ECHOES OF INGEN HOUSZ

We owe this book to ir. A. J. (Jan) Ingen Housz of Enschede –

and so it is our pleasure to dedicate it (twice over)

to Jan Ingen Housz

Jan Ingen Housz (wearing hat) and his manservant, Dominique Tede, preparing an experiment at Bowood House, Wiltshire, in (probably) 1797. From an etching recently discovered at Bowood House. Courtesy of the Trustees of the Bowood Collection.

Echoes of Ingen Housz

the long lost story of the genius who
rescued the Habsburgs from smallpox and
became the father of photosynthesis

NORMAN AND ELAINE BEALE

with foreword by
DAVID BELLAMY

First published in the United Kingdom in 2011
by The Hobnob Press, PO Box 1838, East Knoyle, Salisbury, SP3 6FA
www.hobnobpress.co.uk

© Norman and Elaine Beale, 2011

The Authors hereby assert their moral rights to be identified as the Authors of the Work.

All rights reserved. No part of this publication may be reproduced, stored in a retrieval system, or transmitted in any form or by any means, electronic, mechanical, photocopying, recording or otherwise, without the prior permission of the publisher and copyright holder.

British Library Cataloguing in Publication Data
A catalogue record for this book is available from the British Library

ISBN 978-1-906978-14-3

Typeset in Monotype Bembo 11/13.5 pt.
Typesetting and origination by John Chandler
Printed by Lightning Source

Front cover image: Wedgwood jasper medallion of Jan Ingen Housz. A 1970s version produced from the original plaster mould held by the Wedgwood Museum at Barlaston. The die appears to have been based on a James Tassie engraving of 1778. Courtesy of the Wedgwood Museum, Barlaston, Staffordshire.

Contents

Foreword by David Bellamy		ix
Preface		xi
Acknowledgements		xiii
Preamble		xv
1	De Trecktang	1
2	Higher Education, Family Practice	22
3	Enlightenment London	32
4	Particles	49
5	Foundlings	69
6	Head Hunted at Hertford	92
7	To Vienna	112
8	Royal Inoculations	134
9	Royal Inoculations II	153
10	Paris, London, Paris	175
11	Hunkering Down	193
12	Mesmer & More	221
13	Apogee	250
14	Home – to Vienna	284
15	Intrigues	309
16	Experiments and Distractions	334
17	Shocks	363
18	Retreat to England	396
19	Going Home: Not Going Home	428
20	Putting Down Roots	455
21	A Noisy Quietus	492
22	The Black Hole of History: How Jan Ingen Housz was Forgotten	519
Notes		527
Ingen Housz Publications		585
Bibliography		593
Index of Names		601
General Index		621

Foreword
by David Bellamy

JAN INGEN HOUSZ IS an unknown name to the vast majority of scientists, let alone people who think science is boring. But I have a natural affiliation for Dr. Ingen Housz: like me, his father was an apothecary – of the pharmaceutical ilk – and some of my earlier ancestors were also from the Low Countries.

Jan Ingen Housz deserves to be remembered for many things; not least his aptitude for experimentation, both in the laboratory and in the open air. His experiments became cornerstones of scientific discovery, so ingenious that they also enthralled, enthused, and educated the families and friends of some of the greatest stately homes across Britain and Europe. He was a popular guest at sumptuous stay-overs that could last for weeks, and the ladies loved the country walks with him before lunch and his evening pyrotechnic demonstrations.

It was Jan Ingen Housz who first showed that the power that succours Mother Earth comes from the sun, energy that can be trapped by plants, making food and releasing oxygen. There can be few things more fundamental to the understanding of life on earth than photosynthesis, the engine house of the living world. And there can be little doubt that Jan Ingen Housz would have won the Nobel Prize if it had existed in the eighteenth century.

Sadly, his place in the history of science has long been in deep shadow. The good news is that Norman and Elaine Beale have set the balance aright. Their outstanding work of biography was triggered by them setting off to find his gravestone – which they expected to be in their local churchyard. In the end they have produced the book that Jan Ingen Housz has long deserved. It is a substantial, comprehensive volume, heavy in references and laden with fascinating facts but, please, readers, do not be put off.

'Echoes' is a perfect pie – the lightest pastry, the tastiest sauce and the heartiest meat, all cooked to a turn - easily digestible and satisfying. I couldn't put it down - every sentence has been nursed to perfection. In it you will discover that Ingen Housz also had a gift for being in the right place at the right time – in London when Maria Theresa demanded that her ambassador send her an 'English' smallpox inoculator, a brave man indeed; in Italy when air analysis was a new science; in Paris when the 'new' chemistry was being developed; and back in London when agriculture was moving on from 'muck and luck'. This book is a 'Grand Tour' of Enlightenment science and you will also discover the huge number of important personal contacts that Ingen Housz had – his address book must have read like an eighteenth-century 'Who's Who'.

If you've reached this far in this fabulous book, don't stop here – buy it, read it and enjoy it – I certainly have. And I would certainly have bought it for my father - if he were still with us.

David Bellamy OBE, PhD, Hon FLS.
DSc, FIBiol.
June 2011

Preface

THERE'S a very old saw that a biography is a novel with an index. Though tongue in cheek, the allegation is unavoidably true. Recreating a life in totality is obviously impossible without a lively imagination but one must not fill in the gaps by invention. The task becomes more difficult as one goes back, through generations, into different times, into different cultures and into documents in many different languages, especially if they're scattered and piecemeal. Tackling 'Echoes of Ingen Housz' presented us with all these problems. On the other hand it must be rare for someone so prominent as Jan Ingen Housz to have had such a massive archive left so undisturbed and we have had the joy of making many new discoveries; of correcting many myths. We would, however, have made very little progress, if any, without the most generous help of a large number of people. While it is normally very unwise to pick out any one name from an acknowledgements list we have to make special mention of someone who, with outstanding symmetry, is called Jan Ingen Housz. Six generations on, this Jan, a retired university lecturer living in Enschede in the Netherlands, is a descendant of Louis, 'Uncle Jan's' brother. He launched our project into feasibility. Seeing our despair in the face of stacks, literally, of archive in the Dutch language, he organised himself, his family and friends into a busy co-operative of translators; he acting as recruiting sergeant, adviser, cheerleader and editor. And thanks to Jan we were given privileged access to the complete gamut of relevant papers, pictures and artefacts still in personal possession of the Ingen Housz family. We have no subtle intentions in irony in dedicating our book on Jan Ingen Housz to Jan Ingen Housz. If there's a glaring ambiguity, so be it: we owe our affection to both men.

We have allowed ourselves free rein to the imagination in the short, italicised, introductory sections to each chapter. They are an attempt only to set the scene and are fictional; each is what a musician might call a *leitmotif*. They are, however, based on reliable sources and properly contextual; incidents in the life of our subject or in those of characters he knew. Otherwise, we have tried to produce a worthwhile document of record that will be respected for accuracy and verisimilitude, an academic reference point. At the same time, we have tried, desperately, to make some complex issues digestible, make the prose attractive and deliver an enjoyable read.

Some of the structure of the book is idiosyncratic enough to need explanation. We have translated the titles of articles and books published in other languages into English and foreshortened all such titles after their first appearance. There is also a comprehensive list of publications by Jan Ingen Housz giving full translations into English where necessary. We have appended the modern equivalent (in parentheses) whenever reference is made to antiquated chemical terms. There is a separate names index where the briefest of biographical material for each individual is given. Whenever a monetary value is given in the text, we have appended another figure, in parentheses and converted to sterling units if necessary, representing a present day value of the sum. We have used the National Archives currency converter ('old money to new') for the calculations.[a] This computation of 'inflation' probably underestimates, however, the true worth of the eighteenth-century amounts because of profound and parallel changes in the purchasing power of money over the centuries. We therefore suggest that, as an approximation, readers mentally double the converted values in order to have a better idea of true worth – in other words if we quote a value of five pounds in 1770 and update it, by means of the website, to £318 now, one should think more in terms of £600 at least.

It took two of us to write the book. There can be few individuals, after all, who possess, by themselves, all the talents necessary to get into the way of an eighteenth-century polymath who was a classicist, a physician, a scientist, an author, a diplomat and a complicated human being. If we have successfully managed to combine our backgrounds of arts and science, of education and medicine, of imagination and scepticism, and funnelled them into a text that is worthy of the subject then we are satisfied. Indeed, it may be the need for such a multiplicity of skills, and the luck necessary to find willing translators, that has deterred earlier potential biographers of Jan Ingen Housz. We trust that, at long last, he and his genius can be given an airing. We are not so arrogant, though, to overlook the need to apologise, in advance, for the mistakes and faults. There are bound to be some and the responsibility for them is entirely ours. We can share that too.

Norman and Elaine Beale
April 2010

(a) (http://www.nationalarchives.gov.uk/currency/default2.asp)

Acknowledgements

WE THANK THE BRITISH ACADEMY for their endowment of a small grant that encouraged us to embark, seriously, on our project. We also thank, especially, John Chandler, our publisher. He has always been enthusiastic and encouraging despite our prolixity. Asking us if we had a further 'magnum opus' in mind was clearly hysteria.

The following individuals have all helped us, often in more than one way and often repeatedly and most unselfishly. The order is alphabetical by surname and has no other significance.

Our sons, Alick and Huw Beale; David Bellamy; Bart and Marijn Bijnen; Terry and Diana Dalton; Ptolomy Dean; Jaap de Vries; Robert Dimsdale; Don Emblen; David Falconer; Kate Fielden; Jane Freeman; John Forsyth; Howard Gest; Barbara Gleed; Ian and Jenifer Glynn; Gwendolin Goldbloom; Lorna Haycock; Robin Holley; Karl Holubar; Marijn van Hoorn; Martine Houten-Ingen Housz; Rex Hunt; John Hurley; Bon Ingen Housz; Hans Ingen Housz; Jan and Mia Ingen Housz, Breda; Jan and Caroline Ingen Housz, Enschede; Jan Maarten and Marijke Ingen Housz; Laetitia Ingen Housz; Matthijs Ingen Housz; Nanette Ingen Housz; Noud Ingen Housz; Willem and Mieke Ingen Housz-Coenradi; Hermine Ingen Housz-Menalda; John Jenkins; Ludmilla Jordanova; Marieke Kappelhoff-Noordhof; Bob Kenway; Desmond King-Hele; J K van der Korst; Loes Kuiper-Brussen; The Marquis of Lansdowne; Alfred and Dee La Vardera; Mike Laycock; Julian Litten; Claude-Anne Lopez; Antoine Luyendijk-Elshout; Geerdt Magiels; Scott Mandelbrote; Corrie McConachie; Liesbeth Molen-Reeders; Karl Mühlberger; Sally Northmore; Chris Oosterhuis; Paul Partridge; Peter van der Pas; Margaret Pelling; Robert Perlman; Marie Petz-Grabenbauer; Pierre van der Pol; Jaap Rooijen; Jeffrey Rucker; Annechien Saltet-Ingen Housz; William Shupbach; Pieter Smit; Manfred Staudinger; Audrey Tickner; The Hon. Mrs. Charlotte Townshend; George Twyman; Peggy Valke; Wim Vermaas; Alex Verrijn-Stuart; Timothy Walker; Charles Webster; Agnes Wever-Wittebol; Christopher Wright.

We also wish to record our gratitude to the staff of the following institutions, many of whom went well beyond the call of duty in helping us. It is, again, in alphabetic order rather than any order of precedence or preference.

American Philosophical Library, Philadelphia PA.; Amsterdam University Library; Archieven van Teyler's Stichting, Haarlem; Bath Public Library; Bayerische Staatsbibliothek, Munich; Bergeus Botanical Garden, Stockholm; Birmingham City Archives; Staatsbibliothek zu Berlin; Universitäts- und Landesbibliothek, Bonn; Bowood House Archives; Brabants Historisch Informatie Centrum, 's-Hertogenbosch; Breda Gemeentearchief (now Stadsarchief); Breda Museum; The British Library; The British Museum; Cambridge University Library; Central Library, Ealing; The Library of Congress, Washington DC.; Deutsches Museum Bibliothek, Munich; Devizes Museum Library; University of Edinburgh Library; Biblioteca Nazionale Centrale, Florence; Bibliotheque Publique et Universitaire, Geneva; Glasgow University Library (Hunter Collection); Hampshire Record Office; Hertford Record Office; Hollandsche Maatschappij Wetenschappen Library, Haarlem; Hunt Institute for Botanical Documentation, Pittsburgh, PA.; The Huntington Library, San Marino, CA.; Koninklijke Biblioteek, The Hague; Institut für Geschichte der Medizin der Universität Wien; The Johns Hopkins University Institute of the History of Medicine, Baltimore, MD.; Landt- und StaatsArchiv, Vienna; The Jenner Museum, Berkeley; The Lilly Library, University of Indiana; London Metropolitan Archive; University of London Library; Geerdt Naturhistorisches Museum, Vienna; Nederland Tijdschrift voor Geneeskunde; Katholieke Universiteit, Nijmegen; Österreichische NationalBibliothek, Vienna; Österreichisches Staatsarchiv, Haus-, Hof-, und Staatsarchiv, Vienna; The Royal Archives, Windsor; The Royal Bank of Scotland, London; The Royal College of Physicians of Edinburgh Library; The Library of the Royal College of Surgeons, London; The Royal Commission on Historical Manuscripts; The Library of the Royal Observatory, Greenwich; The Royal Society Library, London; The Royal Society of Medicine Library, London; Semmelweis University of Medicine Library, Budapest; The Science Museum Library, London; Stadsarchief, 's-Hertogenbosch; Strathclyde University Archive; Library and Museum of the United Grand Lodge of England; Universitätsarchiv, Vienna; University College, London; Victoria History of Wiltshire; The Wellcome Library, London; Wiltshire History Centre; The Wedgwood Museum, Barlaston; Yale University Art Gallery, New Haven, CT.

Preamble

The Reverend Thomas Greenwood hovered, quill in hand. 's'. 'z'. That was it. 'h - o – u – s – z'. His normal flowing hand had faltered at the unfamiliar foreign spelling; his last strokes deliberate and laden with ink. But clear enough, he thought. Taking care not to smear the new entry, he glanced back through the pages of the burial register. The names of the recent dead were very familiar to him; Ponting, Pearce, Bayly, Titcombe, Chivers, Bodman, Page; Calne names for generations. He had known them, understood them; their joys, sorrows, kindnesses, foibles, fecklessness, sins. Their world was very limited, but then, so was his. He was their pastor in a shared life. Dr. Ingen Housz had not been one of his parishioners. He had led a much more varied existence, turning in a much larger orbit than he and his flock. Greenwood had come to know him as a friend of Lansdowne, Lord of the Manor. Ingen Housz had taken, particularly, though, to Allsup, the Calne surgeon apothecary. Christopher Allsup, the vicar's brother-in-law, had often, at table, tried to explain why the Dutch doctor deserved to be famous. Greenwood had never grasped the points fully. How can plants possibly eat air? How can it be that we breathe only because of leaves? And why did spreading gypsum on our fields produce better crops? He had always felt uneasy, if less than many other men of religion, about spreading smallpox by inoculation. It had troubled him that Ingen Housz was such a strong advocate. Could this really be God's work? Then there were his demonstrations with blinding lights, the stargazing with the powerful Bowood telescope, the excitement with the moons of Saturn – if that's what they really were. Magic? Pagan perhaps? His mind began to wander at random: Faustus? Surely not; a man so approachable, so considerate. Ingen Housz had lived in palaces but now in two mean rooms in London; confronted loss of a fortune; died courageously. No, a modest, honest Christian. Eccentric, exotic even; but worthy of a proper burial. Sad that his wife was so far away, in Vienna. And, as a physician, Ingen Housz had wonderfully relieved the suffering of poor Davis, the schoolmaster. So, he'd agreed to reading the funeral prayers when Cross, Lansdowne's agent, sought him out. Vault deposition within the church had been the dying doctor's wish, it seemed.

Greenwood started, suddenly alert. Benjamin Bowman might have been standing close behind him for some time but the vicar had been lost in thought. Nodding towards the Register, the churchwarden asked if he could lock it away, now? Greenwood understood:

'Yes, of course – the ink is perfectly dry.'

'I've checked in the North Chancel, Sir. The stone is bedded down right enough. The vault is proper sealed. The sexton and his men were here for some time on Monday.'

'Ah, good.' 'Thank you, Mr. Bowman. Thank you for all your trouble. Sorry to tarry you.'

The warden replaced the register in the chest, distorting the lid slightly as he lowered it, always necessary to engage the locks. Doctor 'Ingerhows' was now dead and buried officially as far as he was concerned. Singular man by all reports. And a Papist. Hopefully the rest of the day – Sunday 15 September 1799 – was going to be as warm and sunny as the early haze had promised.

1
De Trecktang

He frowns. He hopes he's turned away in time; that his men haven't seen. The orders have come up from the hastily assembled officers' meeting in the floor of Glen Shiel. Hussel's Dutch contingent is to stay up on the steep northern slope of the valley and advance westward on rising ground. It means moving up into the sinking sun in close formation. Their ranks will be an easy target for an enemy that's above them and well dug in. From what could be glimpsed of their positions the Highlanders have made good use of their time that afternoon. They're intimate with their own terrain and happy with their sight lines; unlike the Spanish marines who, in their distinctive colours, can be seen easily further along the valley, repeatedly adjusting their ranks in an expression of anxiety and disorientation. Sergeant Ingen Housz also knows, from past experience, that the clansmen are tenacious warriors, tall, strong, determined and bloody-minded; hard men from a tough and uninviting part of the world. There's also the rumour, flying about, that Rob Roy MacGregor is somewhere out there, a guerrilla fighter of fearsome reputation.

As he traverses the slope back to his platoon he feels what hardened N.C.O.s have felt throughout history; that his officers are naive, arrogant and impetuous. His men are tired and it's already five hours after noon. They've force-marched all day on only a poor breakfast and brief stops. Two of his best men are struggling with dysentery; another one has a week-old putrid leg wound after a fall on the wet Inverness flags. Circumstances are stacking odds against them. The only plus is Hussel's clever interpretation of the orders he's been given. Once back among his own officers he's proposed that the regiment makes a diagonal uphill advance. It means an even longer climb and more exposure to fire but it might just create a fear, in their opponents, of being outflanked and then of being attacked by Royalists pouring downhill from their rear. An insecure enemy might break from their positions. The odds might be turned. It's worth a try.

With only two hours of useful daylight left and a mist already congealing, the Dutch infantry prepare themselves for battle. Arnold Ingen Housz insists that his three disabled men go to the rear, to keep up as best they can and then waits in front of all the others trying to look confident. Within twenty yards advance they come under fire from mortars. There are screams from men with shrapnel wounds and several mounds of gruesome, silent, body parts. Then there is musket fire. Men slump to their knees or clutch at limbs half gone. The climb is hard work but as a moving target the soldiers at least feel there is some chance of survival. Landmarks are now lost to the noise and smoke but when the Dutch eventually find themselves looking along a line of enemy trenches, some already empty, they sense the makings of a victory. Now they can avenge themselves. Hand to hand skirmishing breaks out. Increasingly outnumbered, the remaining clansmen become demoralised and begin fleeing further up the slope, towards the peak of Scuir-Curan, turning, at times, to return fire. Sergeant Ingen Housz is trying to gather his platoon, count heads, shield his wounded. He's looking round when . . . blackness and a single, incredibly loud, toll on a cracked bell . . . he falls up the slope, face down, flaccid. The troopers near him hope that, when their turn comes, it will be as sudden, profound and painless.

It's dark when he comes round. He can open one eye. There's a deep-seated pillar of throbbing pain above the other one. He's face down, something fibrous in his mouth. A very heavy weight is across his legs. He can move his toes and his right hand. He has no idea where he is. Something is digging into his left ribcage. With a supreme effort he reaches under his chest to remove a hard object. It feels familiar. He passes out again. An unknown time later it's light. The headache is worse and still he can only open one eye. He can see, now, that his leather-bound prayer book lies near his free hand. Why, when he pushes out, twists his upper half and looks down, he should be pinioned by a dead body, mouth agape and eyes wildly staring, he has no idea. Perhaps he, too, is dead? Many of the pages of his little book are stuck together by blood but he can just push others apart with a finger and prop them up to half read, half call to mind, some familiar prayers. He tries reciting aloud but his swollen tongue is stuck to the roof of his mouth. He has no fear: his thirst is all embracing. As he begins another prayer he passes out again.

A BIOGRAPHY IS a journey: a life will have a beginning and an end. On the other hand, life, the generality rather than the specific, is a continuum. Knowing where to embark and disembark on a life-story is not as simple as it might seem. All this has been said before. So, too, has the fact that home circumstances usually exert a massive influence on children. Foundlings and orphans aside, the biographer dare not omit a portrayal of the parents, however sketchy. Beginning the 'journey' of Jan Ingen Housz is neatly resolved by starting his story with just that – a journey. It was a complex, arduous and perilous one made by a young man called Arnold Ingen Housz, who would later become Jan's father.

Arnold Ingen Housz was born in the small Dutch town of Bommel,[1] a community on the south bank of the Waal between 's-Hertogenbosch to its south and Utrecht, the provincial capital, to the north of the river. He was baptised on Friday 23 October 1693, sixth child[2] and third surviving son of a Catholic lawyer, Louis Ingen Housz, and his wife Hendrica. Besides making certain generic assumptions about his childhood, there are few other details. He had sufficient education to be literate and, as the sons of an advocate, we can expect that all the boys would have been given formal schooling beyond the elementary stages. Arnold certainly followed his older brothers in the next phase of his life; in becoming a soldier. He enrolled in the service of the state at the age of about fifteen, in 1708. Martin and Cornelis Ingen Housz[3] were already serving in the armed forces as Arnold reached enlistment age. They both marched with the allied forces under the Duke of Marlborough in the later stages of the War of Spanish Succession,[4] part of the Dutch contingent that besieged and eventually broke into the French redoubt of Douai in northern France in May/June 1710. Martin and Cornelis had clearly inspired their younger brother but less propitious was the fact that they were both mortally wounded in the action. Cornelis, 20, died almost immediately, from cannon shot and Martinus, 24, during July, from his wounds.[5] They were only two of some 8,000 Dutch casualties sustained between 30 May and 24 June,[6] part of the heavy price paid in forcing the French finally to abandon their ambitions to take over the southern Netherlands (now Belgium). Even this double family tragedy does not appear to have deflected Arnold from his own military career. Nor did the death of his 50-year-old father early in the next year, on Friday 16 January 1711.[7] Presumably Arnold even resisted the entreaties of his grieving mother that he should now consider some other career – pleas that would certainly have been wholly understandable and heavily laden with emotion. Conjecture becomes fact again in 1715. Now 22, Arnold was a member of the standing force available to the 'States-General', the assembly of provincial delegates that was the effective parliament of the United Provinces (as the Dutch Netherlands were then called). This regular army was not large but it was well trained, fit and disciplined. Its courage and skills were soon to be called upon.

Across the North Sea, King Charles II, restored to the English throne in 1660, had died after a stroke in February 1685. Sovereignty passed to his younger brother, James II. Whatever had been the true, but closetted, religious conviction of Charles, James made no secret of his Catholicism. In fact, he was arrogant enough to assure his new court, a hastily appointed cabal of papist sycophants, that the nation would have no objections. He was wrong. Within months of his accession, public opinion was fractured. Another civil war threatened when his second wife, also a Catholic, bore him a son baptised into the same church: there was now the real possibility of a Catholic dynasty. The next sequence of events is also well known.

Politicians pleaded with William III of Orange – married to Mary, James' Protestant daughter by his first marriage – that he rescue England from Catholic rule. William reacted vigorously and landed at Torbay in November 1688. James fled to France and was deemed to have abdicated. William and Mary were enthroned the following year, the high point of what has come to be known as 'The Glorious Revolution'. Their joint reign resolved a crisis. Less well known are some of the preparations that were made in order to reinforce their Protestant rule lest supporters of James – 'Jacobites' after *Jacobus*, the Latin for James – hit back and tried to restore the 'rightful' king. This was seen as so inevitable that part of the treating included clauses to bring substantial numbers of Dutch troops to England to help quell any revolt. These standing arrangements were, in the end, never needed until, almost forgotten, the documents were dusted off in July 1715, long after William and Mary had been succeeded to the throne by Anne, Mary's younger sister, and then by the Hanoverian, George I. He found himself threatened by a substantial Jacobite force assembling in Scotland, the makings of the so-called '1715 Rising'. As Peter Rae was to report, in his history of this rebellion,[8] only 30 years after the events:

> ... the King had, upon the Alarms formerly mentioned, given notice to the States-General, of the Designs of his Enemies at Home, and the Preparations made by the Pretender from abroad; and resolved to demand of the Dutch, the Assistance of 6000 men, Stipulated by the late Treaty of Guarantee for preserving the Protestant Succession; . . . The Dutch came readily into the Proposal, and passed a Resolution . . . that the Assistance his Majesty required should be ready; and ordered their Ambassadors at London to give his Majesty positive Assurances of the most express Terms, that They were always, and would constantly be ready to perform all their Engagements with Great Britain.

By October an attempt, by force, to gain the English throne for 'James

III', the late James II's son, seemed inevitable. George I instructed Horatio Walpole, British Minister at the Hague,[9] and younger brother of Robert Walpole, the first British Prime Minister,[10] to request, immediately, the promised Dutch support. The Dutch were as good as their word and by Wednesday 16 November, 3000 of their troops were already disembarking at Deptford.[11] The other half of the force sailed from Ostend late on the 24th, on a course that was intended to steer them to Scotland. During that evening, however, winds freshened in the English Channel. Not all the Dutch have sea legs and as the ships began to roll, pitch and yaw, the time-honoured recipe for *mal de mer*, the troops became apprehensive, progressively less sociable, and then prostrate with their symptoms. Soon many were vomiting. Worse was to follow. Wind speeds increased and as the fleet tried to tack north a gale stirred up huge waves. A badly ballasted ship capsized and went down with loss of all on board. For the other soldiers the compound terrors of drowning and seasickness led them to plead with their navigators that they be landed; put ashore anywhere on the English coast – even if it meant a forced march all the way to Scotland. Surely this was better than such danger and suffering at sea? Many of the naval officers were successfully bullied by their petrified human cargo, few of whom would ever have been to sea before, and allowed their vessels to be swilled towards the English coast, their warnings of being smashed to pieces on rocks falling on deaf ears. One of the boats beached, eventually, in the outer reaches of the Thames estuary and the men fell onto the muddy ooze with huge relief, among them a certain Cadet Sergeant Arnold Ingen Housz.

The uncomfortable, unscheduled and unceremonious arrival of Arnold's unit is better recorded than any of their subsequent precise activities in the British Isles. Clearly they had embarked for Scotland, their later departure and intended direct transfer by sea implying that they had waited for, and were carrying with them, the bulk of the baggage and munitions for the Dutch expeditionary force. Much of this could only have been abandoned in the disabled ships scattered along the eastern coastline of England. The officers did, however, rally their men and, carrying what they could, they formed up to march the 400 hundred miles north to Leith, their original destination.

Wet, weak and probably still vertiginous, Arnold was one of more than 2000 of the survivors who were grateful, at least, to be on dry land and glad to fall into the reassuring marching routine common to all infantry. Monday 28 November saw columns on the move. By the military standards of the day a long march to meet an enemy was not unusual but for these

Dutch troops it meant trudging all the way to Scotland along squelching roads in deteriorating weather and shortening days. They were soon cold, damp and lousy after sleeping rough night after night: it must have been more arduous than some could sustain. By mid-December, however, substantial Dutch columns were seen descending from the Lammermuir Hills to the Firth of Forth. The final stragglers arrived just before the roads were lost to deep snow.

Long before this time, however, the Jacobite forces had marched in the opposite direction; down into England through the western valleys and coastal plain, to be confronted by loyalist troops reaching Lancashire from the south. Battle was engaged at Preston. This was the first real clash between the two sides, beginning at two o'clock on the afternoon of Saturday 12 November.[12] The Jacobites were defeated by dawn on the Monday, those who could retreat scuttling northwards. And, over the same weekend, if with a less precise onset and outcome, the other rebel army was overwhelmed and scattered on Sheriff Muir,[13] heath land above Dunblane, north of Stirling. So, as Arnold and his countrymen arrived in Scotland, they would have heard that the rising was already over. Many of the defeated troops were now prisoners of war. Those of high rank were taken to London and other English cities for trial and virtually certain execution: there was grim entertainment for many a morbid, jostling mob during much of 1716. Other, miscellaneous, groups of less exalted rebels, many of them Highlanders, fearsome in battle and churlish in captivity, were rounded up and incarcerated in squalid makeshift gaols all over central and southern Scotland. Guarding these surly and difficult men was the task now given to some of the Dutch contingent. Others were used to garrison various Scottish towns where rebellion was still thought to be smouldering – at Perth, Dundee, Dunkeld, Dunfermline, Montrose, Aberdeen and Inverness. Arnold could have lived, therefore, in any of these communities during the late winter and spring of 1716. As a non-commissioned officer he would have needed to acquire some facility in English although many of any prisoners in his charge would have used Gaelic. Nonetheless, he would have gained some of the predominant language as soldiers on a foreign territory always do: after all, the little comforts and consolations that make life tolerable are sourced only by personal contact and by bartering with the indigenous population.

Arnold spent eight months in this foreign service, leaving Scotland sometime near mid-July 1716[14] to disembark at Willemstad, a small port just south of Rotterdam, on the Sunday 19th. From there his unit marched some 50 miles alongside the River Maas to Grave, just south of Nijmegen.

Here he must have felt much more at home – familiar countryside, familiar architecture and language, windmills groaning softly in warm sunshine and the nearest Jacobite clansman a world away. Nor, any longer, was he in conflict with himself - a Catholic fighting a Protestant crusade because of his loyalty to his country. This respite was not, however, the end of his military 'journey' and he was soon to come even nearer to death than he had been in crossing to England.

Although it was accepted, on all sides, that the 1715 rebellion in Britain had been crushed, there were reasons to fear further Jacobite attempts to overthrow the Protestant English court. Several of the rebel ringleaders had escaped to continental Europe where there were endless other supporters of James VIII (of Scotland), the son of the now dead, deposed James II (VII of Scotland). There were also recurring problems, in Scotland itself, from groups of malcontent Highlanders who continued to raid communities and harass the settled population. Among these bloodthirsty marauders was a particularly brave and elusive leader, Robert Macgregor Campbell, better known as 'Rob Roy'. As a consequence, large garrisons of Hanoverian troops remained in Scotland and among them was a Dutch regiment held back from returning to their homeland. Personnel in this Dutch contingent changed with time and after two or more years some of these troops were serving in Scotland on a second posting. Among these, it appears, was Arnold Ingen Housz. In the summer of 1719 he was certainly still in uniform for he was wounded in a battle.[15] This almost certainly indicates that he was in Scotland, the only place in Europe where there was, that summer, any armed fighting involving Dutch soldiers.[16]

The Battle of Glen Shiel, 1719, is not well known.[17] This is partly because it was a single, brief action that put paid to this further Jacobite incursion into Britain so swiftly that the government saw no reason to publicise their victory: it was far from politic to advertise Jacobite activity. This tactic was feasible for there were very sparse records made of the events and these few documents remained in private hands for many decades. Glen Shiel is in northwest Scotland, carrying the River Shiel down to the sea that separates Skye from the mainland at Lochalsh. The glen is steep-sided, rising to high peaks on either side, the slopes of which dominate the road along it, the route inland, to Inverness to the north and Oban to the south. The fighting in the glen, on Wednesday 10 June 1719, was the outcome of the long-feared further attempt, by Jacobite forces, to retake the throne. By the spring of that year Jacobite ambitions had hardened once more. The King of Spain was convinced, by his advisers, that it was time to fight a 'holy' war that would put a Catholic king back on the English throne. Spanish forces were therefore put at the service of the exiled James Butler, 2nd Duke of Ormonde and he conceived a bold invasion plan. Other refugee Jacobite nobility quickly rallied to the cause and before Ormonde set sail for the south coast of England, with 7000 Spanish troops in 27 ships, George Keith, 10th Earl Marischal, set off to create a diversion attack on Scotland. It was integral in the scheme that Marischal would be able to rally Rob Roy's men to his side together with those of other clans, so he confidently set sail with only 300 Spanish marines and a large stock of ammunition. By April, Marischal had arrived off Scotland and was welcomed onto the island of Lewis, a stronghold of Jacobite sympathisers. Then, on 13 April, his force sailed into Lochalsh and landed on the mainland near an ancient castle, Eilean Donan. By now, the presence of the force had been reported to London and within a week or two it was being shadowed by ships of the Royal Navy. Eilean Donan was being bombarded from the sea by the middle of May and there was bad news of Ormonde – his ships had encountered a violent storm off Cape Finisterre and those ships that had managed to limp to shelter had turned tail and returned to Cadiz. It was the second time in history that a Spanish Armada had been pulverised by angry seas. Marischal must have felt very isolated, particularly when the expected highland clansmen failed to appear in any significant numbers. He moved his force inland, out of range of the navy guns, and set up a barricade, straddling Glen Shiel where the drovers' road crossed the river, while he reorganised his plans. Meanwhile, Major-General Joseph Wightman, commander of the Hanoverian garrison at Inverness had also learned of the invasion force. Not wanting to be caught, flat-footed, in the town, he immediately set off for Eilean Donan. Thus it was that on the afternoon of 10 June the 1600 men marching with Wightman, Hussel's 'Dutch Auxiliaries' among them, came upon the entrenchments of rebel troops lying in wait in Glen Shiel. Wightman saw, immediately, that the rebels were well dug in and that many of them were holding dominant ground well above the valley floor, all of which gave them a substantial advantage, irrespective of numbers. Nonetheless, he decided to attack without delay even though it was already five o'clock in the afternoon and darkness was expected by nine. He split his force in two. He placed the bulk of his units to the right of the river, including Hussel's regiment. These troops were sent to attack and outflank the assorted Highland clansmen, including Rob Roy and his Macgregors who were clustered some way up the slope of Scuir-Curan, the peak to the north of the glen. Despite having to advance uphill under fire, this mixture of infantry and grenadiers scattered the clansmen who fled further up the mountain to disperse in the coming darkness. It took longer to clear a way along the valley floor but eventually the rebels and Spanish marines here also gave way and the pass was now open. Wightman sensed victory just as night came on and a thick fog descended. At least twenty of his men had been killed and well over one hundred wounded. Among the fallen was Sergeant Arnold Ingen Housz. Having sustained a frontal head wound[18] that rendered him deeply unconscious, he lay on the mountainside all night. The archives of the Ingen Housz family pick up the story.[19]

> He was severely wounded, such that he was left on the battlefield, amidst the dead. Upon the burial of the killed he was already half-undressed when it was discovered that he was still alive, an old woman giving him a blanket in order to cover himself, after which he was transported to the hospital and recovered. When lying wounded on the battlefield, he exerted himself to recommend his soul to God and gave himself strength with prayers, especially to the Holy Virgin, Mary, prayers he read to himself from a Church booklet in which one sees the printed pages dedicated to the litany of the Holy Virgin still blood-stained.

There is no record of how long it took Arnold Ingen Housz to recover from his wounds. We might assume, however, that as soon as it was deemed possible to do so, he was conveyed back to the Netherlands to be nursed by his mother and sisters at Bommel.[20] The family archives record that Arnold now left military service, refusing the temptation of a commission, 'because of the aversion he now felt for the military life'. Despite the disabilities and disillusionment, Arnold's vigour and determination, his devout faith and the family ministrations saw him through what was likely a long recuperation and one with no early guarantee of success: battle wounds are rarely free of complication. Besides the tissue damage, there is always contamination. Clothing, bone fragments, jagged metal, animal flesh, soil, dung and a host of other materials can all penetrate deep into wounds and fester for months or years. To have been found, semi-comatose and surrounded by corpses, as Arnold had been, with a significant delay before receiving any attention, could only have intensified the risks of organ damage and permanent disability. He was lucky to survive but his recovery was eventually well worthwhile. Veteran Ingen Housz was to go on to live a long and fulfilling life lasting another four and half decades.[21]

Breda in 1731. Painting by de Haen. Courtesy of Breda Museum.

THE NEXT PERIOD of Arnold Ingen Housz's life is not documented. By inference we know that, having left the army, he invested in a business venture – as a *Koopman in leer en huiden*,[22] a merchant in leather and hides. Perhaps he had conserved a legacy from his father. Perhaps he had been extremely thrifty with his soldier's pay. Perhaps he had gambled on commercial success in borrowing some money. A career in such a business seems to have been a new venture for this family. His military service had convinced him, possibly, that there would always be a demand for leather products; for belts and boots, satchels and purses, saddles and straps. Certainly, the skills of the currier were of exquisite value to many, many aspects of human life. By 1726 Arnold was either living in Breda, neighbouring town to Tilburg, where his mother now lived, or at least travelling there very frequently. Breda housed a large garrison and there would have been high demand for leather goods. It was also a port, accessible to the sea by the River Aa, and perhaps he was already exporting his products. Whatever his situation or mission, Breda was to become his new home;[23] his motives for being there soon massively enhanced by meeting a young lady, one Maria Beckers. It was a relationship that deepened and on Monday 5 May 1727 the 33-year-old Arnold Ingen Housz married Maria in the 'Great' church at Breda[24]. His bride was 28 years old, the elder daughter of Abraham Beckers and his wife Allegonda (formerly Gouban).[25] Like her husband, Maria had been baptised into the Catholic faith – in her case, on Friday 8 August 1698.[26] The Beckers family were prosperous. Besides owning his own home, a house called *de Clerbessem* (large clothes brush for outdoor wear) in the Eindstraat – 'End Street', so-called because it led to the southern ramparts and boundary canal of the town – Abraham could afford to rent[27] the next door property, *de Trecktang* (the extraction tongs) for his daughter and new son-in-law. So it was here that Arnold and Maria began their married life, premises that probably also served as the headquarters of Arnold's business. It was to remain the Ingen Housz family home for the next 70 years, allowing for the fact that it was united with *de Cleerbessem* to form one large house in 1761 (now Eindstraat, 3).[28]

Leather production provides the ultimate olfactory paradox. The end product has that seductive smell common to new shoes, garments, bags and belts whereas no one ever asks the way to a tannery. No one ever needs to, certainly down-wind. Indeed, no one with no business there would ever dream of visiting. Fresh animal skins, and certainly those soaked in brine for a week or two, are always obnoxious. The stench is sickening. The word 'hideous' does not derive from 'hide' but it might just as well. Each skin is a spatchcocked silhouette of a recently living creature on one side, sense organs splayed and grimacing; on the other a gelatinous blood-spattered ferment of sinewy remnants and fat globules that all congeal and putrefy as the tanner himself, shunned at work and hardly popular at home, prepares his chemicals. As an established merchant, Arnold traded in leather and the house in the Eindstraat was probably his retail outlet. But his wholesale warehouse was more likely to have been a suitable building outside the town, sensibly close to one or more local tanners. Otherwise Arnold would have been a very unpopular neighbour, not least for his in-laws next door. It is certainly the case that tanners normally worked at premises well outside communities where, in truth, they were encouraged to stay. It was not unheard of, even, for priests to take the Host to tanneries on a Sunday rather than have the tanners in church.

Family life, though, was certainly based in and around Eindstraat where, within two or three menstrual cycles, Maria realised that she was pregnant. This, at least, is a biologically valid conclusion if their first child, Louise, was born after a full gestation period. A traditional *kraam kloppertje*, a baby doll dressed for the appropriate gender, appeared on the front door of *de Trecktang* on Thursday 4 March 1728 and the little girl was baptised in the Roman Catholic Church in the Brugstraat[29]. However, our arithmetic may be misleading us; perhaps she was a premature infant. She was certainly doomed. She survived only into her fourth week, dying on Sunday 28 March.[30] The oscillating emotions of the grief-stricken parents, and adjacent grandparents, can only be imagined: the joy of a first birth but then the free-fall despair of an infant death, the wrenching loss of an *onnozele schaapje*, a little lambkin, all within a month. But in every century until well into the last, human procreation has always been hedged around by tragedy. Parents always knew that early losses would be likely to hound and haunt them. A reproductive resilience and persistence was necessary if parents wanted to see any children grow up. In fact, with no real contraception available, such strengths were mandatory and by early autumn in the same year Maria had conceived again. This time she bore a boy, Louis, baptised, also, in the Roman Catholic Church on the Brugstraat, on Saturday 9 July 1729.[31] His godparents were his paternal uncle, Caspar[32] and maternal grandmother, Allegonda Beckers.[33] Any hasty anxiety to have this child baptised lest he, too, be taken from them proved gratefully ungrounded but the fears were not irrational. Arnold, knowing the sad fate of four of his twelve siblings, would have been extremely tense. Was he transmitting a family blight? Three sons and a daughter born to his parents had all died one to three months after birth. Louis, however, grew into a man of singular vigour who was to die nearly six decades later and then only because of an accident. There was now a gap of nearly eighteen months before Maria was delivered of another live child. It is therefore possible that she both conceived and miscarried during the winter months of 1729/30. Just as feasible is that she breast-fed Louis, stopping after about six months, then 'falling' again on restoration of her fertility.

Her next child, another boy, was also robust and, we presume, gestationally mature at birth. He was born on Friday 8 December 1730.[34] This little boy, baptised 'Jan' during another trip to the Brugstraat, is our subject. His godparents - his mother's brother, Jan Beckers[35] and, once again, grandmother Allegonda Beckers[36] helped to launch him on his 'journey'. It will also be ours.

JAN, LIKE HIS older brother, appears to have been a strong and healthy child. There are no reasons to suspect delay in his developmental milestones nor that he experienced any out-of-the-ordinary infant ailments. It was his emotional development that ran into the buffers. By late November 1731 we can imagine him on the threshold of walking, of articulating a few simple words, of relating intensely to his family. His strongest bond was, of course, to his mother. This was to be severed abruptly and painfully for on Friday the 30th Maria Ingen Housz died.[37] She was buried, four days later,[38] on Tuesday 3 December 1731. It appears to have been a sudden and unexpected demise. She was only 33 years old. No cause of death has come down to us. Having carried three live children, two at least to full term, there is no reason to suspect any serious chronic illness. Perhaps she had an accident – a slip and a head injury? In late November it could easily have been treacherous under foot. Perhaps she died of an overwhelming infection that the rest of her family were able to resist? Maybe she was pregnant again and succumbed to a complication after miscarriage? Graveyards through history have been filled with the coffins of many young women who haemorrhaged to death in pregnancy or died of uterine sepsis. None of these theories, however, help us to embellish a bald record correctly. The facts are simple: Jan Ingen Housz's mother died before he was a year old.

The despair and despondency at the adjoining homes in the Eindstraat defy prose description. Only the poet has the words for agony like this – 'Light breaks where no sun shines'.[39] A husband had lost a wife inside five years of marriage. Devoted parents had lost a daughter. Two infant boys had lost their mother. Their father could have reacted in many different ways. The strength of character, the steel needed to go on with life after an event like this is very different to the courage he had already displayed on the battlefield. His grief could have transfixed him or he might have become embittered. There is evidence, though, that he rallied himself – his boys could not be neglected nor his business abandoned. Presumably his in-laws at *de Cleerbessem*, next door, were an immediate and vital resource. Families have to adapt – even to very premature deaths, an everyday eventuality in the eighteenth century. All we know for certain is that this catastrophically curtailed young family went on to prosper. Both boys matured, adoring and remaining very close to their father until his own death. Arnold's business succeeded. He never remarried. There was to be no stepmother for the boys – wicked or otherwise; no one to vie for their father's affections. We do know, though, that Arnold employed a full-time maid cum nanny – Marianne de Kanter – who 'served him a great many years and took care of and served both his sons with great affection during their minority, like a mother for her own children.'[40] There is also the sadness, even today, that Maria Ingen Housz died before any personal or family mementos accumulated; we have no pictorial image. But these are piffling regrets against the depths of emotion that must have been running raw through the Ingen Housz home on Saturday 8 December 1731, the day the family had expected to assemble in total joy, a day to celebrate, the day of little Jan's first birthday. Arnold seems, however, to have reigned over a nurturing and happy household despite the maternal influence being that of a surrogate. From what would undoubtedly have been a time of boundless young male energy and exploration, twice over, the boys appear to have reached adulthood free of any neuroses despite having lost their mother. We should assume, in fact, that they were a substitute source of joy for Arnold and satisfaction, mostly, for Marianne. The boys' respect for each other, and for their father was, in both instances, life-long.

IN THE MID-EIGHTEENTH century Dutch formal education usually began at the age of six. Louis would therefore have started school in 1735 and Jan in 1736/7. The rudiments of reading and writing were the task of the *lagere school* or lower school. It is at school, in the laying down of literacy and numeration, that we can begin to distinguish the two boys, the one character from the other. Louis was clearly no dunce and he was always adept and successful when interacting in the grown-up world of commerce and industry. But Jan made progress under a star. At the age of ten he was enrolled at the Breda Latin School[41] and was to have outstanding success.

Jan not only displayed an aptitude for languages but he also had a stern and studious attitude to his class work.[42] We find the 'inspiration/ perspiration' recipe so often quoted for any success, a virtuous circle of talent and traction that was driven even faster when Jan came under the influence of inspiring teaching at the school. Drive and hard work were predominant and led to deserved success but being feted at school and praised at home was not all heaven. The cruelty inherent in the young male is legendary. The class swot has always been the butt of his peers. That Jan suffered for his strengths is strongly suggested by a later scenario when he first arrives at University, also recorded in family memorabilia.

The Latin School at Breda, then situated on the Kasteel Plein,[43] would have followed a strict syllabus and timetable. Such things were a matter of national diktat introduced in 1625[44] and only slowly breaking down even a century later. The notion that this allowed boys to make seamless progress if they changed school was a good one in principle but a straitjacket against innovation and reform. For our purpose it is useful: we can deduce, with pleasing accuracy, the rungs in Jan's educational ladder. As a preamble it is necessary to know that classes were stacked in what will seem, to us, reverse order. At ten, already able to read and write in Dutch, boys enrolled at the school and entered the sixth form.[45] By fifteen or sixteen they found themselves in the first form, their last, and conventionally in the hands of the headmaster. He was a man of considerable power and influence, traditionally living in premises attached to the school, a house of sufficient size for him and his family but also capacious enough to accommodate those boys who, far from home, were obliged to be resident at the school. Classes began, each of six days a week, at eight a.m. and, post-prandially, at one p.m. These school routines had changed little by the middle of the eighteenth century. Lessons still occupied six hours every day bar Wednesdays and Saturdays when they were shorter and afternoons were free. Latin was still predominant. On Sundays the boys were compelled to attend services at a local Reformed Church, morning and evening. This would have created conflict for Catholic boys like Jan but they may have been exempted as long as they attended one of the Catholic Sunday schools that had been founded, in Breda, in 1581.[46] There were two long terms each year, beginning in January and July, sessions ending in exams on 10 June and 10 December.

A list of Latin passages and primers were decreed for each year group, some for translation and some for learning by rote. Poetry was pre-eminent. Some texts were used to instil logic and rhetoric, some to preach morality. All was reinforced by tuition in grammar. The approach to Greek, reserved for the afternoons of fourth form onwards, was parallel. The stock authors included Cicero and Cato, Homer and Euripides. The boys were given the *Colloquia Familiaria* (Friendly Conversations) of Erasmus published in 1518; essentially a book grounded in a humanist's appreciation of ancient authors. More Erasmus was studied in his *De Civilitate Morum Puerilium Libellus* (On Civilised Manners in the Young). On the other hand a caustic competitive spirit was inculcated.[47] The boys in each class were strictly ranked. As they took turns to translate, aloud, from unseen texts, they could be challenged and corrected by their peers – all in Latin or Greek – and if the teacher deemed the interruption valid, rankings would change hands. The nearer the top of the class, the more likely a boy's efforts would be attacked. All this seems fair enough in class but the snakes and ladders system extended to a *Stasiland* beyond the schoolroom where the boys spied on each other. Any boy denounced by another, in class and in good quality Latin, for any minor misdemeanour – conversing in Dutch while on school premises, for instance – would slide, ignominiously, down the class hierarchy. Revenge was obtained only by reversing the process – 'sneaks and ladders' perhaps? These by-laws had been the invention, back in the early 17th century, of the Calvinist zealots delegated to the provincial church synods. Predictably, therefore, lessons were sandwiched between prayers.

Only in the last year, at most two, was there a leaven of other subjects. These included history, mathematics, physics and geography. There is good reason to suggest that Jan, just has he had been fired by the classics syllabus, arid and tedious to many boys, was even more inspired by these new subjects, especially physics and geography. From a reference he makes, later in life,[48] to a long-standing ambition to visit there, some of his lessons must have incorporated Roman history and descriptions of Italy. But if going to Rome was a restrained desire, his interests in physics and in the other embryonic sciences were to become an obsession. It was deemed unnecessary to teach the boys any modern languages other than their own for proficiency in Latin, a *lingua franca*, was considered a ready passport to international communication. This was a serious gap in the syllabus in the view of many middle-class Dutch parents who recognised that French, especially, was the real international language, certainly in diplomacy and international commerce. There is no evidence that Jan was ever in difficulty with French at any adult stage of his life and we should assume some formal study of the language. There was at least one school teaching French in Breda from 1566,[49] staffed by Huguenot refugees from France. Perhaps Jan attended one of these institutions on otherwise free afternoons or had the benefit of private tuition from one or more members of the staff.

The most successful aspect of Jan's secondary education was undoubtedly the singular progress he made in Latin and Greek, languages for which he had an innate facility. Then, in his fourteenth year, his skills accelerated remarkably when he came under a new influence. School reminiscences will nearly always include vivid memories of teachers, good and bad. When a teacher is remembered for being inspirational, however, it is often in the context of launching a successful career and for Jan Ingen Housz this is a case in point. A new member of staff joined the Breda Latin School in 1745,[50] engaged to teach Latin and Greek. It was barely two years before Jan left for university but this new teacher was to have an intense influence on him, recognising the depths of his talents and fostering them. His input was therefore as immediate as it was profound. Hendrik Hoogeveen was born in 1712 in Leiden, to parents of modest means. His father's strong views in favour of secondary education held sway in the family home and, though it was a financial struggle for the household, Hendrik attended the Leiden Latin School. At first he did not do well, partly because some overweening teachers tormented him and partly because of recurrent throat infections that put him behind his peers. His third year at school was less traumatic: he had the double good fortune of better health and more sensitive tuition. By his fifth year at the school he was one of the top pupils and was even teaching younger boys in his free time; able, then, to assist his father with the school fees. These coaching sessions also convinced him in his ambition to be a teacher, taking his first full-time post at the age of twenty. After less than a year in Gorinchem, south of Utrecht, he moved to a new Latin School at Woerden, due west of the same city of Utrecht. Hoogeveen married in 1733 but his wife died in 1738. He was left with three children. In the same year he moved some fifteen miles to the southeast, to Culemborg, where, in 1739, he remarried and was to father a further three surviving children. It was early in 1745 that he accepted the appointment in Breda. It was therefore in the few days following that a momentous relationship began between an outstanding teacher and a remarkable pupil. Hendrik Hoogeveen and Jan Ingen Housz not only respected each other's talents but they were to remain in close touch for the next 45 years, years in which the scholar was to have the unique satisfaction of paying back some of the great debt he owed to his instructor. The other momentous figure that entered the life of Jan Ingen Housz in his early teens was someone he met at home, when the school day was over.

A FTER HIS DUET with death, burying a child and then his wife, Arnold Ingen Housz seems to have had a stable middle-age, free of tragedy. He was an exemplary father, an astute businessman, a man of property and a respectable burgher.[51] Particularly sure-footed was his move away from trading in only one product. After the so-called 'Golden Age' when the Netherlands, throughout the seventeenth century, had enjoyed a buoyant economy, when nearly every citizen shared in an era of affluence, the next century was a time of financial stagnation.[52] This relative slump was not universal: some economic sectors continued to boom but many bubbles burst and there was a flush of bankruptcies. One area of decline was agriculture.[53] Arnold would have known very well of the epidemic outbreak of rinderpest, a viral plague of cattle and other ungulates that was endemic in Europe from the Middle Ages onwards. The highly contagious agent is, we now know, related to the measles virus of humans. It had been recognised, though, by the mid-eighteenth century, that quarantine and slaughter of infected herds was the only means, then, of restricting spread. Rinderpest amongst Netherlands cattle herds had peaked about 1720, just as Arnold was entering the leather trade. And, as the number of cattle succumbing to the disease began to climb again in the early 1740s, he would have been uneasy. Fortunately, for him, the outbreak was restricted, finally, to the same provinces, Holland and Friesland, that had been worst affected twenty years before.[54] The adept businessman always carries a contingency plan. It seems very likely that Arnold diversified, becoming an apothecary[55], lest the supply of hides dwindled or became available only at impossible prices. The

Arnold Ingen Housz. Courtesy of the Ingen Housz family.

eighteenth century apothecary was what we would now call a pharmacist; in the sense that he 'made up' medicines prescribed by a physician. Just as often, at a time of self-diagnosis and treatment, he simply supplied what customers wanted to buy. It is impossible to believe, though, that many visitors to the apothecary did not ask his advice even if this was entirely unschooled. Arnold Ingen Housz would have sold herbs and drugs, bitters and tonics – chemicals that were, to him, a mere extension of those he knew well as a merchant in the leather industry. In most apothecary 'shops' the customer could also hope to find other dry goods[56] such as spices, tea and coffee, nuts, sweets and even tobacco. The apothecary shop of the eighteenth and nineteenth centuries was the direct forerunner of the 'drug store' found in every community bar the very smallest in today's United States of America.

Jan may have outshone his brother at school but, at home, Louis seems to have held sway. The older brother followed his father into the family business, no longer only in leather. The solvents and solutes, salts and pickles, bleaches, dyes, tinctures and cosmetics – the effects that they had on tissues, living or dead, humans or hides – appear to have had a direct fascination for Louis, as did the mercantile machinations wielded by the man of business. Arnold may well have had some apprentices under his supervision by now but having a willing and enthusiastic trainee, who was also an elder son, would have been especially fulfilling. Not that all the industry around him had no impact on Jan. But here the sizeable influence was tangential. Experiences with his father – at the abattoir, at the tannery, at the workbench, with mortar and pestle and drug mill, in pill-rolling, in weighing and packaging, with labelling and at the ledgers, manning the retail counter – none of these held any terrors for Jan as he reached his teens. Both boys would have witnessed their father in the wholesale arena; wheeling and dealing with other merchants. Louis successfully took on his father's mantle and Jan was, in later life, very unlikely to take the least line of resistance in academic disputes. His life, however, was to be outside commerce.

In May 1742, Arnold, the oldest surviving son, travelled the fifteen miles to Tilburg to attend his mother's funeral.[57] Of her thirteen children, six were surviving to attend the final ceremony, after which Arnold's brother, Caspar, and his three unmarried sisters[58] appear to have moved to Breda. The centre of gravity of the Ingen Housz genus thus changed and Arnold, Louis and Jan found themselves part of a larger family circle. But new contacts were not only to be among blood relatives. It was at this time that another influence on the life of the young Jan Ingen Housz appeared. In 1742 an English army, 16,000 strong and temporarily led by John Dalrymple, Second Earl Stair,[59] crossed to Flanders and marched northwest across the flat, wet fields of what we now call Belgium to winter in Brabant. This 'preparatory pitch' by an 'expeditionary force' – has been, repeatedly, a dubious, if not disastrous, move in British military history. On this occasion it was the response to threatening gestures, by the French, that they might invade northwards and subsume the country of an ally, that of the Austrian Netherlands (Belgium). The Austrian Emperor, Charles VI, had died unexpectedly after a soaking on a Hungarian hunting trip in October 1740 to be succeeded by his young daughter, Maria Theresa.[60] Despots sat up, or were roused, more likely, by self-seeking advisers. They took the new Empress, only twenty-three, to be naive. She was. Political advisers also assessed her as weak because she was female. In fact, she was far from ineffectual but still her rivals saw the outposts of her empire as ripe fruit hanging at the ends of long branches. Here were the origins of the Wars of Austrian Succession and of the British force overwintering in Brabant. One of the divisions was encamped at Terheijden,[61] a village just four miles north of Breda. It was inevitable that the men, once freed of the chores of installing themselves, would drift into the nearest town in search of extra food, alcohol and entertainment (women) and not, necessarily, in that order. Their officers had more *bona fide* reasons to ride into Breda and their senior physician would have been anxious to locate the local apothecary. It was not long before a foreign officer appeared in the Ingen Housz establishment and announced himself in stumbling Dutch. The family archive is very specific:[62] 'Pringle . . . became acquainted with (Jan) Ingen Housz through the young scholar's father.'

As they became acquainted, Arnold Ingen Housz and Doctor John Pringle must have had conversations in which each described their past, the one with his soldierly smattering of English, the other with a stale, student's knowledge of Dutch. They discovered that they had many things in common. Ingen Housz would have described his military service in Scotland and explained his scars. Pringle would have told Ingen Housz how he had been born around Easter 1707[63] and that his new friend must have passed very near his Scottish birthplace – Stitchill House, the family home near Kelso, alongside the road to Edinburgh. He was from a distinguished Lowlands Scottish family that, like the Ingen Housz forebears, also included lawyers. Until his early teens Pringle had lived at Stitchill, having private tuition. He had then been sent the 100 miles north to St. Andrews to live with his cousin, Francis Pringle, who was professor of Greek at the university. After studying classics and philosophy, John Pringle had moved to Edinburgh University in order to study medicine but he found the teachers uninspiring. After a year he had left the city to continue his studies in Leiden[64] where he found that Boerhaave's excellent lectures and bedside teaching were attracting students from all over Europe. It had taken him only two years in Holland to earn his medical degree but, as is usual in one's student days, he had made some excellent friends, particularly a Dutchman called Gerard van Swieten with whom he had kept up an unbroken correspondence over many years. Piece by piece Pringle's story emerged. He had returned to Edinburgh to start a practice but only after a brief time in Paris where he had attended some further medical courses. He had been honoured with the Edinburgh chair in pneumatics, teaching morals and theology but, meanwhile, he had kept up his medical practice. A greater honour had been the recent invitation from Lord Stair, a patient of his, to take the position of chief physician to the British Continental Army with whom he had travelled, now, back to the Netherlands.

The Dutch merchant, Arnold Ingen Housz, and the Scottish physician, John Pringle, continued to surprise each other by the depth of their mutual interests and shared experiences. Just as striking was their joint realisation that, in young Jan, they had a fluent interpreter when their interactions in broken English or Dutch ran into difficulties. Jan realised that he could converse, rapidly and accurately, with his father's guest – in Latin. In fact Pringle was considerably impressed by the facility in someone so young that a firm bond developed between these two proponents of the Roman tongue – a bond that would be of such importance to the youth's future that these encounters, in the winter and spring of 1742/3 were, for Jan, seismic. John Pringle was to become his career adviser, professional mentor, generous host, life-long friend, counsellor and patron. We have reached a crossroads on the journey, one where the young Jan Ingen Housz was being signposted by very powerful but entirely beneficial influences in the persons of John Pringle and Hendrik Hoogeveen.

2
Higher Education, Family Practice

The space is more vertical than horizontal, a widening funnel of curving steps that channels light down to a circular central dais. An inescapable, revolting reek takes the reverse journey. The stench is dark, dominant; sickly, stale and clinging, hanging about at its nauseating worst in the warmest air up among the high windows. But the nearer one drops to the bottom of the cone, the stronger the visual assault. The block wooden table on which the body lies has a reflecting patina along its edges, polished by repeated greasy friction over several decades. All around it the floorboards are stained despite regular scrubbing. What volumes of blood, of cold body fluids, of globules of fat, of urine, bile and faeces have oozed their way between them, what indescribable organic slag lies beneath defies imagination. In the end the predictable smells from the fresh female corpse in front of them are no real challenge to the hardened medical students assembling, noisily and cheerily for their nine o'clock anatomy demonstration. They jostle each other to gain best vantage point and elbow their neighbours to make room on the narrow desktops where they crane forwards to see. It is a tight amphitheatre in which every cough, gulp or fart resonates.

As the noise reaches a crescendo one of the anatomist's assistants enters the arena from a doorway low down. He pulls off the dank, discoloured sheet and folds it nonchalantly as he backs out of the theatre. The linen had been a tent over the cavernous well that once contained the guts and other organs of the abdomen, dissected and discarded last week. The skin flaps, one with an umbilical eye, drape like segments of wet pastry into a half-eaten pie. In the momentary limbo a student at the front darts forward, eyeing the access corridor and lifts up the arms of the corpse so that her hands cover her naked breasts. He is back in his place as a raucous cheer erupts and desks are banged. The reappearance of the mortician with a tray of glistening knives and hooks ends the hubbub abruptly and he pointedly replaces the upper limbs of the cadaver into the orthodox anatomical position before balancing the tray across the torso. There are mutters of approval. He has always been popular with the students. He understands their need for black humour.

HIGHER EDUCATION, FAMILY PRACTICE

THE NEXT LEG of our journey is as if through a tunnel. There is little in the way of individual record for the later years of Jan Ingen Housz's education, nor of the early years of his career. It is possible, though, to avoid a total hiatus without writing a novel. Generalisations and the sparse, isolated facts can get us through. The next move in Ingen Housz's education should be no surprise. Although there was a family tradition in the law, Ingen Housz opted to study medicine. Like his father, Jan was breaking new ground. But his father's post-military career must have been relevant and germane. Anyone living in the house of an apothecary gained many insights into the multi-faceted interface between health and disease, well-being and misery, life and death. And anyone with a growing intelligence was going to be sensitised to the frustrations of the apothecary himself who frequently regretted that he could not do more to relieve suffering for want of greater depth of knowledge. Jan's ambition was easily nurtured by his father's daily experiences. There can be little doubt that humanity was more on show at the apothecary shop counter than when at home in a sickbed: a physician visited his affluent patients and advised them in a capsule of absolute privacy. Ordinary people, however, stood at the counter of Arnold Ingen Housz's shop and relayed their intimate discomforts for all to hear, accepting advice and a twist or a bottle to the flinching appreciation of those behind them in the queue. Many a grapevine of gossip had its roots in number 17, Eindstraat. And, when customers waiting in a queue knew the remedy they wanted, the sons would 'do' as well as the father to concoct their 'usual.' Familiarity and job satisfaction led Louis into becoming his father's trainee and partner but Jan had the talent and desire to take his homely knowledge of medical 'trade' to another level and become a qualified physician. To an unknown extent, but most certainly of relevance, was the inspiration and guidance of Dr. John Pringle, the Ingen Housz family friend and visitor. In later life Ingen Housz hinted, very strongly, that Pringle was instrumental in his choice of career.[1]

> No man upon earth can have stronger reasons for a due sense of gratitude than I acknowledge to you. You bestowed many civilities upon me, who had never been in the way of doing you any service whatever. You granted me your friendship almost as soon as I was acquainted with you.

The gratitude is palpable and profound but the ambition was littered with steeplechase hurdles. There were many possibilities of stumbling and a soaking. To become anything more than an apprentice barber-surgeon, medicine required admission to a university. It meant, therefore, a facility in Latin and Greek. It required a headmaster's recommendation; a father's consent and financial support; and a favourable outcome at a candidature interview. The last requirement was still outstanding in early 1747 and the sixteen-year-old Ingen Housz[2] travelled to the University of Louvain to be interviewed by the principal of the medical school. Here he took a tremendous gamble. Perhaps the impetuosity of youth is always a chemical display but whatever the physiology, it was a glimpse of the personality yet to mature; unique, determined, and far from docile. The story comes down to us from within the family,[3] from one of his nieces and therefore reliably:

> ... his youth caused some anxiety for the Principal. Ingen Housz was asked to confirm that he had had sufficient grounding in preparatory science and that he knew enough Latin and Greek to be able to follow the courses successfully. Somewhat offended by this, the prospective student snatched a book from the table in front of him. It was a Greek edition of the Old Testament. 'Which part do you want me to translate into Latin for you?' he challenged. After he had very skilfully translated a passage he could sense the Principal's amazement. He also offered to translate some unseen Latin into Greek or, if preferred, into his mother tongue, Dutch. After this, his reputation for exceptional ability was never challenged.

In taking an intellectual lunge at his inquisitors the teenage Ingen Housz was playing for high stakes, playing Russian roulette with his career. Fortunately, the professor allowed him to respond to his self-inflicted challenge, to throw himself at his Gorgon, and he came out the winner. Even he must have seen, later and with a cooler head, that he had been lucky. He was admitted as a student in the medical faculty, a place he probably took in 1747 but we have no actual record of the date of matriculation.

Louvain was in a hotly contested war zone in the middle of the 1740s: the War of Austrian Succession was reaching a climax. The British had withdrawn their troops home to resist a further Jacobite incursion in Scotland – the '45 rebellion', Dr. John Pringle among them. The cities of the Austrian Netherlands had then been overrun, or besieged at best, by the French army. Louvain was no exception and Ingen Housz may have had to wait before being able to go to live there safely. The supposition that he did matriculate in 1747, shortly after his fractious interview, implies that he took six years to graduate. However this is longer than usual for that era and there may be an alternative explanation.[4] At some point in his youth Ingen Housz went to live in France in order to build up fluency and confidence in French, then the universal language of diplomacy and much commerce. Everything seems to point to it being at this time of his life – it was certainly before he graduated. The fact that his immersion in another culture was at Douai is ironic. Did he choose the fortified town because he wanted to pay homage at his uncles' graves or was there some special financial dispensation in their honour? We don't know but for his studies to have been effective he probably stayed in the town at least six months. Why the University of Louvain was chosen as his *alma mater* is subject to much less doubt. Catholics were barred from matriculating at Dutch universities at this time whereas Louvain, closer to Breda than many of the other Dutch cities themselves, had been a papal stronghold for over three centuries. It was one of the oldest universities in Europe[5], founded by Pope Martin in 1425 and was soon held in high esteem, especially for the study of Catholic theology. From early in its existence it was awarded the status of *studium generale*; an institution that had the right to receive students from all over Europe and award degrees that were an international currency. By the seventeenth century it had acquired a particularly high standard for theology and philology and there were now schools of law, medicine and 'arts'. The arts school was a catch-all – students enrolled to study mathematics, physical sciences, philosophy, literature and history. The university had long flourished; size and repute both taking impressive upward curves thanks, in part, to the enthusiastic support of several European royal houses – of Brabant, Burgundy and the Habsburgs. All the while the professors managed to keep their independence of thought – as long as they didn't try to undermine, from within, the ramparts built against the Reformation, the tide of Protestantism that swept across Europe from the beginning of the fifteenth century.

Ingen Housz was, it seems, an undergraduate at a university that was well respected though still, in many ways, a medieval institution with age-old attitudes. This applied to the medical courses as much as to others. Some of the contemporaneous, manuscript pages of the notes that Ingen Housz made in lectures have survived, some 400 folios, all in Latin.[6] The material is dense, didactic and from a long-superseded era of clinical understanding. The physiology and pathology constructs, especially, are antiquated and built on what are now long-abandoned ideas. Ingen Housz learnt on a conceptual framework that was largely medieval although this would disintegrate during his professional lifetime. He was of a generation of doctors who would all face an intellectual dilemma during their careers. Either they could cling on to a hard-won appreciation of what were dark age doctrines or try to adjust their practices to take advantage of new developments in medicine, some cerebral and some practical. In fact, an educationalist would argue that a university degree should be a signal that the graduate has been primed to perform the latter task throughout his life, whatever his profession. But Utopia is a remote country.

In reality it would take several generations of doctors before the 'humoral' theories of disease were finally abandoned. Many physicians continued wearing the symbolic full wig and black silks and persevered with the Hippocratic belief that health was a precarious balance of body fluids, of black bile, yellow bile, phlegm and blood. This equilibrium was even seen to underlie mood and personality – too much black bile made a patient 'melancholic'; too much blood made one 'sanguine' and liable to be referred to a surgeon to be bled. The beginnings of the revolution in clinical ideas had begun in the late sixteenth century when serious observations and measurements of our world began to replace beliefs. Harvey, Galileo and Newton are names familiar to everyone. They represent just a few generations in which extremely rapid and impelling progress was made in science, in a proper understanding of the world based on structure and function and this inevitably spilt over into medicine.

The concepts based on stagnant 'humours' were first translated into visions that health related to fluid movements. Then the mechanics of muscle movement and the obvious irritability of nerve tissue led medical trailblazers towards a modern understanding of physiology. Linked to this intellectual progress was the institution of bedside teaching, notably at Leiden by Hermann Boerhaave, who had been, incidentally, an inspiring figure for the young John Pringle. It may seem bizarre, now, to think that physicians once trained entirely by book-study, in classrooms next to those of their peers learning to cant theology in exactly the same tradition. Boerhaave began a revolution. His students attended his lectures but they then followed him to the hospital ward and afterwards to the mortuary. They learned from actual patients, as did their teachers. Symptoms were back tracked from the gross abnormalities revealed at post-mortem dissection. Disordered function and displaced anatomy were being integrated, observation was replacing speculation and a new nosology of disease was being pioneered, just as Ingen Housz was a medical student. Eventually, medical training took on this model at all universities and it is, in essence, the pattern of modern medical education.

On the face of it, the University of Louvain was not among the pioneers but we may be doing it an injustice for we have no records. Sadly, the ancient city was sacked and burned by the invading German troops in 1914. And the few surviving archives were destroyed by stray bombing in the next world war, particularly in late 1944. We do know, though, that as regards the teaching of anatomy, the one medical discipline traditionally built on real observation, Louvain was rightly respected. One of the alumni of the medical school[7] was the great Andreas Vesalius, the founder of the modern approach to human anatomy.

UNIVERSITY IS A breeding ground for life-long friendships. Sharing hardships and stresses as much as privileges encourages confederacy in people thrown together. Ingen Housz may have made many friends at Louvain but we only know of one. Johann Hendrik Deckers was a peer from 's-Hertogenbosch (Bois le Duc), also a Catholic and a medical student. The ceremony conferring the degree of MD, doctor of medicine, on Jan Ingen Housz was held on Tuesday 24 July 1753.[8] As was the tradition, a broadside[9] poem was published, in Latin, that located the blushing graduate into his ancestral pedigree and childhood community and was recited in the presence of the family. Some over-indulgence can be forgiven: there was an analysis of Jan's erudition and personality that evoked the talents of a triumvirate no less prestigious than Virgil, Cicero and Socrates. And there was express gratitude to his father and to Hendrik Hoogeveen as well as reference to his valiant uncles.

Celebrations over, Ingen Housz took another unexpected path: he remained a student. There have been reports that he went to Paris, attending lectures there as a postgraduate and then, after an unspecified time, travelled the considerable distance to Edinburgh, for further specialist study at the famous medical school there. The story is uncorroborated, however, and there is a danger that previous investigators of his life have dislocated Ingen Housz's whereabouts and activities immediately after qualification,[10] for he certainly lived in both cities for significant periods of his later life; but this was long after studentship. It is certainly the case, however, that the young Doctor Ingen Housz travelled to Leiden[11] where he refined his anatomical knowledge and studied physics, as it is now known. There was no religious embargo directed at Catholic postgraduates provided they had no ambition to acquire a formal teaching post.

Ingen Housz attended lectures and demonstrations given by the famous anatomist then at Leiden, Bernhard Albinus, who held, by the 1750s, the chair of medical practice. His atlas, of 1749, of the human bones and muscles, was rightly famous and he was a very popular teacher. Another ex-pupil of the great Boerhaave was Hieronymus Gaubius, professor of medicine and chemistry – effectively head of the school of pathology. Ingen Housz became, then, a student of Boerhaave at one remove but he was a living associate of another Leiden professor who was to have an exquisitely important influence in his life, Pieter van Musschenbroek.

Born 1692, Musschenbroek was the son of a scientific instrument maker whose ancestors had come from Flanders early in the seventeenth century. Having been awarded a doctorate in medicine at Leiden, his home town, in 1715, he went on to postgraduate studies that included a spell in London at the feet of Isaac Newton. Various teaching posts then led him back to Leiden where, in 1739, he took the chairs of medicine, mathematics and philosophy. He was yet another pupil of Boerhaave to become part of the pedigree, to perpetuate the great man's influence. For the best of reasons, Musschenbroek is remembered for his explorations of phenomena in static electricity. In 1745 he was making repeated attempts to store electrical charge. Friction machines could, by this time, readily generate the mysterious energy, then conceived as an invisible fluid. But the force was fleeting and weak and attempts to stockpile it had all failed. There had been little real progress beyond mechanisation of the original discovery, in antiquity, that a charge could be produced by rubbing dry amber with a cloth – hence 'electricity', 'electron' being the Greek for amber. The breakthrough came in the autumn of 1745 when Musschenbroek, working with a student, Andreas Cunaeus, discovered that electricity could be stored in a glass jar full of water, the energy being led in via a copper rod suspended centrally in the liquid. The energy, they found, went on accumulating until the rod was connected, originally by hand, to the outside of the jar. The resulting jolt could be very unpleasant and the sweaty palm and fingers were soon replaced by a metal chain with a link that could be joined or broken at will. Here we have the 'Leyden Jar'. It was a considerable advance in this new branch of physics, a simple but highly significant piece of apparatus now familiar to every pupil making a decent study of electricity. For Ingen Housz to be able to study with Musschenbroek at Leiden was a singular privilege and must explain his life-long and intense

An early Leyden Jar – the first means of storing electricity.

interest in all things electrical. He would even honour the memory of his mentor by initiating, himself, a major development in the field.

INGEN HOUSZ WAS able to enjoy at least five years as an undergraduate at Louvain and two years more of postgraduate study at Leiden, all at his father's expense. In brief periods at home he may have contributed to the family economy by helping in the business but his education would have been a considerable drain on the family assets. What, then, finally led him back to his roots in 1755 and to start a medical practice in Breda – a few houses along from his father and brother – at number 23 Eindstraat?[12] He was qualified for a higher status in his profession but chose to become, in modern terminology, a general practitioner or family doctor (*huisarts* in Dutch). There may have been an element of duty, guilt even. It was nothing unusual, in the medical profession at that time, for graduates to return home after their training and many practices were dynastic. Deckers, his student friend, is another case in point. He returned to 's-Hertogenbosch and was to remain in practice there for the rest of his life.[13] Medicine also has a long father to son tradition and there are still, today, some general practices, in the UK at least, with unbroken generational links, all in single communities, that have lasted over 200 years.[14] Though they were not qualified medical practitioners, Ingen Housz seemed to have an innate desire to help his father and brother[15] and repay them for their generosity in supporting him as a student. Being close at hand and in a parallel profession certainly had its attractions. Two apothecaries and a doctor in the same family could only be a lucrative symbiosis. A 'customer' needing more than a simple purgative was, rightly, referred to the doctor and, after a diagnosis was made, the 'patient' was given a prescription for the apothecary to prepare. This is certainly a virtuous cycle but open to abuse. However, the citizenry of Breda appear to have respected the Ingen Housz name, that they weren't being duped. It is certainly true that the medical practice flourished so successfully that it became an intolerable burden for one man. The apothecary shop, however, with more restricted hours and delegation of responsibility more available, remained an honourable and valued local business for at least another 60 years.[16]

THE LIFE OF a 'surgeon-apothecary' – a general practitioner – in the middle of the eighteenth century was a lonely, frantic struggle in a

Jan Ingen Housz aged 35. Courtesy of the Ingen Housz family.

vortex of contagious diseases, malnutrition and injury. Armed only with a level of knowledge barely better than common sense, a saddle-bag of simple instruments and dressings, and a plentiful supply of vomifacients and laxatives, the community 'surgeon' was a 'jack of all trades' and always on call. Toothache or worms might wait until next morning but an obstetric disaster or late-night fall would deprive the practitioner of his sleep. Nor was there likely to be respite from teaming up with colleagues. There was a robust tradition of single practice: medical partnerships were rare until at least a century later. Indeed, doctors considered their 'trade' as necessarily competitive. Only the deeply unpopular had an undemanding schedule but being out of favour meant being out of pocket. Neighbouring professional jealousies frequently poisoned the atmosphere between practitioners and there appears to have been an element of this at Breda by the early 1760s.[17] Intellectual isolation was a major frustration for Ingen Housz; wider horizons beckoned. In his limited free time he turned to 'philosophy' (science) for recreation. His keen interest in electricity had survived his 'busy times'.[18] By 1763 he had invented, using homemade apparatus, a revolutionary electrical machine that incorporated a glass disc rather than a sphere[19] and he would test his theories oblivious of time and often long into the night. The vibrant flashes of blue light and the muffled cracks and bangs drew the attention of his superstitious neighbours and many assumed that he had some sort of communion with the devil,[20] that he was 'sitting up in the company of the Father of Evil.'

The growing dilemma in Ingen Housz's life, the need to break out of the backwater that was medical practice at Breda versus loyalty to his father, was resolved in 1764. On Monday 30 April, Louis married the 26-year-old Maria Stuyck,[21] the beginning of a long and productive union. More relevant to Jan's own future, however, was the sad fact that the wedding was only a joyous prelude to something more sinister: Arnold Ingen Housz was ill. The previously robust health, a vigour that had served him well for some 45 years since recovering from battle wounds, broke up and he was forced to seek medical attention. Perhaps to avoid distress, he did not consult his son but one of the other local practitioners. The advice, no doubt sincerely given, was not universally approved. Jan saw it as inappropriate, based, he suspected, on a wrong diagnosis.[22] His inner conflict was all the more painful as his father deteriorated and died, on Thursday 12 July 1764.[23] No longer having to be a bedside diplomat, Jan determined to resolve the diagnostic difference of opinion and, amazingly, proceeded to perform a post-mortem dissection of his own father.[24] Ultimately, however, his father's death presented an opportunity. Within a very short time of the burial, Jan Ingen Housz was making plans for a European tour.[25]

3
Enlightenment London

Sunday 16 December 1764. At three thirty on a day that has been wholly overcast and made miserable by scuds of rain, it is close to dark. As he turns the corner into Pall Mall he is nearly knocked over by two young men running up towards Charing Cross. They are at full tilt and his eyes follow them automatically until they dart into an alleyway and disappear. Only then does he realise what is happening. On the south side of Pall Mall, about thirty yards away, a well-dressed man is being helped to his feet and two women are waving and shouting for help. He hurries to the spot. Others are converging including two liveried servants from the nearest house. The victim of the crime, a man in his sixties, has been attacked from behind. The footpads knocked him off his feet, bundled the women aside and relieved him of his walking cane, his watch and his purse. Ingen Housz explains that he's a doctor and is allowed to make a cursory examination. The man gives his name: he's a titled gentleman but Ingen Housz is still finding English difficult, especially out of context, and the details don't register. Nevertheless, he can announce that his patient is uninjured, though seriously shaken. The two ladies begin to straighten their father's wet clothing and one of them retrieves his wig and carefully replaces it. Ingen Housz senses that he has done all he can and, as he begins to make to leave of what is now a small crowd, he sees, approaching from the other end of Pall Mall, two men who stop immediately outside Dr. Pringle's house. One of them wears no wig, his mane of long white hair giving him a unique appearance. The two of them are obviously in the middle of an intense discussion and stand, for some seconds, before the taller of them, bareheaded and in brown velvet coat and breeches, breaks off, uses the balustrade to help his large frame up the steps to the front door and pulls the bell cord. Despite the activity and buzz of conversation around him, Ingen Housz is transfixed. When Pringle's butler answers the door and greets the guest with a warmth you would expect only from a long-standing friend, he is convinced. He has just seen his hero of many years; a personal champion and inspiration for over a decade and a man he never dreamt that he might ever meet. Dr. Franklin must also be a dinner guest of Dr. Pringle: for Ingen Housz, this is going to be one of the most memorable afternoons of his life. He is so buoyed up and distracted by the prospect of actually meeting Franklin that he walks out into the middle of the street and is nearly run down by a carriage.

ON THE EVENING of Thursday 15 November 1764 the lamp was lit above the Fleet Street entrance to Crane Court. The Royal Society of London[1] was assembled for its weekly meeting, the new President, Lord Morton, in the chair.[2] Fellows were permitted to invite guests to meetings and Doctor Matthew Maty had brought two.[3] One was 'Dr Ingenhouse.' This is the earliest reference to Jan Ingen Housz being in London. The meeting was to focus on astronomy but the attendance of Ingen Housz was based on something more tellurian. Since they were both Dutchmen and both medically trained, Maty and Ingen Housz had much in common. Their keeping company made excellent practical sense. Matthew Maty – pronounced 'Matty' if the various phonetic spellings of his name are compared – had been born near Utrecht in the Netherlands in 1718, the son of a Huguenot clergyman.[4] He studied medicine and took a doctor of medicine degree at Leiden, all against his father's wishes. But, in turn, his father was also unpopular – with his ecclesiastical superiors – so much so that he decided to move himself and his family to live in England.

In London, Matthew Maty may have practised medicine for a time but '. . . to make himself known . . .' he constructed an alternative career as a linguist. He took to translating, into French, serial editions of *Journal Britannique*, which, despite its Gallic-sounding title, was an English magazine published at The Hague. This successful tactic brought him into literary circles in London where his name did, indeed, become familiar. He seemed to carry the same potential for antagonism, however, as his father. For some reason Dr. Johnson said of him '. . . the little black dog, I'd throw him in the Thames . . .'[5] Notwithstanding, Maty seems to have remained popular with, and valued by, the majority of his contemporaries and was elected a fellow of the Royal Society on 19 December 1751.[6] Maty was, from 1762, the foreign secretary of the Royal Society, and from 1765, overall secretary.[7] He was also appointed as one of the under-librarians when the British Museum first opened, in 1753.[8] For all these reasons Maty was a strong representative of a cohort of vigorous, inquisitive and enterprising men who were, in Britain, during the latter half of the eighteenth century, to generate an intellectual and, consequentially, an industrial revolution. But whatever his background and achievements, there can have been few men more immediately suited to bring Ingen Housz to a meeting of the Royal Society. Maty, after all,

Matthew Maty.
Courtesy of the
Wellcome Library,
London.

spoke Dutch. We have no means of knowing how much English Ingen Housz spoke or understood when first in London but without whispered clues in his native tongue it is easy to see how an evening dedicated to complex astronomy would have left him floundering. The astronomer, Neville Maskelyne,[9] opened proceedings by reading a long paper on estimating the distance between the moon and the sun. A discussion on calculating longitude followed and there were then various reminiscences of the June 1761 transit of Venus. Instructive or not, it was very clear that Ingen Housz had been welcomed into an exclusive company. Here, in the very room where Newton himself had once presided, he was among the country's intellectual elite. There was no university in London until well into the next century. The Royal Society was the capital's 'high table' and here was Ingen Housz, already a diner.

Ingen Housz's move to London was planned and deliberate. He was 33 years old but his passage was no impulsive act of early middle age. It is not known, exactly, when he embarked for England but we can isolate the time to the six-week period between late September and mid-November, 1764. During August and September of that year, within a few weeks of his father's death and long before the estate could have been settled, Ingen Housz visited the Breda lawyer, Jacob de Bruyn. He arranged for a deed to be drafted that would give to his brother, Louis, the necessary powers to handle his inheritance ' . . . in the event of my being absent (abroad) . . . .'[10]. Then, presumably taking Louis with him, the doctor called on the lawyer again, so that the deed could be signed, on Monday 24 September. The story within the family[11] refers to Ingen Housz having been invited to London, severally and over a long period, by John Pringle but that he had always refused to leave his ageing father's side. Now his father was gone. Louis had taken over the family business and had recently married. Change was in the air. It is also evident that, after a decade in single-handed medical practice in Breda, ten years in which he could rarely have been off-duty, Ingen Housz was weary and professionally stale. Forever prone to being called to patients, whatever the hour or the day, he knew he could not always maintain the impetus, certainly if he was to have any residual energy for his other pursuits. Of the two obvious options, taking an assistant or a sabbatical, he chose the latter.

It is a reasonable assumption that Ingen Housz had amassed enough money from his medical fees to be able to afford a break in his career and his bank balance was about to be boosted by his share of his father's wealth. There are no clues as to the destiny of his practice but in the eighteenth century medical care was entirely episodic. A doctor could simply abandon his practice though he might try to raise capital from his business by arranging sale of the goodwill as well as any property he owned. Whatever the reasoning and the planning, a sense of relief still pervades the phraseology of a letter Ingen Housz wrote, about a year later, to an Edinburgh professor.[12] '. . . released from confusing and busy pursuits, which I have given up for now . . .' Ingen Housz had an open invitation to London but he had planned a more complex trip, a fact intimated by the same correspondence.[13] '. . . intending to travel through parts of Europe in order to enrich my mind by means of talks with philosophers . . .' He wrote, giving more detail,[14] to Hendrick Hoogeveen; that he intended '. . . to travel through England and France to Italy, and subsequently to return through Germany.' Hoogeveen knew his young friend well enough, however, to suspect that the real ambition of the Ingen Housz tour was to be in Italy. '. . . he, who, after a long absence from here, I would have sought long in Rome.' His 'grand tour', then, was to be spiced with meeting 'philosophers', men we would now call scientists, but culminate, as for so many 'tourists' of the eighteenth century, in a total immersion in Roman antiquity; in columns and obelisks, marbles and mosaics, statues and frescos. Perhaps Hoogeveen had not realised that the young doctor also had a medical mission. Ingen Housz hoped to learn, in England, about smallpox inoculation.[15] This is certainly apposite for it was in England, during the previous decade, that the most dramatic progress had been made in this procedure.[16] This salient fact, perhaps, and the long-standing personal invitation from Pringle, his medical patron, appear to have instigated the first leg of his travels; a visit to London.

MID-EIGHTEENTH-CENTURY LONDON MUST have been alien and initially threatening to a provincial Dutchman. The city was easily the biggest conurbation of the eighteenth-century world. Its residents had defied the royal edict, after the great fire of 1666, that rebuilding should be systematically ordered. Ambitions for wide avenues and more rational street plans were being leisurely considered, at royal request, by fellows of the Royal Society while householders returned, en masse, and got on with reconstructing their homes, regardless, and in the former higgledy-piggledy pattern.[17] The slums had soon reappeared and the bursting of London's medieval seams continued as more walls were breached or devoured by adjacent buildings. The old city gates were demolished or left stranded as frustrating fossils that constricted the flow of traffic. Eighteenth-century expansion pushed out further in all directions, the Thames irrespective. The Fleet, the other main London waterway, was being covered over in 1765 and many new developments would straddle what had become a subterranean sewer, as opposed to an open one. Villages in a rural hinterland were becoming what are now familiar inner-London place names: Spitalfields, Stepney and Bethnal Green to the east; Sadler's Wells and Finsbury to the north; Southwark, Lambeth and Vauxhall to the south. All were swallowed up in a dense, disordered sprawl of overcrowded and insanitary tenements where

> there are such great multitudes of people brought to inhabit in small rooms, whereof a great part are seen very poor, yea, such as must live by begging, or by worse means, and they heaped up together, and in a sort smothered with many families of children and servants in one house.[18]

On the western fringes of the old city developers speculated in high quality habitation. Houses were larger and better appointed. The best craftsmen were allowed the finest materials. There was ambitious planning of generous spaces drawn into elegant squares and imposing terraces. The ploy worked. A 'west-end' address attracted a classier clientele and the bandwagon rolled. The boom pushed up prices so much that a positive spiral ensued - only the most affluent could afford the exclusive addresses of St. James' or Mayfair. This still applies, even today; evidence of the build quality. This was the London to which Ingen Housz came in 1764: a core of soot-stained near-chaos in and immediately around the old city, desirable town houses for the rich in and around Westminster and a burgeoning suburbia spreading, amoeba-like, in all directions.

The best estimates of London's population in the 1760s agree at about 650,000. It was a city of youth and of their restless energy. Large numbers of young men and women were drawn in as recruits to the army of servants but domestic service was profoundly insecure. Employers could hire and fire at will and, without wages, there were only two ways, in essence, to survive: crime or prostitution. In the many cryptic alleys and lanes theft, and flight, were easy. The majority of crime was petty but it was profitable: peddling illicit property was not a problem in the buzzing anonymity of the streets. Punishments were extremely harsh for the convicted but, ironically, most prosperous Londoners rubbed along on the premise that pilfering and soliciting were unavoidable nuisances. The poor, on the other hand, saw selling themselves or swindling as making the difference between starvation and some sort of survival. In *The Fatal Shore*[19], his history of the transportation of British convicts to Australia, Robert Hughes gives a description of eighteenth-century life in London; the unrecorded and largely unacknowledged life of the ordinary people. For Hughes it is relevant because it was the poor who risked this enforced exile and it is a necessary antidote to the romanticized notions of life in Georgian London. His descriptions would be hard to better:

> The rookeries of the poor formed a labyrinth ... choked with offal. Because men had to live near their work, tenements stood cheek by jowl with slaughter-houses and tanneries. London was judged the greatest city in the world but also the worst smelling. Sewers still ran into open drains... Around St. Martin's, St James's and St. Giles-in-the-Fields there were large open pits filled with the rotting cadavers of paupers ... 'Poors holes'.[20]

Living conditions at the lower end of the social spectrum were so bad that, for some, dealing in faeces was the only way to survive. 'Pure-finders [were] old women who collected dog-turds which they sold to tanneries for a few pence a bucket (the excrement was used as a siccative in dressing fine bookbinding leather).'[21] In such circumstances it is no surprise that the gamut of contagious diseases was endemic, that average life expectancy was modest. Only a minority of children ever reached maturity. Here was the substrate of the medical experience waiting for Ingen Housz in London. A milieu in which death and disfigurement from diseases such as smallpox were routine, where illness led to destitution, where the fate of the vast numbers of children, legitimate and illegitimate alike, born to parents who could not afford to feed them, was unspeakable.

London life for the aristocratic and wealthy could hardly have been more different. Stark social inequality in Britain has an extremely long pedigree. For the 'upper crust', over-endowed with money and privileges, the metropolis was a 'seasonal' delight.[22] It held all the attractions. From their solid and serviced houses they sauntered out, elegantly bewigged and accoutred, to coaches waiting to take them to levées, concerts, balls, plays, parties and dinners. They ate huge meals that could take hours, sitting to table in mid afternoon since the unseemly concept of work did not impinge on their days. But 'the devil makes work'. Once replete, they repaired to pleasure gardens such as Ranelagh or Vauxhall; to assembly rooms such as Almanacks or the Pantheon. There was rumour, gossip, intrigue, titillation and more. Entertainment of the most public kind was plentiful, certainly, but so was that of the most intimate. The idle affluent often had complicated relationships and serial sexual adventures. As long as honour was not impugned by blatant dalliance all was well. But they still had their problems; legal entanglements for which they needed lawyers, diseases of over-indulgence for which they turned to doctors. The physician was the archetype of the city's middle-class. London had about a thousand of them in 1765. If successful, personable rather than likely to keep you alive, physicians could make themselves a fortune from a following of rich patients. Being fashionable gave them the assets to lease, or buy, a desirable London address. Of all the possible landing stages in London in 1764, this is the one on to which Ingen Housz disembarked. He moved into a circle that was largely non-conformist while tolerant of beliefs, intellectually inquisitive, unafraid of controversy, champions of the Enlightenment but, at the same time, well-heeled, well-connected and even at court. But in attitudes, interests, fashions and routines, this tier of London society represented a rise from his social status in the Netherlands where, as Schama has reported[23], a much larger segment of the population were embedded in a uniformly providential middle class, the *brede middenstand*. In Britain there was a clear dichotomy between the professional and the tradesman. In time, Ingen Housz would come to see this but not how the friendship and patronage of such influential individuals, especially that of John Pringle, would change his life. Nor that, in socialising with the intellectual descendants of Vesalius and Harvey he was surfing a wave of revolutionary thinking in medicine in which blind dogma was being pushed aside, more and more, by the insights of organised investigation. As was well summarised by one of the circle, in 1794

. . . the art of healing is becoming every day more conformable to what reason and nature require; that the errors introduced by superstition and false philosophy are gradually retreating and that medical knowledge, as well as all other dependent upon observation and experience, is continually increasing in the world. [24]

IN THE TWO decades since visiting the Ingen Housz family home at Breda, Pringle's career had blossomed. When his military patron, the Earl of Stair, resigned from command of the British continental army after the Battle of Dettingen in 1743, he had recommended Pringle to his successor, the Duke of Cumberland, second and favourite son of George II. Cumberland had asked Pringle to remain as 'Physician General to the Forces in the Low Countries and parts beyond the seas'. Towards the end of 1745, the Duke recalled Pringle from Flanders and took him, as the senior physician, with the Royalist force sent to quell the mounting Jacobite rebellion in Scotland. Pringle was therefore attendant on the Duke's army at Culloden on 16

Sir John Pringle. Courtesy of the Wellcome Library, London.

April 1746, but as physician to the force he was probably back at Nairn during the actual battle.[25] The campaign might have been very difficult for Pringle. Here was a Scotsman serving an English royal general who, with a military efficiency that has often been presented as callous, killed some 1000 of Pringle's countrymen. But Pringle, as a Lowland Scot, was as natural an enemy to the Highland clans as any Englishman.

Pringle returned to Flanders for a further two years,[26] retiring to England only after the peace treaty, signed at Aix-la-Chapelle in 1748, ended the War of Austrian Succession. In 1749 Pringle began to build a physician's practice in London. He was hardly likely to fail for he already had the esteem of both the royal family and of the Royal Society. The Duke of Cumberland immediately appointed him his private physician and Pringle was already an intimate of other very important men in the capital. The journal book of the Royal Society records his presence as a guest of the President, Martin Folkes, on Thursday 14 February 1745.[27] By the end of the year he was a fellow. The certificate proposing his election, dated Thursday 25 April, was exhibited that evening and for the next ten meetings and his fellowship confirmed on Thursday 31 October.[28] Pringle was back in London in time to sign the 'Obligations' of the Society and to be admitted on Thursday 7 November. The eight signatures on his election proposal are, besides that of the President, those of some very influential sponsors. The prime mover was the famous London physician Richard Mead, a vice-president of the Society. Here we see, explicitly recorded, the customary patronage of the times. Pringle appears to have deserved his sponsorship for he was soon making important contributions to the medical literature; shrewd and systematic observations made, during his years as a military physician, on the diseases suffered by the incarcerated and by soldiers in the field. These infections were numerous and significant; there were often more burials between battles than after. In 1750 Pringle published, as a letter to Mead, his *Observations on the Nature and Cure of Hospital and Jayl Fevers*.[29] This proved the commonality of the two infections – both were a 'distemper' which, in modern terms, we call typhus. His findings also supported his urgent plea for basic hygiene to be introduced into barracks, prisons and hospitals. And for a series of papers to the Royal Society on infections, works that introduced the terms 'septic' and 'anti-septic', Pringle was awarded the Copley Medal of 1752.[30] All this was crowned by the publication, in the same year, of his most famous work *Observations on the Diseases of the Army, in Camp and Garrison*.[31] Pringle was among the first to suggest that many diseases, capable of killing whole regiments, arose from packing humanity – soldiers, sailors or prisoners alike – into confined spaces. His theory was built on the popular construct of the time, that men became ill because of breathing 'close and putrid air'. Unique, though, among his contemporaries, and inspired by his reading of Bacon,' . . . that whether or not anything can be known – is to be settled not by arguing, but by trying,'[32] Pringle had been able to demonstrate the efficacy of good ventilation. In 'Baconian' science the sleeves are rolled up and progress is made by trial and error, however taxing or tiresome. Ingrained customs are eschewed and the effects of the changes, for good or bad, are honestly observed and recorded. As Physician General to the Continental Army, Pringle had had both the experimental material – the troops – and the authority of rank to follow these precepts. He became a strong advocate for the sanitary needs of the troops being subject to a discipline that was outspoken at the time and often resisted. His 'camp rules' were that indiscriminate fouling by the troops should be punished, that latrines be dug and always used; that they be covered with fresh earth daily, and that if dysentery did break out, camps be moved. It is not surprising that Pringle is often seen as 'the father of public health'. But he was not always interdictory: the remarkable humanity of the man is easily overlooked. As is recorded in his obituary,[33] it was Pringle who was the driving force behind a novel martial agreement to prevent the sick and wounded having to be moved far away from battlefields and therefore away from the care of their physicians and surgeons.

> The Earl of Stair, being sensible to this evil, proposed to the Duke of Noailles (The French Commander), when the army was encamped at Aschattenburg, in 1743, that the hospitals on both sides should be considered as sanctuaries for the sick, and mutually protected.

Here can be seen the true origin of the International Red Cross movement which is usually attributed to the nineteenth century. There is no disputing the humane consequences. The sick and mutilated could be tended in safety and without fear of interference from the enemy when fronts moved. And when the principle was later extended to casualty clearing stations, the assessment and care of the battle-wounded could be immediate and more effective. No longer did they have to be carted away to be tended only after hours of agony, weakened by secondary blood loss and their wounds even more heavily contaminated. It is not fanciful, therefore, to trace modern trauma triage back to Pringle's vision.

Pringle's personal life is barely recorded. We know, however, that he did marry – on Tuesday 14 April 1752 – Charlotte, the second daughter of Bath physician William Oliver.[34] Oliver is now only remembered for devising a recipe for a successful savoury biscuit that remains popular even now – the

'Bath Oliver'. Charlotte was not so enduring and died some two years later. There were no children. Pringle's brief spell as a family man was extinguished barely before it had begun. For consolation he was left with his viola and his poetry. Professional and philosophical matters reclaimed centre stage. He accepted the honour of serving on the council of the Royal Society several times in the later 1750s and he was admitted licentiate of the Royal College of Physicians of London on 5 July 1758.[35] He was not a graduate of Oxford or Cambridge however, and it took another five years for him to be elected as a fellow, and then only *speciali gratia* (by special favour). He was seen as a more acceptable figure by the Royal College of Physicians of Edinburgh and would work closely with its President, William Cullen, in editing the new edition of the *Edinburgh Pharmacopaeia* that appeared in 1774.

In 1761 Pringle was appointed as physician to the new Queen on her arrival and coronation. Presumably being personal physician to fellow Scotsman, Lord Bute, played no small part in this promotion. Bute was at the height of his influence over King George III at this, the beginning of the reign. After the early death of the young Prince's father, Prince Frederick, in 1751, Bute had assumed the role of surrogate father, a position of trust approved by the widow, Princess Augusta. He had the ear of the young man to such an extent that he, alone, advised on the choice of royal bride[36] and the seventeen-year-old Princess Sophie Charlotte of Mecklenburg-Strelitz disembarked at Harwich on Sunday 6 September 1761.[37] She spoke not a word of English and the betrothed couple had never met. George III at least spoke German fluently. His bride was not reckoned to be beautiful and here there was no compromise to be had. Waggish tongues got busy, one cruelly commenting that 'the bloom of her ugliness (was) going off . . .'[38] But the union was an instance of that unexpected phenomenon, an arranged marriage that works emotionally. The young couple developed a very warm relationship and there was soon a succession of young princes and princesses, born almost by annual ritual. Pringle carried the demanding responsibility for the Queen's health during a decade in which she was pregnant more often than she was not. By 1764 Pringle was not only 'the father of military medicine' and an advocate of the scientific approach in medicine – of doctoring built on ordered evidence – but also a top London physician and a palace doctor. He was at the very pinnacle of his profession.

There were few meritocracies, if any, in the eighteenth century and Britain was certainly not one of them. Within the strict hierarchy of royalty, aristocracy and gentry, power and privilege were distributed by personal recommendation. The higher the rank achieved, the greater the expectation that relatives or friends would benefit. Just occasionally, a member of the lower orders might be given a lift out of their humble station, a famous example from an earlier era being Samuel Pepys.[39] So established was the practice of patronage that pleading letters to a potential benefactor were both expected and entertained. But for a patron to use his influence well required care and cunning. Installing friends and family into positions of trust in which they did not succeed could have devastating and very personal consequences. It did not pay to recommend too many bungling nephews, hapless neighbours or, most certainly, bewildered foreigners. At least some of one's promotions should succeed and talent spotting, a necessary skill for the rich and powerful, was easier if you had a large circle of reliable colleagues and contacts. Ingen Housz was not to disappoint his patron.

If Ingen Housz was anxious to meet other 'scientists' here was the ideal conduit. Pringle was in the habit of holding weekly soirées at his home . '. . . his house . . . was a centre of culture and hospitality (where he) held regular . . . Sunday evening conversations.'[40] Unless engaged elsewhere, the members of his circle rarely missed. Here was a leaf that stood proud of the rest of the vegetation. Cockchafers chirred in and chirred out, vying by luminescence. But they weren't just clattering carapace. They were powerful men, men of

Crane Court, Fleet Street, London. Home of the Royal Society 1710-1780.

means and men of influence. They were virtually all fellows of the Royal Society or destined to be so. Here was a club within a club. Within days of reaching London Ingen Housz crossed two more thresholds, all at once: those of Pringle's home and of his friends. Here were men who were to have a massive impact on his subsequent life. No doubt Pringle would have circulated, introducing his young friend to, among others, Andrew Kippis, Richard Huck, William Watson, George Baker, William Heberden, and Donald Monro. Ingen Housz must have been stunned by the quality of the assembly. Politics, religion, theatre, literature, medicine, history, architecture, music, travel; science would have been only a small part of the cross-chat. But in eighteenth-century thought, science was merely a technical term for knowledge – that which bore any possible relation to natural phenomena. Some of the information sought by Ingen Housz would have been more mundane and Pringle's friends were certainly experienced in advising strangers to London. He often extended invitations to foreigners visiting the city, a point made by James Boswell in his *London Diary*.[41] Confounded though he might have been, Ingen Housz could have had no better motive, nor practice slope, for rapidly pushing up his fluency in English nor occasions for seeing the value of his French and Latin. Indeed, it is easy to appreciate how his natural ability for languages would have impressed the whole circle, as it had John Pringle many years before, and hastened his full acceptance into the society. If Ingen Housz had resumed his European tour after only the few weeks he originally envisaged, he could have been congratulating himself on its memorable and successful beginning. His inquisitive, acquisitive mind had been stimulated and his address book filled up.

INGEN HOUSZ HAD been a family doctor, *huisarts*, at Breda; known today, in the UK, as a general practitioner. Although he was a university graduate, having a doctor of medicine degree, his practice had been a mixture of primary medicine and surgery commensurate with many who had served only apprenticeships. It seems clear that Pringle intended a climacteric for his protégé: that Ingen Housz should update and improve his knowledge of diseases and their treatments. He could then consider himself a physician – a doctor respected for his specialised knowledge and consulted by other less-qualified colleagues nonplussed by patients with complex diseases. The first clue to these medical machinations is the fact that Ingen Housz, sometime in late 1764 or early 1665, embarked for the five-day coach journey to Edinburgh. It is almost possible to hear Pringle's advocacy for Caledonian medicine and his Dutch protégé was happy enough to go ' . . . where the pleasure of conversing with experts led me'.[42] One wonders, though, whether Ingen Housz was conscious, painfully perhaps, that he was retracing the journey his late father had made, on foot, some fifty years before.

Ingen Housz found, at Edinburgh, a city undergoing profound social and structural changes. The dark, forbidding old city, multi-storey and hugger-mugger, high on its volcanic outcrop, consisted mainly of seventeenth-century tenement dwellings separated by very narrow streets and by the 'tofts', tunnel-like alleyways. It was now in the process of being abandoned by the affluent middle classes, for airy new buildings to the north of Nor Loch. Rather like in Westminster and Mayfair 500 miles to the south, these were being aligned in regimental squares approached by wide, tree-lined avenues, served with safe water and drained by efficient sewers. Princes Street was the new, out-of-town, shopping mall.

Edinburgh was certainly then a medical Mecca; successor to Leiden. It was the biggest and the best medical school in Britain, if not in the world; its teachers renowned. We know something of its daily routines and administration from a contemporary document, a first letter home from a new Edinburgh medical student to his father in Yorkshire. Thomas, the 21-year-old son of the Reverend Joseph Ismay, Vicar of Mirfield, near Huddersfield, went up to Edinburgh in November 1771.[43] He tells his father of finding 'very convenient' lodgings with the widow of a late professor for £10 (£640) per quarter including two hot meals a day and candles – 'the lowest one can get Boarded in a genteel manner.'[44] To justify the expense, he was obviously anxious to convince his father that he was the most diligent of students and details all the courses and ward rounds. We therefore have a comprehensive timetable and although six years had elapsed since Ingen Housz was in residence, it is very likely that routines had changed little.

The Colledge bell generally rings to every Class; as at 8 in the morning Dr Home's which continues till nine – From 9 till 10 Dr Cullen; From 10 to 11 Dr Black; From 11 to 12 Dr Gregory; From 12 to 1 Dr Russel upon Natural Philosophy; From 1 to half past 2 or 3 Dr Monro; From 3 to 4 to get their Dinners …' and so on until seven in the evening when Dr. Monro's dissector gives you the ' . . .opportunity of examining every Part of the Subject yourself.

The medical students also attended weekly lectures on surgery, chemistry and botany and '. . . from 12 to 1 o'clock each Day they walk the Royal Infirmary along with the Physicians and Surgeons.' It is not known how long Ingen Housz spent in Edinburgh that winter term nor which courses he attended. Lecture registration rules at the time were surprisingly informal and there are no reliable archives. There were some five to six hundred medical students according to Ismay's letter.[45] When courses were advertised tickets were put on sale and no student was admitted to any of the lectures without one. Alexander Monro's lectures were the most popular, attracting audiences of over 300. Ismay couldn't help highlighting to his father, with a sense of amazement, that this activity, alone, would have earned the professor some 900 guineas (£60,200) a year. Presumably Ingen Housz selected only those courses that he saw as relevant to himself. After all, he was nearly a generation older than most of the students and had gained considerable professional experience. We do know, though, that he spent the three guineas (£200) necessary to buy a ticket to attend the daily ward rounds at the Edinburgh Royal Infirmary since he wrote, again to his friend, Deckers.[46]

> I am at the hospital every day, where I admire the care and attention of the physicians, all doing their best both for the patients and for the instruction of the students. One writes down all the symptoms and the effects of the treatments in a journalistic way and at the least difficulty the doctors discuss the case among themselves. I am making notes of everything unusual.

INGEN HOUSZ HAD travelled back to England by May 1765, writing another letter to Deckers, postmarked London, 31 May.[47] Once again he was able to take a passive role in the London medical scene, attend Pringle soirées, visit the coffee houses and otherwise mingle as an intimate in the glittering circle he had met before going to Scotland. It was all propitious and, just as important, fascinating, enjoyable and recreational. Ingen Housz faced a dilemma, however. He had never planned a stay in London that would be pure vacation In marshalling the sparse clues we have, by intruding on his thought processes, we can see that Ingen Housz had four objectives on leaving Breda. Most pressing was a respite from the exceptionally busy life of the young doctor. He had an urge to travel but not just to sightsee: he was anxious to meet 'learned' men of science and medicine which made London an attractive first stop, particularly as he held a long-standing personal invitation. Fourthly, he also hoped to be initiated into smallpox inoculation; hence London again. By the early summer of 1765 he had achieved three of these four declared ambitions and going up to Edinburgh to upgrade his medical knowledge had made good sense. However, he had rather neglected one main motive for coming to England – to learn about the latest techniques in the fight against smallpox and he had now been away from Breda for over six months. His schedule for visiting other parts of Europe was slipping as he became more enmeshed in Enlightenment London and eventually it would require a totally unexpected external influence, a very strong centrifugal force, to pull him away from England.

Remaining in London meant problems at a practical level too. Where Ingen Housz had first lived on moving to London is unknown but being a guest of Maty at the Royal Society so promptly after his arrival may be a clue. Maty's position at the British Museum gave him and his family imposing and generous quarters in one of the wings of Montague House in Bloomsbury, where the collections had been installed.[48] Perhaps Ingen Housz was a house guest of Maty, at least for a short time, until he travelled to Scotland. On returning to London Ingen Housz took rooms and every time his departure was delayed he needed to extend the tenancy of his lodgings. He moved into rented accommodation and lived independently for he refers, in correspondence,[49] to domestic difficulties where he quotes his address to be 'Messrs. Dunn and Taylor at the Temple Bar, London'. He was living, therefore, at 6, Fleet Street.[50] This located him at the west end of this main thoroughfare, at the very point of one of the old city gates through which people and traffic had to pass to enter the City of London. By the 1760s this was just an arch stranded on an island in the middle of the road, its name deriving from the fact that it was next to the Temple Law Courts. A feature of past ceremonies and processions, it could be a gruesome edifice. It used to display, on spikes, the heads and other body parts of traitors who had been executed, the last such occasion being after the Jacobite rebellion of 1745.[51] For Ingen Housz it was a convenient address, halfway between the old city nexus of shops and coffee houses and upper crust Whitehall and Mayfair where Pringle and many of his circle had their houses.

His longer than expected settlement in London would also have made Ingen Housz yet more beholden to the Dutch Ambassador, Count Welderen, to whom he had been given a letter of introduction.[52] Welderen was an anglophile of the most ardent kind: he had an English wife. He was also most hospitable to his young compatriot. 'The ambassador's kindness towards me is such that he has invited me to dine quite frequently; uninvited, too, I am free to dine with him as often as I have an opportunity.'[53] It is also clear that Ingen Housz became particularly close to the legate's secretary, Collard. 'I also have a confidential and friendly relationship with . . . Collard, who is his (the Dutch ambassador's) secretary.'[54] These umbilical attachments obviously helped him in what was still a largely alien society. Not least it gave him a reliable means of communicating with home, his letters being carried in the diplomatic bags. London routine was, however, becoming more diverting and more cloying by the day. In just nine months, despite being away in Edinburgh for much of that time, Ingen Housz had come to know, and be known by, a host of new friends. These were men with diverse backgrounds and experience, wide interests and activities. They all had a magnetic attraction for the young doctor who had obviously, in turn, become popular with them. Like the wispy, wafting and paralysing tentacles of a jelly-fish, a succession of shared projects that he found irresistible would now hold Ingen Housz back from his onward journey; and the longer he remained in London, the more intellectually ensnared he became. The only blemish of the medusal analogy is that he was a willing victim.

4
Particles

He finds the shop easily enough. It's in Whitefriars Street off Fleet Street. This area of the city is still the epicentre of the English publishing trade despite the recent lifting of the restrictive practices favoured by its guild. The door puts up some resistance but finally gives and a bell rings. By the time he's negotiated the steps down into the interior, side-stepped some rough timbers propping a crossbeam and flinched from the noise and the dust, a man has appeared from a back room through a doorway, reducing the clattering noise, at least, by drawing across a curtain. Ingen Housz looks around him, quickly noting that the room is lined by shelves and that many of them bear stacks of unbound books and pamphlets. They are dishevelled and, from the dull patina of fine dust, some of the piles have obviously lain undisturbed for a long time. Ingen Housz squares up to the man behind the counter and apologises for being distracted. He reckons him to be about five feet four inches tall – the same height as himself. Can this, then, be Mr. Bower? He has been warned that Bower is remarkably short, that he is very self-conscious of his stature and that it is best to feign indifference.

'Mr. Bower?'

The man hesitates and then nods, wiping his blackened hands in a cloth. Ingen Housz launches into his story.

'One of my old schoolmasters, from Brabant, has written a book on Greek particles, a sort of dictionary translating the many random Greek idioms into Latin. He is hoping to have his book published in England and . . .'

The curtain in the architrave is pulled aside and a much smaller man appears. Despite his size he has an air of authority, is bewigged and well dressed. He moves in on the conversation and introduces himself as William Bowyer. Ingen Housz is confused.

'But I thought this gentleman was Mr. Bower, sorry, Bow-yer' *correcting the pronunciation.*

'This' *the little man says,* 'is Mr. Power. He has worked here since his apprenticeship to my father and manning the presses has left him deaf. He knows more about the trade than I do and would have answered your enquiry but now that I'm here perhaps I can ask you to start your story again?'

Joseph Power retires, re-tying his leather apron.

Ingen Housz begins his account of Hoogeveen's 'Particles' once more, trying, this time, not to stumble over the description of its contents. Bowyer listens him out, recognising that this must be the young Dutchman mentioned to him by several of Doctor Pringle's circle. He has plenty of experience in rejecting authors diplomatically: authors' agents normally give him no qualms.

'Unfortunately such a book will be very expensive to produce, demand a high price and be unlikely to sell. It could so easily end up as one of the piles of unloved paper that you see all around you.'

He makes a flamboyant gesture to the unsold books that line the shop.

'Even if I accept the manuscript and the copyright for nothing, it will take years, if ever, for any profit to appear and though just seeing his work in print may be reward enough for your old teacher; I, Sir, must run a business, pay my staff and make a living.'

However, for a friend of a friend, he throws in a sweetener.

'The book sounds, though, to be worthy and unique and it might be worth making a more singular assessment of its prospects. Would your countryman permit us, perhaps, to have a few sample pages for us to peruse together? We should have, also, a better knowledge of the number of likely pages.'

Ingen Housz reads between the lines perfectly well. Mr. Bowyer is saying 'yes, maybe' but meaning 'no, thank you'. But he has no choice but to follow his suggestions and return, perhaps, with some specimen pages. He thanks Bowyer for his time, apologises again for mispronouncing his name and, outside, climbs up to Fleet Street, crosses the road and makes his way east to St. Paul's Churchyard in the shadow of the Cathedral. On a Thursday there might be people he knows at St. Paul's Coffee House: the 'Honest Whigs' meet there fortnightly and often call in at other times.

AFTER SIX MONTHS in England and Scotland Ingen Housz was probably having few problems with speaking and understanding English; a reasonable assumption, knowing his prowess with other languages. Indeed, it was likely to have been his rapid progress in ability to communicate that would have impressed his new friends in those early months. And since they were a group in whom literary talent was highly valued we can emphasize the difference between average and gifted translation using an appropriate text. We could say that Ingen Housz had become *noscitur a sociis*. This Latin stub translates simply as 'he is known by his companions'. Whilst this is an acceptable rendition of the phrase, it is bland and lacks the flair of the talented linguist. More meaningful, and the way that we believe Ingen Housz would have translated it, is 'you can tell the quality of a man by the company he keeps'. Here the point of the quip can be grasped immediately and could serve as the theme of this chapter in the life of Ingen Housz: premise and motto unite. It was a time in which he was to meet and impress a whole host of contemporaries by undertaking a series of activities that were tangential, at best, to the medical world.

The result of what was probably the first unexpected project that Ingen Housz embarked on in London still exists in the extensive family archive at Breda.[1] It was more a transcription than a translation but no less significant for that. In an octavo, calf-bound and clasped book we still find page after page of what are best described as remedial recipes, 'receipts' as they were then usually known, for concoctions designed to alleviate a wide range of symptoms. They are entitled, in Latin, 'Formulae which are used by the very famous Pringle'.

Worm Powders
Take of tin reduced into a fine powder, an ounce; ethiops mineral,[2] two drams. Mix them well together, and divide the whole into six doses. One of these powders may be taken in a little syrup, honey, or treacle twice a day.

Tincture of Black Hellebore
Infuse of the roots of black hellebore, two ounces, bruised, in a pint of proof spirit,[3] for seven or eight days; then filter the tincture through paper. A scruple of cochineal may be infused along with the roots to give colour.
A tea-spoon of this tincture may be taken in a cup of camomile tea twice a day for obstruction of the menses.

Liquid Laudanum
Take crude opium, two ounces; spirituous aromatic water, and wine, of each ten ounces. Dissolve the opium, sliced, in the wine, with a gentle heat, frequently stirring it; afterwards add the spirit, and strain off the tincture.
As twenty five drops of this tincture contain about a grain of opium, the common dose may be from twenty to thirty drops.

Foetid Pill
Take of assafoetida,[4] half an ounce; simple syrup as much as is necessary to

form it into pills. In hysteric complaints, four or five pills, of an ordinary size, may be taken twice or thrice a day. They may be likewise of service to persons afflicted with the asthma.

Soon after his appearance in London, Pringle gave Ingen Housz the permission and the privilege of copying out these medicinal preparations, ones he was in the habit of recommending to his patients. The few examples shown above are typical of that era. They are fundamentally different to modern prescribed medicines, having no theoretical basis except that of trial and error.[5]

Georgian therapeutics did not proclaim there was a pill for every ill . . . After all, contemporary medicine did not view disease itself as 'specific', but as 'dis-ease' or 'dis-temper', marking a general imbalance of the humours, a disorder of the digestive system, nerves or blood. Hence drugs were typically expected to perform indirect services. They might . . . purge or fortify the body's own resources by cleansing, sweetening and purifying the blood. Some drugs would warm, others cool the system, important for counter-balancing the effects of chills and fevers.

We also find balsams and boluses, clysters and confections, decoctions and draughts, gargles and juleps, as well as tinctures, pills and potions. The fact that this fascinating little tome has found its way back to Breda suggests that it was among the few personal effects still in his possession when Ingen Housz died, for a box of private effects was returned to the family at Breda. In other words, this little book became one of his most treasured possessions. It was a singular honour to have been allowed to reproduce all Pringle's personal remedies, the outcome of a lifetime's clinical experience. Ingen Housz would have found many of the actual ingredients very familiar. He was, after all, the son of an apothecary and shopping for herbs and drugs in the Ingen Housz premises at Breda was the very reason that Pringle had come to know the family. This little book can also be seen as a tangible signal of Pringle's ambition to turn Ingen Housz, the general practitioner, into a physician. And it was in this context that Ingen Housz appears to have accompanied Pringle on his daily visits to patients, imbibing his bedside skills and his mannerisms, noting the assurances that he always gave to patients and the due deference he gave to their confidences. Since many of the famous physician's patients were distinguished, if not titled, his young Dutch 'assistant' would not always have been invited to the bedside itself. Waiting outside imposing buildings, some of them palaces, was nothing less than confidentiality in action. We know, at least, that Ingen Housz did accompany Pringle to the autopsy table:

the dead do not stand on ceremony. For instance, Pringle and Ingen Housz watched, with several other colleagues, a Mr. Bayford, anatomist, dissect the body of a 60-year-old patient whose death had been a diagnostic enigma. He had died, it transpired, of a brain tumour the size of an orange, possibly a secondary growth from a primary lesion suggested by some suspicious tissue in the wall of the stomach.[6]

ANOTHER TEXTUAL ACTIVITY, which Ingen Housz began after returning from Edinburgh, was to attempt a translation of a new medical volume[7] written in English, into Latin. This would have made it accessible to a much wider readership; Latin was familiar to most physicians throughout the educated world. The book's author was Robert Whytt, 52-year-old professor of medicine at Edinburgh. He was physician to George III should the King visit Scotland, President of the Royal College of Physicians of Edinburgh and a fellow of the Royal Society. He was an expert on diseases of the nervous system and it is very likely that Ingen Housz had gained an introduction to him while in Edinburgh: Whytt was a personal friend and regular correspondent of Pringle. Ingen Housz, anxious as ever to employ his facility for languages, had taken up and translated, unbidden, some pages of the newly published book by Whytt; perhaps Pringle's personal copy. This was *Observations on the nature, causes and cure of those disorders which have been commonly called nervous,* published in Edinburgh in 1765. It was a considerable tome, 520 pages in octavo – some 80,000 words. Nevertheless, Ingen Housz pounced. He wrote to Whytt sometime during late 1765:[8]

> ... fortuitously...I translated into Latin some pages of the work published by yourself... the work about the disorders of the nerves (which) is so keenly desired by Physicians not acquainted with the English language.

His Latin translation of Whytt's text, he went on

> has not only been approved but has also been praised, though I say so who shouldn't... I shall be very grateful indeed for your giving me permission to translate the whole book into Latin, and equal gratitude will be felt by the learned world.

Whytt's reply was favourable so the quality of the Latin must have been remarkable. Whytt, himself, was renowned for his Latin prowess.[9] He knew of no other translator working on the book and raised no objections to Ingen Housz's proposal. Ingen Housz, wearing his enthusiasm on his sleeve, wrote again in February 1766,[10] enclosing some of the translated pages. He asked Whytt to approve them and, if so, prepare a short contextual preface such that the Latin version of the book would be seen, by readers, to be '. . . by no means a bastard or an aborted child, but a true descendant of an illustrious father.' Ingen Housz admitted relief that Whytt was in agreement with the whole project for, he now confessed, he had already translated over 100 pages. He had even conceived a business plan – that the finished manuscript be sent to one of the Dutch publishers because of their ' . . . extensive business contacts with all European countries'. He signed off on a more personal note that, cleverly, reinforced his pedigree. '. . . the esteemed Pringle, whom I met last night in good health, wishes to pass his greetings to you.'

Unhappily, this project aborted when Whytt died on 15 April 1766. With sad irony the cause of death was an obscure disease of the peripheral nervous system that no one has ever been able to diagnose retrospectively. But, for Ingen Housz, there was some residual compensation in having to put aside this massive undertaking. Ploughing through the translating process could only have improved his English, grammar and vocabulary alike. For many weeks the intellectual mechanics must have involved a triangulation between Whytt's text, Ingen Housz's native Dutch and the target Latin. As further consolation there were now other embryonic projects.

THE NEXT ENTERPRISE in our artificially sequenced list of Ingen Housz's London literary activities also remains unfinished, even today. No-one died but it ran into the quicksand so long associated with the publishing industry. However, it was to be crucial in establishing an important and long-sustained friendship for Ingen Housz; such a close and life-long liaison that the destiny of the project itself was almost superfluous. In December 1764, a major player in eighteenth-century history had returned to London. He knew the city well for he had worked, as a young man, in the printing trade in and around Fleet Street in 1724/6.[11] Back in America he had pioneered research into electricity and had made groundbreaking discoveries that brought him to the attention of the Royal Society and a fellowship in 1756. He was back in London in a diplomatic role in the late 1750s[12] and now he was welcomed, for the second time, as the 'Agent' of the American colonies. This time he was to be less popular, having been sent by an increasingly disgruntled and disputatious Assembly at Philadelphia. Independence was on the wind.[13] We are, of course, talking about Benjamin Franklin. When describing Pringle's circle we have been referring to men with individual skills, talents and aptitudes that made them outstanding but Franklin was a

Benjamin Franklin FRS. Courtesy of the Wellcome Library, London.

polymath in a higher league. His life spanned every decade of the eighteenth century. Born in 1706, he was the thirteenth child of a Banbury soap-boiler who had decamped to America, like thousands of other Dissenters affronted by the popish leanings of Charles II, arriving in Boston in 1683. From the smallest and the last came the greatest: he was the only 'father' whose signature featured on all three 'founding' documents of the United States of America – the Declaration of Independence, the Peace Treaty with Britain and then the American Constitution. Franklin truly was 'The First American'. He had left Boston for Philadelphia when he was fifteen and clawed his way into the growing but highly competitive printing trade. He proved to be a very successful businessman and had achieved financial security by his late forties. This gave him the ability to move out into many other activities combining, always, inspired inventiveness with sublime pragmatism. Franklin is given almost two pages of text in the late Alistair Cooke's popular history, *America*[14] and we can enjoy the acme of reportage.

More than any other in the long history of the colonists, he is the one most often instinctively recalled on many daily occasions of American life. A siren whines up Third Avenue in New York: Franklin organized the first volunteer fire brigade. The New York Public Library announces a microfilm room: he founded the first free public library. The middle-aged man who finds himself looking up from the newspaper to refocus on his wife is told by his oculist it is time for bifocals: Franklin was the first to wear them. The northeast wind drenching down from Maine: Franklin first spotted it as the characteristic storm signal for the North Atlantic seaboard. The latest heating system boasts it is an intelligent, inevitable development of the Franklin stove.

Ingen Housz probably met Franklin very soon after their almost simultaneous arrivals in London. It was not long before the two men were involved in a joint project. In his correspondence with Whytt, Ingen Housz had referred to a series of conversations '. . . in the home of Dr. John Pringle.'[15] It had been suggested that Ingen Housz translate, again into Latin so that their contents would be accessible all over Europe, a substantial collection of personal papers and letters, mostly on electricity, that were in Franklin's possession.[16]

. . . letters on electricity . . . that have now been amended and in some places provided with notes. A number of (other) letters by Franklin . . . describing the scientific dealings with his friends and treating of various subjects in the field of natural philosophy such as whirlwinds, gales, hot winds etc. interspersed with some essays on diverse matters e.g. on the increase of the population in a region, on magic squares, on cooking pans . . .

If Franklin's papers were still only a random bundle of manuscripts we have to ask how Ingen Housz accessed them. Did Franklin pass them over, a few at a time perhaps, or did Ingen Housz go to the colonist's rooms at 36, Craven Street just off Charing Cross[17] to copy them or work on them? Perhaps they had already been copied and the replicas deposited somewhere secure. We can only guess but Ingen Housz certainly translated 387 pages of tightly written script into Latin. These 'interesting letters', all in his unmistakable hand, are now in the Austrian National Library, high above the Josefsplatz, in Vienna[18] still waiting, in a way, to be handed over to a publisher. Perhaps problems arose concerning copyright or confidentiality, or the project was simply unfinished when other events in Ingen Housz's life suddenly took over. But though they were shelved, the life in the pages had a very personal significance for Ingen Housz. Nothing cements a growing friendship better than joint working on an activity of strong mutual interest. The young Dutch doctor and the somewhat older colonial polymath spent many hours together, and on many occasions, during the translating process. Many of the technical terms describing discoveries in electricity and other scientific developments were newly invented words or phrases and converting them into Latin would have been inherently difficult and often arbitrary. Franklin was renowned for his relaxed and genial personality but it masked a brilliant brain and a diamond-edged wit. Solving conundrums with his 'translator' must have been intellectually satisfying, emotionally rewarding and, knowing Franklin's reputation, amusing – for both men. It is not difficult to see how often the conversations would have raced off into new territories, novel hypotheses and apparatus adaptations. It was electricity, particularly, that had kept Ingen Housz's scientific bent alive during his frantic time as a doctor at Breda. The intense closeness and mutual respect that clearly developed between Ingen Housz and Franklin, was to survive, very tangibly, until the older man died, in 1790. And since Ingen Housz was not English, there was to be none of the frigidity and ostracism that developed between Franklin and many of his other 'English' friends, during the next decade. Franklin foresaw that a rift was coming between England and her American colonies and, long before the outbreak of hostilities, he revealed his strongest loyalty to be to America.

By 13 March 1767 the translating of Franklin's papers was '.well-nigh finished . . . (and) in his preface, Franklin will highly recommend my translations so that there is hardly any doubt that the book will meet with a ready sale.'[19] The conception and the pregnancy were safely over but, true to form, the birth would be long, tiresome, difficult and delayed. Lambertus Bicker, a Rotterdam physician whom Franklin and Pringle had met during a journey of the previous summer, had written to say that there were two Dutch publishers willing to print the manuscript – one in Rotterdam and the other in Leiden.[20] Their terms, however, were equally ruthless. They would pay only 1575 Dutch guilders (£9,000) for the manuscript, want the illustrations printed in London at no cost to themselves, and would provide only a very small numbers of free copies for author and translator.[21] They also threatened a price review if their offers were not taken up without delay and the manuscript delivered promptly. Franklin had always been a very shrewd businessman and knew the publishing trade well. These were not men with whom he wanted to contract. Ingen Housz must have felt some intense embarrassment, particularly as a Netherlander, and perhaps this is more the reason that the project was shelved.

INGEN HOUSZ WAS no mere reporter of the world of electricity. He was a very active participant and must have been very excited to be able to demonstrate and discuss his own experiments with Franklin. Reminiscences that he was to incorporate into an invitation lecture to the Royal Society in 1778[22] tell us that Ingen Housz had taken the development of the friction electrical machine to new heights of efficiency around 1763.

> I began to make use of flat glasses instead of globes or cylinders, to excite electricity. Finding that a greater quantity of electricity would be excited upon a flat piece of glass, when rubbed on both surfaces, than when it was exposed to friction on one side; . . . I also thought another material advantage might be derived from a plate of glass, as the form of it admits of placing cushions or rubbers upon different parts of it.

In other words, he had built, during his last months at Breda, a prototype electrical machine that used a rotating glass disc rubbed by a number of pads

Disc electrical machine that generates a strong static electrical discharge.

to generate the charge. This was entirely novel. The inventor's hunch proved correct – as the circular disc of glass was spun on a central axis, also of glass or wood, it produced more electricity than earlier, spherical, machines and was capable of a most dramatic discharge. He takes up the story of his new apparatus:

> In this imperfect state I showed it to Dr. Franklin, who approved much of the scheme, and advised me to pursue it. Soon after, I shewed it to several of my acquaintance, and in a short time I found such machines ready made at Mr. Ramsden's and some other mathematical instrument-makers.

Here is a conundrum: was Ingen Housz the sole inventor of the glass plate electrical machine? As can often happen in science at the fulcrum of a breakthrough, other 'electricians' in England may have conceived ideas parallel to his own and pitched to the professional instrument maker first. If this was the case, no one has staked a claim for recognition and we know from Ingen Housz's own words that, isolated in Breda, he had worked entirely alone, reliant on his own genius. If someone had stolen a march on him, and this seems to be the most likely scenario, his trip to Jesse Ramsden's instrument shop in the Strand[23] would have been very upsetting. Ingen Housz learned, in the most brutal form possible, the lesson for all inventors: that intellectual property is easier to steal than watch or wallet, and often with more long-term consequences. He disguises any chagrin or anger in the account published by the Royal Society but this was probably edited anyway and appeared some thirteen years after the events themselves; a significant cooling-off period. The design of his glass plate electrical machine, about which he had had no hesitation in describing to many in his new 'acquaintance' had been plagiarized. He had been unwise, perhaps, to broadcast his invention and then go to Scotland without seeing it through to fruition. Perhaps only Jesse Ramsden could have solved the riddle by revealing his trade secrets. This seems untenable for he had a supreme motive for not doing so and in the end, the glass plate design came to be known, colloquially, as 'The Ramsden Machine'.[24]

IN THE EARLY months of 1766, Ingen Housz received a reply to a letter he had sent to his old teacher and friend, Hendrik Hoogeveen. Hoogeveen had just taken up the post of Principal of the Latin School at Delft. He had brought with him, from Breda, a massive manuscript. Now, during precious spare moments in the Headmaster's house on the Schoolstraat, in the morning shadow thrown across the canal by the Oudekirke, he was wondering how best to arrange publication of his 'magnum opus'. *Doctrina particularum linguae*

Graecae[25] had taken him sixteen years of devoted and determined study. His *Particles*, as it came to be known, is a comprehensive lexicon giving the Latin equivalents of those minor parts of the Greek language that derive their meanings only from the context of the whole sentence. Particles are usually short and impossible to classify: the odds and ends that are often missing from language manuals but which make for fluency. For instance, in the phrase 'the plane took off', 'off' is a particle, giving a new meaning to the verb 'took'. In 1829 the English scholar of Greek, John Seager, released an English language and abridged version of *Particles*,[26] referring to the original author as 'celebrated'. As Seager wrote: 'One of the principal difficulties of language arises from their 'particles'. That difficulty is perhaps greater with the Greek language than in any other.' He goes on to describe how Hoogeveen's book rationalizes and explains their use in Greek. '. . . to show that they were all originally formed from words separately significant; to teach the various ways in which they affect other words and to ascertain their proper meaning.' The fact that Ingen Housz had written to him from London presented Hoogeveen with a unique opportunity. He wrote back, presumptively enclosing an outline sketch of his book: perhaps Ingen Housz could arrange publication in England? This was surely a tactic that would result in wider distribution and better sales? Owing so much to his former tutor, Ingen Housz seized the chance to repay him and accepted the challenge. But, in a way, Hoogeveen was about to inflict on his pupil an experience that was going to be as invaluable as any previous lesson; the stark realities of publishing.

Naturally Ingen Housz consulted his new friends in London – the '. . . *eruditi* . . . the learned gentlemen whom I consider good acquaintances . . .' at the first opportunity. 'Each sent me to the other for advice . . .' he reported back to Delft.[27] The work would certainly be valued but there were obvious problems: the book would be expensive to produce and could not be expected to sell widely, even in England. Neither the author nor the publisher was likely to make a profit. Indeed, it would probably be difficult to find a publisher willing to handle the project even if he were presented with the manuscript free of charge. But a consensus grew that Ingen Housz should take professional advice. The obvious port of call was a common friend, the 'respectable' publisher, William Bowyer, who was known to have Greek typefaces in his printing room. Bowyer was certainly sceptical but suggested that Ingen Housz try to obtain, from the author, some copied manuscript pages. A few of them could then be printed ' . . . so that the value of the contents and the future size of the book could be judged.'[28]

Some draft text duly arrived in the Dutch diplomatic bags and Collard had it delivered to Ingen Housz. By the end of June Bowyer had printed some of the specimen pages and was showing them to potential customers. The view remained, however, that the book would be so large and expensive that there would be few sales.[29] Ingen Housz was forced to admit to Hoogeveen that the prognosis for his book was still gloomy, doing his best to sweeten the bitter pill with some pleasing news. John Pringle had been made a baronet by the King on Thursday 5 June 1766.[30] It was now high summer, the Queen was 32 weeks into her fourth pregnancy and was well. So the new knight could leave London, in the company of his great friend, Benjamin Franklin, for a short holiday in Holland and Germany. Ingen Housz also took a holiday; a break from his growing commitments, to Hoogeveen, Franklin and others.

> In the month of July I completed a walking tour, of about 600 miles, which took me through Yorkshire, Lancashire and Derbyshire and I saw many of their most memorable sights. I went into two very famous, but also very frightening, caves one of which extends 1978 feet into the earth and the other 2250 feet, both under awesome rocks.[31]

A vacation may be a wonderful respite from the 'daily round', a helpful distraction from looming problems, but may solve nothing. On returning to London, Ingen Housz was confronted, anew, by the difficulties facing his friend's embryonic book, Hoogeveen's *Particles*. He immediately went to see Bowyer. Physical exhaustion was compounded by continuing discouragement: after trawling for further opinions on the specimen pages sent from Delft, Bowyer was of the opinion that there was 'hardly any hope of success.'[32] The size and, therefore, the likely price of the book were consistently off-putting. Volumes of similar bulk had a sale price often as much as two pounds sterling (£127) even before paying the binder. In effect Bowyer was turning the manuscript away. Ingen Housz did not give up. Driven, perhaps, by the debt he still owed to Hoogeveen, Ingen Housz was 'impelled by such feelings of loyalty to you, to whom I feel obliged for so many reasons that no difficulties can stop me from what must be undertaken on your behalf.'[33] He went to see his friend Matthew Maty and we sense with what aplomb this nonpareil combined the important post of secretary to the Royal Society with that of librarian of the British Museum. 'As soon as he hears the name of the author . . .' Maty was upbeat. But we can also feel the *sang-froid* of the experienced manager. Surely, he said, this enterprise could succeed only by having different financial arrangements. The book '. . . should be offered to the 'learned' world by subscription.'[34] This was a useful tactic in the publishing business and still considered in some circumstances today.

Expected purchasers of a book were persuaded to put money 'up front' so that some or all of the costs of publication were covered, reassuring both author and printer that they weren't gambling on a book's success. Maty also displayed further business acumen. The book should be printed in Holland under the close supervision of the author himself; Hoogeveen acting as his own proof-reader. This would save money and the imprint was far less likely to carry printing errors. All within a critical half-hour the project was revitalized. Ingen Housz wrote, excitedly, to Hoogeven on Friday 17 October 1766.[35] There had been sudden and significant progress. Someone, perhaps Maty himself, had told the Reverend John Foster, headmaster of Eton College, that subscription to an important new textbook on the Greek language was about to open. Foster certainly wanted to subscribe for himself but he had also written to old Etonian, Dr. George Baker, one of the Pringle 'circle' now familiar to Ingen Housz. Baker, in turn, had told some twenty 'learned friends'. Out of mere speculation, a subscription price of two guineas (£134) seemed to have become accepted. Ingen Housz then described how a cascade of recommendation had ensued and that Baker had told him, on 15 October, that success for the book was now guaranteed. Baker had also wanted him to implore Hoogeveen to tie up any loose ends in the manuscript so that it was ready for the presses: subscribers of significant sums of money could become disenchanted if the product did not appear promptly. The drudgery and anxieties of the enterprise were thus re-exported, to Delft, leaving Ingen Housz with all the fun. His name was now on the lips of a growing number of London society, and for the best of reasons. He was introduced to many of them. His resourcefulness and dogged perseverance were acclaimed. He had demonstrated maturity and an ability to act independently of Pringle and the other luminaries. No project could have been better designed for enhancing and promoting his name and reputation.

Now the occasional correspondence with Hoogeveen became a flood, a neap tide in which Ingen Housz tried to ensure actual progress. Hoogeveen must settle on terms.[36] Could the price of subscription be accepted as one guinea? Could a further guinea be paid on delivery of the actual books? The latest prediction was that there would be some 1700 pages of printed text if produced in octavo (eight pages per sheet of printed, pre-cut, paper). This would be an excessively bulky tome. Better, therefore, that the pages be stitched into two volumes. Buyers could, after all, have them bound into one volume if they really wished. What was now obvious in London may not have been so apparent at Delft: the project now had a life of its own.

Word was on the street. People were stopping Ingen Housz and wanting to subscribe there and then, to hand over a guinea and be given some sort of receipt. After all, men bought books for different reasons, only one of them being that they intended to read them. Books were more treasured than now. A valuable library was an enormous status symbol and a matter of honour; a display of worth in embossed leather decked with gold leaf. Books were usually acquired unbound and taken to a favourite binder to be 'dressed' in the house style – leather, lettering and gilding to personal taste. And in a modish age ownership of a certain, fashionable, book could be as vital as parading the latest style of wig. The nature of the contents might sometimes be less than relevant but *Particles* had the distinct advantage of advertising that a man was 'learned'.

Soon there were further positive developments. William Heberden invited Bowyer, Baker and Ingen Housz to a meeting over a 'sumptuous' dinner: perpetrator, publisher and host were at table together.[37] Further advertising of the book was considered and it was agreed that 250 'prospectuses' would be published. Bowyer promptly offered his services in producing these; he was anxious to nullify any air of disapproval stemming from his initial scepticism and reluctance to publish the book. He was even happier when Heberden said that he would meet the printing costs. Each prospectus would describe the book – in English – and would also specify the terms of subscription. The same flyer would leave a space for Ingen Housz's signature to acknowledge receipt of the cash. An announcement was to appear in the newspapers. Perhaps, now slightly under the influence of the good food and wine, Heberden boldly announced that he would ' . . . lead a *decemvirate* . . . he would generate, nay, guarantee, ten subscribers'[38] and immediately handed over ten guineas to Ingen Housz. Baker trumped him. He would guarantee that all of his friends would subscribe and he had already recruited twenty. But trumpery it wasn't: Baker and Heberden were both as good as their word. They went on to produce both the subscribers and the money.[39] Heberden even insisted on paying for a new set of prospectuses when, later, some printing errors were discovered. Ingen Housz did the proof-reading this time.

By early January 1767 there were 30 known subscribers. Prospectuses had been sent to Oxford and to Cambridge; a batch sent to Dublin and another reserved for distribution in the House of Lords when it reconvened. There was also interest, of an additional kind, from Franklin. He was intrigued by the typescript that had been used, in Holland, to produce some sample printed pages sent over by Hoogeveen. The American had been a

Dr. William Heberden
FRS. Courtesy of
the Wellcome Library,
London.

printer in his first career and had seen nothing like this before: the pages gave the appearance of the text having been handwritten. Could Hoogeveen buy a set of the engraver's dies and send them across for Franklin?[40] Ingen Housz thought they would probably have come from Haarlem, which Hoogeveen confirmed, and reported that some were on the way.[41] Ingen Housz's friendship with Franklin deepened with this further small success though it was only a brief distraction.

Meanwhile the Hoogeveen project was progressing by leaps and bounds. Having started with Pringle's own subscription – he had paid both guineas at once – the number of subscribers had climbed[42] through 44 (23 January 1767) to 61 (16 February), to 92 (13 March). Ingen Housz was able to report, to Hoogeveen, the actual names of some of the subscribers and they were remarkable. One was no less than The Most Reverend Thomas Secker, Archbishop of Canterbury. There was David Hume the Scottish philosopher and historian who had just returned from Paris to introduce his friend, Jean-Jacques Rousseau to London society. There was Joshua Reynolds, society portrait painter at the height of his fame. And besides the now familiar individuals among his new London friends, Franklin, Watson, Baker, Heberden and Huck, the tally of peers was approaching a dozen. By late March another subscriber was William Hunter, gynaecologist to the Queen and famous for his public lectures on anatomy, some of which Ingen Housz had attended.[43] By this time Ingen Housz was concerned for the security of the 108 guineas (£7,200) secreted in his lodgings, lodgings in which personal property was far from safe. 'I must move house in order to escape from the thieving clutches of a servant girl.'[44] His move took him to rooms where he was to be much happier – with a family called Pitters in Northumberland Court, just off Charing Cross.[45] Nevertheless, he wisely arranged transfer of the considerable cash he was holding to a bank in Amsterdam. He also hinted to Hoogeveen that though subscription had proceeded very quickly until then, progress was tangibly slowing and he should hurry to finish the book. To animate the author he reminded him that most subscribers were assuming that the finished manuscript was already at the printers.

In the next letter from London to Delft, in May,[46] any fatigue and anxiety was blown away. Only five more subscribers had been found but one of them was no less a person than the King. His majesty, approached by Pringle, wanted two copies and for one of them to be adorned with his name, a privilege rarely bestowed on any book written by a foreigner.[47] Now Ingen Housz's friends were advising that he should have no less than 200 copies pressed and that the paper used should be of the highest quality. And, suddenly, the vaguely hinted notion that subscribers would be pleased if a list of their names appeared at the front of the book[48] became an imperative. Even Ingen Housz was anxious not to miss out and hastily subscribed rather than rely on goodwill for a copy; 'I do not want my name missing from the register.'[49]

By the summer of 1767, Ingen Housz had become so popular in London intellectual circles, his life so crowded and complicated, that the end of the season brought a flurry of invitations. As an instance, Lord Morton, President of the Royal Society arranged a dinner party for the afternoon of Saturday 23 May.[50] Pringle had already said that he would attend and Morton hoped that Ingen Housz and Franklin could both join them, particularly if able to arrive early – about eleven o'clock. There would then be time for the two 'electricians' to demonstrate some experiments before food was served. Ingen Housz and Franklin arrived, we can assume, carrying various items of apparatus and in time to demonstrate their joint efforts; experiments they had, presumably, rehearsed together. We have, therefore, a true eighteenth-century occasion – refined dining after a 'scientific' cabaret; experiments before food. It certainly slots Ingen Housz into his newly found status. No longer was he merely the young Dutch doctor, the close friend of Pringle. He was moving on a par with men whose social standing could not be higher. He and Franklin were at table with both the current President, and a future President, of the Royal Society. And the opportunity was not missed. In the next surviving letter to Hoogeveen,[51] recording the latest number of British subscribers as 126, Ingen Housz reported that 'Earl Douglas of Morton, President of the Royal Society, has joined us. He is a Knight of a very ancient order and may cause some person or other in Scotland to subscribe.'

Hoogeveen's *Particles* now had a life spirit and a safe future. Ingen Housz was able to leave it to the printers and the booksellers and the first edition of *Doctrina Particularum Linguae Graecae* appeared in 1769.[52]

WHETHER INGEN HOUSZ conceived a summer holiday for 1767 or not is unknown. He did manage to escape the summer oppression of London, however, to visit the university towns of Oxford and Cambridge. A motive, if needed, had appeared in the guise of a letter from a Dutch friend, Nicolaas Jantzon.[53] Jantzon was a Leiden-born and Leiden-trained lawyer who had moved to Breda in 1745 having married locally and become a vice-principal of the Latin School.[54] Having discovered that a former pupil was in London, Jantzon wanted to know if Ingen Housz could find, in England, some definitive manuscripts that would relate Hebrew laws to their Roman successors. Ingen Housz assumed that Oxford would have the best archives and spent a week there, beginning at the Bodleian Library. After '... searching high and low among the old manuscripts, ... a very time-consuming job ...'[55] the hunter found little quarry. A few, uncertain, references to the ante-Justinian jurisprudence, were unearthed with the help of the librarian at the Bodleian, the Reverend Humphrey Owen, principal of Jesus College '.. a gentleman as friendly as he is erudite'. Owen was also thrifty and, perhaps, supercilious, for he declined to subscribe to *Particles*,[56] even on behalf of his library. Finally Ingen Housz found nothing that was going to help his friend from home, but it is inconceivable that he did not take the opportunity to wander through the colleges. Having found no news for Jantzon at Oxford, Ingen Housz travelled to Cambridge to search the libraries there. Once more he had little success except in recruiting more subscribers for Hoogeven.[57] Back in London, Ingen Housz sought the advice, once more, of Matthew Maty, visiting him at the British Museum. As one of the museum librarians, Maty was able to show him a '... a treasury of all sorts of antiquities ...' but still not the elusive legal codices, despite considerable effort. Ingen Housz had to draw a line. He had other pressing commitments, he told Jantzon, after searching ' . . . as long as time allowed.'[58]

In a similar vein to the appeal from Jantzon, Ingen Housz received, in early 1767, a letter,[59] bundled up with other papers and books, from another compatriot, Gerard Meerman. Meerman was a high-ranking state official living at Rotterdam. His plea was self-explanatory.

> I understand from Mr. Hoogeveen, Rector of the Latin School at Delft, that you are a veritable advocate of men of letters and . . . that you will not refuse me your help. I have had printed a work on the origins of typescript, in two quarto volumes. I am the first person to have reconciled the dispute between Haarlem and Mainz on the invention of printing having proved that moveable wooden type was conceived and used first at Haarlem, and that only later was it that molten metal type was invented at Mainz. Among the many appendices . . . there is a dissertation in which I amply refute . . . with proof, that printing wasn't carried to England, in 1470, by Caxton but earlier, about the year 1460, by Corsellis. This book, which I have had printed at my expense, has cost me at least 500 pounds Sterling . . . I have contracted with three booksellers for its sale, one at The Hague, one in Paris, and one at London. The last is Thomas Wilcox, bookseller in The Strand . . . However much I might have flattered myself that I could sell 40 or 50 copies in the first year in England, he sold only seven or eight.

Meerman obviously stood to lose a considerable sum of money and begged Ingen Housz that he recommend the book[60] to all his new London friends, imploring them to call into Wilcox's bookshop, near St. Pauls Cathedral. As an incentive, Meerman arranged for Ingen Housz to receive a 'beautifully bound' edition of the two-volume work. The recipient immediately reported his reactions to Hoogeveen.[61]

> Meerman's letter is very friendly indeed but I think he has too high an opinion of me as a promoter of men of letters. (He) is well known to the Britons and they speak highly of his learning . . . my recommendation might be beneficial . . . I shall be pleased to speak words of praise to my friends . . . recommending his work to learned people is an activity that appeals to me.

Unhappily, Ingen Housz's salesmanship did not prove successful on this occasion, reporting to Hoogeveen, later in 1767, that '. . . the results of my efforts on behalf of Meerman do not match up to my unceasing zeal. I have left nothing untried . . .'[62]

Literary activities, on his own behalf, and in promoting the enterprises of others appear to have consumed much of Ingen Housz's time and energy during his second and third years in London. It is also difficult not to create the impression, by the restrictions of story telling, that his efforts with the works of Pringle, Whytt, Franklin, Hoogeveen, Jantzon and Meerman were each compartmentalised and strictly sequential. This, of course, is not true: it would be naive even to consider it. His dabbling in translation, his seeding a book subscription, his researching antiquarian vellums and pointing potential readers to otherwise neglected books all ran together in a hotchpotch, serving a common cause. Ingen Housz was no longer a cipher, a fleeting contact, an inquisitive tourist. He had come to be seen, by all his new friends in London, as a pleasant companion but also a valuable colleague with infinite drive and huge potential. Pringle himself must have been pleased to see how his upbeat predictions of twenty years before had been realised. Nonetheless, one should not assume, because of the temporary neglect of the topic, that Ingen Housz had totally subsumed his clinical activities to literature, particularly his interest in smallpox inoculation. By July 1766 he had had cause to visit a medical instrument shop.

5
Foundlings

Saturday 17 October 1767. A man in early middle age is idly watching the dull distorted leaves spinning slowly down from the nearest trees, musing that the air must be undisturbed in the lee of the building. Further away, though, there's a breeze and the anchorage of the mottled foliage is being tested relentlessly. Turning away and standing clear of the window so that he can see, he lines up, afresh, the articles on the table in the middle of the room. The nurse has gone to ask for fresh water and new linen after cleaning up: the last child had vomited on the floor. With the door left open, the corridor to the scullery is in full view and a simple nod summons in the maid when she approaches. She places a clean bowl and towels on the table and he thanks her. Even so, the house is now tainted by terror, by the permeating smell of sick. The nurse reappears clutching, firmly, the upheld left hand of the next four-year old; number 21, the last of the girls.

Being last in the line doesn't help. Having seen her peers return, one by one, apparently unharmed, has been consoling but not enough to counter the unspoken dread on the little girl's face. Bereft of all bonds, she is already at an emotionally low ebb. In the last two weeks she, as all the children in the group, has been abruptly wrenched away from the only mothers, fathers and siblings they've ever known. Total strangers had come to scoop them up and forcibly escort them, in a hail of pleas and tears, on long journeys, sometimes even through the night. At their common destination they've all been stripped, pawed over, washed vigorously and slotted into the scratchy new components of a uniform. Individuality obliterated, the buoyancy of childhood lost, they've spent their last few days sobbing or in sullen silence, shunning each other and fearful of the impersonal surroundings. There are new routines where there'd been cuddles or a crust. Here there are no treats, no warm familiarity. Those whose bowel or bladder function have regressed are scolded, further worsening control. It's as though they've been taken hostage. But as the days pass they've become aware of the distant voices of older children who sound happier. Being, at least, warm, fed and organised, their instincts slowly begin to intimate survival and, that assured, they've started to take an interest in life again. They've begun to adapt and to play, as children can and do. Now, suddenly, here's fresh trauma. They've been lined up, boys and girls separately as always, and marched in two columns, away from the tall buildings, through some large gates, left past some fields and a farm or two and then right, into the claustrophobic clamour of some city streets, to one of the houses around a square. The rooms, here, are on a more human scale but are far from homely. There's still only essential and rudimentary furniture, bare boards and walls doing nothing to offset the chill. They've been shown into single-sex dormitories and have heard the doors locked behind them.

This little girl, the last for this month, seems composed and courageous but, moon-eyed, her gaze darts from the seated man to the items on the tabletop and back again. She then fixes on the chair's occupant, watching him intently. Though leaning forwards and smiling warmly he speaks strangely. She makes no response. Nor does she react when the nurse removes her newly-issued blouse before lifting her onto her lap. Her name is announced and the man turns away to write it in the book on the table. Too late she realises that the nurse's embrace is now a vice: when she tries to squirm she can't. All she can do is tense as the man examines her eyes, her teeth, her torso. He then turns and takes into his hand something from the table. She watches him stretch the skin of her puny arms and make a firm scratch on each with a cold piece of metal. It didn't actually hurt and was very quickly over. She relaxes; as do the two adults. She has been a 'goot girl' and the man strokes her hair as the nurse lowers her back to her feet. The nurse says nothing as she leads her back to the dormitory. Some minutes later the child's courage finally fails her. Tears roll down her cheeks as she hears, from below, a boy's screams of protest and the sounds of furniture scuffing resonant boards. Why has her mummy let her be taken away? What has just happened? Is she ill? Is she going to die? The little girl's questions are precocious for a four-year-old but far from irrelevant. In fact, what has just happened will reduce her chances of a premature death. She has just been inoculated against smallpox, one of the most feared diseases of mankind.

S MALLPOX OR, MORE specifically, the earliest means of fighting the disease – by so-called 'inoculation' – was going to be the cause of such a titanic and unexpected sidestep in the life of Ingen Housz that it is easy to underestimate its significance. Ingen Housz had an ambition to seek tuition in inoculation when he came to England. Just how significant a part, though, that inoculation would play in his life story could not have been foreseen, far less planned. It was both fortuitous and fortunate for Ingen Housz that he came to a city that had the morbific substrate – a dense and impoverished population – and a device – smallpox inoculation – that were to combine in such a propitious way.

In describing smallpox and its consequences, the eloquence of Macaulay has never been bettered.[1]

> The most terrible of all the ministers of death . . . the smallpox was always present, filling the churchyard with corpses, tormenting with constant fear all whom it had not yet stricken, leaving on those whose lives it spared the hideous traces of its power, turning the babe into a changeling at which the mother shuddered, and making the eyes and cheeks of the betrothed maiden objects of horror to the lover.

All diseases are unique but smallpox is extraordinary in several respects. It was the first infectious agent for which a prevention strategy was developed and came to be the first human disease known to have been eradicated by man's ingenuity. We can even dare to talk about it in the past tense. Unlike most contagions, smallpox was not socially selective: it presented equal danger to all classes of society; to kings and emperors as much as to tradesmen and labourers. This indiscriminating threat led to a third reason that it was exceptional: with powerful people being prone we have good records of its impact, both clinical and political. Indeed, the list of prominent names that this often-fatal disease suddenly eradicated from history is immense. From the late Middle Ages, physicians began to distinguish the 'small' pox from the 'great' pox (syphilis). And the number of times that royal and political lineages took a sideways lurch because of fatal cases of smallpox diagnosed in palaces and great houses is a hugely important aspect of our past.[2]

To understand the full impact of the disease we need only quote the case of Queen Elizabeth the First. In 1562 she was in the fourth year of her reign, 29 years old, fit and able. She began to feel unwell during the weekend of 10 to 11 October, at Hampton Court Palace.[3]

A day or two later she became feverish, faint, and began shivering. On 15 October . . . she concluded a letter to Mary, Queen of Scots, saying 'The fever under which I am suffering forbids me to write further'. Elizabeth's advisors summoned a highly respected German-born physician, Dr. Burcot, to the palace, and although the Queen's skin was still clear, he told her 'My liege, thou shalt have the pox'. Hours later the Queen became incoherent and soon sank into a coma. She was unmarried and without a designated heir. Here was the grave crisis many Englishmen had feared since her accession. The Queen was desperately ill, and England's fate depended on the outcome of her illness.

Ultimately, Elizabeth I recovered and was more than grateful that her face was virtually unscarred. But she was lucky to survive. Eight out of eleven rulers or potential rulers of the next royal dynasty, the House of Stuart, were not. One of them was Queen Mary, joint regent. She fell ill at Christmas 1694. The immediate verdict of smallpox, pronounced by the renowned but blunt diagnostician, John Radcliffe, proved to be correct. William III, her husband, was distraught. He had already lost, to the very same disease, both his father and his mother. Now he was widowed and childless. When he died, in 1702, he was succeeded by Mary's sister, Anne. After eighteen pregnancies, most of which had ended in early miscarriages associated with her porphyria,[4] she, too, eventually died childless; three of her live-born children having succumbed to the dreaded smallpox. The eradication of the Stuarts left an empty throne to the House of Hanover. Such were the well-recorded ravages of the disease at royal palaces and therefore on the twists and turns of dynastic succession. From the overcrowded tenements of surrounding London we only have impersonal statistics but they are equally daunting. During the same period smallpox was killing about 10,000 Londoners a year: from a population averaging half a million this represents one person in every fifty. Similar court tragedies and population statistics can be repeated from nation states all over Europe and, though the records are patchy, probably from elsewhere in the world.

Smallpox was probably at its most destructive and most widespread in the eighteenth century although it is difficult to be absolutely sure. The conundrum very familiar to epidemiologists may apply. An apparent rise in the incidence of a disease can be deceptive – better records can give the impression of higher numbers and cause a false alarm. Nevertheless, this highly contagious disease thrived on the huge increase in travel across, and between, countries during a period in which Europe was, for long periods, at peace. Smallpox has a ten to twelve day incubation period – time enough for an unsuspecting, and unsuspected, 'case' to convey himself, and his lethal contagion, long distances before actually falling ill. But, paradoxically, it was as a by-product of travel and international commerce that doctors came to learn of their first effective weapon against this enemy.

In 1708, Edward Wortley Montagu, a young man of 30, with a fortune made from coal and with political ambitions, was forging a name for himself in Parliament. A well-connected Whig, he held, by 1714, high office in the Treasury. Powerful and jealous colleagues marked his progress,[5] among them Robert Walpole. Walpole was clearing a path towards taking the posts of both First Lord of the Treasury and Chancellor of the Exchequer, a power combination that would establish him as Britain's first elected chief executive, the post later named 'Prime Minister'.[6] Knowing glances were exchanged and, when a vacancy arose in the British Embassy at Constantinople, Wortley Montagu was offered the post of Ambassador to the Ottoman Empire. At first he refused the appointment but eventually felt obliged to accept. On 1 August 1716 he left London with an entourage that included Lady Mary, his young wife,[7] their young son, Mr. Cross the embassy chaplain, and a surgeon, a Scot called Charles Maitland.

Mary Wortley Montagu was far from the prim, passive consort that society might have expected. She is viewed, now, as a trailblazer of feminism.[9] A talented and biting poet, she was a confidante of Alexander Pope and others on the literary scene in London. She was not one to spend days hunched over the embroidery sampler. Once installed at Adrianople,[10] where the plague was less endemic than at littoral Constantinople,[11] she chafed at her isolation whilst her husband's diplomatic duties took him away. She regularly visited the local communities, noting their dress, routines, mores

Lady Mary Wortley Montagu. Courtesy of the Wellcome Library, London.

and institutions, especially those of her own gender. She entered homes, mosques and even harems. Using a Greek friend as 'interpretress', she talked freely with the local population. Primed by prior knowledge, she sought out, and was soon able to testify to, the local method of fighting smallpox. The practice was remarkable and entirely a lay activity. Within three weeks of arrival, she placed in the diplomatic bags to London an astounding letter, dated 1 April 1717, to her childhood friend, Sarah Chiswell.[12]

> A propos of Distempers, I am going to tell you a thing that will make you wish your selfe here. The Small Pox, so fatal and so general amongst us, is here entirely harmless by the invention of engrafting. There is a set of old Women, who make it their business to perform the Operation. Every autumn in the month of September, when the great Heat is abated, people send to one another to know if any of their family has a mind to have the Small Pox. They make partys for this purpose, and when they are met (commonly 15 or 16 together), the old Woman comes with a nut-shell full of the matter of the best sort of Small Pox and asks what veins you please to have open'd. She immediately rips open that you offer her with a large needle (which gives you no more pain than a common scratch) and puts into the vein as much matter as can lye upon the head of her needle, and after binds up the little wound with a hollow bit of shell, and in this manner opens 4 or 5 veins. . . . The children or young patients play together all the rest of the day and are in perfect health to the 8th. Then the fever begins to seize 'em, and they keep their beds 2 days, very seldom 3. They have very rarely above twenty or thirty in their faces, which never mark, and in 8 days time they are as well as before their illness . . . Every year thousands undergo this Operation, and the French Ambassador says pleasantly that they take the Small Pox here by way of diversion, as they take the Waters in other Countrys. There is no example of any one that is dy'd in it, and you may believe I am well satisfy'd of the safety of the Experiment since I intend to try it on my dear little Son.[13]

Here we have an excellent verbatim description of the process of 'inoculation' against smallpox. The descriptive term 'en-grafting' is noteworthy for its botanic imagery – we all know of the grafting of buds from one tree to another – of scion to stock. And there is nothing more arcane in the analogous Latin: *in-oculum* – the transferring of an 'eye' as from a chitting potato – gives us 'inoculation.'

It should be no surprise that Mary Wortley Montagu was intensely interested in all aspects of smallpox. In 1713 it had killed her 20-year-old brother after he had suffered terribly[14] and in December 1715, only six months before they had left England, Mary herself had laboured under the disease. Her crisis came over Christmas. Though she had contracted the most severe, confluent, version of the disease and her pocks ran together and suppurated so that she was unrecognisable, she pulled through. Not surprisingly she was badly scarred, losing her eyelashes such that she developed a lifelong 'fierceness to her eyes.'[17] Being rich and influential, her husband had consulted top London physicians. Their patient, when recovering but still 'enjoying' their visits, would not let the doctors go without their satisfying her innate curiosity. Sir Samuel Garth, Richard Mead and John Woodward were cross-examined on all aspects of the disease. Garth and Mead were fellows of the Royal Society[16] and Garth a royal physician. They must have been astounded to be interrogated on medical minutiae and especially by a young lady. Nevertheless, they seem to have co-operated and she learned of a letter that had been read to the Royal Society, by Woodward in May 1714. It had come from an Italian physician, Emanuele Timoni, practising in Constantinople and reported the practice of inoculation in Turkey.[17] In other words Mary Wortley Montague was not really surprised to witness this practice when she arrived in Adrianople, the reaction often surmised. She had, in fact, been primed to look for it.

Edward Wortley Montagu was not always away and in midsummer 1717 Mary realised she had conceived again. Their second child, a daughter, was born on 19 January 1718. Timoni, whom she had now met, became her physician for the pregnancy. Mary was on her feet within six days of the delivery. Though short, her successful confinement seems to have raised her determination to have her son inoculated. She appears to have acted on it independently of her husband,[18] who was travelling in Turkey once more. Maitland was despatched to find, in the neighbourhood, a suitable smallpox victim and one of the 'old women' and, around the middle of March 1718, Edward junior, two months short of his fifth birthday, was inoculated. Maitland wrote a contemporary record that was later published and though the very simplicity of the method is readily apparent, so is his compassion.[19]

> After a good deal of Trouble and Pains, I found a proper Subject, and then the good Woman went to work; but so awkwardly by the shaking of her Hand, and put the child to so much Torture with her blunt and rusty Needle, that I pitied his Cries . . . . and therefore inoculated the other Arm with my own Instrument, and with so little pain to him, that he did not in the least complain of it. The Operation took in both Arms, and succeeded perfectly well . . . . betwixt the Seventh and Eighth Day . . . . the Small Pox came out . . .

This private, but now famous, medical incident is usually reported as courageous and controversial; redolent of Mary's personality. Resolute it may have been but it was not entirely unprecedented. The previous British ambassador and his wife had had their two sons inoculated during their time at the embassy and just as successfully.[20] Where we should particularly value Lady Mary Wortley Montagu is in her provocative evangelism for inoculation when, later in 1718, the family and their staff returned to England. Back in London, she was prepared to challenge what she sensed would be a stubborn conservatism among the medical profession. 'I am patriot enough to take pains to bring this useful invention into fashion in England (and) if I live to return I may, however, have courage to war with 'em (English physicians).'[21]

In London, in 1721, Mary Wortley Montague junior, the daughter born in Turkey, was old enough to be inoculated. Maitland was called on, not only to perform the task, but also to admit two physicians as witnesses and permit maximum publicity. Inoculated in late April or early May, the little girl progressed well.[22] This, the first recorded professional inoculation in England, succeeded clinically but also politically. Caroline of Ansbach, Princess of Wales, wife of the future George II, was impressed and begged Sir Hans Sloane, royal physician and closet protagonist of inoculation, to organise further 'experiments'. On the prospect that they would be pardoned and freed, six convicts, three men and three women, waiting on the hangman at Newgate jail, were invited to submit to inoculation. Not surprisingly, all of them grasped at this unique route to release. On 9 August 1721,[23] Maitland performed the inoculations, this time with 26 potentially hostile physicians breathing down his neck, one of them Sloane himself. All five criminals who responded to the treatment recovered perfectly well and all six were freed. A further, similar experiment on six foundlings was also successful and finally convinced the princess. On 17 April 1722,[24] two of her daughters were inoculated, by Maitland again, with the consent of their grandfather, George I. This was the seal of approval for smallpox inoculation and many aristocratic parents immediately sought to have their children so protected. But the clamour soon died down in the face of adverse publicity. Deaths in some well-known families, both above stairs and below, were attributed to the practice by those who were still highly sceptical. With modern insights we can see that these conclusions were probably unjustified but when prejudices are deeply ingrained facts are only an inconvenience. It took another forty years before smallpox inoculation became commonplace in England and even longer on the European mainland.

BY THE 1750S reputable, university-educated doctors were beginning to inoculate routinely in England. Here was a significant break with tradition – physicians who were willing to get their hands dirty. They were also 'enlightened' in trying to refine their techniques and some of them, for the first time, were keeping records sufficient to establish true success rates and make risk measurements. In the interim the practice had been kept alive largely by unscrupulous practitioners and itinerant quacks who saw it only as a lucrative source of income. Many of them had double-wrapped the basics inside two requisites that, though entirely spurious, were used to justify extortionate fees. In preparation, patients were obliged to undergo compulsory bleeding, starving and purging to 'lower' their constitutions. Then, after at least a week of being enfeebled under close supervision, they were inoculated while being administered 'essential' secret concoctions that were unproven, unpleasant and prohibitively expensive. The whole miserable experience was close to torture:

> He was bled, to ascertain whether his blood was fine; was purged repeatedly, till he became emaciated and feeble; was kept on very low diet, small in quantity, and dosed with a diet-drink to sweeten the blood. After this . . . . he was removed to one of the inoculation stables, and haltered up with others in a terrible state of disease, although none died.

Such was the actual experience, in 1757, of an eight-year-old Gloucestershire schoolboy at the hands of a local doctor at Wotton-under-Edge.[25] It typifies the process at that time. The patient himself would, some forty years later, revolutionise smallpox inoculation. His name was Edward Jenner. But as he was being inoculated, as a small boy, more subtle improvements were already being introduced.

In the 1750s an Essex family of rural surgeons, the Suttons, effectively resurrected the simple, unencumbered technique brought back from Turkey by Charles Maitland. But even they persisted with a 'protective' medication, given on the day of inoculation, the recipe for which was a jealously guarded secret within the family. Nevertheless, the impressive size of their case pool, 20,000 patients safely inoculated over five years,[26] was witness enough that theirs was one of the best practices available; that many technicalities had been mastered. For instance, it had come to be recognised that the incubation period of smallpox after inoculation was only eight days, some four days shorter than when the disease was contracted naturally. But a great deal of what we now understand of the pathology of the condition was then totally mysterious. Eighteenth-century physicians floundered when this dreadful disease struck. They had no concept that the cause was a complex virus particle that invaded the victim's cells and turned them into factories for turning out more viruses. Neither did they understand, of course, that in inoculating someone they were provoking an antibody response that would endow permanent protection – 'immunity'. How could they have had this knowledge in an era long before powerful microscopy and other evidence had led to the 'germ' theory, the existence of bacteria, in the mid-nineteenth century? And it was another century yet before the reality of viruses like smallpox was proven.

Nonetheless, pioneer inoculators had made considerable progress in the absence of theoretical foundations and had led their profession away from medieval dogma. They had established that close contact with cases transmitted the disease, as did 'foments' – contaminated clothing and bedding. They knew that the onset of fever and rash marked the beginning of infectivity and that this then continued until the very last scab had dried and separated. They therefore knew the mandatory minimum isolation period for their inoculated patients if they were to avoid fuelling an epidemic rather than averting one. Canny observation had also taught them that the best 'matter' to use in inoculating was that taken from fresh pocks on a patient who was suffering a mild version of smallpox. Likewise, they had discovered that the watery droplet from a vesicle kept its potency on drying but only if it didn't go 'putrid'; in other words if it remained uncontaminated. Trial and error had also taught them to inoculate through the most superficial of scratches on thin skin such as the upper outer arm. Even today, however, we are not sure why it was that inoculated patients risked a mortality that was only a tenth of the disease if it had been acquired naturally; that the risk of dying was reduced from 30% to 3%. Exposure to a much smaller number of virus particles and the fact that they entered the body by an unnatural route are probably the two factors which combined to make the procedure relatively safe.[27] However, there is an impressive pedigree of physicians who have remained baffled by many aspects of this terrible disease however sophisticated their practice; investigators that stretch back as far as the generation of Dr. William Watson.

IN THE THREE years and more that Ingen Housz was in London there would have been many fleeting chance meetings that were insignificant. Others were to be the reverse. Among the latter was his introduction to Dr. Watson. William Watson was to be an extremely important contact for Ingen Housz; a conduit to a new life. Relatively unknown today, he was an exemplary figure of the Enlightenment. He was born near Smithfield

Dr. William Watson. © The Royal Society Library.

in the very heart of the city on 3 April 1715 and was to remain, always, a Londoner.[28] This, though, was the only life feature that could have seen him accused, with any justification, of having near horizons. His father had died prematurely but had been successful enough as a young tradesman to afford Merchant Taylor's School for his son. The boy became an apprentice apothecary in 1730, an obvious career choice for someone already fascinated by plants and their potential. So outstanding was his boyhood knowledge of the natural history of vegetation that his career as an apothecary flourished from the outset.[29] In 1738 he married and established himself as a general practitioner near Aldersgate. From this unassuming launch he was to rise to the top of the London medical hierarchy, able to afford an imposing house in Lincoln's Inn Fields. It was in 1759 that he was recast from apothecary to physician, from dispenser to diagnostician; approved and licensed after satisfying the examiners of the Royal College of Physicians. He had already long been a fellow of the Royal Society; his interests and aptitudes extended far beyond medicine. By the late 1740s he was a renowned expert on electricity generating machines and the static electricity they discharged, demonstrating his experiments to fashionable visitors to his home, among them the Prince of Wales and the Duke of Cumberland.[30] He did much to promote a scientifically-based classification of species, being an early and strong advocate of Linnean taxonomy. Watson accepted many honorary posts in the London establishment including positions at the Chelsea Physic Garden and at the British Museum. At the Royal Society he served on the council and as a vice-president.[31] Small wonder that he should prove to be an attractive personality to Ingen Housz. The two men seem to have had similar relentless energy, physical and intellectual stamina, and affable temperaments. They had both 'risen from the ranks' – successful men in a profession that didn't often recruit from 'trade' and they found themselves sharing many interests. Watson was a regular attendee at Pringle's Sunday soirées as well as an habitué of the Mitre coffee house in Fleet Street where the Royal Society Dining Club assembled.[32] In fact, Watson had been an inaugural member of this London intellectual roost.

Ingen Housz had probably met Watson very soon after arrival in England but it was a year or so later that another of Watson's honorary responsibilities pulled the two men from pleasing acquaintance to joint activity; a rare case of similar poles attracting. Watson was physician to the London Foundling Hospital. Busily preparing a book on the nuances of smallpox inoculation,[33] he needed to delegate some of his clinical commitments. He recruited Ingen Housz to help him at the Foundling Hospital and must have taken him there some time in the summer of 1766. As the carriage ricked and rattled its way north from Lincoln's Inn to the city extremity, Watson presumably told his junior colleague what to expect at the institution and something of how it had come to be built. It may also have been the occasion of a private conversation that must, somewhere, have taken place. To work as an inoculator, to collect serum from active smallpox cases, handle the material and be in close contact with inoculated patients, was dangerous, foolhardy indeed, for anyone who was not immune. This required that an individual had knowingly survived an attack of the natural disease or had been, themselves, successfully inoculated. Watson would have discussed Ingen Housz's history in this context and since inoculation was then rare in the Netherlands it is a fair assumption that Ingen Housz had suffered and survived the disease.

The story of the London Foundling Hospital was one of cussed perseverance and quiet heroism. On 1 March 1715 the House of Commons had debated the sad fate of ' . . . a great many poor Infants and exposed Bastard Children . . . suffered to die.'[34] But it was to be a full 24 years before anything was done about the babies regularly abandoned on London's streets, thought to be some 1000 a year. And it was no thanks to any elected politician. 'Captain' Thomas Coram was a retired shipwright born in modest circumstances at Lyme Regis in 1668.³⁵ He had emigrated to Boston as a young man and, having been moderately successful, returned, in the early 1720s, in middle age, looking forward to a comfortable retirement on the outskirts of London. Though able to afford to dress suitably and mingle in society he was no minor aristocrat nor even from the gentry. His speech and manners were those of the craftsman he had been. He was endowed, however, with one noble characteristic – a stubborn determination. A contact once described him as ' . . . a man of that obstinate, persevering temper, as never to desist from his first enterprise, whatever obstacles lie in his way.'³⁶ And he had a focus for his pertinacity.

Though childless, Coram had a passion for children; for particular children. There was nothing sinister or sordid in this philanthropy, nothing

'Captain' Thomas Coram.

perverted or threatening. He simply felt driven to do something about the appalling social phenomenon of abandoned infants. London, a youthful city, suffered particularly from this horror. The bulk of the population were both fertile and very poor, a coincidence that generated unwanted pregnancies; children that spelt economic meltdown for the parents. Unmarried mothers, from the moment their 'blooming' began to show, lost their jobs together with the associated income and accommodation. Even married parents, sooner or later, could be confronted by more mouths than they could afford to feed. The daily discovery, on London's streets, of tragic post-natal packages, bundles of rags encasing babies found to be comatose or dead

The London Foundling Hospital. Courtesy of the Wellcome Library, London.

from dehydration and hypothermia, was a shameful scandal. Equally odious was the entrenched view that this unnecessary wastage of human life was the consequence of wilful promiscuity that would only be encouraged if the children were taken in at parish cost. Refusing to see how he could be condoning immorality and irresponsibility, Coram determined to do something about the poor mites found, many a morning, outside churches and workhouses and even on spoil heaps. He spent his retirement pounding the city streets trying to cajole moneyed and influential people into supporting his cause. Perhaps because of his business experience in a less hidebound America, he was not deterred by rank. Somehow he broke through the class barriers and taboos to gain personal access to aristocrats, usually via their more charitable wives. The enterprise succeeded imperceptibly at first but a major leap forward came when King George II signed a royal charter in 1739. This sanctioned the opening of a foundling hospital in London

'. . . for the education and maintenance of Exposed and Deserted Young Children.'[37] Coram knew of such institutions in continental cities such as Lisbon, Paris and Rome but, being in Catholic countries, these were mostly funded by religious foundations and run by nuns. Coram broke new ground in extracting, somehow, promises of capital from the British aristocracy and a volunteer management from among a particular intellectual and professional London élite – non-conformist lawyers, doctors and bankers. He used paid employees to man the institution, arm-twisting 'celebrities' to contribute to regular revenue-raising activities to pay the wages. The best-known example is the generous regular input of new music and concert performances by George Frideric Handel, famously organising and conducting an annual performance of his *Messiah* in the newly completed hospital chapel from 1749.[38]

The purpose-built and imposing hospital, on the northern edge of the city where present-day Bloomsbury was then the beginning of the countryside, was completed in 1752.[39] Having been the oft-shunned missionary, the dogged fund-raiser, the draughtsman and, finally, the worn-out project manager, Coram had been able to bask in the glow of some success when the first children had been admitted to temporary premises more than ten years before, on 25 March 1741.[40] But with painfully acute irony he soon suffered a fate parallel to 'his' children. No longer wanted, he was abandoned. The governing body that he had helped to recruit, embarrassed by his lower class speech and manners at their weekly meetings, conspired not to elect him to any of the committees or subcommittees and effectively banished him from the board. Coram was left to use, as he wished, the begrudged concession of visiting 'his' hospital to console and play with the youngsters, ' . . . distributing, with tears in his Eyes Gingerbread to the Children.'[41] He died on 29 March 1751, but was granted, at least, his wish to be buried under the altar of the hospital chapel.

A decade later, by the early 1760s, the London Foundling Hospital had a well-established record of success. The governors had continued to recruit from among their own and William Watson had been chosen as the new hospital physician, giving him a seat on the board, in 1762. Visiting the Foundling Hospital had become part of Watson's weekly routine and as his carriage swung through the gates and approached the now fully completed buildings, Ingen Housz was able to admire the imposing three wings of the institution. The etiquette required of the hospital physician in engaging an assistant is enigmatic. Perhaps Ingen Housz was presented, in person, to Watson's fellow governors at a general committee meeting. Maybe he was taken to one of the weekly subcommittees where inoculation matters are known to have been discussed.[42] Watson probably felt, however, that the secondment of a colleague to help him with strictly clinical matters was entirely within his gift. He had certainly been handed a very clear mandate to organise the inoculation of the children. '... the children who are proper to be inoculated, be inoculated as soon as conveniently may be, under the care of the Physician of this Hospital.'[43] This may explain why there are no official references to Ingen Housz among the surviving archives but there can be no dispute that Ingen Housz inoculated the children of the Hospital for, as he himself says, in a letter of 1768, he performed this task[44]

> for two years at the Foundling Hospital under the instruction of Dr. Watson ... who had the kindness to confer on me the responsibility to inoculate almost all of them with my own hands and to attend them during the whole course of their illness.

Ingen Housz also left his 'mark' in the surviving medical records of the hospital. Among these is a substantial collection of small, leather-bound books in which inoculation details were assiduously recorded;[45] dates, gender and names of children, results of treatment and, sadly, deaths. The entries made for 27 August 1766 are, for the first time, in the distinctive hand of Ingen Housz. He records the progress of three of the 21 boys inoculated on the 19th of that month. Barnabus Norton, John Carpenter and Maximillian Short may well have been his very first inoculees, his actions supervised, presumably, by Watson himself. The boys would all have been about four years old. This was customarily the age at which the children were taken back from the rural foster mothers with whom they'd been placed, as tiny

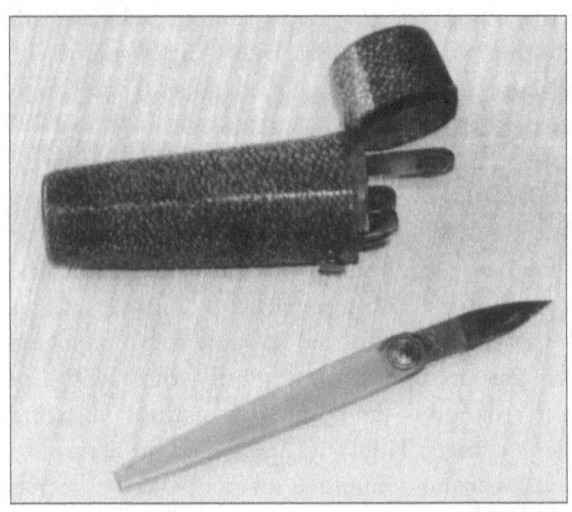

Inoculation lancets and shagreen case owned by Jan Ingen Housz. Courtesy of Breda Museum.

infants, immediately after initial acceptance into the hospital. Now they were escorted back to the hospital to be institutionalised and begin their formal education. Being fitted for their first foundling uniforms – grey serge with red flashes as designed by William Hogarth, the famous artist[46] - was no consolation for having been wrenched away from intimate family life,[47] and now they were lined up to be punctured by yet more strangers.

The children who developed the mild version of smallpox that was the intention of inoculation, created a serious threat of spreading the active, vigorous infection. It had always been management policy, therefore, that the groups of children being inoculated should be isolated, not merely within the hospital, but physically remote from the main buildings. Various properties had been rented for this purpose. From the early 1760s, this was a house in Cold Bath Fields,[48] a residential development on a square about a quarter of a mile to the east of the Hospital, also on the very edge of the north London conurbation. The general committee had minuted, on 9 January 1766 '. . . that the Apothecary do see that the Inoculating House be immediately put in a proper condition for taking in the said children, and to prevent the Distemper from spreading.'[49] The Hospital apothecary, from 1755 to 1797, was Robert McLellan[50] and he would have engaged nurses and other carers, all smallpox survivors or successfully inoculated individuals, to look after the children during their incarceration. McLellan held one of the few salaried posts at the hospital. Watson would certainly not have expected any remuneration for his own responsibilities, considering his position an honour. He had no need for such income – ample remuneration was secured from the wealthy patients that consulted him in his central London practice. Indeed, far from being paid, the post of hospital physician appears to have burdened Watson with considerable costs. The secretary of the hospital subcommittee reporting, for instance, that he had[51]

> . . . received of Dr. Watson for the charge of the maintenance of a Person in the Inoculating House at Cold Bath Fields for 10 weeks, £4 -4 -0; more, for Medicines furnished, £1- 11 -6; which with other Moneys, he shall pay into the Bank, for the use of this Hospital.'

Small wonder Watson should see fit, as it seems, to appoint whomsoever he chose to assist him at Cold Bath Fields.

The carefully tabulated statistics of the monthly inoculation routines continue in Ingen Housz's handwriting until December 1767. During that time he records his puncturing of 219 children, boys and girls in equal proportions, some 25 to 30 each month.[52] This was intense skill acquisition and experience and even more valuable than it appears at first sight. Watson was making a detailed analysis of every aspect of the inoculation technique, the conditions for best selecting and preserving the 'matter', the best sites and techniques for puncture, the need, or not, for ancillary medications, for starving and purging, and had already acquired enough consistent observations to begin the book that was published in 1768.[53] Ingen Housz had the privilege of tuition from an inquisitive and experiential clinician. Ingen Housz learnt quickly in the circumstances and became established in the routines of the inoculation house at Cold Bath Fields.

Ingen Housz would presumably plan a day for the inoculations with the resident staff at the 'Cold Bath' house so that the children newly returned to the Foundling Hospital could be transferred across for their isolation. Then, for each appointed day, he would arrive early bearing some fresh smallpox matter and some lancets. The children would be brought in and, once the child's identity was recorded, the young Dutch doctor would instil a small quantity of his inoculum into a superficial scratch drawn on each of their tiny arms. It was probably not quite this simple for the children were a very stressed group and some would be unco-operative: the inoculator had to be empathetic but, at the same time, very determined. Inoculation itself was, of course, only the beginning of the treatment. The children needed careful on-going supervision, especially around the eighth day after puncture when mild fever and pocks could be expected. This, at least, must have been the time for a routine follow up visit by Ingen Housz. He had to satisfy himself that the disease had been provoked in each child and that he could enter the case as a success. This was not merely a clinical nicety. The hospital intended all the children for careers in service, military or domestic according to gender; and both soldiers and servants were, in the eighteenth century, value-added if they had survived smallpox, naturally or by inoculation. Otherwise they might bring smallpox into your home or your regiment.

Ingen Housz carried a considerable responsibility: besides mastering the practical skills of inoculating, he had to be a competent diagnostician and satisfy Watson of such. After all, clinical medicine is ever full of pitfalls. A child with a fever eight days after being inoculated did not necessarily have smallpox. Spurious cross-infection always threatened. Some of the 20 to 30 children, packed together for two weeks or so were almost certain to develop intercurrent illness and a false positive diagnosis of smallpox could have dire consequences. On the other hand there were exquisite difficulties when a child failed to respond to inoculation and remained symptom-free. If there was absolutely no trace of pock, fever, or glandular enlargement, the honest clinician had to face some alarming possibilities. Perhaps the child had already survived the disease as an infant and was already immune? But perhaps the inoculator had not been dexterous enough? Had the smallpox 'matter' been taken from a misdiagnosed subject?[54] Had the inoculum putrefied, losing its potency? Here were many issues that the novice inoculator would want to discuss with his mentor and, no doubt, Watson was able to teach his acolyte many lessons at the bedside.

Despite all the potential problems and dangers, Ingen Housz became a remarkably proficient inoculator. Only 15 of his 219 little patients failed to react to inoculation in his hands and only two died – two boys in October 1766.[55] This remarkably low mortality is significant: these heavily disadvantaged children were far from robust and, in the mid-eighteenth century, inoculation, even in the most favourable circumstances, was acknowledged to carry a usual mortality risk of some three to four percent.

As HIS EXPERIENCE as an inoculator grew and as his clinical confidence matured, Ingen Housz began to think ahead. His intended tour of Europe seems to have been delayed repeatedly by entanglements in England and, now, in the summer of 1767, after three full years in London, he was contemplating his immediate future. There was now the possibility of Ingen Housz taking his new skills as an inoculator to the Netherlands and trying to convince his countrymen of its value. Already, in a letter of 31 May 1765,[56] he had reported to Deckers, his friend at 's-Hertogenbosch.

> I have seen inoculation practised, here, to my great satisfaction. No-one doubts, here, the usefulness of the practice and the single example of the Foundling Hospital is demonstrative for there used to perish, every day, four or five children of this malady whereas less than three percent die at present.

And, then, more than two years later, in August 1767,[57] he wrote again.

> The smallpox inoculation has never been more in vogue than it is at the moment . . . . I hope that the eyes of my compatriots will be opened, at last, on this subject and that they will accept a practice that is so beneficial to human kind.

Pringle probably reminded Ingen Housz, though, that his clinical experience as an inoculator was, to date, rather narrow: his practice had been restricted to the Foundling Hospital, valuable though that was. Once again his mentor was anxious to broaden and deepen Ingen Housz's bedside experience. He was also hardened enough, as a professional, to know the dangers of sending an ineffectual missionary into alien territory. If Ingen Housz were to start an inoculation campaign in his homeland and make too many clinical mistakes, the whole mission could abort and reactionary forces would hold sway. Perhaps Ingen Housz should also observe inoculation as best practised by someone in the general community before trying to pioneer the practice in the Netherlands? And he certainly needed absolute confidence in diagnosing childhood illnesses – after all inoculation was mostly performed on children.

Ingen Housz was advised to spend some time with Dr. George Armstrong of Hampstead village, another outstanding young doctor patronised by Pringle. George Armstrong[58] had had to abandon any hopes of graduating in medicine from Edinburgh when the family finances collapsed. Denied the means of qualifying as a physician, he became a general practitioner. On starting a family he and his wife had discovered that she could not breast-feed their children. The young couple were forced to invent, by trial and error, methods of artificially nourishing their offspring. This had led Armstrong into the whole field of children's medicine, a discipline largely ignored by eighteenth-century physicians: to them infant mortality was a 'natural wastage.' Armstrong had recoiled from this misguided notion and was inspired enough to record a decade's worth of clinical observations and write a book that was to be a landmark in medical history.[59] He was well supported by the Pringle circle. The names Baker, Huck, Watson and Hunter, besides that of Pringle, are all recorded as sponsors for his next medical move, the opening of a dispensary for sick children at 7, Red Lion Square, Holborn. Beginning on Monday 22 April 1769, Armstrong attended here to see sick children, for free, for two hours daily on five days a week.[60] Here is the birth of paediatrics. Visiting sick children in the slums of Holborn, in the company of Armstrong, would ensure that Ingen Housz's knowledge of all childhood medicine was sound and Pringle was glad to arrange the introduction.[61]

All the time, Ingen Housz seemed to have been deliberating with his friends at the Dutch Embassy whether he should sacrifice the rest of his projected tour through Europe and now return directly to the Netherlands. The themes in the conversations ran parallel to those with Pringle. Might he not be the person best placed to establish widespread smallpox inoculation in his fatherland? After all, the practice had now been given the seal of approval of the Royal College of Physicians of London[62] and not one of his high-ranking close contacts in English medical practice harboured any continuing doubts of its efficacy. Its safety had undoubtedly improved thanks to the minimalist approach fostered by Robert Sutton[63] and his sons. The credit due to them was well merited for their impressive successes with thousands of grateful patients were crowned by a mortality rate reported to be less than one percent. The Suttons had created a benchmark. If Ingen Housz was going to inoculate large numbers of his compatriots his results were going to have to be equally convincing. Leaving large numbers of patients falsely reassured by less effective treatment or, on the other hand, scarred by serious disease was going to make him unpopular and invite a critical backlash. Such failings, relished by critics and highlighted in the newspapers, could leave yet another generation of the Dutch exposed to the terrors of the 'natural' disease. There was, however, an undercurrent of suspicion attached to the Suttons. Robert Sutton senior had served a conventional apprenticeship to a surgeon, John Turner of Debenham in Suffolk[64] and began inoculating, seemingly, for the best of motives. Some early disasters had led him to adapt the usual practices, again for 'honest' reasons. There is no evidence, though, that any of his sons had any rigorous medical training. They had learned inoculation, and only inoculation, from their father and its practice became an exclusive family business. The 'Suttonian' method, pioneered by their father, finally eliminated many of the customary routines. But the family still owed their successes, they claimed, to a secret oral concoction they had developed and which they administered to patients around the time of inoculation. Whether or not there was any real pharmacological benefit conferred to the patient, this 'unique' remedy gave them a distinct advantage in the market place. Only they or their 'agents' were able to give patients this 'security' and it gave them a profit premium. Competitors might mimic their extremely light scratch technique, adopt their policy of using fresh, liquid matter from new vesicles taken only from victims of the mild version of the disease, and allow patients plenty of fresh air while discouraging lying abed; none of this methodology could be obscured. But only the Suttons had the necessary 'nostrum', the ingredients and proportions of which could be kept confidential.

In December 1767, Daniel Sutton was in London. Collard organised a meeting, perhaps at the Dutch Embassy, between Sutton and Ingen Housz to which he must have acted as observer.[65] Perhaps Collard thought that bringing the two men together might help resolve, in the mind of Ingen Housz, whether he should take inoculation to the Netherlands as an agent of the Suttons, ensuring safety and success. Decisions were crystallised but not, perhaps, as Collard had hoped. Ingen Housz was finally convinced that the Suttons were motivated more by profit than by genuine altruism; that they were, in fact, protectionist racketeers. The nub of the conversation, recalled later by Collard[66], reveals the now finely honed facility in English, and the forensic intellect, of a sceptical Ingen Housz:

... I put to mind our conversation with Mr. Sutton in December. When you asked him, out of curiosity, whether he would reveal all of the secrets if you were to tie yourself to him, he answered at once not to have any secrets and to inform you of everything without any reservation. As you thought this answer to be somewhat ambiguous and that you were not satisfied, you rearranged your question as follows: 'whether he would indicate to you the name and quantity of each constituent part of the medicines so that you could prepare them for yourself after the supply he gave you, as his agent, was used up and you needed a new supply. On this he said a simple no, because it is not necessary for his agents to know the singular composition of the medicines that render, in the preparation of the human body (for inoculation), such excellent services, just as it is sufficient for a physician to know that jalappe has a purgative action when he needs such with a patient, without it being important whether that action is specific to one of the components of the jalappe.

Ingen Housz turned elsewhere for the further inoculation experience he saw as essential before launching himself into a solo career. After all, as Pringle had reminded him, he had been inoculating only young children and these in a very particular circumstance. Everyone in the London 'circle' knew of Thomas Dimsdale from his book, published earlier that year.[67] It was clear that here was a man who had also adopted an uncluttered inoculation technique that gave patients freedom of movement within their geographical isolation and although he, too, administered some drugs, the regime used familiar ingredients in customary quantities, full details of which he had now published. Ingen Housz determined to seek an invitation to visit Dimsdale whose practice was at Hertford. He knew him to be a friend of the Quaker physician, John Fothergill, yet another of the Dutchman's London acquaintances. The hoped-for invitation from Dimsdale was forthcoming. Ingen Housz was welcome to delay his return to the Netherlands and travel to Hertford. He was still in the firm grip that England held on him as he reported, with discernible homesickness, to his friend, Jantzon: 'I shall not revisit hearth and home at the time that I had in mind; I shall be detained in these regions longer than I had thought ... please persevere in the affection towards me, even though I am absent.'[68]

WE HAVE TO imagine Ingen Housz mounting the Northern Mail in Lombard Street in the heart of the city[69] sometime in late December 1767. This coach connected with the York Mail at Waltham Cross and thence Hertford. He was well wrapped, we trust, against a day likely to have been cold, dark, and heavy with the threat of snow. 1768 was going to be as exciting a year as could be imagined for both Dimsdale and his imminent guest. But the slightly apprehensive passenger, settling himself into his seat and making inspired guesses about the nature of his fellow passengers, couldn't know this. The only consideration was arriving in Hertford before nightfall, before the threatened snow obliterated the roads.

6
Head Hunted at Hertford

Three figures lean forward in their saddles to cut the bite of the blizzard. They struggle on through a white world of darkness. The older man, a local physician, knows the way and leads but there are no horizons, no landmarks, no lights. First one mount slips and falls and then another sticks fast. The riders help each other, in turn, and finally reach the outskirts of a silent, deserted village and, then, in the lee of the churchyard yews, its largest house. They dismount. One man holds the horses and the others scramble up the tiered ramp of snow to the porch and knock at the door. Snow has stuck to them and though the light and the warmth are inviting they prefer to stay at the threshold rather than thaw out and risk a soaking. Words are spoken, some abrupt. There are some impatient gestures. The man summoned to the doorway disappears again. Eventually the Reverend Richard Levett emerges in hat and heavy cloak and points out the stable: the horses are stalled. Levett leads through the blanketed graveyard and high steps his way down a near-obliterated lane to a small, sad cottage. We are in the village of Berkhamsted, three miles south of Hertford. It's the first week of a new year – 1768.

An hour later Dr. Thomas Dimsdale has done all he can for the ten-year-old boy inside. George Hodges is dying. He has an extreme fever, is delirious, his face so engorged that he will never see again and he mouths a scream when touched. His parents have lost the courage to move him any more. Livid vesicles cover his body, some running together on his face and arms. There is a compound stench of stale, steaming bedding and of rotting flesh. Smallpox. Confluent smallpox – the worst kind. Dimsdale takes charge: he wants warm water, clean clothes and dry bedding. His servant knows the routine, having seen the humanity of his master before, and helps the numbed parents prepare. Dimsdale and his new assistant, Ingen Housz, then gently wash the boy, soak the crusts from his eyes and mouth, and unfurl clean bedding under him, rolling him, first one way and then the other, noting the large bruises on the back and buttocks that signal inevitable death. The choked cries and groans don't deter them and parental terror turns to gratitude. The horrified rector prays silently.

Dimsdale and his servant, Ingen Housz and Levett, all make their way back across the silent village. This time the visitors go into the rectory. Over hot drinks

Dimsdale introduces Ingen Housz properly, explaining that the young Dutch doctor has come up from London to gather experience from his practice. As for young Hodges – his death will be a blessing for him but his case could presage hundreds. A third of the village could die – a dozen or more burials each week between now and Easter. A silence measures the impact: it's only a decade or so since a previous epidemic struck down sixty inhabitants and twelve died.

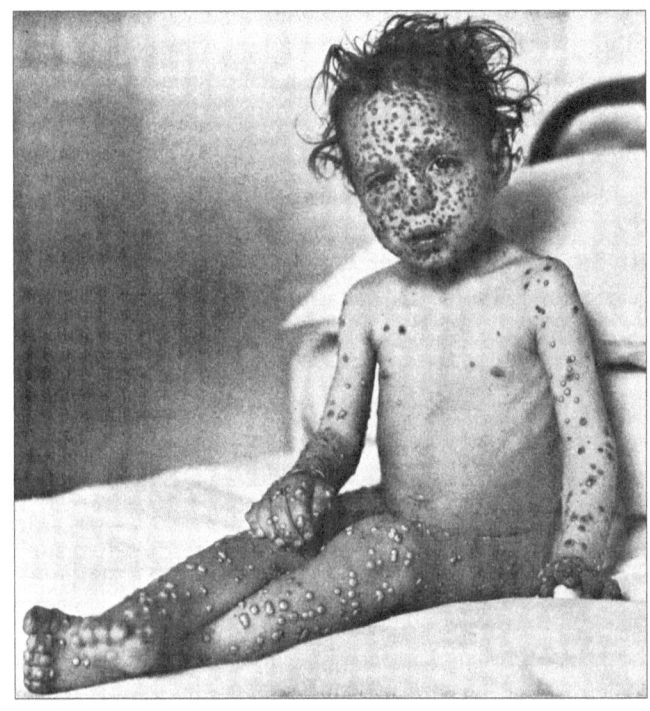

Child with smallpox. Courtesy of the Wellcome Library, London.

WHEN THOMAS DIMSDALE'S book on smallpox inoculation had appeared in 1767,[1] Ingen Housz had read it avidly. With a growing ambition to take smallpox inoculation to the Netherlands, he was anxious to digest all the latest developments. He was pleased when circumstances led to an invitation to Hertford, to work with Dimsdale for a few days. He had been warmly welcomed into the household; it had been a long time since he had shared a family hearth. Dimsdale had had a substantial family with his second wife, Ann. A strong bond formed between the two doctors that was to be lifelong. There would have been visits to Dimsdale's patients, to his inoculation house at Bengeo[2] and, no doubt, long conversations well into the night; until embers and candles were exhausted. One evening, however, domestic comforts were interrupted by a messenger at the door. Despite a snowstorm, a villager from Berkhamsted, three miles from Hertford, had come into the town to call Dr. Dimsdale to a very sick child.[3] From the details of the illness, Dimsdale suspected smallpox and began to prepare for an immediate expedition to the village. Ingen Housz was invited. As his host explained, the boy lived in an isolated community, a situation where an outbreak might be confined and an ideal circumstance to try a mass inoculation. If it was smallpox, perhaps Ingen Housz would be prepared to help him with all the inoculations? Ingen Housz appears to have agreed without hesitation.

Here was a whole new aspect of smallpox inoculation – its potential for limiting the extent of a new outbreak. Dozens, perhaps hundreds, of lives could be saved. The two men agreed that Ingen Housz would do the bulk of the inoculating[4] in order to gain more experience. Ingen Housz's midwinter trip from London, to see how inoculation was managed in a country town, had been intended to be brief because he was anxious to return to the Netherlands. Here, however, was an unforeseen but priceless opportunity. It would, in the end, keep him in England for another three months, a time in which destiny would catch up with him.

Dimsdale was a vigorous 55-year old. His father, a knight of the neighbouring shire of Essex, had sent him to St. Thomas' Hospital in London to study medicine. He had qualified in 1734.[5] His first marriage had ended tragically early when his wife died in 1744. The following year the young widower had volunteered for the force taken north by the Duke of Cumberland to quell the 1745 Jacobite rebellion.[6] After the liberation of Carlisle, Dimsdale had returned home, to the practice he was building up at Hertford but not, perhaps, before meeting John Pringle, senior physician on the military expedition. Dimsdale and Ingen Housz were not automatically destined to like each other. They were, in some respects chalk and cheese, the one a Quaker, the other a Catholic. But both had returned to their boyhood communities after formal medical training and set up successful practices. And they were each very pragmatic men with a 'hands-on' mentality. Not for them the aesthetic, coldly intellectual approach to their patients that so many physicians cultivated. Neither man had had any hesitation in becoming inoculators. Dimsdale had put into practice his scepticism that much of the flummery surrounding the inoculation practices of many doctors, whose strongest motives were high profits and who needed to vindicate their steep charges, was quite unnecessary. Like the Suttons, he, too, had pioneered a very simple approach to the whole process, one that was highly successful. Ingen Housz had already extracted the details from Dimsdale's book and reported them to Deckers.[7]

> The principal point consists of making the very lightest incision; in taking fresh, fluid matter; in giving, as preparation, two or three light purges with a preparation of mercury; and in insisting that the patients are exposed, throughout the course of their illness, to the fresh air.

Dimsdale had also developed the notion that if everyone in a circumscribed community was inoculated simultaneously and without exception, no-one need be isolated. The imminent danger of an inoculated patient infecting an unprotected contact was eliminated. Dimsdale had revealed all this to Ingen Housz but not in a formal seminar, of course. Friends have free-ranging and disordered conversations. In the chaos of chatter they touched on techniques, on efficacy, on problems and uncertainties, and on the history of inoculation. Ingen Housz already knew, from Watson, of Lady Mary Wortley Montagu. Dimsdale had not supposed the progress of inoculation in England to be unique and Ingen Housz was able to call on personal details of American practice that he had learned from Franklin. Smallpox inoculation was now common practice in America but there it had developed from an independent source. A Puritan minister in Boston, Cotton Mather, had learned how to impregnate skin incisions, as was performed in Africa, from one of his slaves and promoted the practice in the colony.[8] Franklin had opposed the notion at first but, after having changed his mind, he had always regretted not having his son, Frankie, treated before the four-year old had succumbed to the dreaded disease in 1736.[9] In fact Franklin was unable to talk about inoculation without emotion.

> I long regretted bitterly and still regret that I had not given it (i.e. smallpox) to him by inoculation . . . . Franky, whom I have seldom seen equalled in every thing, and whom to this day I cannot think of without a sigh.[10]

It is remarkable that the fundamental techniques of inoculation from two widely separated parts of the globe should be virtually identical: that matter from the lesions of a mild smallpox case was transferred into one or more superficial cuts made, with a sharp blade, on the skin of the recipient. Perhaps these customs had spread both into the Levant, and into Africa, from a common source, but not from the Far East. The Chinese practice of introducing dried smallpox matter via the nose – by using it as a kind of snuff – was fundamentally different.[11]

Dr. Thomas Dimsdale. Courtesy of the Wellcome Library, London.

Dimsdale, one of his servants and Ingen Housz had set out on the mercy mission to Berkhamsted and had eventually struggled back into Hertford through deep snow very late in the evening. There was little to be done until morning except retire with a warming pan. Some time over the next forty-eight hours Ingen Housz accompanied Dimsdale back to the village. They carried with them some matter recently drawn from some of the vesicles on another current smallpox victim, one who had been suffering only a mild attack, someone more fortunate than young Hodges. They also had with them their personal collections of steel lancets. In Berkhamsted and the proximate hamlet of Bayford the two doctors were surprised by their reception. Far from having to persuade the villagers to be inoculated, they were implored on all sides to begin at once.[12] There were many who could remember, only too well, the dreadful results of the last outbreak in the village – in 1754. Among the first group that Ingen Housz punctured were very young children and some who appeared to be far from fit. He hesitated each time. He was not accustomed to inoculating so indiscriminately and blindly. He had been used to his subjects being chosen carefully, rejecting them if they had even the mildest symptoms and for them to be prepared with purgatives and a special diet. Dimsdale assured him that he should go ahead. But should the patients be returning, immediately, to their work? Dimsdale was again adamant – it was quite unnecessary for them to rest. For some days Ingen Housz must have felt very anxious and exposed – after all he was doing the actual inoculating and, if things went wrong, would he not be blamed? Dimsdale refused to be displaced from his stratagem: there were nearly 300 inhabitants of the village[13] and the priority was to have them all inoculated in as short a time as possible. Here, at least, the continuing bad weather was helping their work by confining the population to its parish. Their real anxiety, Dimsdale reminded his Dutch assistant, should be the gruesome possibility of the disease reappearing naturally in the community, in someone not yet protected by inoculation. Imperceptibly at first, and then with increasing satisfaction, Ingen Housz began to share the confidence of his senior colleague. He was clearly amazed by the eventual results of their mass inoculation for, as he later reported:[14]

> There were many children of tender age and of imperfect health (and) there were twelve complete families inoculated at the same time and in which they nursed themselves, one and another, through their consequent illnesses. I cannot express to you, Sir, what pleasure I obtained in seeing how the inhabitants of these two villages, although all being attacked simultaneously by the smallpox, were able to go about their work without interruption. Instead of finding patients in their beds, I would meet them busily repairing their clothes, cutting down trees, etc. etc.

By the end of January Ingen Housz had, himself, inoculated almost 250 villagers, and Dimsdale 50 others.[15] They had treated five expectant mothers, one already in the seventh month of her pregnancy, and many women who were still breast-feeding their own infants, or being hired as wet-nurses. They had treated all ages – the youngest being only three weeks old, the oldest at 70 years.[16] Back in Hertford, in the evenings, the inoculators discussed their progress. Ingen Housz had many questions for Dimsdale; a colleague he was coming to respect very rapidly for his stamina and clinical acumen, ' . . . a physician of great merit and humanity.'[17] And Dimsdale was probably anxious to know all about the experiments that Ingen Housz had been performing, with Watson, at the London Foundling Hospital.

Towards the middle of February, and in the absence of any new cases of natural smallpox in the villages, Ingen Housz was able to share, with Dimsdale, the satisfaction of such success that he was able to record, with irresistible irony;

... the first person attacked was the victim (George Hodges). The second (the only other victim), who had withdrawn unnecessarily from the village in the hope of avoiding the contagion, survived but only after the loss of the left eye, and begins, at last, only now to get better: in contrast, of nearly 300 to whom we gave the illness by inoculation, none died, none were even in danger.[18]

In Innsbruck, Upper Austria, on Monday 19 August 1765, the most powerful woman in the world was sewing. It was a shroud – for her husband, Francis Stephen. Maria Theresa, Empress of Austria, Queen of Hungary, Queen of Bohemia, was suddenly, at 48, a widow,[19] after 29 years of a happy marriage. She had brought her crowded household, and her court, to the Tyrol for the marriage of her eighteen-year-old second son, Archduke Leopold, to Maria Luisa, the Infanta of Spain. The family meant everything to Maria Theresa. Her people called her 'landsmutter', mother of the nation, a term meant endearingly: she was held as a shining example of marital loyalty and fecundity. Her eleven surviving children had just lost their father.

The expected euphoria at Innsbruck that August had already been interrupted. The groom, Leopold, had spiked a fever early on the very day of his wedding and had been carried to bed with pleurisy. After a few days he began to recover and as they watched him rally, his parents, and his fiancée, were thankful that the festivities could resume. Maria Theresa decided, however, not to attend a gala performance at the theatre on the afternoon of the 18th, leaving the duty to her husband and Joseph, their eldest son. On their way back to the palace, Emperor Francis suddenly staggered and collapsed. His son ran to his aid but the unknown pathology was catastrophic. Joseph found himself propping up a corpse.[20] There were to be no last rites for this Holy Roman Emperor.

If summer 1765 was a low-point of Maria Theresa's personal life it was only one of many in her middle years. Charles Joseph, her sixteen-year-old son, a particular favourite, had died, of smallpox, in 1761. Her daughter, Joanne Gabrielle, had died of typhus in 1762. In a crescendo of brutal losses during the 1760s, these were only the beginning, deaths that we might number one and two. Isabella, the Empress' first daughter-in-law, married to Joseph in 1760, had nursed the fourteen-year-old Joanne Gabrielle personally; mindless of her own safety and that of her baby daughter. She outran the

Maria Theresa, Empress of Austria.

exposure to typhus only to be caught, by smallpox, a few months later.[21] Archduchesses Marie Christine and Maria Antonia[22] were the first cases in the palace. Both had survived. Isabella did not. After three days she was in a labour precipitated by her illness and Joseph's second daughter was stillborn, eight weeks premature. Isabella herself died five days later, in November 1763;[23] deaths three and four. Joseph was bereft: an otherwise aloof man, he had lost the one love of his life apart from his mother. His next wife, Princess Maria Josepha of Bavaria, was a match hastily arranged by the court; for political motives as usual. 'She is twenty-six. She has never had smallpox . . .'[24] was the nearest Joseph came to being emotional when describing her. But how portentous! The death of Francis Stephen at Innsbruck in the middle of 1765 was the 'fifth' in this tragic sequence of bereavement. Deaths 'six' and 'seven' were both in the next six months. Although these were Haugwitz[25] and Daun,[26] two old favourites among Maria Theresa's personal courtiers rather than actual family, they deepened the well of her loneliness.

Life at the Hofburg, in central Vienna, was showing some signs of returning to normality by early 1767. During fasching, in February, the sounds of carnival in the city streets were audible in the palace. And in May, at the Schönbrunn palace, southwest of the city, the Empress shared with her people the celebrations of her fiftieth birthday. But within hours she was confronted by news of death number 'eight'. Her daughter, Archduchess Marie Christine, married to Albert of Saxony and living at Pressburg[27], had had a near-fatal first labour. She survived but the child did not.[28] The grim reaper now returned to Vienna. Joseph II, newly joint regent with his mother, was growing ever more involved and powerful around the court, glad to be distracted from his new wife, Marie Josepha, whom he callously ignored. It was his mother and sisters who first saw that she was ill. Maria Theresa was there when the diagnosis of smallpox was pronounced and stayed with her, trying to compensate for her son's distant disdain. The Empress paid the penalty and contracted the disease. Now Joseph attended – but only at his mother's sickbed. He did not visit his dying wife and was not with her when she died on 28 May. Nor did he pay respects to her corpse or escort it to the tomb: he was constantly with his mother who was also expected to die.[29] The Viennese churches were packed with her incanting subjects, praying for a miracle. Providential or not, Maria Theresa did survive, though heavily scarred. Now another Habsburg wedding was looming – and yet more tragedy. Another Marie Josepha, Maria Theresa's 9th daughter, was betrothed, at sixteen, to Ferdinand, King of Naples.[30] This odd young man[31] had been, in 1762, engaged to Joanna Gabriella before she had died. Now, days before the planned wedding to his second Habsburg fiancée, the intended bride contracted smallpox and died[32] – death ten – but not before infecting her older sister, Elizabeth. She survived.

There was now a smallpox epidemic in Vienna and there were hundreds of deaths. Not all the statistics were anonymous. Leopold Mozart had brought his talented children from Salzburg, hoping that they would attract attention during the Royal marriage celebrations. Now the wedding was a funeral. The 'speckled monster'[33] was at large in the tall tenements of the crowded city. The Mozarts fled; but too late. Wolfgang, 11, and his sister, Nannerl, 16, were scarred but survived, the family rescued from the roadside by a local aristocrat.[34] Of the thirteen members of her immediate family and courtiers that Maria Theresa had lost, or nearly lost, in just six years, eight had been smallpox victims, of which five had died. Here, once again, is the epidemiology of smallpox with flesh on the bones: a highly contagious and often fatal disease occurring in outbreaks among dense human populations.

Royal and aristocratic pedigrees had long been shaped and re-routed by this dreaded disease and Maria Theresa's own life story was a prime example. She had married at nineteen and had always loved, to distraction, her husband, Francis Stephen, even when he had 'wandered'. She bore him 16 live children. The supreme irony was, however, that it was smallpox that had given her all these children as well as taking many of them. Maria Theresa's father had originally fixed on Francis Stephen's older brother, Clement, as her husband-to-be, but smallpox had killed him in boyhood.

The autumn of 1767 saw an important development at the Hofburg Palace in Vienna. Initially devastated and demoralised by her early widowhood and recurrent loss of children, Maria Theresa had lapsed into very uncharacteristic passivity. Her former trenchancy had gone; many tasks and decisions of the kind that she had formerly devoured energetically were sidelined or delegated to Joseph. But surviving her severe attack of smallpox was, she interpreted, a message from God. He must still see a purpose in her life.[35] Her former vigour and flair for incisiveness began to return and among the new ambitions was a resolution to fight smallpox; to confront the contagion before it could annihilate any more members of her family. Brushing aside the strong prejudices of her medical advisers and the conservative certitudes of her priests, she took a unilateral decision.[36]

I T WAS NOT a long carriage ride from Hertford to London but Ingen Housz would have spent most of it in deep thought. He had no idea why his presence was suddenly so vital in London. Sir John Pringle's letter had said that he was to come immediately for there was urgent business of the highest importance. There were no clues because there had been discussions only in the highest possible diplomatic circles and therefore at the highest levels of discretion. Nine months earlier, in May 1767, John Pringle had written a letter[37] to Dr. Terence Brady, physician at the Austrian court at Brussels. It had described the present state of smallpox inoculation in England. Ingen Housz may have known of it. He may even have seen the text – the letter was one intended for circulation – but neither Pringle, Ingen Housz or Brady could have foreseen the consequences of this correspondence. By autumn 1767 the letter, or a copy, was in Vienna. It proved to be an inspiration to a grieving Empress who was no longer willing to be passive in the face of smallpox. Pringle had written of very high success rates for inoculation and that its safety had come to be accepted. Maria Theresa could see no reason why an experienced inoculator from England should not come to Vienna to introduce the practice among her subjects. Very shrewdly, she saw that the best chance for general acceptance of the practice would be that her own family be among the first to be treated. She commanded Prince Kaunitz, her Chancellor, to write to London: to ask that His Majesty the King give consent that his personal physicians send their consensual recommendations on smallpox inoculation. And, if they saw no reason for it not to be introduced into Austria, that they ask His Majesty to approve a suitable person to travel to Vienna to inoculate among her family and teach the skills to Austrian doctors. Kaunitz knew that the Viennese medical establishment was against the practice and he himself had a constant terror of the disease. But he did not need his renowned and considerable diplomatic skill to know that it would be unwise to deter Her Majesty in her present frame of mind. He therefore wrote to Count Seilern, Austrian ambassador in London, on 17 December 1767.

So it was, that in very early January, 1768, Seilern hurried to St. James' Palace for an interview with the King's Chamberlain. After seven years on the throne, George III had matured into a disciplined and efficient manager of the nation's affairs. He immediately accepted the charge: his advisers knew that he would be looking for a prompt and judicious response. The Royal doctors convened. A reply to Vienna was formulated – a version appeared in *Gentleman's Magazine* in February.[38] It was dated 23 January and must be only a truncated version for it makes no mention of a person being chosen to go to Austria to start inoculation there. However, the doctors had

> no doubt, but that the method of inoculation, practiced in England with such universal success, would be as successful at Vienna, provided the inoculation was performed with the same skill and prudence . . .

The latter phrase certainly chimed with Maria Theresa's instinct – that the successful introduction of smallpox inoculation into her empire would require demonstration by, and tuition from, an experienced and successful practitioner. But who was the inoculator to be? Who should be recommended to His Majesty? David Middleton, surgeon to the King, was the first name to emerge. It was agreed that someone with the dignity of a court appointment and, therefore, exposure to protocol was needed, whatever their skill as an inoculator. But Middleton was 64 years old. He did not relish the rigours of the long journey to Vienna nor the immense responsibilities on arrival there. From the known timings of these deliberations it would seem that he asked for some time for thought. But he was adamant in his final decision. The mission was for a younger man. The job needed someone who was less likely to become incapacitated or die – outcomes that would frustrate the Empress' ambitions and embarrass the King. This was a powerful and entirely selfless argument and the others had to agree. John Ranby, the next ranking royal surgeon, was also well over 60 and for this and other equally valid reasons, it transpired that neither he nor any of the other royal doctors were suitable. They had to look outside the establishment; hope to graft on to a younger candidate the necessary arts of diplomatic discretion and the demands of protocol. It is not known how many names were proposed. One, or more, of the Sutton family might have been suggested but the family were from a humble background, lowly educated, and had, in any case, turned their inoculation practice into a highly remunerative 'trade' frowned upon by some qualified doctors. Watson might have been mentioned but he was also an older man, as was Dimsdale. Whoever it was had to be independent, educated, a skilled physician and an experienced inoculator: respected and respectable.

Eventually, it was Pringle who proposed that Ingen Housz be recommended to the King as the best person for Maria Theresa to consider.[39]

> The Court of Vienna has at last consented to the introduction of inoculation and being desirous of a proper person from this place . . . I have recommended Dr. Ingenhousz of Breda, who, after having practised nine years at home, came here for further improvement and in particular gave attention to inoculation. He was some time with Dr. Dimsdale, has inoculated above 300 or 400 with his own hand in the country and has improved upon his master. He is a man of study, good understanding, modesty and honesty, and with all these qualifications, being a Roman Catholic, I thought I could not better acknowledge the honour which was done me by naming so fit a person; his age is about 37.

Not all those present knew the Dutch doctor personally but all had to agree, on common knowledge, that he could be a good choice. He was young, energetic, single, respectable, skilled, trusted and likeable. It was felt unlikely that the King would refuse this suggestion, even though he might have preferred an English doctor to be so honoured. However, the arguments in favour of Ingen Housz continued. He had fluent French, the language used at the Austrian court, as well as outstanding Latin and Greek. He had been a first-class apprentice in inoculation, both to Watson and to Dimsdale. And he had been an excellent ambassador for his old teacher's book on Greek particles: there were now nearly 150 paid-up subscribers including the King himself. Moreover, the Empress seemed to like Netherlanders. Van Swieten, her personal medical adviser, was Dutch, as was de Haen, the professor of medicine at Vienna. Although the case for sending Ingen Housz to Vienna appeared overwhelming, the team of royal doctors were thrown back on to the fact that the young Dutch doctor was not a courtier. He could not just be seconded and it appears to have been agreed that his permission should be obtained before his name was submitted to the King. It was necessary that Pringle send a message to Hertford, requesting that Ingen Housz return to London immediately. It seems extremely unlikely that Pringle did not arrange to speak to him personally and privately. Such was the sequence of events that the young doctor, in a quandary, in a coach, between Hertford and London, could not yet know.

INGEN HOUSZ'S REACTION, when Pringle made the proposition, is another phenomenon where the only evidence is circumstantial. He would certainly have been shocked and perturbed. Here was another unexpected circumstance that would delay, again, his intentions to finish his tour of Europe and his return to the Netherlands. But the last three years in England had been so rewarding, so successful, thanks in large part to Pringle. And should he refuse, what would all his other new friends think of him? But some of this verges on the imaginary. All we know for certain is that Ingen Housz accepted the massive responsibility and that, having explained the need for absolute confidentiality to his protégé, Pringle requested an early appointment at the palace in order to inform the King.

The next day Ingen Housz attended a post mortem[40] but was probably deeply distracted. He was uncertain how many of his friends, some of whom were with him and Pringle around the corpse, were privy to his agreement that his name be sent to Vienna. Having committed himself, he would have begun to conceive, as the day progressed, the nature of the challenge. Suppose something should go wrong? Suppose one of his royal patients developed a severe form of smallpox? They might even succumb under his hands. Later in the day he called on Matthew Maty who had just received copies of some important letters about inoculation published at The Hague. The subject was hardly a suitable one to divert Ingen Housz from his anxieties about Vienna. But here, in the rooms of Matthew Maty at the British Museum, Ingen Housz took another significant step in his life, the second in twenty-four hours. Whether or not he foresaw this as its destiny, he wrote a letter that would become his first printed document in a long life of publications. The opening letter in the sequence that Ingen Housz was about to embark upon was from Charles Chais, dated The Hague, 4 January

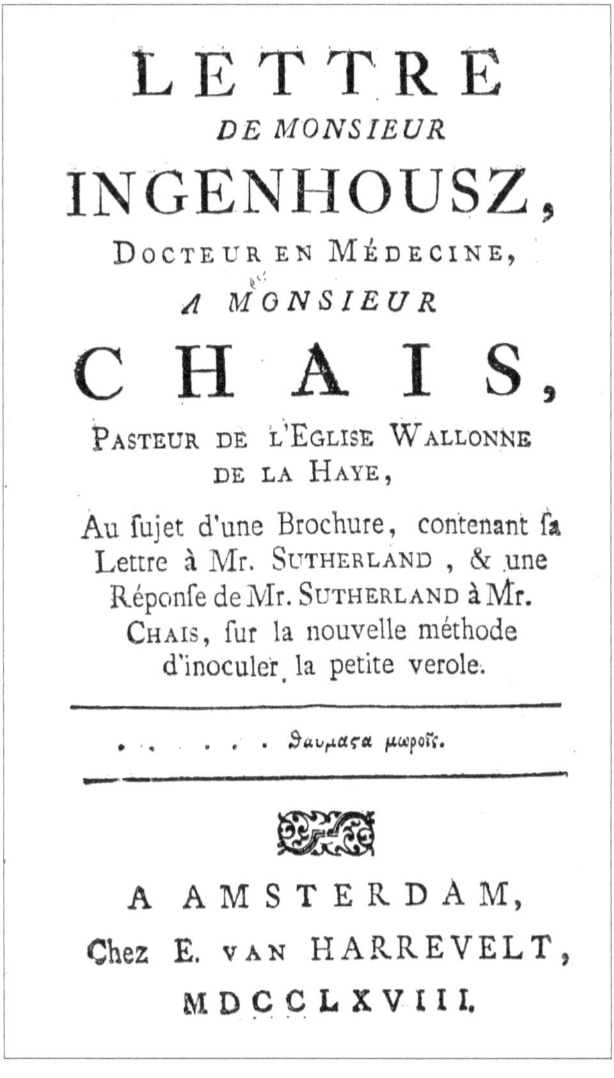

> # LETTRE
> ### DE MONSIEUR
> # INGENHOUSZ,
> ### DOCTEUR EN MÉDECINE,
> ### A MONSIEUR
> # CHAIS,
> ### PASTEUR DE L'EGLISE WALLONNE
> ### DE LA HAYE,
>
> Au sujet d'une Brochure, contenant sa Lettre à Mr. SUTHERLAND, & une Réponse de Mr. SUTHERLAND à Mr. CHAIS, sur la nouvelle méthode d'inoculer la petite verole.
>
> . . . . θαυμαςα μωροῖς.
>
> A AMSTERDAM,
> Chez E. VAN HARREVELT,
> MDCCLXVIII.

Letter to M. Chais – Ingen Housz's first publication.

1768.[41] The Reverend Chais was the preacher of the Walloon Church in the city, the religious 'home' of French Protestant refugees. The addressee was a Scottish physician, Alexander Sutherland, who was in Holland on his way to Berlin.[42] Chais had obviously met him and talked to him about smallpox inoculation. There had been tantalising and sometimes conflicting tales, at The Hague, of very significant recent improvements in the 'English' practice and here was someone who could bring the Dutch pastor properly up to date. He took Sutherland to be a reliable source. Chais related his long-held passionate advocacy of the practice and his conviction that it must come to the Netherlands. In this context, he persuaded Sutherland to respond, in kind, to an open letter so that knowledge imparted could be printed and distributed. For a man of religion Chais was amazingly free of biblical prejudice and prepared both to publicise and act on his own beliefs. He had already had his own children and grandchildren inoculated and expounded so convincingly to his neighbours and friends that many of them, too, had lined up to be treated by Dr. Thomas Schwenke, professor of anatomy.[43]

From detailed study, as well as from personal observation, Chais had a firm grasp of smallpox inoculation; its history, technicalities, efficacy and safety. In fact, he had a 155-page dissertation to his name, published at The Hague in 1754. In French and Dutch in tandem, it was a very detailed and reliable record of smallpox inoculation during its 30-year gestation in Europe.[44] There could have been few people in the Netherlands who had a better intellectual grasp of the process, including surgeons and physicians, and Chais's reputation had spread well beyond Holland. Matthew Maty, who had long taken a personal interest in the history of inoculation, knew of his prowess in these matters. [45]

> The ardour with which you have fought the worst prejudices of my compatriots by your excellent defence of a practice which could have been taken up there some 20 years ago, and the wise reserve which you have held in the midst of that torrent even more ferocious and more noisy recently, deserves the recognition of any Dutchman . . .

Now, it appeared, Chais had heard of the latest refinements as developed by Robert Sutton and by Thomas Dimsdale. Whilst not seeking overall opinion or recommendation, he wanted definitive answers to twelve itemised and very specific questions.[46]

Sutherland's reply was dated 1 February 1768.[47] He responded fully to the forensic interrogation but the answers were evasive in places. This may have been inadvertent, indeed unavoidable, but the reek of mercenary opportunism was less easy to forgive. Sutherland could not resist the chance to present himself as the best man to establish a private inoculation practice in Holland. He claimed to have obtained, from the Sutton family, their secret remedy, the ancillary treatment that supposedly gave their extensive practice in East Anglia such safety and efficacy. Preparing and dispensing this special but 'essential' recipe was not cheap, he implied, so he would be obliged to charge those who wished to be treated. Nevertheless, he was sure that his fees would be money well spent by those who would be, effectively, saving themselves from a very common cause of mutilation and, even, death.

Ingen Housz was incensed when he read of Sutherland's venal attitude and took up a pen. The resulting letter,[48] a thousand words in French, begins with his personal experiences as an inoculator. He reported how he became, in 1766, Watson's assistant at the London Foundling Hospital and then the frantic short winter days with Dimsdale, near Hertford, trying to combat a smallpox epidemic. His credibility was secure but the letter was clearly intended to be more than a *curriculum vitae*. Perhaps buoyed up by the private knowledge of his expected appointment at Vienna he was aggressively assertive. He was certainly supportive of Dimsdale whose inoculation philosophy and practice appeared to have been impugned by the correspondence that Maty had shown him. Dimsdale, he wrote,

> has inoculated many workers and peasants (free of charge), revealing his humanity for which he should be respected by the whole world

And as for the odious claim by Sutherland:

> I do not know of any reputable medicines which are unique to Mr. Sutton and I am surprised that Mr. Sutherland should want to tell such a story. . . . What name should one give to those for whom the truth only makes itself clear by the dangerous help of the lie, the mystery, the disguise?

The letter was despatched across the North Sea. It joined a sequence that would eventually give good insights into the issues on the periphery of inoculation as it was then being practised in England: the striving for best technique, for risk-free practice and, for some, huge profits.

INGEN HOUSZ PRESUMABLY left London the moment that Pringle intimated that the King was content for the name 'Ingen Housz' to be sent to Vienna. There was work to be done at Hertford. As the winter stronghold relaxed, but before a consistent thaw swamped the lanes, more people were on the move again in rural England. Word of mouth soon distributed the news: the events at Berkhamsted and Bayford were firing a surge in demand for inoculation that was already apparent before Ingen Housz had been called away by Pringle's message. Dimsdale had been assailed from all quarters and was delighted that his assistant reappeared so quickly. The confidential news from London was astounding but there was little time to celebrate it. The local aristocracy, gentry and middle classes were now also convinced that they and their households should be protected from the smallpox and were clamouring to have precedence according to their class and status. During the next three weeks Dimsdale and Ingen Housz inoculated, between them, over 700 'county' people[49] including a woman of 92 years and many suckling children with their mothers. And once again, despite the absence of any preparation with diet, drugs or bleeding, all the patients 'did well' and there was not a single death. Ingen Housz must have been fully occupied, treating and then following the progress of some fifteen to twenty new patients per day, scattered in and around Hertford. Nevertheless, he would have been increasingly tense. Powerful and remote strangers were determining his immediate future. He had only his friend, Dimsdale, as a foil. Even by the fastest post then known, in the diplomatic bags that were carried across Europe lashed to the best horses and riders, he could not expect a verdict from Vienna for about a month. In fact, the reply arrived in better time than that. There had been no protracted deliberations or hesitation at the Hofburg: Maria Theresa was functioning efficiently once more. Count Seilern found the following letter[50] in a bag that arrived in London on or about 12 March 1768.

27 February 1768
Your Excellency,
I collected together, for presentation to Her Majesty the Empress, all the various papers concerning the perfect state of the art, over there, of the smallpox inoculation. Her Majesty has therefore seen, with the most tender sentiments, the compassion of His Majesty the King and how, day-by-day, he wishes her well. And as the convocation of English Royal Doctors assert that smallpox inoculation can be performed just as well here as in England, Her Majesty is also assured that it could not be better performed than by any of the King's Physicians who have learnt the technique and practise it themselves.
Her Majesty also noted that the Dutch-born Doctor Ingen Housz, who arrived in England three years ago, has been given the highest testimonials as a smallpox inoculator, and could be the person to introduce it over here .
. . . Her Majesty has deigned to command me, therefore, Your Excellency, herewith, that you immediately make a proposal to Dr. Ingen Housz that he come here to undertake the smallpox inoculation; that you agree with him to come here as the best-qualified person, that you confirm his acceptance and despatch, and that you proceed as quickly as possible in view of the advancing season.
Your own activity will therefore bring about the highest of all important commands and you should not forgo the opportunity to express to His Majesty the King particular feelings of gratitude, in fitting expressions of

friendliness, and thank, also, Doctors Middleton, Pringle and Duncan and other such counsellors for their specific advice.
Kaunitz.

Once more Ingen Housz was urged back from Hertford and, on the afternoon of Tuesday 15 March 1768,[51] found himself at table with the Austrian ambassador. Count von Seilern revealed, formally, that 'Ingen Housz' had been approved at Vienna. He would have judged, correctly, but probably not admitted, that taking smallpox inoculation to the very heart of the Holy Roman Empire was an awesome task. Ingen Housz was told that speed was of the essence and it had already been assumed that he would travel through the Netherlands to Brussels where he would be given documentation and further assistance. If Ingen Housz were able to leave London by the end of the month he could be expected in Vienna by '. . . the beginning of May.'[52] This was a demanding schedule but permission had been given for all possible assistance on the journey. Count Reischach, Austrian ambassador at The Hague was being requested to lend all support after Ingen Housz had arrived in the Netherlands and officials at the court in Brussels would also be at his disposal. They would advise him, specifically, on the best onward route to Vienna. There were no objections raised to the doctor spending one or two days with his brother at Breda and Ingen Housz was welcome to arrange, for himself, his passage to Holland. All his expenses were to be paid by the court and, in order to offset any immediate costs, Sielern had been instructed to furnish him with sixty pounds sterling (£3,800). Ingen Housz can only have left the embassy with his mind in turmoil.

It must have been difficult for Ingen Housz to organise the logistics of his next few days. After arranging an early passage across the North Sea he had a long list of tasks thrown up by the sudden turn of events. He may have returned, very briefly, to Hertford to say goodbye to Dimsdale and his family – there is no surviving thank-you letter, suggesting that he may well have gone in person. There were also the other obligations of a sudden departure: restoring personal possessions to their rightful owners; settling outstanding accounts with traders; negotiating a rebate on the rent for his lodgings; writing to his brother, to Hoogeveen, to Jantzon and to Meerman. His real need was for time, which he did not have. In London he had made friendships of such intensity that it would likely offend if he did not say goodbye to a long list of individuals in person. Unless, by good fortune, he was able to see them congregated at one of Pringle's soirées he was, presumably, forced to scurry around Westminster and the City knocking on doors. In particular there was Pringle, Collard, Franklin and Watson. We know that he found Matthew Maty at home for Ingen Housz invoked the help of his compatriot to round off some 'business'.

Chais had already replied to the 19 February letter and although we no longer have the document itself, some of its contents can be deduced. At The Hague, Sutherland was still claiming privileged knowledge for which he could justify a monetary dividend. He was also denying that Ingen Housz could know better. These matters could not be left in mid-air but Ingen Housz was frantic with his leaving arrangements and, in any case, now in a situation where controversy was best avoided. Matthew Maty was persuaded to write a reply on his behalf and posted a letter to Chais on 27 March,[53] five days before Ingen Housz left London. Maty had no need for diplomatic reticence; he could treat discretion with disdain but did explain the substitution with some delicacy.

> Mr. Ingen Housz is called to introduce inoculation into one of the principal Courts of Europe and will no longer need to care . . . about scrapping, even with honest men, who dispute his reputation with him.

He reassured Chais, most emphatically, that Ingen Housz was an unimpeachable medical witness.

> He is one of those in our circle in London. Pringle, Heberden, Baker, Watson, Franklin and others hold him in high esteem, as much as for his virtues as for his friendship.

Chais was told to treat the contents of the Ingen Housz letter of 19 February as entirely trustworthy, particularly the statement that Sutherland was not privy to the actual formulation of the secret Sutton concoction. This fact was, he pointed out, something the Dutch doctor had learned from Daniel Sutton himself and would be corroborated by Collard. Chais must be wary of charlatans, even medically qualified ones, for

> an enlightened doctor will have, even if he is easily satisfied, advantages over those who have only a blind routine and over those who are not content unless they are paid exorbitantly for nonsense that they dare to purport as 'mysteries' while repudiating those modest practitioners who donate their knowledge and assistance freely.

The other, perhaps final, conversation that Ingen Housz must have had in London would have been with Sir John Pringle. Even at this late stage, Pringle was in a position to provide valuable support. He was able to remind his protégé that Gerard van Swieten, the Empress's personal physician at Vienna, had been one of his corresponding friends since student days together at Leiden and he gave Ingen Housz a letter of introduction to his old colleague. Moreover, Richard Huck had been in Vienna, observing medical practice there, only five years before, between June and November 1763.[54] This was recent enough for his personal reminiscences to be of great value. He, too, was able to recall, nostalgically, the affability of van Swieten. Not all of the briefing was reassuring, however. Huck had also met Anton de Haen, Professor of Medicine at Vienna. Though Huck probably tried to minimise the vehemence with which de Haen always expressed his opposition to inoculation,[55] he could not have been frank and honest without revealing that some antagonism was inevitable in the Austrian capital. It is almost certain, however, that Pringle would have been more optimistic: with the royal imperatives behind it, he could not afford the mission to Vienna to abort and Ingen Housz would have been sent on his way with a positive parting flourish.

Ingen Housz left London, for Harwich, on Friday 1 April 1768. Having reached the Essex coast in good time, he was unlucky. The winds were totally unfavourable and his ship was confined to the harbour for a week. After a day on the road from London, and then this frustrating and lonely incarceration in Harwich, he must have been relieved to hear, at last, the rattle of anchor chain and the clap of filling sail.

7
To Vienna

For some moments he's confused. His head has flopped onto his chest; there's stillness but for a splashing sound that he can't explain. The tension in his neck unwinds and as his eyes focus he realises where he is. There's just enough diffused moonlight for him to see the empty seats opposite and realise that the coach has stopped: that the coachman is relieving himself. He pulls his coat collar forward, shudders and yawns. This is the third night on the road and the craving to lie flat – somewhere, anywhere – is almost overwhelming. The horses, it seems, are in need of a rest having climbed a long hill – at least that's the story according to the driver as he turns from adjusting his clothing. He speaks a heavily accented German but Ingen Housz understands well enough. He thought the man had been drinking heavily when they changed horses at Feuchtwangen and his foul breath confirms it. It's a quarter before five and they should be in Ansbach by full light. The driver swigs from a flask and offers it to his passenger with a half-hearted gesture that's ignored. The coach reels crazily as the heavy man climbs up to retake the reins and they move off. Now wide-awake, the sole passenger is left with his thoughts. He has had six days and almost three nights to try to master how and where to sit so that he might be safe and comfortable. There's no real answer. It will be many days before his bruised ribs and swollen knee will be comfortable. He'd been unguarded when the wheel hit a particularly deep rut on the road down into Mannheim and might have suffered less damage if the coach had continued rolling and fallen on its side rather than bounce back on him. Every other lurch now gives him pain and only sheer exhaustion allows him to doze off but there is little hope of that now that his mind is racing.

The sun has risen and the coach has stopped yet again when he next wakes. He listens to the voices, some of them loud and irritable. He recognises that of his driver protesting, ' his important passenger is on a very urgent mission' and the rough reply that if he is in that much of a hurry he can 'bloody well get out and walk.' Once outside he sees the problem. They are in deeply wooded countryside and the trees come right to the edge of the road on both sides. A wagon loaded with freshly cut timber has a broken wheel. It's at a crazy angle and has disgorged its load all across the road. The lumberjacks are sitting on their spilt cargo, smoking. They raise their hats politely but Ingen Housz senses that they are enjoying the power they now wield: the

Elector of Bavaria himself would be powerless to proceed. When they feel that they've exasperated the travellers enough, the loggers own up to one of their number having gone for another wagon and help: with luck the road will be clear in about two hours. Ingen Housz clambers back into the coach and winds himself into a blanket. This is going to be the most comfortable sleep for three days and nights.

AFTER NEARLY TWO hundred unsettling nautical miles on the North Sea Ingen Housz reached the Dutch coast[1] around 11 April 1768. At least, here, he was among familiar sights and sounds, the native returned. He was able to command, very quickly, the necessary help to get himself the three or four miles past the dunes and into the centre of The Hague. Here he was taken aback[2].

> I was two days at The Hague and was very surprised that my presence caused such an instant commotion.

Having left the Netherlands an obscure doctor from a provincial town, he was returning to the capital a celebrity. He was now someone people were anxious to meet, to fraternise with, to talk to, to flatter. Having had only fleeting small-talk, conversations with anonymous strangers for over ten days, he was hurled into a frantic period of intense and important social interaction with famous and powerful people. In the future he was going to come to value time to himself, hours out of the public eye, genuine privacy and friendship where protocol didn't interpose. But for the moment the novelty was exciting and he found himself, immediately, having *tête-à-têtes* at the very top of the social ladder[3].

> I dined one day at the home of Mr. Fagel . . . . (who was) more polite than I deserve.

Ambassador Welderen, or his secretary, Collard, would have indicated to their young friend, on his leaving London, that it would be good manners, and certainly diplomatic, to pay his respects to Hendrik Fagel, the Pensionary. At this time in their complex history, the United Dutch Provinces all sent their own representatives to sit in the 'States-General' (assembly of delegates, national parliament). The 'chief executive' of the state, particularly for foreign affairs, was the Pensionary, the man at the very top of the pyramid of power and influence at The Hague, the seat of united government. He was certainly no mere mandarin and this top post was secure for life. In other words, and excepting the Prince of Orange, Ingen Housz was, in Fagel's state apartment looking out onto the Binnenhof,[4] at table with the highest in the land. It appears that the Pensionary had a mandate to recruit his guest.[5] There seems to have been a groundswell of aspiration, at The Hague, that Ingen Housz would return to the Netherlands and introduce smallpox inoculation. 'Very advantageous terms' might have been offered to him[6] but both men knew, of course, that they were dabbling in a rock pool after the tide had gone out. Ingen Housz was committed to the Empress; indeed, on his way to Vienna. It may be, though, that he and Fagel were really shadow-boxing with the notion that, after a few months in Vienna, Ingen Housz would return home. Tides turn. At this time, Ingen Housz himself can hardly have held any other concept for his future than returning to the Netherlands in some capacity or other: it was surely his tacit destiny. Perhaps he made a diplomatic half promise to Fagel. Inoculation had been the cause of quite a furore in the Holland press in the previous weeks and Fagel would have been desperate for inside information from someone with, now, unique personal experience; to have, at first hand, confidences on Pringle, Dimsdale, Watson and Sutherland. It is also very likely that he conveyed to Ingen Housz the message that the Reverend Chais wished to see him. No doubt the reverse was also true. Finally, there were, no doubt, offers of diplomatic assistance and a gentle enquiry to be sure that Ingen Housz was planning to call on the Empress' representative at The Hague; a courtesy that should not be overlooked.

The prompt was superfluous: Ingen Housz had already been to the Austrian Embassy to present a letter of introduction. He had received an invitation to dine at the earliest convenience. Once again, a very high official saw him as someone important enough to want to meet at length and at leisure; across table rather than in study or library. Baron Thaddäus Reischach, Austrian ambassador at The Hague, was probably relieved that his young guest had actually materialised. He'd been expected at The Hague for well over a week. Ingen Housz had an impeccable reason for his delayed arrival, of course, but there was some astringency, a tension that even a skilled diplomat could not disguise. The plans for Ingen Housz's passage to Vienna had hardened. A letter had arrived, from Brussels, two weeks before, enclosing more details. Everyone knew, of course, the difficulties that befell travellers but an air of anxiety in the corridors at Brussels had pervaded the text. A passport for Ingen Housz had also arrived together with a royal dispensation that customs officers should not open his bags; all such unnecessary delays were to be avoided. Everything pointed to a degree of impetuosity at Vienna, a desire for immediacy. Reischach understood that his guest had already been granted permission to see his old teacher and his brother but were there going to be any other delays? He must have been reassured to find that, after calling on Chais that evening, Ingen Housz intended an early morning departure for Delft and Breda. He should be in Brussels within five days. Reischach was able to relax, duty done, and report so to the Viceroy on 15 April.[7]

> Your Highness,
> For three days we have had a Doctor called Ingen Housz from London. . . he being called to Vienna, in order to establish there the art of inoculation . . . he shines out as a very reasonable man in possession of much learning but a wholly modest man to be with.

Time, distance, season and expectations at Vienna were beginning to coalesce into an uncomfortable reality for Ingen Housz. Still, there was a happier meeting to come.[8] 'That evening I was at the house of Mr. Chais'. Meeting Chais was wholly pleasurable: here there were no difficult undercurrents generated by officialdom. The two men raced along on a shared hobbyhorse. But within minutes a flustered servant appeared on the threshold to interrupt them.[9] ' . . . his Highness, The Prince of Orange sent me a message that I must, under no circumstances, leave (The Hague) without speaking to him.'

It is difficult to see how the easy-flowing conversation with Chais can then have been continued. The Netherlands did not really have a royal family. The Prince of Orange, despite his title, was officially only the 'Stadthouder', the figurehead of the court. But if not, definitively, a monarch, he was far more than just first minister or commander-in-chief and was marriageable to other European royals. Wilhelm V of Orange-Nassau was twenty years old in 1768. His mother had been Anne, Princess Royal, daughter of George II. The messenger had presented Ingen Housz with a dilemma. Just hours before he had been assuring Baron Reischach that his departure was almost immediate. Now he had a commitment to which he should give precedence. By inference he must have accepted the request but also sent a very apologetic rider that the Prince should know that he was expected, by no less than the Empress, to be on his way early next morning. The result was an immediate invitation brought back by a breathless servant.[10]

> That evening, at 9 o'clock, he (The Prince of Orange) sent for me, by messenger, calling me to The Court, where I had, at once, a gracious audience; in brief I was ashamed of myself, seeing how more was being made of my case than I would have made of it myself.

No details of this particular interview have survived; no hints or inferences, even. Ingen Housz was already taking on the heavy mantle of discretion assumed by wise courtiers, a leak-proof confidentiality beyond anything known in medicine. At court it was often unwise to reveal confidences even to close colleagues. It is not difficult, however, to guess the drift of the conversation and the fact that the Prince would have been anxious to give his personal blessing to the enterprise, that one of his subjects should be seen to have his support in this remarkable undertaking.

IT IS ONLY a few miles from The Hague to Delft – not time enough for a bemused traveller to regain his composure before arriving at the Latin School where Hoogeveen was now head teacher. Now the debts owed to each other were self-cancelling. The younger man's outstanding education in the classical languages had led him to unforeseen life opportunities. Meanwhile, the older man's erudition and enterprise had been promoted so well, by his former pupil, that it was now guaranteed success. The name 'Hoogeveen' was also going to be famous but there was little time for celebrating the new circumstance. From the located dates that we know of his journey, Ingen Housz must have left to travel the 40 miles further south to Breda within a few hours. Hoogeveen would have been delighted to hear, though, that there were some sample pages of *Particles*, and a bundle of prospectuses, in his young friend's baggage. The prospect of selling his book at Vienna was an unexpected fillip.

At *de Trecktang* Ingen Housz found a dramatically new situation. When he had left Breda, in autumn of 1764, Louis and Maria had been married for only six months.[11] Maria must already have been pregnant but when Jan left she would have been only in the early stages. Now the 'bulge' was a very active little girl of three – Jan's first niece, Maria Catharina, born in March 1765.[12] She was the first of a whole new generation in the family. Then came Arnold Joseph, Jan's first nephew. He was now a month short of his second birthday. And just beginning to walk was Peter Joseph Wilhelm born in April 1767. Maria herself was already heavily pregnant again – child number four was due in July. In a way Louis' life had changed almost as much as Jan's and the demands of the family business superimposed on those of a rapidly filling home probably left little time for fraternisation. But sharing practical matters, profit margins and small domestic crises were a welcome respite for Jan. All too soon, though – in two to three days – Ingen Housz was on a coach to Antwerp and thence to Brussels, arriving on or about Wednesday

20 April.[13] He would have made his way, immediately, to the court.

The Austrian Netherlands, almost contiguous with present-day Belgium, had become part of the Habsburg Empire with the Treaty of Utrecht in 1713. From 1744 to his death in 1780, it was ruled over, as Viceroy, by Prince Charles Alexander, the younger brother of Francis Stephen, Holy Roman Emperor until his death in 1765. Therefore Prince Charles was brother-in-law to Maria Theresa: twice, in fact, for he had married the Empress's sister, Archduchess Maria Anna. Charles's instinctive fondness for the arts and the Enlightenment were fostered by the long-standing and progressive head of the country's administration, Count Philip Cobenzl. These were the two men waiting for Ingen Housz at their capital, Brussels. Here was an outpost of Vienna and, for Ingen Housz, a foretaste of the Austrian court and an occasion to rehearse his rusty French.[14] Details for the onward journey were finalised in briefing him. The essential paperwork was validated and double-checked and his best route rehearsed. The immediacy of the bureaucratic machine impressed on Ingen Housz, yet again, the high priority of his mission and the unspoken censure that he was running late. That speed was of the essence seems to have been reinforced. A generous sum of money was handed over – frequent changes of horses and travelling through the night, as seems to have been suggested, were going to be expensive. He was advised to travel overland to Ratisbon (now Regensburg) where he could embark on one of the regular boats that sailed down the Danube to Vienna. When Ingen Housz eventually reached Vienna he began a 'financial' daybook,[15] using the first page to describe, in his own words, his journey to, and reception at, the Austrian capital. The quotation below is the first sentence of this 'log'. 'I left Brussels for Vienna on 22 April 1768, in a travelling coach that I had bought, at Brussels, for 49 Ducats.'[16] There are no details, however, of the staging posts to Ratisbon. The most direct route would have taken him through Liege, Trier, Mannheim, Heilbronn and Nuremburg, some 420 miles. Presumably, at least two coachmen were engaged and fresh horses harnessed in every ten or so miles, the exhausted team then taken back to their stables by a postillion. All of this was then standard practice[17] and may have a romantic resonance if it calls to mind the familiar scenes of British royalty gliding comfortably along The Mall on ceremonial occasions, smiling and waving serenely within their padded cocoon. The eighteenth century reality was far from being so sedate even though some stretches of road were being improved by turnpike schemes. No roads were metalled, of course, and even in late spring the slightest hollow was a foul, gelatinous quag. If the coach didn't stick fast it moved in crazy lurches or slewed from one deep rut to another. Over drier stretches on high ground the rigid wheels fought unarmed combat with proud rocks. The suspended cage that was the passenger compartment was too large for the occupants not to be thrown around inside and too small to seal off from fresh air and the weather. Even at only four miles an hour passengers were shaken like the pea in a whistle with every bump and then, if the road was smooth and even, nauseated by the pendular rocking and yawing. Add boredom and sleeplessness to this sensory nightmare and one can see how long journeys in the eighteenth century were a test of both stamina and purpose. Obviously Ingen Housz had a powerful motive for his journey and was aware that he was running behind a schedule that had been concocted behind his back. He tried to catch up by travelling non-stop for part of the journey, blessed by a full moon. 'I arrived at Ratisbon on the morning of 7 May, having passed four entire nights on the journey without sleeping in a house.'[18]

ALL THE TIME that Ingen Housz had been making his way from London to Vienna, from west to east, two other men had also been travelling across central Europe, in their case south to north. Plotted on a map of the continent the traces of the two long journeys were bound to intersect. The fact that the crossing of the ways also coincided on the clock so that the travellers would meet was a remarkable concomitant. It was a fateful meeting, a very long-odds liaison. The 50-year-old Johann Winckelmann and slightly older Bartolomeo Cavaceppi had set out from Rome on Sunday 10 April 1768 and were heading for Berlin.[19] They were travelling together for their mutual convenience. They were both taking a vacation, a respite from their individual commitments in Rome but both had obligations north of the Alps. Winckelmann had been born in Prussia and moved to Italy at the age of 38. He was fluent in several other contemporary languages besides German. This was of huge benefit to Cavaceppi, an Italian who floundered outside his mother tongue. Sharing the costs of their extensive journey was also prudent: they each moved and worked in circles where there was indescribable wealth whilst they, themselves, were both of modest means. They enjoyed privileges and high status rather than the bounty of rich rewards. Their journey had been planned carefully so that it took in a sequence of cities important to one or both of them. They had been at Loreto, Bologna, Venice and Verona. After 25 days on the road, they had arrived at Ratisbon in Bavaria on Thursday 5 May. By this time there was serious friction between them.

Johann Winckelmann.

Winckelmann had been born on 9 December 1717 in the Prussian country town of Stendal, a settlement devastated during the Thirty-Years' War.[20] His father was a poor cobbler in this run-down community. The child grew up in one room that served as shop, kitchen, living room and bedroom. His parents had risked total penury in allowing their son to continue to study into his teens rather than contribute to the household economy and both had then died before he was fully-grown. Young Winckelmann's talent for languages was recognised at school and led to an early career of translating and tutoring in the classics. He became particularly fond of the Greek language. Winckelmann was an isolated figure, his loneliness exacerbated by his insecure employment and an impelling sexual preference that placed him among a persecuted and shadowy minority. He was unequivocally homosexual and his promiscuous experiences, deviously hidden from public gaze by necessity, seemed to make him emotionally labile although his career path did not seem to suffer. By his mid-thirties he was a valued cataloguer and librarian, moving between various aristocratic seats around Dresden where he discovered, within himself, an intense appreciation of the classical art that adorned the houses in which he was working. Greek statuary was a particular love and inspired him to study its history. Born and brought up a Lutheran, he converted, in 1754, to the Roman Catholic faith. This opened new doors and, the next year, he found he could afford to visit Rome. Here he was very surprised that the inhabitants knew virtually nothing of the countless art treasures that surrounded them. Indeed, they showed no respect for the continual stream of artefacts that was being unearthed as the city developed. The only common motive was obtaining the best price in a viciously venal market place. Winckelmann was appalled by the abuses and was impelled to challenge them, slowly changing from horrified visitor into resident authority and conservationist. In turning a hobby into a profession he effectively founded the science of archaeology at the very time that private collections, museums and galleries for classical antiquities were being established all over Europe and the desire for exhibits exploded. It was Winckelmann who instilled a discipline into digging and collecting, insisting that the exact location and nature of antiquarian treasures be carefully recorded. Winckelmann lived in Rome for the next thirteen years, increasingly respected by genuine collectors of antiquities. Among these was Cardinal Alessandro Albani, a powerful Vatican influence and it was he who recommended Winckelmann as papal antiquary in order to

> keep track of all antique statues, paintings, engraved gems, coins, mosaics . . . in the Papal States, make sure that no such objects were smuggled out of the country, and dispense permits to undertake excavations . . .[21]

Winckelmann had, at last, a secure and high status occupation and to him, the ultimate and most welcome power – to control the quality and locations of archaeological activity. The position also carried other responsibilities, however; ones which he also appears to have relished and in which he was eminently successful.

> The Papal Antiquary . . . acted as a guide whenever prominent visitors were taken on a tour of the city . . . (occupying) the top rung of a professional ladder that extended all the way down to the thousands of nameless guides who offered their services to newly arrived travellers[22].

Winckelmann seemed to be able to cope, consistently and tactfully, with all types of personalities visiting Rome, many of whom had less interest in architecture and antiquity than in carousing with prostitutes. However, his standing as a guide had become such that a conducted tour with 'Abbé'

Winckelmann formed the high point of many a traveller's visit to Rome, royalty among them[23].

In fact, Winckelmann's successful progress in Rome was so outstanding that the real reason for his decision to return to Germany, even temporarily, has always been a subject of intrigue.[24] His correspondence hints that he hankered to meet some of his old friends again and, once his half-hearted intention was discovered, various agencies in Rome engaged him as a secure messenger. One of those hoping that he could carry confidential correspondence was his employer, Cardinal Albani, who was Maria Theresa's representative at the Vatican.[25] This made it particularly difficult for Winckelmann to back down but it was his private life that gave him the real dilemma. Homosexuality was blind-eyed more in Italy than anywhere further north in Europe whatever the letter of the law. Established in the upper reaches of Rome society, Winckelmann had made no secret of his sexual orientation. He had enjoyed several stable, openly homosexual, relationships with important men, nobility among them. Indeed, today's homosexuals proclaim his courage in being so forthright, a fact that should be placed, they assert, on a par with his fame as an archaeologist and art historian. In Rome his high status and respectability were not challenged. People valued him for his professionalism and accepted, or chose to overlook, his intimate proclivities. But none of this social tolerance was guaranteed outside broad-minded Italy and, in Germany in particular, Winckelmann had some very strongly motivated enemies. In his homeland his homosexuality would leave him a hostage to fortune; and justice, were it administered, could be extremely harsh. He was a very vain man, however, and perhaps he simply wanted to indulge his conceit and sartorial flounce in former circles. He was certainly now a significant world figure, famous even, and expected, perhaps craved, attention. Quite what Berliners would make, though, of his fully pleated silk shirts and tight leather trousers was not so clear cut as his narcissism. Was he not the thrill-starved persistent criminal tempting capture?

Cavaceppi, on the other hand, seems to have been a less complex, more stable character. He had trained as a sculptor and eventually established a workshop in Rome that came to enjoy a deserved reputation for the restoration and selling of antiquities.[26] His address was one of those most sought out by 'Grand Tourists' visiting the holy city and shopping for artefacts. Not only did he have genuine antiquities available for sale, he could restore them sensitively if they were badly damaged or copy them in pristine stone if that was preferred. Stately homes all over England contain examples of his artistry although, as he was reluctant to sign his works, even his own excellent copies, their provenance is often unrecognised. Like Winckelmann, Cavaceppi was patronised by Cardinal Albani and, having sensitively restored much of the Cardinal's personal collection, had been appointed restorer for the Museo Pio-Clementino at the Vatican.

The friction between Winckelmann and Cavaceppi was not because they were lovers – there is not a shred of evidence for this. The conflict could be traced to a different cause and back to a specific point on the journey. It was as their coach was being pulled up the southern face of the Alps above Trento, towards the Brenner Pass and Upper Austria, that Winckelmann showed the first signs of an emotional breakdown.[27] Increasingly fixated by the steep stone roofs of the houses in the high valleys above Bolzano he became paranoid and obsessed. To him the pitch of the gables and the dark colours spelt certain doom. Trapped in their travelling coach hour after hour, Cavaceppi began to find his companion interminably irritating. There was no reasoning with him, no means to convince him that they had no need to go back downhill to the Lombardy plain and back to Rome. More used to repairing and renovating inanimate statuary than crumbling intellects, Cavaceppi was out of his depth. There was little else he could do except stubbornly insist they continue their journey and hope that Winckelmann's sanity would return.

Cavaceppi showed amazing tolerance for many difficult days, until they arrived in Munich. Here Winckelmann seemed to improve somewhat, at least by day, and both men were able to grace receptions and attend meetings they had planned. At night, however, things were far from normal. Winckelmann was unable to sleep, terrorised by 'grey shadows' moving around his room at the guesthouse. He would not be left alone and Cavaceppi was unable to relax; far less, sleep. However, he managed to persuade Winckelmann to continue northwards with him, at least as far as the Danube and they entered the city of Ratisbon on 5 May.[28] Winckelmann seemed, again, to remit from his illness. Logical thought, coherent conversation and conventional behaviour returned and Winckelmann coped, appropriately, with a request for an audience from the local Bishop. And both men slept well; the threatening nocturnal shadows of the Munich nights did not return. However, on the 7 May Winckelmann suddenly announced that he was returning, instantly, to Italy. He would catch the next boat down the Danube to Vienna and then take the quickest means to Trieste, thence Ancona and Rome. This time Winckelmann was not to be persuaded otherwise. Cavaceppi could see from the blank expression in his eyes that his friend was again unharnessed from reality. The only practical response was to leave Winckelmann to make his own way home: to book a berth for him on the next Vienna-bound boat.[29] This, he discovered, would leave at noon the next day. The remaining twenty-four hours that these two men spent together appear to have been superficially convivial, the premium of a truce. But it must have been very difficult for Cavaceppi, setting his friend, translator and guide adrift, knowing how disturbed he was. What we do know – and Cavaceppi did not – was that these difficult last few hours with Winckelmann were the very ones that saw the heavy coach bearing an exhausted Ingen Housz approach Ratisbon and, by crossing the old stone bridge over the Danube, lumber into the city. In place, and in time, Ratisbon was the meeting place of the two expeditions. And it was on the bank of the Danube that the three travellers were re-sorted: Winckelmann and Ingen Housz both boarded the next morning's boat to Vienna while Cavaceppi continued north, overland to Berlin.

'I LEFT RATISBON, by boat, at midday on Sunday the 8th. The boatman promised to deliver me to Vienna in four or five days but we only arrived at midday on Saturday the 14th.'[30] Ingen Housz and Winckelmann were to be in each other's company for six full days. The boat that left Ratisbon that morning, 8 May 1768, was scheduled to arrive at Vienna on the 12th but it was delayed and arrived two days behind schedule. Ingen Housz and Winckelmann were thrown together; two isolated travellers under stress, the one going to confront a challenge, the other running away. Here we have two men in a boat: not alone, certainly, but effectively cut off from the other passengers by their unusual stations, backgrounds, educations and interests. They must have faltered, for a few minutes, until they found a common spoken language in which they were both comfortable but went on to have extensive conversations during the trammel of their river journey. Ingen Housz may have heard the name, Winckelmann, during London conversations and actual identity merely required confirmation. On the other hand, the name, Ingen Housz, was enigmatic but his mission to Vienna could not be kept a secret in the circumstances: both men were on their way to meet the Empress.[31] Their common ground would have opened up many conversational avenues. They discovered that they were both enthusiasts of the classical cultures: that they both loved Latin and Greek. At some stage of their drift down the Danube together, Ingen Housz, as an experienced clinician, may have spotted the signs of an ongoing emotional fragmentation in his new friend. Coping with a psychotic fellow passenger in the middle of a river voyage would have been a fearsome prospect. Ingen Housz may also have been sensitised to Winckelmann's homosexuality. There is no reason to think, however, that he would have responded to any advances or propositions. Despite the voluminous and detailed surviving archive pertaining to Ingen Housz, some of it personal and revealing, there appear to be no doors that open into his sex life. All one can say is that at the age of 37 he was sexually unattached. There appear to have been no intense relationships and certainly no commitments beyond those at the level of friendship and intellectual common interest. Motherless from early infancy, though with a stable live-in nanny, Ingen Housz had been educated in an exclusively male environment until the age of 24. At that time these were not unusual circumstances: aristocrats and many affluent middle-classes, throughout Europe, seconded the care of their children to surrogates of various kinds and only boys were formally educated outside the home. However, even sexually blinkered schoolboys go on to form loving relationships with women and start families. On the other hand, that a man clawing his way up a professional hierarchy or building a business remains unmarried, unattached even, when nearing the age of 40, is nothing at all unusual. But whatever the personal chemistry, the two men had some rational discussions at some point on the journey. They discussed, for instance, Hoogeveen's *Particles* subscription in some detail. Winckelman, whose first love had been the Greek language, enthusiastically agreed to act as Hoogeveen's 'agent' in Rome.[32]

> I established friendly relations with him. After being informed of the matter and after reading the prospectus, this learned man and patron of savants spontaneously offered to be the writer's publisher in Italy.

EVENTUALLY, ON THE morning of Saturday 14 May,[33] the boat listing Ingen Housz and Winckelmann on its manifest neared Vienna and swung to starboard into the Danube canal. Ingen Housz was probably expecting transport to be waiting at the dockside. From now on, as a guest of the court he would find that servants would cater for every item of his daily needs. He had had a foretaste of such privilege at Brussels. He was almost certainly scooped up and taken, by monogrammed carriage, into the heart of Vienna, to rooms in the Julianischer House,[34] a small palace just a stone's throw from the Hofburg, the royal palace itself. He would soon learn that the nerve centre of the court was six miles to the southwest of the city at the Schönbrunn Palace. This had been, formerly, the summer residence of Maria Theresa but she had now taken to spending more and more time at what was her favourite home. Secured within its protective walls, mid-eighteenth-century Vienna was a tightly corsetted city often struggling to breathe. The mildest dry warmth, and the least traffic, raised clouds of fine chalky dust that settled as gossamer on the densely packed dwellings. Accommodation pressures had long forced city builders to think vertically. It was reckoned that in the 1760s the 52,000 inhabitants living within the fortress walls of central Vienna occupied only some 1,300 structures – an average of 40 people per dwelling.[35] And most of this densely packed population resided above ground level, over shops, taverns and stables. Ingen Housz would have noticed that many tight streets ran off the squares into gloomy canyons that could never admit direct sunlight even from a high, midsummer source. He would also have noticed that the light that did penetrate the metropolis was grainy and translucent; already in mid-May the atmosphere bore the suspension of powdery particles that mired every sweat globule and irritated every eyelid. Vienna was a much smaller anthill than London but even in his first trajectory across the city Ingen Housz would also have seen that its 'insect' life was much more varied. It was memorably described by Johann Pezzl in 1786[36]:

> Here you can often meet the Hungarian, striding stiffly, with his fur-lined dolman, his close-fitting trousers reaching almost to his ankles, and his long pigtail; or the round-headed Pole with his monkish haircut and flowing sleeves.... Armenians, Wallachians and Moldavians, with their half-Oriental costumes are not uncommon. The Serbians with their twisted moustaches .... Greeks smoking their long-stemmed pipes in the coffee-houses .... bearded Muslims in yellow mules, with their broad, murderous knives in their belts. The Polish Jews, all swathed in black, their faces bearded and their hair all twisted in knots ... Bohemian peasants with their long boots; Hungarian and Transylvanian waggoners with sheepskin greatcoats; Croats with black tubs balanced on their heads – all provide entertaining accents in the general throng.

Vienna was one of the most eastern of the recognised European cities and certainly more accessible to the Orient than others. Some 800 miles east of London it was one of the last Christian outposts for the traveller heading east, a defensive bastion for this very reason. Eyed from the opposite pole it had frequently been in the first bite of aggressive empires of the Levant, Ottomans especially. Significantly further east than Prague, well north of Venice and Trieste, south of Berlin and Krakow, Vienna was disposed to the Middle East; it lay on a regularly contested crossroads to another continent.

One might suppose that once Ingen Housz had been installed in his rooms, had been introduced to his personal servants,[37] Louis Mathis and Franz Mochet, and had had some rest, he would have been able to draw a veil over the traumas of his journey, feel confident that the worst part of his mission to Vienna was over, and immediately get down to the reassuringly familiar routines of inoculating. Far from it; he soon felt trapped and under pressure. Within ten days of his arrival he was feeling so isolated, alienated and abused that he was on the point of resigning from his appointment.[38] Generosity in kind, luxurious accommodation, servants, transport and friendly greetings were poor compensation for his sudden and significant loss of liberty. He had grown so fond of his London life, its freedoms, informality and enlightened attitudes. He had been able to go where he wanted, say what he liked, virtually do as he wished. The tedious confinement of his coach and the claustrophobic voyage with Winckelmann had been tiresome enough but the coercive quarantine that he found at Vienna made him feel victimised and angry. Something finally deterred him but he certainly poured out his caustic reactions to his circumstances in a letter to Sir John Pringle, copying it to Richard Huck. Perhaps expressing his feelings in strongly worded letters to London was therapy enough for him to hesitate, not to over-react. The letters themselves, which arrived in London on or around 14 June 1768, no longer exist but we can unpick, from the joint reply,[39] the issues that proved to be so tormenting; the 'devils' that had put all the 'doubts, jealousies and fears' in his mind. Huck and Pringle were clearly distressed to receive such letters. If their 'envoy' behaved recklessly there could be a difficult diplomatic 'incident'. Huck was delegated to spend some time penning an immediate long reply and Sir John Pringle appended, in his own hand, six lines of blunt advice. 'Remember (that) when I gave you the recommendatory letter to van Swieten, it was expressly on the terms that you were to be his friend, that is have confidence in him.'

THE NEWS OF Ingen Housz's arrival at Vienna would have spread around the court and the city in hours and there was soon a whirlwind of activity around him. His servants took him to a series of introductions. Ingen Housz recorded the names in his cash book, in a more matter of fact entry, perhaps, than others.[40]

> I found everything to be in order. Baron Van Swieten welcomed me very politely as did Prince Kaunitz, Prince Starhemberg, Baron Reisach the Chamberlain, Count Ferdinand van Harrach and Count Ernest van Harrach, Count van Degensfeld the Dutch Ambassador, and others.

The most appropriate welcome was certainly that of van Swieten and a substantial personal relationship developed over a series of occasions when Ingen Housz was a dinner guest in the van Swieten household.[41] The most powerful personality Ingen Housz met, however, was Wenzel Kaunitz. He was Maria Theresa's chancellor, her most senior and most regarded political adviser and on the face of it his task was simple enough; to reassure the new arrival that he would want for nothing and that he would have the full cooperation of the whole court. As the highest-ranking dabster in court ritual it was, for Kaunitz, a matter of pure routine to announce to Ingen Housz his formal appointment as body physician to the Empress[42] and her family and that he would be initiated in matters of protocol and dress before meeting her. And it was probably Kaunitz who offhandedly informed Ingen Housz of Her Majesty's strong preference that courtiers should be married.[43] These singular slavish arrangements carried massive implications for Ingen Housz. They strongly contradicted his assumptions that his role at the court was to be transient and soon completed, that of a tradesman on call. It must

Wenzel Kaunitz.

then have dawned on him. No one, up to now, had disabused him of his own notion that he would be at Vienna for a few weeks at most – perhaps for the duration of the summer. He had thought that inoculating the royal children and supervising their recovery should take about a month. And teaching the technique to other doctors in Vienna so that the practice took root would, he had thought, easily be dovetailed into his circumscribed royal responsibility. These had been his concepts but he realised that he had not actually discussed them with anyone, tested them, even with any of his friends in London. Now it was apparent that the court at Vienna had preconceived a substantially different scheme, using a very different time frame. But Kaunitz was a man of such high status that one did not dispute with him any more than one would with the Empress herself. The inability to protest, however, meant that there was tacit agreement to the terms of engagement, terms that made no mention of remuneration or reward. This, too, was on the list of irritations sent to London as the Huck/Pringle rejoinder makes clear.[44]

> . . . you should not be overscrupulous in making a bargain. An appearance of disinterestedness has great weight with a generous mind. The Empress has one. If you are successful, and I think you can scarce fail, she will load you with favours.

We can certainly be sure that Kaunitz made no mention of smallpox, not even couched in analogous terms. His deep-seated abhorrence and fear of illness was well known, even outside Austria: it was second nature in those who worked with him never to trespass into medical matters in conversation. Ingen Housz would have known, anyway, that inoculation was a topic for discussion with van Swieten.

His very famous compatriot, Gerard van Swieten, was someone Ingen Housz had been looking forward to meeting. In the event he appears to have been disappointed. There was a coldness that went beyond the expected formality of office. It was not that van Swieten was not perfectly polite[45] and hospitable but for some reason Ingen Housz felt an unease.[46] Van Swieten was a man in conflict with himself – 'the situation he is in makes him the object of envy' is the phrase used by Huck. As old as the century, van Swieten had arrived at the Habsburg court in 1745. Like Ingen Housz himself, he had been recruited from afar and with gilt-edged recommendations. He had been practising as a physician but in the Austrian Netherlands rather than in his homeland. Also a Catholic, he had been spurned by the academic medical establishment at Leiden even though he was among the most respected of the pupils of Boerhaave.[47] His bedside empathy and diplomatic finesse when Maria Theresa's much-loved sister, Maria Anna, developed a fatal illness at

Gerard van Swieten. Courtesy of the Wellcome Library, London.

Brussels in 1744, had attracted Van Swieten to the young Empress[48] and nothing, over the next two decades, had slighted her belief in him. Indeed, her faith in his potential had taken him far beyond the clinical sphere and he also did the work of three other men. She had entrusted him to modernise the Jesuit-dominated university of Vienna. Van Swieten had also embarked, enthusiastically and with great energy, on reorganising medical education in Austria: it was he who sowed the seeds of what would become the famous Vienna medical school of the next century. And it was Van Swieten who had reorganised and upgraded the Imperial library. A family man who probably would have wished to see more of his wife and children, he had suffered the loss of his second son, Jan, from smallpox, in 1750.[49] Despite this, and the fact that he himself bore scars of the disease, he had remained very conservative in his treatment of smallpox, trusting to therapies that were seen, even then, to be little better than useless. He had continued to agree with his dogmatic disciple, Anton de Haen, whom he had brought from the Netherlands to be the professor of medicine, that inoculation was counter-intuitive, counter-productive, anti-canonical and probably dangerous. Then, having been a pivotal but helpless figure among the royal victims of the disease in 1767, van Swieten had faced a dilemma when Maria Theresa took heed of one of her new personal physicians, Anton Störck. He had helped her through her own assault by the disease[50] and backed her decision to invite an inoculator to the court. De Haen, it is clear, remained in violent opposition and took no pains to disguise his revulsion.[51] Van Swieten was in a more tricky position, being unable to settle, in his own mind, his predicament – should he remain loyal to his Empress or to his beliefs? He opted, finally, to cooperate with Maria Theresa's wish. Presumably he would otherwise have had to tender his resignation and suffer a catastrophic fall from grace. How this affected his relationship with de Haen and the rest of the very conservative medical fraternity at Vienna is easy enough to project. His pragmatism isolated him and some of his regret and resentment were probably still gnawing at his usual affability even as Ingen Housz arrived at Vienna. He, in his turn, could not have been aware of these unspoken tensions in the court; that van Swieten was struggling with a loss of face and authority. After only a few days in Vienna, however, the war being waged among the physicians came rampaging into his life. Ingen Housz began to feel the wrath of the opposition to inoculation and was subjected to some spiteful verbal defilement of van Swieten and perhaps, even, threats. This tension, too, had been relayed to London.[52]

> Listen not to the insinuations of de Haen or (other) enemies. If you confine yourself to the practice of inoculation you will interfere little with other physicians and therefore be less the object of jealousy.

But van Swieten's inner conflicts appear to have affected his manner and put up a barrier; to have made Ingen Housz uncomfortable and doubt his new mentor's sincerity. Huck and Pringle tried to reassure and rebuild broken trust.[53]

> . . . you may be sure . . . Baron van Swieten . . . is your friend. He has such influence over the mind of the Empress that no man will be appointed a Court Physician without his approbation.

From the timing of subsequent events, Ingen Housz must have acted in accordance with the Huck/Pringle advice, but by intuition and long before it actually arrived. He put his trust, finally, in van Swieten. A more compliant Ingen Housz was initiated into the correct greetings, the expected way of parting and taking one's leave; the response to make if offered a royal hand; that in formal situations one always spoke in French. There was a provision of ready cash to meet any of his immediate needs. He should not leave the coins lying around and should keep a careful account of his spending. He should use some of them to pay his servants, the standard court wage of twelve florins (£80) monthly, and get them to sign receipts.⁵⁴ His equipage and coachman were paid directly by the court.⁵⁵

There was another immediate problem. He had so enjoyed the company and esteem of his many medical friends in London that he had been looking forward to meeting Austrian colleagues. He was to be profoundly disappointed. As soon as the news of his arrival reached the medical fraternity at Vienna, the grapevine hummed. The agent of the royal rebellion had finally come among them. Their almost universal aversion to inoculation was about to be challenged. Later in the year, van Swieten reported to Pringle that he had known of only one case of smallpox inoculation in Vienna before 1768.⁵⁶ Even that had not been conducted on a native. The English ambassador had arranged it for his daughter and although it had been successful it had not fired any enthusiasm. However, the glacier was beginning to melt. In February 1768, Dr. Collin, physician in charge of the hospital at Vienna, had approached van Swieten, asking if inoculation was actually permissible in the city or if there was an embargo.⁵⁷ Whether he was genuinely uncertain of the legal situation or craftily seeking van Swieten's support is not clear. Collin was, in any case, probably prodded into activity by Thomas Houlston, a younger, English doctor visiting Vienna. Houlston seems to have been an inoculation enthusiast for he travelled on to Linz later in 1768 and treated a series of patients there.⁵⁸ With van Swieten's approbation, at least, Collin and Houlston inoculated a number of Viennese children in February 1768. A month later, another team also took up the practice. Dr. Locherer was 'chief' at the Vienna Maternity Hospital and he and the royal physician, Störck, inoculated twelve newborns.⁵⁹ Then, later in the spring, they punctured a number of older children at a city orphanage. Suffice to say that they all fell into a ferment of confusion and three children died.

Ingen Housz found himself having to advise in these difficult circumstances and he was soon reporting, in disparaging detail, to Deckers.⁶⁰ What he had found horrified him. The Viennese doctors didn't seem to realise that the incubation period in inoculation – the time between puncture and the appearance of pocks – was considerably shorter than in natural smallpox. Therefore, they had been looking for positive reactions to treatment too late after puncture. Worse still, they didn't seem to realise how important it was to examine the children at the outset, newborns apart, for evidence of previous smallpox affliction – for scars. Smallpox was, after all a common disease, and many mild cases survived without conspicuous stigmata. Assiduous clinical inspection was necessary because survivors of the disease would fail to react to inoculation. This was disconcerting for patient and practitioner alike because it led to uncertainty – was the patient still vulnerable to smallpox or not? It must have been very difficult for Ingen Housz. While trying, desperately, not to sound arrogant he was forced, by honesty, to point out the failings of men with very strong personalities and in established positions. How was it possible to be diplomatic whilst demonstrating old smallpox scars on 'failed' inoculees? And how invidious it must have been to be shown copies of Thomas Dimsdale's book, well thumbed but clearly not understood. If there was any common feeling, any warm camaraderie, it was that he and the very few colleagues who had already tried inoculation in Vienna were all risking professional ostracism and that every slightest setback would be transformed into a predicted and well-advertised disaster. 'You can easily believe that my appearance should ignite jealousy . . . a trait I was able to discern because the Germans do not so much tend to conceal it.'[61]

Ingen Housz could not have foreseen that the few trials of inoculation at Vienna would have been so mishandled. Nor could he have known that he was going to be so unpopular. He had imagined, until now, that the taxing journey to Vienna might have been the worst part of his expedition. Now he was not so sure. He had badly underestimated the strength of opposition to inoculation in Austria and that he was going to be such a lonely missionary. He certainly had a gloomy prognostication for his cause 'I don't think that inoculation will be general here, even in ten years time, because many of the doctors and the clerics are against the practice.'[62]

Ingen Housz seems to have been increasingly irritated to find that so many other matters were taking precedence over the intended royal inoculations. He had to accept that no decisions could be taken until he had met with and had the final approbation of the Empress. But he had expected to find an anxiety to proceed as soon as possible, before smallpox could strike again; otherwise what had motivated the pressure forced on him to reach Vienna at all possible speed? Van Swieten presumably tried to reassure his new colleague in the face of all the feedback but was probably caught off balance. A courtier himself, now for well over twenty years, he had failed to see how a new recruit to Her Majesty's household, especially one who had never been a courtier and never in Vienna before, would have no knowledge of the normal arrangements, that there was no such beast as a semi-detached courtier. The evidence all points to a sudden and profound change of atmosphere; that the whiff of rebellion and threat of a complete detachment turned into a soothed compliance. The lure was Maria Theresa herself.

8
Royal Inoculations

It's just getting light. There's a knock on the door and one of the women recruited to help with the children asks, through the panelling, that the doctor come to look at one of her charges. Quickly. He dresses rapidly and goes along the corridor to the dormitory that has been made of one of the end bedrooms. All the boys are sitting up or out of bed except one, where two of his staff are hovering. He sees immediately that the boy is ill – very ill. His breathing is very fast and shallow, he's flushed and there's a foul smell of fresh faeces. His eyes are sunken into dark hollows and Ingen Housz notes that his skin stays creased when pinched. There's no sign of awareness, let alone response, when the boy is spoken to and there are muscular spasms when he's turned. There's no rash and his two-day old inoculation sites show no reaction. His bowels open again in a torrent. Ingen Housz takes charge of the situation. His calm exterior is hiding a seething internal horror and his own pulse is racing. He orders that all the other children be taken downstairs to be dressed and fed and that the sick child be transferred to one of the small dressing rooms at the other end of the corridor. He should be cleaned up, put in a dry bed and all the soiled linen burned. He asks for the name and home details to be brought from his room downstairs and that someone be sent to ask that van Swieten come across to Meidling urgently though the messenger mustn't enter the palace. That the boy might die and that his death be linked to having been inoculated is the nightmare that Ingen Housz fears most. His whole mission to Vienna is on the brink of disaster.

By mid-morning four-year-old Franz Felbiger is in a deep coma, his lips and fingers are blue, he's breathing in fits and starts and it's obvious that he's not going to recover. Ingen Housz has already sent for the parents and the father arrives at midday with the village priest. Van Swieten appears at the same time. On being shown into his son the father takes one look and folds into a chair, sobbing loudly. Van Swieten notes how Ingen Housz gives him time, speaks softly to him, puts a hand on his shoulder and hides any signs of his own distress. Then, outside in the corridor, the priest tells van Swieten of a double tragedy. An infant, the boy's sister, died only yesterday, after vomiting and diarrhoea for only twenty-four hours and the mother has been sick this morning. Van Swieten is pained by the disastrous circumstances but relieved beyond measure that the family distemper is very obviously nothing to do with inoculation. When he calls Ingen Housz away from his patient and relates what he's learned, his young colleague is almost overwhelmed and has to turn away.

The priest performs the necessary offices and the poor lad dies shortly after two o'clock. By now his father has seen how the young doctor is almost as bereft as he is but listens carefully as Ingen Housz explains that the little body should not be exposed because of the risk of smallpox. Van Swieten has sent to the palace mews for a wagon. Franz is wrapped in two tightly bound bed sheets and his father carries the little body out to the yard. In about an hour the poor man will have to confront his wife with the heart-rending parcel. Tomorrow the Felbiger family will have a double funeral, all the time praying that their mother recovers and that the other three children remain well. Van Swieten has now understood why his royal colleagues in London were so glowing in their recommendation of Ingen Housz. Not only is he a skilful inoculator, he is a sensitive human being and an excellent physician.

T HE HABSBURG RULERS had long owned a hunting lodge, the 'Katterburg', in wooded hills to the west of Vienna and when an artesian well was discovered the name was changed to 'Schönbrunn' – the 'Beautiful Spring'.[1] The Turks ravished the area in 1683 but the new name survived and the land became the site of one of the great palaces of Europe. Schönbrunn is sometimes referred to as the 'German Versailles' but the

Schönbrunn Palace, Vienna.

Austrian architect was mainly influenced by Italian concepts, having been a one-time pupil of Bernini. His initial plans for the new palace were seen to be far too extravagant but building to his second set of drawings began in 1696. In outline, there are three freestanding facades enclosing a central courtyard orientated on the centre of Vienna some six miles away. Building work stopped in 1711, only to resume in 1736 when the newly married Archduchess Maria Theresa chose to make it her family home and grew to love it. When her father died in 1740 and she became Empress, she began a long series of improvements and enlargements, both to the buildings and to the gardens. Her palace certainly needed many adjustments if it was to accommodate the court, even if only in summer. By the 1750s Maria Theresa and her husband had some dozen children growing up at the Schönbrunn, each of whom had five rooms at their disposal and there were almost 1000 courtiers and servants. Even when the palace had grown to over 1000 rooms the demand for space was such that many of the state apartments served multi-functional roles but all were magnificently appointed as befitted the dignity of the Habsburg dynasty.

Maria Theresa also initiated massive redesigns and extensions of the gardens. Here the Versailles model was used as a template. All parts of the park came to be connected by symmetrical avenues to form a geometric and optically pleasing design. Statues or fountains at the intersections of the paths complemented the ornamental flower gardens, shrubberies and trees. The ground rises significantly to the rear of the palace and the height was exploited to create cascades. Then, later, a maze, a rose garden and a zoo were added. The last major overhaul of the garden supervised by Maria Theresa began in 1765 and was not completed until the late 1770s, during the Empress' last months of life. The environs of the palace were therefore in a state of some disturbance when Ingen Housz came to know it. On Whit Sunday 1768, ten difficult days after his arrival in Vienna, he was taken out to the Schönbrunn for an audience with the Empress.[2]

> On 22 May, I had the honour of a private audience with Her Majesty, kissing her hand with bended knee. There and then she spoke to me in a most friendly manner. Saying that she was most pleased to see me she immediately showed me the smallpox scars on her arms saying that the illness had violated the members of her family so terribly and that she hoped I would be able to save her two Archdukes by inoculation although this would be the subject of the approval of Baron Van Swieten (who was with us). She asked me ... whether I was ready to perform inoculations and would I be allowing patients into the fresh air. She said to me 'would you like to see my family?' on which

she guided me to the next room where I found four Archduchesses and two Archdukes to whom the Empress presented me and with all of whom I was given the honour of kissing the right hand with bended knee.

Ingen Housz's first meeting with the Empress was clearly significant. It is possible to sense the genuine human warmth of Maria Theresa, just as he did. Here, in these few momentous minutes, Ingen Housz felt real empathy. It was, perhaps, the first time since his arrival at Vienna that he had not been confronted by a veneer of formal stiffness and even antagonism. He could see why Maria Theresa was so popular with her people; how the grief and distress of her subjects would have been so universal and genuine when she had been nearly extinguished by smallpox the previous summer.[3] He may not have been completely spellbound – he was too tense for that – but at last here was someone inspiring and genuinely appreciative that he had made such a long, tiring and selfless journey.

It appears that the audience may have triggered something in van Swieten: events now proceeded at some speed. Perhaps taking his cue from the conversation they had had with the Empress, van Swieten appears to have revealed the proposed scheme for the inoculations. The two royal children who were to be inoculated were the sons of Maria Theresa, Archdukes Ferdinand aged 14 and Archduke Maximilian aged 11. But first, van Swieten explained, it had been agreed that Ingen Housz should demonstrate his inoculation techniques on some volunteer subjects. Ideally, these would be children so that the 'trial' would be identical, in as many respects as possible, to the proposed royal inoculations themselves. It had been resolved that he should inoculate a number of children brought in from among the poor of the parishes surrounding the Schönbrunn.[4] Ingen Housz may well have seen this as implying a lack of faith in his skills. After all he had now inoculated, personally, nearly a thousand patients and in all sorts of circumstances. Two deaths, at most, could have been attributed to his technique and he could have referred to the fact that royal children in London were now inoculated as a matter of routine.[5] He would also have seen that the stratagem inevitably prolonged his time at Vienna. It protracted his unwanted role as a courtier and intensified his exposure to professional jealousies and fanaticism. But in proposing the plan to Ingen Housz, van Swieten would have pointed out that he was not acting alone.

The notion of running 'rehearsal' inoculations had evolved under the eye of supreme authority and was already formulated in some detail. The Empress had even offered to help in finding parents willing to offer their children for inoculation,[6] presumably by arranging for letters to be sent, in her name, to local officials or parish priests. The children were to be fed and nursed at court expense[7] and there is one report[8] that the Empress would give each of them a ducat. She had also authorised the renting of a vacant house near the Schönbrunn where Ingen Housz could treat the children in suitable isolation. Despite the valid argument that such superfluous rehearsals were tempting fate – active smallpox might, at any time, afflict the royal children waiting to be inoculated – Ingen Housz had little option but to agree to the plan. Van Swieten then revealed that the number of children the court scheme assumed was as many as one hundred.[9] Here Ingen Housz moved away, perhaps, from objections in principle to protest on practical grounds. A hundred children was too many to inoculate and supervise all at once and, in any case, how could so many children be confined without a significant risk of other infections. So the fortunate circumstances that Ingen Housz had enjoyed at the London Foundling Hospital appear to have been used as a template. Repeated smaller batches of children might be inoculated, making the whole process – recruitment, examination, inoculation and recovery – more manageable.

Having met Maria Theresa and learned, from van Swieten, his immediate destiny as inoculator, Ingen Housz appears to have settled down to preparing for his task. But he was about to learn that he would confront one further responsibility. This came to light when the Emperor, Joseph II returned, in late May, from a trip to Hungary.[10]

> Yesterday the Emperor was in my room for two hours and behaved in a very familiar way with me, so much so that I often forgot that I was talking to the most important man in the world. In talking I forgot about protocol, which he didn't take badly. On the contrary, being free of all the splendour of his ancestors, he prefers familiarity. It is a far-seeing Prince who doesn't allow himself to be duped by flatterers. He said he was glad that an honest man had arrived, free from the beliefs of gypsy women.[11]

Joseph's manner and conversation were friendly and direct. He explained that he had a daughter, his only child, six years of age, Archduchess Maria Theresa.[12] He had lost her mother to smallpox four years previously and wished for his daughter to be inoculated together with his two younger brothers, Archdukes Ferdinand and Maximilian. Ingen Housz agreed; he was hardly in a position to refuse. Then, after extensive and informal conversation, the Emperor suddenly asked whether Ingen Housz was happy with his equipage, particularly the horses.[13] Ingen Housz had no reason not to be but Joseph was not content with this. He wanted, he said, to inspect them personally. The verdict was instant and condemnatory: the whole team was to be replaced.

Emperor Joseph II with his younger brother, Leopold.

Within a few days, Ingen Housz learned that the court had rented a large furnished house at Meidling,[14] then a pleasant village just to the east of the Schönbrunn.[15] Access to fresh air was an essential part of the inoculation scheme[16] so the house presumably had a large walled garden in which the children could play. And there had been a very positive response from the Empress' subjects: Ingen Housz could now prepare himself by obtaining some smallpox matter and moving into his 'inoculation house'. If many of the Viennese doctors had not grasped the fundamentals of inoculation, van Swieten certainly had. He understood that although the children were free to play in the open air after having been inoculated, they should only mix within their own group for about two weeks, until they were no longer able to pass on the smallpox by the natural route. The selection of resident staff at Meidling would have been given careful consideration. They were to be chosen because they were smallpox survivors and willing to remain strictly resident at the house – happy to be quarantined during the infectious periods. In fact the Empress had suggested that the inoculator himself, his assistants and all the children be fitted with special coats so that they could be easily identified. The 'uniform of Ingen Housz' would be blue jackets edged with gold braid.[17] He and van Swieten must have deliberated on the exact numbers of children to be inoculated and finally agreed on four batches of about twenty each time.[18] With each group taking about three weeks to be processed, it meant that Ingen Housz would not be inoculating the royal children and leaving Vienna until October at the earliest.

BAD NEWS ARRIVED at Vienna on or about June 11. Although details were confused at first, it was immediately clear that Johann Winckelmann had been killed, at Trieste.[19] How much this thunderclap upset Ingen Housz personally will never be known. If van Swieten had discussed personal safety with him, here was an immediate, sobering example of what could happen to those who were known to move in royal or papal circles. After a few days, the Emperor, on an occasion when the court was assembled, read out a letter that relayed the whole story[20]

> (It was) about 11 a.m. that Winckelmann was stabbed by an assassin in a tavern in the town centre (of Trieste). Five wounds were inflicted . . . and at 5 pm on the same day he breathed his last. The murderer is said to be a cook from that town (who) when Winckelmann showed him some gold coins that the Empress had given him this evil man stabbed him.

Ingen Housz, through a veil of shock, realised that his scheme for promoting *Particles* at Rome had died with his late friend and wrote of the sad news, with its implications, to Hoogeveen around 20 June.[21] 'You will understand that this unfortunate event has annihilated the possibility of distributing prospectuses in Italy.' True or not, various sordid embellishments of the Winckelmann story were now filtering through to Vienna. Winckelmann, making his way back to Rome alone, had put up at the Locanda Grande, Trieste's largest hotel.[22] He had hidden his true identity until his death throes. Francesco Arcangeli, his assassin, had been a fellow guest and known to be 'a man of Italian tastes' (a homosexual). With no immediate boat to Venice, Winckelmann and Arcangeli had been seen absorbed in each other's company for several days, shopping and eating together, relaxing in each other's rooms. Arcangeli had discovered that Winckelmann had a collection of silver and gold medals given to him personally during a private audience with Maria Theresa. Then, mid-morning on 8 June, astonished witnesses had seen Winckelmann stagger into the dining room of the hotel with a bloodstained noose of rope around his neck. He was faint and bleeding heavily from stab wounds to the body, which probably explained the fiction that he had disembowelled himself. Conscious and well aware of his fate until the very end, it had taken six agonising and dyspnoeic hours to die. Arcangeli had been apprehended and made a full confession. He was publicly executed at Trieste on 21 July,[23] strapped to a wheel and torn apart limb by limb. No wonder there was a feeding frenzy of rumour and counter-rumour in the humming stairwells and corridors of an attentive Vienna.

BY THE MIDDLE of June 1768 the first batch of volunteered children for Ingen Housz to inoculate were being received into the house at Meidling. Some of the first twenty must have failed to turn up or others, unwell on arrival, were returned to their parents. The remaining sixteen[24] were prepared for inoculation by giving them a mild purgative. Ingen Housz had decided to follow all the other exact methods of his friend, Dimsdale. He must have been pleased, delighted even, to be back to practising what he had learned in London and at Hertford. On the appointed day he dressed, we imagine, in his blue coat. It had been agreed that he shouldn't actually begin before a party arrived from the Schönbrunn,[25] led by the Empress and van Swieten. Ingen Housz himself would have greeted the Empress and introduced his staff and the children. He explained that they had already been administered a mild purgative and that everything was ready. Ingen Housz had certainly never inoculated under this degree of scrutiny before but he obviously made a good impression. Van Swieten was to write to Pringle:[26] 'Ingenhousz proceeded to use inoculation of the pustules; he commends himself to everyone by dint of his modesty and knowledge.' Maria Theresa would have been delighted to see that none of the children suffered the slightest distress and that as they were dressed, each donned their little blue jacket that signified that they had been inoculated. In less than a quarter of an hour all the children had been treated. As Ingen Housz escorted Her Majesty back to her carriage, there must have been a warm glow of approbation. If there were those in the party who were not impressed by what they had seen, they were not about to voice their doubts in the presence of the Empress. But something might still go wrong; and something did.

Ingen Housz inoculated further batches of children at Meidling. The timings and the final total of 65 children[27] suggest another three groups. One of the children was to die during treatment, a fact that no one dreamt of trying to conceal from Maria Theresa. Indeed Ingen Housz had predicted this possibility. The Empress was not deterred. She continued to visit the inoculation house frequently, as did Joseph II, both continuing to give their approval of all they saw. All in the immediate circle were convinced that the death was not linked to inoculation[28] and excepted it from their overall conclusions. However, van Swieten and Ingen Housz must have been acutely aware how this might stir up the opposition in the city. Otherwise '. . . all of the children overcame their treatment without problems, none being put to bed and spending their days happily playing in the open air.'[29] To have accommodated the four batches of children at Meidling without their overlapping, without exposing the uninoculated to the already treated, needed consistent organisation and impetus. To be ready to inoculate the royal children by September, the target date that was now being aired, Ingen Housz would have had very little spare time. The remaining groups of children must have followed in quick succession. Nevertheless, the inoculator had some pleasing distractions.

AT SOME POINT during his first month at Vienna Ingen Housz wrote to his friend, Dimsdale. The Hertford physician was not at home and the letter took some time to find him. He had particular business in London and replied only on 10 July. Dimsdale was brimming with confidence.[30]

> I make no doubt that you will give perfect satisfaction & have the pleasure of seeing your patients recover as well as those in England.

Dimsdale was able to report that there had been no problems with any of the several hundred patients they had divided between themselves and inoculated around Hertford that spring – 'at least I don't know to the contrary'. But his letter did contain one 'surprise' as he modestly called it.[31]

> I am now setting out for St. Petersburg to inoculate the Empress (Catherine the Great) and the Grand Duke (Paul, her son). The Russian envoy Monsr. Putsckin sollicited me extremely to accept of the commission & wrote most politely to acquaint me that he had consulted the most eminent Physicians in London as well as others who united their suffrages in me . . . I long refused but at length my friends advising me strongly to go & assuring me it was an Employ of the greatest honor that could be conferred on a man I have consented and my son who you may remember to be a Physician Pupil at St. Tho's Hospital . . . goes with me. . . I believe that I shall get out from hence on the 27th inst. by the Pacquet to Helvaetsloes & thence go by land through Berlin, Dantzick and Rigen . . . and am in expectation of returning to England in December.

Ingen Housz would have suffered very mixed emotions with this astounding news. He must have been delighted for his great friend and, in one sense, for himself: he was no longer so alone. His teacher was now also embarked on exposing himself to the risks that unavoidably haunted the inoculation of prestigious patients. On the other hand he now saw how he, too, should have been more circumspect and negotiated, more precisely, his terms of employment. Dimsdale had obviously managed to strictly limit his commitment to that of a transient visitor providing short-term expertise; the very circumstance that Ingen Housz had probably assumed he was going to find at Vienna. But Dimsdale had been less successful in avoiding unwarranted jealousies and ill-will, even before leaving home. There had been gratuitous gossip that his initial reluctance to accept the commission in Russia had been a negotiating ploy to win better remuneration and such tittle-tattle had found its way into the London newspapers. There were even rumours, on the Continent,[32] that he had pushed an initial offer of a hundred thousand pounds sterling back across the table. Ingen Housz knew Dimsdale better than most and would have been relieved, but not surprised, to read the truth.[33]

> . . . you know full well that my situation and practice was so very happy and advantageous that nothing but a full assurance of a very important Commission could have induced me to leave England . . . (and) . . . you would be mistaken in supposing that I made any sort of stipulation for either the manner in I should be received, or the gratitude with which I should be rewarded for my time and trouble – I could not possibly think of making terms with the most generous as well as most powerful Court in Europe, nor indeed have the least idea of what will be given me when I take my leave.

CATHERINE THE GREAT's experiences with smallpox at St. Petersburg had been far less tragic than those of Maria Theresa at Vienna.[34] The Russian ruler had not lost any members of her immediate family but there had been some frightening encounters and smallpox was still a constant dread. Her husband-to-be had been stricken in December 1744,[35] some eight months before their wedding. The betrothed couple had been immediately parted when Grand Duke Peter developed the tell-tale symptoms and Catherine had escaped infection. But she had been horrified when, a month later, she saw him again. She found herself repelled by his 'swollen, pockmarked features.[36] Her revulsion from his 'hideous disfigurement'[37] was probably a trigger in the rapid disintegration of their subsequent personal relationship. Catherine

Catherine the Great, Empress of Russia.

herself had, twice before, in adolescence, been abruptly quarantined because incubating smallpox was suspected but, unusually for that era, she had, by spring 1768, reached early middle age without contracting the disease. She knew she had been very lucky and though still prone, her anxiety had now been transferred to her thirteen-year-old son and deemed successor, Grand Duke Paul. She had thought of having the boy inoculated against smallpox in 1764 – her royal physicians were all non-Russian and knew of the process.[38] Nevertheless, she allowed these same advisers to dissuade her on the grounds that the boy was not in sound enough health. As a compromise she issued an edict that any of her subjects in contact with smallpox, or any other kind of rash for that matter, must not attend church services and certainly not approach the court. Then, in the late spring of 1768, Countess Anna Sheremeteva, one of her closest advisers, caught the disease.[39] Although Catherine's son, Paul, was immediately separated from his *oberhofmeister*[40] – Anna's fiancé – all Catherine's old anxieties were rekindled, especially when she herself became febrile on 14 May. Twenty-four hours later she was fully recovered but only to hear that Countess Sheremeteva had died and there were reports of a growing epidemic in St. Petersburg. Catherine was galvanised into an action long promoted by her regular correspondent and advocate of inoculation, Voltaire;[41] hence her letter to Pushkin, her ambassador in London.

The Dimsdales, father and son, had their first audience with the Russian Empress on 28 August and went on to dine with her.[42] Officially their presence was to go unrecorded but, according to the English ambassador, Sir George Macartney, their arrival and their purpose was an open secret. Dimsdale was impressed by Catherine's knowledge of inoculation and despite her poor English and his stumbling French they eventually agreed a plan. The Empress had wanted to proceed to her own inoculation immediately but Dimsdale persuaded her to wait a few weeks while he checked that his routines were safe with 'Russian' smallpox and in the Baltic climate. And so Catherine was inoculated on the evening of 12 October 1768 and the next morning she left St. Petersburg to incubate her disease in rural seclusion at Tsarskoe Selo, fifteen miles inland.[43] There she did as Dimsdale advised, spending two to three hours daily outside obeying his 'cool' régime and was fully recovered from her sparse pocks within three weeks. There were public celebrations on her return to St. Petersburg on 1 November and Paul was inoculated the next day; he, too, recovered perfectly. The transport, secretly arranged at staging posts to Berlin to whisk the Dimsdales safely away if something had gone wrong, was stood down. Facetious congratulations arrived from Voltaire who, reiterating the simplicity of the process he had long advocated, described it as less troublesome than for 'a nun taking an enema.'[44] Like Maria Theresa, Catherine now utilised the political capital of her courage for all it was worth and 20,000 Russians were supposed to have been inoculated by 1780; two million by 1800.[45]

O N 7 AUGUST, about the time that his third batch of children were recovering from their inoculations, Ingen Housz stole an hour to write to Cardinal Albani at Rome.[46] He was trying to restore some possibility of selling Hoogeveen's *Particles* in Italy.

> Your Eminence will ask yourself, in surprise, why a man whose name is perhaps not known to you comes to bother someone whose workload is too heavy as it is. Winckelmann's grievous fate, about which Her Majesty the Empress, depressed about this sad incident, spoke to me and the interests of a friend who is well worth being honoured by me for his great merits, are the reasons why I dare to apply to you. I met Winckelmann after I had been

summoned here to practise inoculation against smallpox . . . After being informed of the matter and after reading the prospectus (of *Particles*) this learned man and patron of savants spontaneously offered to be the writer's agent in Italy. He took some prospectuses with him . . . and he mentioned you, Your Eminence, as the protector of savants par excellence . . . Deep down he cherished the hope that through your strong mediation and influence he might induce the Pope to add his name to the honourable list of subscribers after the example of the King of England. I do not doubt that the documents . . . have come into your hands because I heard that Winckelmann had appointed you as executor of his last will and testament. This offers hope that you will be prepared to undertake, perhaps only in part, the task that Winckelmann had been willing to fulfil but could not . . . The highly learned Baron Van Swieten has ordered two copies for the Imperial Library . . .

By the end of August the 'Meidling' children were all safely back with their parents bar the one single tragedy. After the difficult start, all was now clear and convincing – in the skilful hands of Ingen Housz, at least, inoculation was simple and safe and he was preparing to move his personal belongings across to the Schönbrunn. The royal children would be the next set of patients.

Archdukes Ferdinand and Maximilian and little Archduchess Theresa, their niece, were examined by Ingen Housz on Friday 9 September[47] and found to be fit. Permission was given for them to be administered the small usual dose of calomel[48] and for their suppers to be withheld. Presumably they were also despatched to a part of the Schönbrunn palace that could hold them in quarantine. Ingen Housz probably slept fitfully in the same part of the palace and inoculated the three royal children next morning, 10 September. Maria Theresa and Joseph II may well have been present and perhaps gave more than just moral support to their children. The day would also have been a very important one for van Swieten who, no doubt, supported Ingen Housz in turn. Happily, the three children did well.[49] On the 5th day after inoculation they each developed slight fevers and the inoculation sites became flushed.[50] Three days later small, scattered areas of reddened skin appeared and next morning there was a solitary pock in the centre of each. Fortunately, they were few in number on each child: none of them involved the eye membranes, none ran together and there were no enlarged axillary glands. The next week must have been seven very long days for Ingen Housz but as each day passed, he visited his patients with the royal parents and van Swieten. A justified optimism turned to a euphoria when, on 25 September, fifteen days after puncture, there were no new pocks detected on any of the children. Generally, the three patients were hale and hearty, indeed they had suffered no distressing symptoms at any time, fully able, as they were, to walk in the palace gardens every day and go for secluded carriage rides. By Thursday 29 September the two Archdukes and their little niece had each been examined for the last time.[51] Bar the scattered and sparse dry scabs, now shrinking rapidly, they bore no signs of any continuing malaise and Ingen Housz was able to declare them fit and fully recovered. The inoculations had been successful. The two last-borne sons of Maria Theresa and her first grandchild were now safe from smallpox.

Memorable celebrations were soon under way. The Schönbrunn gardens were thrown open to the public late in the afternoon of 29 September and a *Te Deum* sung in the palace chapel during the evening. The official court newspaper, the *Wiener Diarum* reported these events two days later.[52]

> On Thursday last, 29 September, 'God we praise thee' was sung in the Schönbrunn Palace Chapel in order to give thanks to the Almighty for this fortunate cure and to offer due thanks. In the evening everyone was granted free admission to the gardens where the buildings, in which the Regal Highnesses lived, were lit up in a very fancy way and various choirs sang praises; there were musicians and a general air of joy permeated the air. The whole town participated and all gave thanks for the undertakings that had occurred and everyone was very happy about the outcome.

Maria Theresa also ordered preparations for a party in the great hall of the Schönbrunn[53] to which all of the little 'Meidling' children were to be invited, with their parents. It was a gesture of gratitude so typical of her and of her appreciation of the part they had all played in the build up to the inoculations of her own children. At the Schönbrunn party, on Monday 3 October, Ingen Housz and his staff would be, like the volunteered children, guests of honour and as a further signal of appreciation, the Empress and her family would serve at table. The happy occasion was an event that showed Maria Theresa at her best; an occasion that no one present would ever forget.

Life for Ingen Housz after the royal inoculations and all the celebrations could well have been anticlimactic but it was certainly not dull. Just as he had seen happen at Hertford, the aura of success, and particularly here, with royal patients, created immediate demand for more inoculations. Following the example set, very astutely, by Maria Theresa, a queue of high-ranking Viennese nobility sought introductions to the new royal physician; van Swieten, a familiar figure among the local aristocracy, serving as a conduit. Meidling was not a suitable inoculation house for the larger numbers that were clamouring to be treated and was particularly unsuitable for adults. Van

Swieten aired these emerging problems in council and the court issued a decree, the Empress in full agreement, that Ingen Housz should take over a vacant royal palace as a new inoculation house. Ingen Housz seems to have accepted this somewhat open-ended new commitment. He therefore closed down his activities at Meidling, and set off in his carriage to inspect a much larger building where he would be inoculating all-comers and where the throughput could be higher.

Schloss Sankt-Veit[54] (Castle St. Vitus), some three miles to the west of the Schönbrunn, had come into the hands of the Empress in 1762.[55] It had undergone a recent restoration under the supervision of Nikolaus Pacassi, the same architect who had been involved with the Schönbrunn. Schloss Sankt-Veit had formerly been the summer palace of the Archbishops of Vienna and would, some years later, resume this function, so there was some irony in using it as an inoculation house knowing the depth of clerical opposition. In early October, Ingen Housz moved in and rapidly adapted the building so that it could receive patients from Vienna and surrounding towns and villages.

The court had also deliberated on how Ingen Housz should be rewarded and in early October he was able to write to Dimsdale with astounding news.[56] Presumably Ingen Housz had been invited to a meeting with Kaunitz, van Swieten and perhaps other officials, just a few days after the garden party. Van Swieten said that he had already written a letter of thanks to Sir John Pringle for his excellent recommendation.[57] There would have been some 'diplomat-speak' informing Ingen Housz of the great gratitude of the Empress who had commanded that he be promoted to the rank of *Hofrat*, the equivalent of Privy Councillor.[58] The council had submitted a scale of rewards, to Her Majesty, for Ingen Housz's services and these had been immediately approved bar one exception. Ingen Housz was to receive, from the time of his arrival at Vienna, an annual salary, for life. But the customary annuity for royal physicians had been increased, by the Empress, to 5,500 florins (£35,000).[59] She hoped that this would entice him to stay in Vienna.[60] Nevertheless she would not expect him to be in constant attendance; rather, that he should be available for difficult cases. His immediate rewards were to be a payment, now, of 3000 Ducats (£86,000) and that he continue to have his court lodgings, servants and equipage which were valued at 1000 Ducats a year (£28,300). He was to receive 200 Ducats (£5,650) to repay him for the expenses of his journey to Vienna and 1200 Guilders (£7,600) for his time in Vienna so far. The Empress also desired to present Ingen Housz, personally, with a diamond ring worth 1000 Ducats (£28,300). And

Inoculation medallion that celebrated the successful inoculation of Maria Theresa's sons and grand-daughter. Courtesy of Breda Museum.

the Emperor had also ordered, for him, a gold snuffbox bearing a portrait in enamel. Ingen Housz was now substantially rich. He was going to be able to bank more than 15,600 Austrian florins (£99,000) as well as deposit at Stametz,[61] for safe keeping, an encrusted diamond ring and a gold snuffbox, both of enormous value. And he would not need to call on this capital for his daily living expenses. His annual salary and free lodgings and transport were worth almost ten thousand florins a year even after tax (£64,000). At that time a typical middle-class Viennese family in a central city apartment and with a carriage and pair would have a household budget amounting to a tenth of this.[62] In addition, Maria Theresa had instructed that the reputed artist and sculptor, Lazar Widmann, be commissioned to design and cut medals to commemorate the safe inoculation of her children, some to be cast in gold and others in silver. Ingen Housz, as inoculator, was to be presented with some of them. There was also news of a different designation. Her Majesty hoped that Ingen Housz would agree to travel to Italy, to inoculate her second son, Leopold, Grand Duke of Tuscany,[63] perhaps in the coming spring when the Emperor himself was planning a visit. Ingen Housz was in no position to cavil. Yet again, he could see his return to London or the Netherlands receding still further into the distance. Nonetheless he had always had an ambition to go to Italy,[64] and here, at least, was an opportunity. He may have wondered, though, what had happened to the second purpose of his journey to Vienna – to teach inoculation. Someone, behind the scenes, had recognised, perhaps, that expecting the medical fraternity in the city to attend demonstrations given by an interloper, someone who had been promoted over their heads and showered with huge rewards, was too provocative. The alternative was certainly a shrewd tactic. Ingen Housz was told that a royal edict was being drawn up[65] that would give him the authority to organise teaching sessions on inoculation in each of the Austrian cities that he would pass through on the way back to Vienna from Italy. A head-on confrontation in the capital, a 'noisy' dispute with de Haen, was thereby avoided, at least until objections could be quashed by sheer weight of evidence.

IN THE THIRD week of November Ingen Housz received a letter, in familiar handwriting, from St. Petersburg. Dimsdale's expedition had also succeeded: Catherine the Great had been safely inoculated.[66] In fact, as Dimsdale had intimated only in a letter to a close friend in London,[67] there had also been some demonstration runs at St. Petersburg: progress had been parallel in many ways to that at Vienna, if on a smaller scale and on a more urgent schedule. Dimsdale's son, Nathaniel, had done the first inoculations – on two military cadets. Unhappily, one of them fell ill two days later and although the Dimsdales knew this could not be attributed to the inoculation, there were serious anxieties that the Russians would immediately lose their confidence in the English visitors. The Empress herself was not disillusioned, however, and when Dimsdale had, himself, inoculated another nine cadets she instructed him that her turn had come even though four of the young men did not respond to their punctures in the expected manner. She arranged a secret, nocturnal tryst. One of the cadets who did develop pustules was used as the donor of serum for Catherine who

> desired I would fix the time and bring one of the boys to the Palace in the night and go up stairs in a bye way that I had been acquainted with where I should find her prepared. I could scarce believe my eyes for it seemed absolutely improbable to the last degree, her resolution and constancy. However, in obedience to the commands I was carried in a coach to the palace. Was met by a nobleman and conducted to a little room where her Majesty waited and was inoculated alone – all this being done in the night no one knew of it and I returned to my lodgings . . . She has had the smallpox in the most desirable manner – a moderate number of pustules and complete maturation. Now, thank God, it is over and I find an inexpressible load of concern removed from my breast.

Grand Duke Paul, Catherine's 13-year-old son, was then inoculated and had

'had the distemper in a very favourable manner'.[68] And, as in Vienna, mass celebrations were being planned, for 24 November, just as the aristocracy were jostling to be the next patients.

> The nobility in general are mad to be inoculated insomuch that, although I have my son with me who was at Edinburgh while you was at Hertford, yet we cannot properly attend the number which at present amounts to about 100 and all have very happily had the disease . . . I perceive that the practice will extend here, as universally . . . But you will be surprised to hear that I am applied to on account of one of the Archbishops who is determined to be inoculated & this I consider as the greatest proof of the extensive influence and infinite good sense of the Empress. This will strike at the root of all the religious simples and have a stronger effect to remove objections than all the declamations that could be made.

By 10 December[69] the Dimsdales had safely inoculated about 140 Russian nobility and were preparing to travel on to Moscow before returning to England – the roads across northern Germany would be frozen hard in January and February; it was a good time to travel so long as they were well-wrapped against the cold. Dimsdale was also plied with substantial rewards.[70] He was created a Baron of the Russian court, body physician to her Majesty, state councillor with the rank of Major General and presented with £10,000 sterling (£650,000). He was also awarded a life pension of £500 (£32,000) per year, portraits of his patients – the Empress and the Grand Duke – and many other gifts.

For Ingen Housz, the final outstanding events of his momentous year occurred on Tuesday 23 December,[71] two days before Christmas. He was called to the palace for a formal presentation. Maria Theresa received him with great warmth and gave him two gold, and thirty silver, versions of the celebration medal of 29 September, the day her two sons and granddaughter had been pronounced successfully inoculated. And later in the day Prince Wurtzburg, a grateful inoculee, personally gave him a present of a silver toilet set of 24 pieces.

The two friends, Ingen Housz and Dimsdale, who had started 1768 with an altruistic mass inoculation of a humble and obscure English village population, had ended their year in resplendent palaces and endowed with considerable wealth. They had both gone out on a perilous limb of faith in their clinical skills and had, jointly, knocked down the resistant ramparts to smallpox inoculation across substantial swathes of Europe. The eradication of smallpox, finally achieved 209 years later,[72] had taken a huge leap forward. That both doctors had been trusted by and supported by imperious royalty is certainly relevant and perhaps it is also significant that their sponsors had both been very powerful and determined women. But ultimately the success had been that of the reliability and efficacy of inoculation, essentially a long-known procedure that had returned, in a sense, eastwards, whence it had come in 1719 with Mary Wortley Montagu.

9
Royal Inoculations II

From the palace at Caserta, it's fifteen miles south to San Giorgio a Cremano where he'd agreed to meet William Hamilton at eight o'clock. He'd requested a carriage for five thirty but the servants were up even earlier to pack him a hamper of bread, cheese and olives with two bottles of wine. After a very sociable and pleasant breakfast in the first warmth of the day, all the while gazing up to their left towards the still shrouded peak of Vesuvius, the two explorers travel on the last mile or so, in the ambassador's coach, to the site of the excavations. At Resina they're met by a guide called Turco, a man warmly greeted by Hamilton for they've shared previous outings and 'Cesare can be trusted.' Ingen Housz has brought some old garments to cover his suit and, as advised, is wearing his strongest pair of shoes. Hamilton gives him a linen hat and a candle lantern. He himself fills a canvas bag with spare candles, some flints and emery, some fruit and a hammer strung together with some cold chisels. They all make their way down a path to the right, towards the sea, Vesuvius behind them. Beyond a row of low hovels under terracotta tiles, many of them cracked, they come to a pile of fresh spoil garnished with small broken pieces of sculpted stone, some of it marble. And a few yards further, the entrance to the workings is under some timbers weighed down with particularly large pieces of smashed statuary. It takes the three of them just to roll these aside while Hamilton is explaining that an unfortunate local child who recently fell down the hole was found dead and horribly contorted.

The Roman town of Herculaneum, entombed in petrified lava for 1700 years, had been found by accident on the digging out of an intended well and to the untutored eye the revealed shaft looks just like another wellhead. Hamilton's two coachmen, having fed and watered the horses, reappear. Ingen Housz is surprised to see that they're now armed. Hamilton explains the need for their protection: no-one in the vicinity, he's sure, would mean them any personal harm but the theft of the valuable rope ladder would leave them stranded deep underground and in obvious danger. Cesare leads, climbing down one-handed so that he can show Ingen Housz the way with a lantern. Seventy feet is a long climb down and one that will certainly seem much longer on the way up. At the bottom they're in a sizeable chamber that leads off to several others through narrow openings. Everything is wet and there's an overwhelming smell of damp rock. Water is dripping on them and shadows move around them when a candle flickers. None of this bothers Ingen Housz – he is back in the Italy of antiquity. He can already see frescoes through an obvious doorway and a Latin numeral carved into a pillar. Little did he ever imagine, in the classroom at the Breda Latin School, that he would, one day, be standing in an ancient Roman street. With all three lanterns alight they set off to William Hamilton's latest dig site – he's half way through exposing a female statue in marble – ' he's having an affair with a rather frigid woman', he jokes to his companion.

THERE HAD BEEN a real breach in the resistance to smallpox inoculation at Vienna by the end of 1768. The successful treatment of the royal children and the ensuing celebrations were a public demonstration, ultimately, of Maria Theresa's personal courage and resolve. Her gamble might have been low-risk but failure had always been a possibility: fortunately the speculation paid off. Inoculation now had the highest possible seal of approval and the opposition was showing signs of fracture. Some, at least, of the Vienna physicians had been convinced by Ingen Housz's successful campaign and were now *au fait* enough with the essentials of the process. They now saw it as mandatory to examine their potential patients thoroughly before puncture and then to reassess, again with obsessive care, from about a week later until full recovery, keeping them isolated throughout. It was now evident, and credible, that such vast numbers as were reported could have been inoculated safely in England. Maria Theresa knew well, though, the obstinacy of the entrenched opposition that would remain, even in defiance of the evidence. In November she had arranged a sideshow, a further eye-catching manifestation of her faith in inoculation and of her trust in her new body physician. She had twenty of her army officer cadets (twenty who had never had smallpox) brought from the Wiener Neustadt Military Academy[1] to the Schönbrunn where Ingen Housz successfully inoculated them[2]. And all the time Ingen Housz was responding to pleas for treatment from the Viennese *prominenti* and inviting them to Castle Sankt Veit. He was able to record, in a letter to Sir John Pringle on 14 December,[3] that he had safely inoculated over one hundred of the Viennese nobility since early October. The only problem, unique to the area, had been the frequency of intercurrent eye infections provoked by the dusty atmosphere, he himself having been a sufferer on two occasions.[4] Otherwise, he was only afflicted with praise, gratitude and gifts.[5] He had also taken receipt of a travelling coach.[6] In the ensuing five years he was going to spend a great deal of time in it.

Not all of Maria Theresa's children were yet safe from the threat of smallpox. Her second surviving son, Leopold, still only 21 years old, but establishing himself in Florence as the Grand Duke of Tuscany, was still vulnerable to the disease[7] and smallpox was no less a threat in Italy than in Austria. Sending Ingen Housz to Tuscany was the obvious ploy and could be combined with him also going to Naples to inoculate the young King there[8] – he, likewise, had never had smallpox. These plans carried enormous implications for Ingen Housz. When he had first arrived in Vienna he had been a professional with no official position, having a break in his career, travelling for enlightenment and intending, eventually, to return to his homeland. If one of George III's appointed physicians or surgeons had gone to Vienna what happened next would not have been possible – they, after all, were already serving another court. Ingen Housz, however, had been installed as a courtier and was now given other responsibilities. He had been appointed as a royal physician and then *Hofrat* (Privy Counsellor). He could not but play his part in the new inoculation scheme. His availability to the Viennese *bonne temps* was cut short sometime in late December, by which time he must have been party to the court's intentions for him: that he was to travel around the empire, first to the country we now call Italy. In actuality he would be visiting the Venetian Republic, the Papal States, the Kingdom of Sicily and then Tuscany. He only had a short time to make arrangements for his households, servants and finances at Vienna to be supervised in his absence. Fortunately, he was able to call on a new friendship made at court – with the Alsatian courtier, Baron Johann de Fries,[9] director of the Imperial silk factories. In early January Ingen Housz set out for Italy. To travel here on this, his first ambassadorial trip, at least provided some personal consolation. As he wrote to his friend, Deckers, from Florence later in 1769.[10]

> Here I am in a country that has always been, as you know, the object of my curiosity. In the middle of the tranquil peace that I enjoyed in London, I was always tempted to make an excursion to this country.

INGEN HOUSZ WAS part of an advance party that travelled to Italy in convoy in early January. Joseph II was to follow, leaving Vienna secretly on 3 March.[11] The Emperor intended to be at Florence to support his brother through his inoculation but he had reasons for first visiting Rome and then travelling on the further 300 miles down to Naples to see his sister Maria Carolina. There were anxieties, at the Hofburg, about her ability to cope with her new situation as wife of the King of Naples. She was a headstrong girl and although she had obeyed her mother's wishes and complied with the marriage, it had not been without protest – 'you may as well throw me into the sea'[12] – and the King was not a man likely to reassure any mother-in-law. On a less regal plane there were also some misgivings about Ingen Housz's difficult mission and he was not sent off alone. He was accompanied by one of the other, more established royal physicians, the 49-year-old Frenchman, Dr. Alexandre Laugier. Another Leiden-trained doctor, he had been recruited to Vienna by van Swieten over twenty years earlier and was seen, presumably, as seasoned enough to serve the role filled, up to now, by van Swieten himself: that of wise counsel and prop should anything go wrong for Ingen Housz. A very portly figure,[13] he certainly had the gravitas necessary for the responsibility but there may well have been an ulterior motive played out by van Swieten in arranging for Laugier to be away from Vienna for a substantial time. Laugier's younger brother, Robert, had also come to Vienna, at van Swieten's behest, to be professor of botany and chemistry at the university in 1750.[14] The appointment had turned out to be a disastrous one, the younger Laugier proved to be not up to the post despite his older brother's high-flown advocacy. Amongst other flaws, he had insufficient Latin to be an effective lecturer. After many years of displeasure, van Swieten had just had the younger Laugier summarily dismissed from his chair, denying him, even, a pension.[15] Packing the older brother off on a foreign trip was timely for lowering the tension at court.

After at least a week on the roads of Styria and Carinthia (now Slovenia), Ingen Housz stopped at Venice,[16] his first real taste of Italy. It was not the sunny and expansive experience he would have expected. As all visitors, he would have lost himself in its dank and disorientating labyrinth where everything is especially confusing in January when the city is not in its finery. Carnival is yet to come and winter hangs in the saturated cold air. All is chilled, still, misty, mysterious. The Doge's Palace, St. Mark's Cathedral, the Grand Canal and the Rialto Bridge were all much as today, as were the shops. Nothing prevents trade. After walking the alleys and the bridges, and having himself rowed across the lagoon to the Santa Maria Della Salute, the church celebrating the end of the plague, Ingen Housz went shopping. He noted[17] the purchases he made in his new 'finances' notebook. Typically eclectic, as Venetian shopping tends to be, his purchases included ribbed pompadour,[18] a strung lorgnette, and a variety of silver spoons.

Ingen Housz and Laugier next broke their journey at Rome, surely the culmination of an Italian pilgrimage for any classics scholar. There are no surviving records of their days in the city but the experiences of the thousands of 'grand tourists' to eighteenth-century Rome are very well documented and can be generalised. Unlike in Venice, founded only in the ninth century,[19] Ingen Housz was surrounded in Rome by the buildings and relics of the ancient civilisation that had captivated him at school. 'In a culture dominated by the classics, Rome was the focus of interest . . . (and) . . . offered classical and Baroque art, sculpture, architecture and painting, and many tourists treated it as the cultural goal of their travels.'[20] It was also, by pure chance, an interesting time to be in Rome in historic terms, and especially for a Catholic – the Pope, Clement XIII, had died on 2 February[21] and the holy city would have been abuzz with the election of his successor.

Ingen Housz had personal business in Rome, however, and was not just another overwhelmed 'tourist'. In the bags he originally packed in London before heading for Vienna, Ingen Housz had obviously packed a portrait of himself. At least, this image must have accompanied him to the Austrian capital if we assume that the artist, cryptically referred to as A.L.L., was Anna Louisa Lane.[22] She was related to the more famous William Lane of Pall Mall, also a London portraitist, and she, too, exhibited at the Royal Academy.[27] Perhaps Ingen Housz also carried with him from London a recommendation that should he ever visit Italy, Rome was by far the best place to have an engraving made from the portrait, therefore allowing copies. Alternatively, someone in Vienna gave him this advice or, of course, it could have been Winckelmann. Whatever the antecedents, Ingen Housz sought out, in Rome, the intaglio engraver Domenico Cunego. At the foot of the result, the best surviving image of Jan Ingen Housz (see p. 158) we find, in Latin, the attributions 'A.L.L., drawn from life' and 'Cunego, cut in Rome, 1769'. Cunego may have suggested the design and layout but the garland of oak leaves adorning the elliptical frame would have had an obvious attraction and double meaning for a classics scholar like Ingen Housz. The so-called *corona civica* of the ancient Romans was traditionally awarded to any soldier who had saved the life of a comrade in battle. It was clearly an appropriate representation to accompany the otherwise simple Latin text under the head and shoulders image.

> J Ingenhousz, appointed as Royal Physician in order to protect the children of the Empress against smallpox by means of inoculation.

However, inscribed under the text block itself, easily missed among the dense cross-hatching, and in a different font, is yet more Latin – *utinam citius* –

Jan Ingen Housz, Rome, 1769. Courtesy of the Wellcome Library, London.

'would that it had been earlier': 'about time, too' would be more colloquial. This is very intriguing for there are at least two possible interpretations. According to the Ingen Housz family archives,[24] the new royal physician was rather hurt that the celebratory medals Maria Theresa had had cut in the autumn of 1768 did not record his name and the central part he had played in the inoculations. Now, in Rome, a few months later, Ingen Housz may have been unable to resist a minor but satisfying reprisal. Still smarting, perhaps, that his personal contribution had not been recorded on the medals, he now had his own memento inscribed, one that justifiably highlighted his role. Or, less to do with self-aggrandisement, but no less provocative to the Hofburg, he may have meant the added inscription to imply that inoculation should have been started sooner at Vienna despite the opposition, thereby possibly saving at least three more of the Empress' family. We do not know how much the etched copper plate cost but we do know its features since it has survived.[25] It is about one and half millimetres thick, 214 millimetres high, 153 wide and still lives inside a leather sheath that is probably original. It presumably accompanied Ingen Housz throughout the rest of his Italian travels that summer and it was with him in London in 1779.

By March 1769 Ingen Housz had left Laugier at Rome and rattled on down to Naples. Joseph II recorded Ingen Housz's presence when he himself arrived there on 31 March.[26] It made sense for Joseph II to visit Naples before descending on Tuscany where the Grand Duchess, Maria Luisa, was heavily pregnant and about to be delivered of her third child. The baby arrived on 6 May 1769, another healthy boy, baptised Ferdinand, a younger brother to Francis. He made the Habsburg dynasty more secure. There was certainly joy at Vienna but Maria Theresa's pleasure was, this time, more restrained than when little Francis had been born in February 1768 – she had then burst into the theatre at the Hofburg in the middle of a performance, shouting repeatedly from the royal box, 'My Poldy's got a boy'.[27] Spring comes early in Naples. It must have been very pleasant for Ingen Housz to have dismounted his coach after 800 miles on the road, gaze at Vesuvius in the warm sunshine and know that the return journey to Vienna would not be for some months and in many easier stages. He had not gone to Naples to stave off death, as so many travellers did. He was there to try to prevent a death. He was to inoculate Ferdinand IV, King of The Two Sicilies (Naples).[28] His patient was just eighteen years old but had been put on the throne by his father ten years before. King Charles III had been obliged to give up his kingdom at Naples in 1759 and succeed his late brother to the throne of Spain, an international treaty disallowing Naples and Madrid to share a monarch.

The inoculation of King Ferdinand was a political act, as was his marriage. Archduchess Maria Carolina, aged 16, had become his wife at a proxy wedding in Vienna on 7 April 1768[29] and then travelled to Naples, so bonding an empire and a kingdom. There was something of the conveyor belt in the way that Maria Theresa used her daughters as political pawns. Formerly, Ferdinand had expected to be married to Joanna Gabriella and then, after she died, to Maria Josepha but, as we have seen, she succumbed to smallpox only days before her expected marriage in October 1767. Maria Carolina was a convenient replacement who came with the profound advantage of having survived her own attack of the dreaded disease. Now, as a young wife, it was vital that Maria Carolina produce male heirs, bonding by blood the Habsburgs and the Bourbons. As the potential father, the life of Ferdinand was at a premium, at least until he had sired a son or two. Seen from the perspective of Vienna, inoculating him had become urgent and critical. Ingen Housz, with his clinical skills, was now part of the political scheme of things.

Strictly, Ferdinand had no right to be the king. He had a living older brother, Philip, but the boy had developed into an imbecile.[30] He was so deranged that it was obvious from a very early age that he should be sidelined. To defend the decision, the regency, by a means both callous and dubious, even for those times, would have him conveyed into public view occasionally so that the Neapolitans would not dispute the unorthodox order of the royal succession. Not that Ferdinand himself was wholly suitable: he, too, was far from stable in the view of many, including his wife. She freely described him[31] as *ein recht guter Narr* – a right good fool – but managed to settle her feelings just short of actual disdain. Ferdinand's education had been very narrow, serving his father's deep-seated opinion that the only way to offset the constant threat of madness in the line was exhausting physical exercise. Hunting and killing game more days than not, winter and summer alike, was seen as ideal. Ferdinand needed little persuasion. There were very few days in his long life that were not to be spent slaughtering, or butchering alive, purely for his own amusement, anything that moved in the hills and forests around Naples or in the waters of the bay. At home his manners were indistinguishable from those of his dogs who, like him, roamed the palace freely and unpredictably. It was all far from appealing. He would fart in company and drag hapless guests, Joseph II among them on one occasion,[32]

Vesuvius 1776. Courtesy of the Wellcome Library, London.

to continue conversations in the latrine while he sat and defecated.[33] And, having finished at the 'close stool' he relished the panic in guests' eyes as he dragged out the chamber pot, threatening to use it as a close range mortar. His impish immaturity also expressed itself in playing fatuous tricks on the servants; tripping them up or slopping jam into their hats. Effectively, his kingdom was being managed by his First Minister, Tannucci. Negotiations between him and Count Kaunitz, the Imperial Ambassador at Naples, had eventually finalised the marriage contract between Ferdinand and Maria Josepha in 1767. When the designated bride died it took little intellectual effort to change only the name of Ferdinand's intended. The clause that gave the Queen, from the birth of a male heir, a seat on the ruling council, survived. When the time came, Maria Carolina eagerly took the reins and was not the least shy of exerting power. She did not find it difficult to dominate her bumpkin husband and became a successful, long-serving ruler. Her older sister might also have become an adept monarch but we shall never know: smallpox had once again diverted a dynasty.

Ferdinand and Carolina lived twenty-two miles to the north of Naples at the royal palace of Caserta. It had been the preferred site of Ferdinand's father – at a good remove from the squalor of Naples and well inland from the coast and the threat of Barbary pirates. It was, and still is, a huge palace.[34] There are well over a thousand rooms, a large library, a theatre and vast gardens with statuary and an impressive cascade. It is said to be the inspiration of the lampoon that 'the smaller the royalty the larger the palace'. Versailles had been used as a template but in 1769, sixteen years after building work had started, Caserta was still far from finished. Ingen Housz arrived in early March and having presented himself and his letters of introduction, was presumably allocated one or more rooms near the state apartments of the King and Queen. He would have needed the help of the local royal physicians to find a suitable new case of mild smallpox as a donor of serum. This was not difficult perhaps; Naples was then one of the largest conurbations in the world. But in its teeming back streets he would have needed some substantial protection and guidance. He must have succeeded, however, and Ferdinand was inoculated sometime around 15 March.[35] This left Ingen Housz with some free time for about a week and there were many local attractions – some high above, and others under, the ground.

I F INGEN HOUSZ did not already know of William Hamilton,[36] the British envoy at Naples, he soon would. The two men were almost exact

Sir William Hamilton.

contemporaries: Hamilton had been born only five days later than Ingen Housz, on 13 December 1730. As Envoy Extraordinary and Plenipotentiary to the two Sicilies, to use formal diplomatic parlance, Hamilton had presented his credentials to the thirteen-year-old King Ferdinand on 25 November 1764.[37] Hamilton, after active service in the British Continental Army, had been able to resign his commission on marriage in 1758: his wife came with a small fortune. His only contribution to their household economy was from his meagre income as equerry to the Prince of Wales (a sinecure found for him in 1751 by Hamilton's scheming mother) although this did increase when the Prince became King George III in late October 1760. The poor health of Catherine, Hamilton's wife, dictated the next significant event, and eventually the rest of his life. She suffered from chronic asthma and the very severe winter of 1762-3 – carriages were able to cross the ice on the Thames at Twickenham – nearly killed her. Responding to a rumour that Sir James Grey was returning to London from Naples with the intention of resigning as ambassador there, Hamilton began a series of manoeuvres and string pulling to secure himself the post. His motives were not wholly those of personal ambition and advancement. The climate of Naples sold itself as the one thing that might ensure his wife's survival. Hamilton was eventually appointed ambassador at Naples but it had taken a year's lobbying and a new administration under George Grenville.

William and Catherine left England, fleeing the prospect of another dangerous winter, in September 1764. Hamilton was to find that Naples engrossed him and set him on the path of an unexpected career. Bar London and Paris, it was the most populous city in the world in the mid-eighteenth century.

> Travellers from the north felt hit, on arrival, by life lived with more show, noise, gaiety and violence than anywhere else . . . the city assaulted their senses, all of them. . . . There was . . . an anarchic energy and ebullience.[38]

But, for all this, Hamilton found himself in charge of an embassy of only minor importance to London. He had plenty of spare time once he had settled into the routines, even for an ambassador. However, the low status of his post hardly mattered at first. As he and Catherine set about furnishing the Palazzo Sessa,[39] their official residence, they were already overjoyed by the rapid improvements in her health.[40] Whereas the general prediction in London had been that she would be lucky to see thirty, she was to live on, at Naples, into her forty-fifth year. So, with the prime ambition for moving to Naples fulfilled, and having mastered the job of envoy, Hamilton was able to adjust his focus. Naples might be an idiosyncratic community but it was also in a unique location.[41]

> Many thought the site – the sea in a bay, the amphitheatre of hills, the volcano – the most beautiful in the world (but) the vicinity of Naples . . . is deeply unstable. The land can't stay still . . . The instability when Hamilton was there . . . was extraordinarily evident . . . Vesuvius was more active in (his) day than at any time since Pliny's.

Known in London for his love of paintings, Hamilton now developed two new passions. The first was Vesuvius: in June 1765, five years after a rumbling overture in pianissimo, the volcano began belching thick smoke. The locals predicted an imminent eruption and they were right. Hamilton was high on the snow-clad mountain in January 1766 when it started to spit superheated debris into the sky. One stone, 'three times the size of his head' fell within a yard of the naive British ambassador.[42] By Easter there was lava flowing at the 'speed of the Severn Bore'[43] and pushing a boiling bow wave of ice, snow and mud in front of it. Hamilton had the disciplines of the best observers of nature and he responded as a worthy exponent of the Enlightenment. He systematically recorded all the phenomena he saw and began a long correspondence with the Royal Society. His impressive observations earned him such approbation of the fellows that he was elected to fellowship himself on 6 November 1766.[44] The following year he was able to report on a massive eruption, provoking another spate of letters and mineral samples despatched to London. But besides pioneering the science of vulcanology, Hamilton was seduced by a second Neapolitan passion.

In Paris, on his way to Italy, Hamilton had met Abbé Galiani, the Secretary at the Neapolitan embassy. Galiani had given Hamilton a book, newly published in French. It was a copy of Winckelmann's account of the excavations around Herculaneum. In 1711 some labourers had been digging a new well near the former royal palace at Portici,[45] six miles south of Naples. Some fifty feet down, in petrified mud, they discovered an intact Roman statue, a discovery soon followed by many others at that depth. Within a couple of decades the palace was filled with a multitude of relics retrieved from the long-buried Roman town of Herculaneum. The tight shafts and seams, a civic honeycomb, became a subterranean tourist attraction. One privileged visitor was an appreciative Horace Walpole. 'There is nothing of this kind known in the world . . . a Roman city entire . . . that has not been corrupted with modern repairs . . .'[46]

By the middle of the century there had been more discoveries twelve miles along the coast, at Pompeii, another time capsule from the first century BC. Here the finds were nearer the surface but even more dramatic and macabre. It had been engulfed so instantaneously that hardened ash imprints of many of its petrified inhabitants were found in foetal postures of agonising death. And, from sarcophagi under the lava to the north and east of Naples, came twice-entombed grave furniture including well-preserved 'Etruscan' vases. Hamilton was particularly intrigued by the many fine clay vases that had been disinterred. Although they are still known as 'Etruscan'; it was Winckelmann who had first recognised them as Greek, from around the 5th century BC.[47] Within a short time Hamilton owned hundreds of these fine artefacts, some from raiding tombs personally but many bought from local 'dealers' who were, in truth, only other grave robbers who had stolen a march on him. His enthusiasm for the vases had two underlying motives – that of art appreciation and scholarship on the one hand and, on the other, as a potential source of income. He was finding that the expenses of his rank in Neapolitan society far outstripped his official emolument of five pounds sterling per day (£375). In fact, his stock of vases never proved to be a source of any wealth. His reward was to be the justifiable admiration for his self-taught scholarship when many of his finest vases were shipped back to England in 1771. They were acquired, for £8400 (£530,000) but at a net loss to Hamilton, by the British Museum.

The vases were instrumental in the change of emphasis of the exhibits at the museum.[48] Formerly a collection of books, manuscripts and natural history artefacts the accent changed towards the great collection of art and antiquities that it has become today. This, then, was the interesting character that Ingen Housz was about to meet at Naples in early 1769. However, he could so easily have arrived to hear stories of a foolhardy, ex-ambassador.

On 19 October 1767, some eighteen months before, Vesuvius had erupted massively and suddenly, its most dangerous discharge for over a century. Hamilton happened to be at the very foot of the volcano at that precise moment and was unable to resist climbing to a better vantage point. He and his guide were almost chargrilled.[49]

> ... on a sudden, about noon, I heard a violent noise ... the mountain split and from this new mouth a fountain of liquid fire shot up many feet high, and then like a torrent, rolled on directly towards us. The earth shook at the same time that a volley of pumice stones fell thick upon us; in an instant clouds of black smoak and ashes caused almost a total darkness; the explosions ... were much louder than any thunder I ever heard, and the smell of sulphur was very offensive... I must confess that I was not at my ease.

Hamilton may have been master of the understatement but all around him was a cauldron of chaos; primeval forces way out of his control. His guide having taken to his heels, he suddenly realised that he was alone and, molten lava pouring towards him, his adrenaline finally preserved him. He too ran for his life, a red-hot avalanche pursuing him for three miles: not that his fright stopped Hamilton from walking on the mountain on many later occasions. He happily acted as a guide to visitors and would have taken Ingen Housz, in all probability, to the very edge of the crater. After dinner the ambassador and the royal physician were able to discuss volcanoes, vases, lava, minerals, Neapolitans, Greeks, Romans and Winckelmann. Hamilton, too, had met the author of the book on Herculaneum – during Winckelmann's fourth visit to Naples, in 1767.[50] Ingen Housz's and Hamilton's paths could also have crossed before: in 1747, aged only sixteen, subaltern Hamilton had been garrisoned near Breda with a contingent of the British Continental Army.[51] Did Hamilton, like Pringle, visit the leather shop/apothecary in Eindstraat? It's certainly possible but probably unimportant.

Ingen Housz was at Naples long enough to be able to supervise the King through the 21 days of his recovery from inoculation. Having only one patient[52] was an undemanding commitment in terms of time and he would also have had ample opportunity to visit Pompeii and Herculaneum. The Emperor certainly visited the former site, Hamilton acting as his guide, on 7 April. King Ferdinand was among the party and therefore, we know, fully recovered from his inoculation and released from quarantine.[53] With his intense interest in all things classical, Ingen Housz would have found these newly discovered Roman sites irresistible, probing his way around Pompeii and burrowing down into Herculaneum. He was probably guided, again, by the ever enthusiastic William Hamilton. Here, as if preserved in aspic, was the world of his schooldays – Roman houses, shops, stables, streets, lead water pipes, under-floor heating flues – an advanced civilisation from two millennia before. However, Ingen Housz was soon heading north again, to Florence.[54] Yet another very high-ranking patient was waiting.

INGEN HOUSZ WAS at the Tuscan court in Florence from late April,[55] until the end of June, 1769. Here he caught up with the Emperor and Doctor Laugier and was formally introduced to his patient, the twenty-one-year old Leopold, Grand Duke of Tuscany. As he had found at Vienna a year earlier, two apartments had been prepared for him.[56] One was at the vast Palazzo Pitti on the south bank of the River Arno, about a hundred metres from the Ponte Vecchio. It had been the former home of the powerful Medici family and was now the official residence of the Grand Duke of Tuscany. The other was on top of a hill some two miles south, a cooler summer retreat for the royals, the Villa Poggio Imperiale, also with Medici connections. It took Ingen Housz a couple of weeks, presumably in concert with local physicians, to find a suitable early case of mild smallpox among the population of Florence. He appears to have received less than full and enthusiastic cooperation.[57] Eventually, the necessary preliminaries had been completed and Grand Duke Leopold, morally supported by his older brother, was inoculated on or about 14 May.[58] The actual surgical procedure was as routinely simple as always but Ingen Housz had to insist that the Grand Duke would be sure to isolate himself from his family, especially from his young children. This did nothing to make the patient less phobic. Ingen Housz would have seen that he had to adapt the approach to his patient: after the brief and unnecessary description at Naples, he had to give considerable and persuasive reassurance in Florence. The Grand Duke was a very different patient to the King. Whereas the Neapolitan was a lithe and restless 'carnivore', boorish and carefree, whose only reaction to inoculation was to resent his quarantine and the fact that he couldn't hunt, the Tuscan ruler was a starchy salon ruminant, a sensitive harpsichordist and introverted neurotic who questioned everything. Ingen Housz and Laugier, subjected to a gruelling cross-examination, would have been desperate not to contradict each other and fuel the fire. Even Leopold's own brother, the Emperor, normally bluff and thick-skinned, was on edge.[59] And, sure enough, Leopold's anxieties gave him specific food intolerances only two days after having been inoculated[60] – at a time when his symptoms could only have been psychosomatic. Ingen Housz had to resort to prescribing *laudanum* as a sedative.[61] After a very tense week, the timorous Leopold, perfectly well but for a slight fever, developed a scattering of pocks that appeared during 23 May, the due indication that his inoculation had 'taken'. Only some of them were blistering or troublesome, according to a now more optimistic and confident Joseph, writing to his mother the next day.[62] There were fourteen about the face and scalp, about twenty on the body and five on the feet. Joseph was hardly the sympathetic bedside relative one might have expected after his own tragic encounters with the disease.[63]

> . . . my brother is so frightened and such a hypochondriac that if the infection had been only slightly more violent, I don't know what we might have started . . . the least thing he experiences, he begins to imagine a thousand possibilities that are without any reason.

Ingen Housz was, no doubt, more empathetic and professional but would probably have found his task impossible without the authoritative presence of the Emperor – something that Joseph II and his mother had probably foreseen. Even Doctor Laugier, someone Leopold would have known and trusted throughout his childhood at Vienna, might have been an insufficiently reassuring presence per se. Finally, everyone was hugely relieved that, by the first days of June, Leopold was out of all danger. Ingen Housz's fifth royal inoculation was safely over. Again he was very generously rewarded. The grateful Archduke gave his inoculator a very fine enamel-plated gold snuffbox and a richly jewelled ring.[64] And within a few days a particular packet was found in the latest diplomatic bags from Vienna. It contained a personally handwritten thank-you letter from Maria Theresa.[65]

> After God, I owe to you alone the saving of three sons and you must judge the strength of my gratitude to you by the tenderness I have for my children.

The letter was accompanied by a bank draft for 8,250 Florins (£48,000).[66] Sadly, and paradoxically, in the glow of this medical success and material boost, Ingen Housz was physically drained and emotionally bereft; grieving, as he had done the previous summer, on his arrival at Vienna, for his lost independence and privacy. A letter to his friend, Deckers, was surprisingly intimate and revealing.[67]

> Having been dragged from the obscurity I had before, I am now doomed to end my days placed, I find, in a light that is too bright for me to suffer with

impunity . . . I have exposed all my reputation to the important enterprise which has been entrusted to my hands. A single unforeseen accident . . . could have plunged me into an abyss of ruination . . . The inoculation of the Grand Duke of Tuscany reminded me of those ineffable anxieties which haunted me during the inoculation of the Archdukes (Ferdinand and Maximilian) making me so over-wrought that my health was impaired. Although the illness was of the mildest kind and his Royal Highness didn't have to take to his bed or to his room for a single day, the mere thought of the possibility that some accident could occur would shake any man called upon to take on, alone, these responsibilities . . . I no longer have any ambition to inoculate other Princes; I have only that to finish my days in the midst of tranquillity.

He was clearly hoping his royal inoculations were at an end: 'I hope that she (Maria Theresa) will free me.' It was not to be. But the mood

Abbé Felice Fontana.
Courtesy of the Wellcome
Library, London.

changes, even in the course of the letter. There was the prospect of scientific investigation, always a potent recuperative agent for Ingen Housz. Suddenly he was cheerfully referring to his instant affection for a new friend, Abbé Felice Fontana, '. . . a savant who makes my principal amusement here. His conversation is pleasing to me and instructive.' Fontana was another exact contemporary of Ingen Housz. He was born on 15 April 1730 at Pomarolo, an Italian village near the north end of Lake Garda. After schooling he was a student at the University of Padua.[68] He trained for the priesthood but also attended lectures in medicine, chemistry and physics. He was comfortably at home in religious robes and was usually addressed as 'Father' even though he never accepted any formal position as a priest. His inquisitive mind held no limits and he was to make many original contributions to anatomy, physiology, botany, chemistry and experimental pathology. He collaborated with Leopold Caldani in researching, at Bologna, the sensitivity of various parts of the animal body, a topic proposed to scholars by Albrecht von Haller. It was in reporting this work that he gave the first accurate description of a nerve fibre, the very genesis of electro-physiology, the study of nerve function. The workings of the iris of the eye, the features of red blood corpuscles and the actions of viper venom were all topics in which this young genius made fundamental discoveries.

By 1765 Fontana held professorships at Pisa but was summoned, in 1766, to Florence, by the Grand Duke who made him court physician. He obviously met Ingen Housz, therefore, just before or during the Grand Duke's inoculation. Otherwise, Fontana was already very busy beginning to reorganise, for the Duke, the 'philosophical cabinet' of the Pitti Palace and sorting the surviving instruments and artefacts of the Medici collection into a science museum. These, together with many of the other treasures scattered around the Duchy, were eventually organised and put on display in the Palazzo Torrigiani. In 1780 the exhibits were to be moved to a purpose-built home,[69] 'La Specola' ('The Observatory'), one of the most important museums of its day, especially for its scientific exhibits. Fontana was its first director. How natural that Fontana and Ingen Housz should become friends – how much they had in common was remarkable. And how apt that their camaraderie was sealed while passing to each other neglected artefacts that had once been in the hands of the great Galileo. What enthralling hours they must have spent fossicking among the heaps of dusty relics below stairs in the palace. However, Fontana was to become much more than a knowledgeable and fascinating friend. His enthusiasm for investigating gases seems to have thrown a switch in the mind of Ingen Housz. Here was a new contact that was to further, and fundamentally, change Ingen Housz's life within the next decade. The two men would meet again – in Paris and in London: Fontana became an inveterate traveller, being sent all around Europe, by the Grand Duke, to collect exhibits for his museum.

In his letter to Deckers, Ingen Housz had indicated that, in late June, he was to leave Florence for Lombardy,[70] the next part of the empire in which he was to demonstrate and teach inoculation. However, in the event, he was delayed and diverted.[71] 'I made a journey from Livorno to the Isle of Elba with their Royal Highnesses where I visited the famous iron quarry, known to the ancient Romans, and the great diamond mine.' This interesting excursion, in close company with his royal masters proved to be both fascinating and recuperative. Elba was part of Tuscany and the Grand Duke was touring part of his domain; making a diplomatic visit. But Ingen Housz was clearly more interested in the geology of the island than in its politics or demographics. In later years he was to advise his young nephew, Josef Jacquin, that he should certainly not miss a trip to Elba.[72]

> You will do well, as I did, to send two or three mine workers onto the mountain for 2 to 3 days for them to dig out and make available a good quantity of diamond stones from as deep down as they can reach. I wish very much that you will use a couple of sequins for me in this dig.

On the point of leaving Florence, in early July, Ingen Housz received the wonderful news that he had been elected a fellow of the Royal Society.[73] Perhaps he had known that his recommendation had been written by Watson, handed to the secretary and posted on the meeting room wall, on 13 February. Perhaps not; but he must at least have given permission, at some earlier time, for his name to be proposed. In the end eleven fellows were to sign the proposal that

> John Ingen Housz, Doctor of Physic, now residing at Vienna, physician to her Imperial Majesty, A gentleman of great merit and well versed in natural knowledge, being desirous of being elected a Fellow of the Royal Society upon the inland list, is recommended by us on our personal acquaintance with him, as highly deserving that honour[74]

The recommendation remained posted in Crane Court for the next ten meetings, as was required, and he was balloted and elected on Thursday 25 May 1769. The supporting signatures that had been added were of people now already familiar to us – Pringle, Franklin, Huck, Baker, Heberden, Maty and, in addition, Gowan Knight, William Watson junior (William Watson's son), the Revd John Blair (historian and friend of Pringle) and James Parsons, a London physician who was a contemporary of Watson and who had been a protégé of Hans Sloane. Richard Huck paid the fee of 31 guineas (£2000) on his behalf and Ingen Housz's name was entered on the rolls on Thursday 16 November – on the 'inland list', a particular honour for a foreigner.

INGEN HOUSZ WAS back in Florence by 9 July and, accompanied by Laugier, left almost immediately, heading north for Lombardy. After Lucca[75] they followed the coast road to Genoa. Then, turning inland to Pavia and Milan, Ingen Housz began the activity that would occupy him for the rest of 1769. He would have learned, probably from court officials at Florence, that an Imperial Decree was about to be published, at Vienna, that gave official approval to the practice of smallpox inoculation.[76] It referred to Ingen Housz by name. Here was the official authority he might need if he was to convince highly sceptical colleagues to learn, at least, the practicalities of the procedure. In the scattered outposts of a huge empire, far removed from the institutional antipathy of powerful men like de Haen, Ingen Housz was more likely to convince colleagues of the efficacy of inoculation. But the schedule he had been given gave him very little time in each of the eight cities he was to visit even if the new protocol gave him a captive, compliant, attentive and at least partly respectful audience.

At Pavia, Milan and Turin he would have introduced, by lecturing, presumably in Latin or French, the background to inoculation before demonstrating the practicalities, instructing specifically on the selection, preparation and isolation of patients. And, of course, he would also have needed to demonstrate how the best – the safest – type of smallpox matter should be obtained and conserved. In Florence where, we may assume, he had given physicians the benefit of his tuition on inoculation, he had had the powerful support of the Emperor and the Grand Duke. Moreover, he was familiar with the city and was not pressed for time. But teaching in a succession of alien surroundings and to pupils who were full of scepticism must have been tense and taxing. Time and again he would have been confronted by an audience who were, frankly, deeply antagonistic and reactionary, many of them anxious not to abrade their local church authorities. Ingen Housz's personal courage and conviction must be admired. It took nerve, also, to defend his preference for the philanthropic posture taken from Dimsdale: that inoculation was simple and should not be an expensive device for an affluent minority who could afford extortionate fees justified by shrouds of mystery. Many of the doctors turning up for tuition would have been dismayed to find that inoculation was not going to turn out to be the income generator they expected. It took altruism to stay the course.

From Lombardy the two royal doctors turned east towards Verona and, eventually, on to roads that would lead them back to Vienna. Their route back took them through Brescia, at the southern end of Lake Garda. Here Ingen Housz was witness to scenes that had a lasting impact on him.[77] He and Laugier crossed the Lombardy plains and reached the town in late August – if they had been there ten days earlier they could have been killed. On the afternoon of Friday 18 August, during a summer storm, a bolt of lightning struck the San Nazaro church.[78] Unfortunately, the building was housing 200,000 pounds of gunpowder in its vaults and a huge explosion resulted. A sixth of the town was destroyed and the death toll was well over 3000. When the two royal physicians did reach the town it was probably too late for them to offer any real help but Ingen Housz certainly never forgot the lesson to be learned among the carnage and ruins: that arsenals and munitions stores should never be housed in tall buildings in densely populated areas and, as his friend Franklin rightly advocated, be well-endowed with lightning conductors. Neither was the lesson lost on higher authorities. The Vatican promptly dropped its scriptural opposition to lightning rods.[79]

During September and October the peripatetic teaching sessions continued. Ingen Housz and Laugier zig-zagged their way back to Vienna leaving further newly-skilled but tentative cohorts of inoculators at Gorizia, Trieste, Ljubljana, Klagenfurt and Graz. They arrived back at Vienna on Saturday 4 November,[80] completing a round trip of about 1600 miles, having spent at least 50 days on the primitive, punishing roads of southern Europe, many of them in sweltering heat and choking dust. Exhausted or not, Ingen Housz found many tasks urgently awaiting his attention: some were financial manipulations and he resumed responsibility for his affairs and his servants from Baron de Fries. Otherwise, he seems to have given priority to arranging, with the bookseller, Graeffier, the sales of some more copies of Hoogeveen's *Particles*. He had the names of new subscribers from most of the Italian cities he had visited and The Grand Duke had ordered two copies.[81] He and Graeffier settled on a price of 14 florins (£82) per copy, 8 Florins more (£127) for better quality paper editions. In April 1770 Ingen Housz was able to send Hendrik Hoogeveen a bill of exchange that would boost his bank balance at Delft by 500 Dutch Guilders (£2,900),[82] a bonus that Hoogeveen could not have expected without the further help of his ex-pupil. For Ingen Housz, not all his homecoming business concerned such sublime matters. Having been away from his Vienna apartment for ten months and having dreamt, no doubt, of sleeping in his own bed, he found it far from the comfortable haven he anticipated. He had to pay 28 florins (£170) for a new straw mattress 'of the best quality'.[83]

Ingen Housz settled back into the routines of a courtier and began, again, to accept patients for inoculation at Castle Sankt Veit.[84] With de Haen still in post and still vehemently opposing inoculation, he probably found himself still to be just as reviled as ever by the Viennese medical establishment. With the obduracy of those holding such firm convictions in the face of the perfectly credible evidence, his further successes inoculating elsewhere in the empire probably made him even more resented. There were clues, however, that the ramparts of resistance to inoculation were beginning to crack. Very many of the Viennese nobility had now been inoculated and, in Florence, not more than a day's journey from Rome itself, a local bishop had even circulated a pastoral letter recommending the practice.[85] Ingen Housz also knew, by letters from Dimsdale, that high-ranking Russian clerics were also beginning to take on board the objective evidence in favour of inoculation[86] and abandon their canonical objections.

Perhaps the stubborn ostracism by the medical fraternity in Vienna was partly why Ingen Housz now came to build a close friendship with Nikolaus Jacquin, Robert Laugier's replacement as professor of botany at the University. He was also Dutch, only three years older than Ingen Housz and also a physician (though no longer practising medicine), so there was obvious fraternal feeling. Jacquin had been a far more successful 'van Swieten recruit' to Vienna. In 1755 the Emperor had sent him to the West Indies to find plants for the Schönbrunn gardens. His fieldwork in the Caribbean, the plants he brought back and his subsequent cataloguing had made him many admirers. In 1759 he was made professor of mines and minerals but was clearly the professor of botany in waiting. His expertise in mineralogy was certainly a draw for Ingen Housz – he now had a great deal of money to invest. Speculation in the recovery of precious metals and gems from the earth's crust was growing in popularity as more and more prospectors emerged from deep underground bearing sizeable fortunes, especially in Transylvania and Prussia.

Having now safely inoculated three of Maria Theresa's sons and her oldest grandchild and having pump-primed inoculation in Austria, albeit mainly in the outer reaches of the empire, Ingen Housz felt it timely to apply for leave of absence from his new responsibilities at court. However, new plans had already been tabled for him to travel to Bohemia in order to introduce inoculation there. It appears that his request for permission to visit England was granted but only after he had agreed to initiate another tranche of physicians into the mechanics of inoculation. And once again his mission was to be away from Vienna. By 14 May, Ingen Housz had travelled the five-day journey north to Prague.[87] He then spent four hot months touring Bohemian and Moravian cities repeating his exertions as in Italy and southern Austria, teaching and advancing inoculation by personal tuition and demonstration. Wiesner reports that he inoculated mainly the children of the poor, seemingly without any unfortunate losses or setbacks.[88] Mozart was to find the Bohemians less reactionary and more open-minded, adventurous even, when it came to musical tastes[89] and perhaps this also applied to preventive medicine, giving Ingen Housz an easier time than he might have feared. His inner thoughts, however, were probably very often of London and all his friends there.

10
Paris, London, Paris

Saturday the 21st of April 1770. In the Hofburg palace, Vienna, a young, newly married woman wakes in the dawn chill and shrinks further down in the bedcovers. She is alone. There never had been a husband in the bed; she's been wed, by proxy, for some 36 hours, to a youth 750 miles away. At the age of 14 it's difficult for her to grasp what is happening except that today is the day that she has to say goodbye to her mother, her nurse, her boxer dog, everything. She withdraws totally under the bedding where there's warmth but no escape. She hears the voice of Frau Weber, her nurse, her second mother.

'Good morning, my sweet. It's necessary . . .'

The choking voice tails away to nothing and as Maria Antoine emerges she sees 'Weber' hurrying out of the room with her apron held to her face.

Breakfast over – it was never started – the young Archduchess makes her way back to her bedchamber out of habit but only to find the servants stripping the room, emptying chests and boxing all her dolls. They stare at her, aghast, and as she stares back she remembers that, this morning, she is not supposed to return to her own apartments. Still nothing is said as she starts to weep and turns and runs, lifting her skirts and screaming silently. Without looking where she's going in the familiar corridors, she runs full tilt into her mother and is swallowed up in powerful arms, just as she has been able to do at times of distress all her life. But this morning both women realise what must come next and grip each other as they walk out into the courtyard. A train of coaches is waiting, horses nodding and breathing heavily; the postillions trying to be invisible and impassive. The Empress is talking softly.

'. . . . I am sending them an angel.'

There is a last, long hug, and then another one, and they kiss each other's wet cheeks as they part. As if seeking distraction, they both turn to look at the alert and questioning dog held on a long lead. Maria Antoine can't bear to give her pet a last hug and snuffle – that would be too painful. She looks for Weber among all the assembled staff but can't see her. She starts to ask for her but impersonal hands motion to her and help her up into the coach. She collapses in sobs. As the vehicle lurches away she leaps up and hangs perilously out of the window to take a last look at her mother, at her dog, at the familiar palace buildings, at all of her life up to this moment. Two weeks hence she will even have a new name – she is to be called Marie

Antoinette. Within a month she will be the Dauphine at a palace in France. There will be a husband in her bed. Achingly alone in her carriage, she breaks down again.

A SENSE OF urgency dominated Ingen Housz's return from Bohemia. There must also have been considerable forward planning. Having arrived back in Vienna on Tuesday 21 August 1770[1] he was gone again within a month.[2] It was as though he feared being within the claws of the court for any significant time lest he be given yet another royal commission, another commitment that would prevent him, again, leaving for London. The conscious ploy, if that's what it was, failed. There was a new task already waiting for him. However, his journey to England was integral to the new mission for which he was briefed: his immediate ambition was not to be frustrated and Ingen Housz was probably relieved to avoid outright conflict between desire and responsibility. The task itself, though, proved to be very delicate and he would, for his first time as a privy counsellor, fall short of success. In the meantime he caught up with various domestic issues. He made contact, again, with Nikolaus Jacquin and acted on his latest recommendations for investing in mineral explorations. Ingen Housz decided to sink considerable sums into mines in Transylvania, 'putting the money between the hands' of his new friend because the purchase of the actual kuxes (mining shares) was a rather drawn-out process. He converted much of the rest of his new capital into fixed interest bank bonds at Stametz and by the time he set off west he had certificates worth 16,750 florins[3] (£98,000). And it was with Stametz that he entrusted all his material valuables – presentation snuffboxes, toilet sets and so on, having carefully packed them all in a wooden box. He also honoured his friend, Jacquin, by agreeing to take with him,[4] on his journey, some early printed pages of the professor's intended 'magnum opus', his impressive *Hortus Botanicus Vindobonensis*.[5] Though far from finished, this luxuriously illustrated catalogue of all the plants in the Vienna botanical garden, was already substantial enough, at over 100 plates, to provide adequate sample pages for inciting interest, if not promises of subscription, at various stations on Ingen Housz's imminent journey and in London. Being an impromptu book salesman was a fate that seemed to haunt Ingen Housz but any successes certainly helped to cement friendships.

STRASBOURG WAS THEN a fortnight's coach ride from Vienna and Ingen Housz, travelling, this time, with an Antoni Schitar[6] as personal manservant, reached the border town on 3 October.[7] He was aiming for Paris. The evidence is lodged in the so-called 'secret' correspondence between Maria Theresa and the French court at Versailles; in letters that were only fully published in 1933,[8] well over a century and a half after they had been written. The Empress, we now know, had given Ingen Housz some very specific instructions or, rather, questions to answer. He also bore a letter to the new French Dauphine, her youngest daughter Maria Antoine, now known as Marie Antionette.[9]

Then aged 12, Archduchess Maria Antoine had been among the four Archduchesses introduced to Ingen Housz, by her mother, after his first private audience with the Empress on Whit Sunday, 22 May 1768.[10] Ingen Housz had learned, then, that she would not need inoculation; she had survived smallpox, virtually scar-free, as an infant. 'Madame Antoine', as she was known in the family, had recently become the focus of her mother's attention. At the tail end of a long sequence of daughters (the seventh to reach adulthood), the Grim Reaper and political necessity had brought her forth from the shadows. The 1756 Treaty of Versailles, a year after her birth, had united Austria with France in a defensive pact against the recurring threat from Prussia. This unexpected alliance was a 'monstrosity' according to Voltaire[11] but, he had been forced to concede, 'necessary and therefore natural'. Now, more than a decade later, the bond needed reinforcing. The French and the Austrians still despised each other. In the political chess played by Maria Theresa, many moves took the form of arranged marriages, the pieces being her children. There was nothing unusual about the tactic. It was barely two years since Maria Carolina had been matched, by proxy, and despatched to Naples. But now the stakes were higher. Maria Theresa saw that Madame Antoine, her only remaining marriageable daughter, was her last chance of binding her dynasty more firmly to that of the French. Maria Antoine must be bride to Louis-Auguste, grandson of Louis XV and now first in line to the French throne after the premature death of his father, Louis Ferdinand, from tuberculosis, at the age of 36. When Ingen Housz had first met the young Archduchess in May 1768, the Marquis de Durfort, French ambassador at Vienna, had already been sidling in and out of the Hofburg for over a year, representing Versailles in the marriage contract negotiations.[12] Then, on 7 February 1770, Antoine had her first 'Générale Krottendorf', the nickname for menstrual bleeds among the Habsburg women.[13] Here was the signal, to Maria Antoine's mother, that full womanhood had arrived. An Imperial whistle was blown on the diplomatic dalliance, the contract settled forthwith, and Antoine, not yet fourteen and a half, was escorted to the altar of the Church of the Augustine Friars in Vienna on 19 April[14] by her elder brother, Archduke Ferdinand, to be married, by proxy, to the 15-year-old Dauphin, Louis, a total stranger at another palace 750 miles away. Two days later the naive and gauche teenager said an agonising goodbye to her mother. Her entourage made its way to an island in the middle of the Rhine near Strasbourg. It was from this symbolic, inter-national launch pad, two and a half weeks after leaving Vienna, that she was propelled into her new life — as a French princess. In a wooden pavilion, erected specially for the purpose, and even then partitioned down the middle, she was divested of all her Austrian apparel and passed through to French ladies-in-waiting to be dressed in French finery. Even her name was changed — to Marie Antoinette.

After another week on the road she arrived at a secluded spot in the forest near Compiègne where, under the intense gaze of hundreds of courtiers, the young couple first cast eyes on each other.[15] Louis was slightly more than a year older than his bride. He was just as emotionally immature although characteristic Bourbon bulk made him look further on in adolescence than he really was. The new Dauphine discovered that she had wed a human tortoise; round, retiring, lethargic and clearly lacking the notorious sexual appetite of his uninhibited grandfather, Louis XV. Perhaps, then, it should be no surprise that this shy and inexperienced couple failed to consummate their marriage, especially knowing that their every reaction to each other in public, and very probably in private, was being scrutinised for reassuring signs of attraction. Penetrative sex probably never occurred for several years[16] and was, even then, too fleeting for impregnation to occur. Although this may be clear in retrospect, at the time the problem presented only as a failure to conceive. It was a serious disappointment. The marriage, politically motivated, was meant to produce male heirs, sons that would unite an Empire and a Kingdom, Austria and France. The mixing of dynastic bloods was meant to generate a geometric progression of power and influence.

By autumn 1770, Maria Theresa was already fretful. Although the couple had only known each other a few months, successive letters from her daughter reported yet another 'Générale'. The poor Dauphine was having to tell 'Mama', again and again, that she was not pregnant. She certainly could not reveal to her in a letter why it was not possible and neither woman

The Palace of Versailles

seemed to appreciate that having had her first period only earlier that year, Marie Antoinette might not yet be ovulating regularly anyway. The no-nonsense approach to these matters at the Hofburg led to an unavoidable conclusion – if Antoine wasn't fecund, she must be ill. She should be medically examined; but by whom? An answer suggested itself: wasn't Dr. Ingen Housz soon going on leave to London? There seemed no reason for him not to travel via Paris.

Ingen Housz reached Versailles around 20 January 1771.[17] The palace, some eleven miles south-west of Paris, was the largest in Europe. Even after the Schönbrunn and Caserta, Ingen Housz must have gasped as he approached, through freshly fallen snow.[18] Renovated and massively extended in a succession of stages during the second half of the seventeenth century under Louis XIV, the facades, the gilded roofs, the sumptuously appointed apartments and horizon-wide gardens could not but impress. However, there were problems.[19]

> In the eyes of the world, Versailles was the enchanted palace in which all the arts had contributed to its perfection: in the eyes of those responsible for its upkeep, it was a fabric on the verge of ruin.

Maintenance of the palace had been much neglected during the reign of Louis XV and Ingen Housz would find, on closer inspection, that there were several areas that were now roped off as dangerous; century-old structural timbers having been allowed to rot.[20] It was also poorly administered: many of the domestic staff were several years in arrears with their salaries.[21] Whether or not Ingen Housz saw the unsightly defects as allegorical, or winced at the sloppy organisation, is nowhere recorded but he was to have a far from happy time at Versailles. But any problems he may have had with the buildings' shortcomings or the palace staff were as nothing compared with his frustrations with protocol. The Dauphine, the future Queen, had still not conceived. A letter, of January 1771, from Count Mercy-Argenteau,[22] Imperial Ambassador and Marie Antoinette's surrogate parent, continues the story.

> Dr. Ingenhouse, having arrived a short time ago, has made known to me the intention of Your Majesty that he will be allowed to see and examine closely the state of health of Madame La Dauphine, of Monsieur Le Dauphin and of the royal family. I am busy arranging the means for the doctor to achieve his objective . . .

However, there was a problem with protocol – as so often at the French court.

> . . . but I have come across an obstacle on the part of Madame the Comtesse, in which the initial reaction has been that Dr. Ingenhouse is not of the rank to have access to the royal bedroom.

The Comtesse de Noailles was 'Dame d'Honneur', the 'Mistress of the Household'. Nothing in the life of the Dauphine escaped her scrutiny; nothing happened without her permission. Rigidly obsessed by court etiquette, a skewed straitjacket of manners unique to Versailles, 'Noailles' and her husband were at the very top of the pecking order and determined to stay there. Ingen Housz, as far as she was concerned, was just another physician and therefore a mere worm, instructed by an Empress or not. Mercy-Argentau tried to squirm under the barrier by organising a personal appeal to Marie Antoinette herself, clearly confident that she would respect her mother's wishes.[23]

> I have charged the Abbé de Vermond to propose to Her Royal Highness that a man sent for the explicit reason of Your Majesty is to be excepted from all the rules of etiquette, and that the least reflection would be sufficient to persuade Madame the Archduchess.

The ruse worked. Marie Antionette saw Ingen Housz in private and he was at least able to report some progress to the Empress.[24] The Dauphine had

Queen Marie Antoinette.

'treated him very well' but it is clearly recorded, if between the lines, that the meeting was seen as only a preliminary one.

> Ingenhouse tells me that he has found you to be very well and grown up, that he has seen all the family and found you all in good health, but that he has not had an opportunity to examine you because of etiquette . . . I can only believe that a man of our court will not be denied access to you . . .

It is then quite impossible, from the subsequent linguistic somersaults of Mercy-Argenteau's letters, to know what happened next except that there was such a huge outcry at the defiance of protocol that the matter proceeded no further. Maria Theresa had to be content, for the time being, with a series of platitudinous observations from her simpering Ambassador.[25]

> . . . when her Royal Highness manages herself with proper sentiments, she always moves in a way to be admired and adored. There is no point in the day where she doesn't give proof of judgement, of a singularly just manner, and of a good character, generous and compassionate.

Maria Theresa must surely have seethed with annoyance. She was not the least anxious to know of her daughter's deportment or manners. Why was she not having babies? Ingen Housz, for his part, must have been very troubled. Not only was this the first royal commission in which he had failed, but to anyone not appreciating the serpentine writhings of the status-obsessed courtiers at Versailles, his failure seemed pitiable. Presumably, he made a second and, this time, entirely unsatisfactory written report to the Empress. And more aware, now, of the subterfuge and paranoia at the French court, he may well have kept this about his person until he had access to the diplomatic bags that went direct to Vienna from Brussels, where he travelled next. He consoled himself, however, before leaving Paris, by buying a handsome chiming clock,[26] for 850 Livres Tournois (£2,300) from the shop of the famous Ferdinand Berthoud.

G IVING PRECEDENCE TO royal affairs and politics has broken the sequence of the story: Ingen Housz had taken some diversions on his journey to Versailles. He spent a couple of weeks at Strasbourg before following the Rhine upstream to Basle in Switzerland, arriving there on 13 November. Then, 60 miles further southwest, at Berne, he met Pringle's very old friend and regular correspondent, Albrecht von Haller. Ingen Housz was in the company of a trusted confidante and well beyond the tentacles of the Austrian Empire for the first time in more than two and a half years. He seemed to enjoy the licence to talk frankly. Haller may have drawn him out but Ingen Housz was certainly bluntly honest in giving his opinions of the Vienna physicians[27] – working among them had been like 'being in a wasp nest' despite the hospitality and protection of van Swieten. After taking a parcel of books for Sir John Pringle into his care,[28] Ingen Housz left the internationally famous physician and spent four more days travelling west in his coach. The highest peaks of the Alps loomed up across lake Geneva on his left as he approached the city of Geneva itself. Here he sought out Horace-Bénédict de Saussure, professor of philosopy at the university. Saussure's interest in local botany had taken him up into the snow-clad heights of the Alps and his focus of study was already changing to that of mountaineering and geology. Nonetheless, he was a likely customer of Jacquin's *Hortus Botanicus Vindobonensis* and Ingen Housz was able to sell him some sample pages at cost price.[29] And it was at Geneva that Ingen Housz spent his fortieth birthday, on 8 December 1770, before setting off on the next major leg of his journey, a long diagonal across eastern France to Versailles.

After his mission at Versailles and having passed through Brussels, Ingen Housz was able to spend some of February 1771 at Breda with his brother. Louis and Maria now had six children ranging down from the age of six. Jan remembered Maria Catharina, Arnold and Peter from his very brief visit, in April 1768, but now there was also Henry, born in July 1768, little Jan in July 1769 and Louis junior who was just six months old.[30] 'Uncle Jan' was only able to stay for a few days but this welcome respite, once again ensconced within his family, appears to have been his first real interruption of a long period of highly intensive and stressful activity in a very public limelight lasting nearly three years. Even the unexpected trip to Elba had been in the company of royalty. After visiting an old friend, the physician, Salomon de Monchy, at Rotterdam and Hieronymus Gaubius, his old teacher at Leiden, Ingen Housz crossed to England in early March 1771, arriving at Harwich[31] on Tuesday 12th. There had been many significant events in his life since he was last in the Essex port, delayed, anxious and impatient. There was no reason to linger in the port this time and within forty-eight hours he and Antoni were in London. Ingen Housz was delighted to find that his old rooms, with the Pitters family, of Northumberland Court, Charing Cross, were immediately available. The rent, including coal and candles for them both, was an acceptable £1.00 (£64) per week.[32] Antoni himself, while in London, was paid half a guinea (£33) a week as wages banked, on his behalf, at Vienna but also eight shillings (£25) a week ready spending money.[33] To be able to pay his way, Ingen Housz went into Drummonds Bank, also at Charing Cross, and handed in two letters of credit, one from Wolfort van Hemert of Amsterdam and another from Stametz, his bankers at Vienna, depositing, effectively, some £100 (£6,370) sterling from each.[34] Immediate needs organised, he then, no doubt, took off in search of Pringle, Huck, Watson, Franklin, Heberden, Collard, Maty and Baker; all his old London friends.

THURSDAY 21 MARCH, was a momentous day for Ingen Housz. In the middle of the afternoon he accompanied Pringle to the Crown and Anchor Tavern on the south side of the Strand. Here he was introduced to members of the Royal Society club.[35] They dined here each week throughout the year before going on, during the winter 'season', to the regular Thursday evening meetings of the Society. The members present that evening[36] included John Smeaton, Neville Maskelyne and Henry Cavendish. Pringle, Smeaton and Cavendish were all holders of the Copley

Medal, the ultimate annual accolade of the Royal Society and Maskelyne would be awarded his in 1775.[37] The company was most certainly top quality and the food likewise; generous in quantity and variety though not unusually so for the eighteenth century. The table groaned with cod and baked tench; there were whole roast chickens, a lamb shank, roast beef, bacon and greens, a calf's head and a selection of pickles. Afterwards there was plum pudding and apple pie.[38] The liquids available are not mentioned in the archive – a discretionary oversight, perhaps, throughout these records. Ingen Housz had had only a few days to practise his English but probably coped reasonably well – certainly better than he had when he first came to London. As evening wore on the members all rose from table but only to decamp together to Crane Court where Ingen Housz was formally admitted as a fellow of the Royal Society,[39] signing the charter book at the invitation of the President, James West. Ingen Housz was to attend, as a guest, 19 of the next 34 meetings of the club.[40]

Except for one of his attendances, when he was introduced by Richard Huck, Ingen Housz was consistently the guest of Sir John Pringle.[41] In fact he was seen in Pringle's company so frequently that word was put about that he was Pringle's houseguest and perhaps even a relative.[42] After a few weeks Ingen Housz's confidence in English would have returned and he was able to make significant contributions to conversations again. Then there was the 'Honest Whigs' discussion group that met at the London coffee house in the shadow of St. Paul's Cathedral,[43] Pringle's domestic soirées, the latest experiments, interesting medical cases and autopsies, new exhibits at the British Museum, private dinners and concerts. He would have found himself in some demand when he revealed that he had been an eyewitness of the ruins at Brescia. The news had caused alarm in London, as elsewhere; anywhere that munitions were stored. Benjamin Franklin, his very close friend still serving, in England, as the agent of the American colonies, and inventor of the lightning rod,[44] would have been particularly intrigued. He, together with William Watson, Henry Cavendish, John Robertson and another Royal Society fellow and electricity expert, Benjamin Wilson, would serve on an advisory committee to the Board of Ordnance, in 1772.[45] The Board were to approach the Royal Society, anxious to know how they should best install lightning rods on their new gunpowder magazine at Purfleet on the Thames. The committee was formed following the general alarm, right across Europe, generated by the Brescia explosion.[46]

Then there was shopping. Ingen Housz bought some very expensive fine cloth at a drapers in Cornhill.[47] The fickle summer weather in England was now something of which he was well aware and he also picked out some shalloon woollen lining material, sent, no doubt, to a quality tailor now that he was a man of substance. Nevertheless, the equable spring weather invited relaxation and he was obviously enjoying himself.[48]

> You can imagine how happy I feel among old friends, because you know how dearly I love this blessed Isle of the Britons. Released from the restlessness of the Court, I am spending quiet days here, free from cares, far from jealousy of any kind. I live in the house of the same friends as I used to, I am like a member of the family and I feel very much at home.

Ingen Housz's life in London was clearly less fraught than in Vienna but it was not entirely frivolous. On Thursday 2 May he was in Berwick Street in Soho, at the premises of a Mr. Jackson, a brazier.[49] He paid 12 guineas (£800) for a 'Pennsylvania Fireplace', one of the heat-efficient stoves invented by his friend, Franklin, and a further 27 shillings (£86) for it to be packed and despatched to Brussels.[50] The stove was for His Royal Highness Prince Charles of Lorraine, Viceroy of the Austrian Netherlands.[51] The head of Apollo that served as adornment on its front face was therefore not inappropriate. This was a simple enough commission but there were on-going and serious problems with the still-delayed final publication of Hoogeveen's *Particles*.[52]

Despite having received his first-quality edition of the book before leaving Vienna and having been very pleased with it,[53] Ingen Housz discovered, to his dismay, that none of the subscribers he had recruited in England, four or more years earlier, had received theirs. Worse still, he found that they had all been served a demand for more money. He was forced to administer massive reassurances to an indignant constituency whilst trying to find out what had gone wrong. There was a widespread suspicion of fraud but he only discovered incompetence – in a merchant middleman who had allowed costs to spiral and then refused to import the printed volumes without more payment. His London friends sent Ingen Housz along the Strand to Peter Elmsly, a bookseller and publisher who specialised in foreign books. Elmsly agreed to try to sort out the muddle, taking a fifth of the profits on any further sales of *Particles* for his trouble. But he clearly had a good grasp of the cheapest freight routes, taxes and other relevant matters and Ingen Housz immediately recommended him to Hoogeveen despite the erosion of profit margin. Order was restored, the patient subscribers soon had their books and Ingen Housz was hugely relieved. But somehow, just being in London seemed to generate spadework.[54] 'I had not foreseen how much business there was waiting for me to settle among old friends. The vastness of this place causes some insignificant matter or other to take up almost a whole day.' There was an accelerating cycle of increasing 'busy-ness' that needed breaking and there were plans.

FROM HIS FIRST arrival back in London, Ingen Housz and Franklin had been planning a vacation and Franklin wrote to John Canton[55] their mutual friend and expert on electricity.

> Dr Ingenhauss and myself purpose to set out . . . to visit Birmingham and some other manufacturing Towns, intending to be absent about 10 days; a young kinsman of mine accompanies us. Will you make a fourth, and so reduce our Triangle to a Square?

Franklin's 'kinsman' was Jonathan Williams, a son of one of Benjamin Franklin's nieces. He would have his twenty-first birthday during the trip. The 'square' set off, from London, at nine o'clock on the morning of Saturday 18 May.[56] The over-excited repartee probably took two or three days to abate by which time they had plenty of new sights and sounds to talk about. Their journey described a huge figure of eight through the heart of England at a time when the Industrial Revolution was just nascent. Heading northwest to and through the Peak district to Manchester they then crossed back, west to east, over the Pennines to Leeds and Sheffield. Turning southwest to Derby the coachman was then given orders to take the road to Birmingham and the last lap to London was a long, tiring day on the road, Thursday 30 May. They visited a silk mill, a marble mill, a mortar mill, a brickworks, lime kilns, Leeds Cloth Hall, a silver plating mill at Sheffield, an iron smelting works, a pottery for both crockery and figurines, and a button manufactory. One of their first lengthy visits was on Monday 20 May, to a natural phenomenon in the Peak District called the 'Devil's Arse'. This large cave at Castleton, previously visited by Ingen Housz in the summer of 1766 and perhaps his recommendation, was a hive of industry. Whole families lived in cottages just within the windy, echoing entrance that gave it its name, eking out a dingy living by making ropes and string. The cave dwellers had also realised that there was money to be made from tourism as visits to the Peak District grew in popularity in the eighteenth century. A guide cost a half-crown per person (£8) and candles a further shilling (£3).[57] At the back of the entrance cavern there was a 'squeeze', a low-ceilinged passage virtually filled with water. Visitors were pushed through, prostrate, in straw-lined punts. Negotiating this internal sphincter of the 'arse' was known as 'crossing the Styx' and was not for the faint-hearted. But it gave access to the marvels within the inner chambers, the 'Flitch of Bacon' stalactite among them. Safely back above ground the foursome then travelled on to Manchester where, after a night's rest, a very long day awaited them.

After rising at five a.m. they were on the streets of Manchester by seven and meandered around the city to see all the sights. Then, at nine, they set off to tour all the admirable enterprises of the Duke of Bridgewater. They were conveyed a third of the way along his famous canal, cut between the centre of Manchester and Worsley, a mining village ten miles to the northwest, their longboat drawn by a single horse. They were 'entranced' by the engineering feat of the waterway, constructed by James Brindley ten years earlier. It snaked through the undulating landscape and there was a 200-foot aqueduct that crossed the valley of the Irwell, 39 feet above the river.

Bridgewater Canal near Manchester.

What is almost incredible is that this Canal goes over a river . . . (and) a road . . . so that a boat may be sailing in the river, a Coach going by the side of it, and a Boat immediately over their heads.

The canal led them to the Duke's coalmines which were served by side arms from the canal. Even narrower boats could be drawn deep into the shafts to bring out the coal, seven tons at a time, with very little effort. The visitors accepted an invitation underground, crouching in one of the coal barges.

Dr. Joseph Priestley. ©
The Royal Society.

We went under the Mountain . . . got out of the boat and walked many yards, stooping till we came where the Miners work . . . it is exceeding hard Work: as fast as they make a cavity they shore it up to prevent it falling in.

One can imagine the mutual amusement when, on surfacing, they saw each others' blackened faces, coal dust having settled on their perspiration. 'We returned to Manchester' Williams wrote, 'highly pleased with our entertainment.'

To stave off the physical exhaustion and intellectual indigestion they indulged in all the regional fare, all backslapping in hearty agreement that Buxton Ale was particularly fine. But perhaps the most significant event of the whole trip, for Ingen Housz, was their meeting with Joseph Priestley, on Thursday 23 May. Priestley was just 38 in the spring of 1771. A native Yorkshireman, he was a man up to his gills in irony. A heavyweight in the history of science, in real life he was a twitchy prairie dog of a man.[58] A preacher by profession; he was burdened with a stammer. Adored and absorbed by the establishment early in his career he eventually became a pariah and was forced out of the country. And for an extremely intelligent man with an exquisitely tuned, inquiring mind he could be agonisingly stubborn. After an education at a Dissenting Academy designed to train him for the Ministry, his first appointments had been unsuccessful. A flair for teaching had been spotted, however, and he moved to a Dissenting Academy in Warrington in 1761 where he introduced 'science' into the curriculum. With the apparatus bought by the college he was able to embark on a subsidiary career; that of self-taught scientist. His outstanding success as a teacher resulted in a Doctor of Laws degree awarded by Edinburgh University in 1764. In 1767, he tried pastoral work again, taking over as minister to a Presbyterian congregation in Mill Hill, Leeds, which is where the four travellers caught up with him.

Priestley already knew Franklin and Canton well and was able to greet them as old friends. With William Watson, they had been among the sponsors for his fellowship of the Royal Society in 1766.[59] But this had been only the culmination of the help they had given him in writing his book, and the basis for his election, *The History and Present State of Electricity*. This had been published in 1767[60] and contained many details of his original electrical observations. By 1771 he had moved on to chemistry experiments, fired, again, by self-taught appreciation. Here his discoveries would make him a household name. When first at Leeds he had lived next to a brewery and was intrigued by the gases that bubbled up from the fermentations — 'fixed air' (carbon dioxide). In researching these 'airs' he first repeated the experiments of Stephen Hales, also a priest, but was soon doing such original work that it was clear that he was an even greater genius. It must have been some of his earlier findings that Priestley was able to demonstrate to his visitors in 1771; a brief preview of his first report on 'airs', published as a book in 1774.[61] This would point up another Priestley irony. His religious dissent was trenchantly held throughout his life. He was a Unitarian, rejecting The Holy Trinity of 'Father, Son and Holy Spirit' or any other notion that split God up into constituent parts. Now he was saying that all around us, the supposed location of God, the very air we all breathe, was anything but a unity. For Ingen Housz a seed was sown: another one alongside that already germinating — the one planted by Fontana.

By the twenty-fifth of the month the four tourists were at Sheffield admiring the production processes of plated silver. They watched in awe as large sheets of the metals were pressed and rolled together, smelted and then polished, noting that some of the labourers were women. Then there were visits to stately homes. On Sunday 26 May, with local industry subdued for the Sabbath, they toured Chatsworth, home of the 5th Duke of Devonshire. They particularly admired the spectacular cascade down the hillside east of the house and the many fountains fed, underground, from lakes atop the hill. But they had found the interiors more impressive at Wentworth Woodhouse near Rotherham, seat of Lord Rockingham, where there were '40 very elegant apartments and many fine paintings'. The next day, at Derby, they visited a pottery and then made their way south to Lichfield. At Soho, just north of Birmingham, the travellers visited the new and splendid 'manufactory' of Matthew Boulton. They saw his production lines for watch chains, silver trays, kitchen utensils and, especially, his famous buttons. On Thursday the 30th they headed back for London.[62] The costs of their tour, calculated by their log keeper and accounts clerk, Jonathan Williams, came to sixpence short of £70 (£4,460) and were divided equally between the 'square'. Ingen Housz happily paid his share. He also repaid Franklin who had lent him, at Sheffield, just under thirteen pounds (£828) to buy silver plate. But within days he was again having the pleasure of spending someone else's money.

A LETTER FROM Vienna, from Count Dietrichstein, had been awaiting his return to London. The Count was chief equerry to Joseph II and wanted Ingen Housz to buy some English bed linen for the Emperor. The impish thought may have occurred to Ingen Housz that even though he could not find out what was happening between the Habsburg sheets, the family at least trusted him to buy some new ones. He took advice and set off for the premises of a merchant called John Davenport.[63] He paid £54 8s. 0d. (£3,460) for what was obviously very high quality cloth but a sum that must have bought a huge quantity of linen for a widower who presumably slept alone. It was despatched to Brussels, on a sloop going to Ostende. From Brussels there were well-worn roads to Vienna. He also had further business with Elmsly whom he now saw as a prospective agent for selling, in England, the great work of his new friend, Jacquin, his *Hortus Botanicus Vindobonensis*. Elmsly seems to have been impressed by the very fine illustrations and took the commission so that some, at least, of the exemplar drawings were put on sale in England.[64] However, Ingen Housz did not allow commerce to fill all his days: there were friendships to foster, especially with his great mentor, Pringle. Here the trade was free. The two physicians, both now having royal patients, exchanged new 'receipts', medical treatments they had come to know in the three recent years they had been apart. Ingen Housz detailed various therapies he had seen being used across Europe: for jaundice,[65] diarrhoea,[66] coughs,[67] tapeworm,[68] sciatica[69] and for the eye inflammations[70] that were such a prevalent problem at Vienna. And he was able to describe, first-hand, the customary clinical practices of van Swieten, Pringle's old friend.[71]

A day or two after 19 September,[72] Ingen Housz took the now familiar road to Hertford, delighted to have received an invitation to stay with Dimsdale and his family again. He was their guest for nearly a month.[73] Dimsdale was able to finish the story of his adventures in Russia and, no doubt, be less circumspect than in the letters that had travelled between St. Petersburg and Vienna. For his part, Ingen Housz also had a full portfolio of experiences having exceeded his mentor in the score of successful royal inoculations. But they both agreed there was still much to learn in this field of study and the two friends went to see some of the families that Ingen Housz had inoculated in the bitter early weeks of 1768. They were especially interested in the children born to those mothers who had been inoculated in early pregnancy – were the offspring safe from smallpox, was inoculation effective across the placenta? Frustratingly, the picture was unclear.[74]

Thursday 7 November saw the first of the new season of weekly Royal Society meetings and Ingen Housz was back in London in good time. He also resumed his regular dining, as a guest, at the Royal Society club and, on Thursday 21 November,[75] he found himself at table with another guest, someone who would become a household name – Captain James Cook, circumnavigator and 'discoverer' of Australia. But, by the end of the month, Ingen Housz was preparing his return to the Continent. Although he was expressing regret for 'an unwelcome departure' almost as soon as he had arrived at London,[76] at least he was going, this time, to a Vienna that was no longer an enigma and with the knowledge that he now had, according to Nikolaus Jacquin, genuine friends there.[77]

> I would like to think, Sir, that you have a few friends among our 'savants'. I don't know if I have any merit to my own name but I can assure you very sincerely that one of those things that would give me the greatest pleasure will be always to cultivate the friendship of which you so honoured me and of which I shall take to be a particular tribute and, I dare say, pleasure of my life.

Of many farewell meetings, we know, specifically, of one very interesting dinner party. Benjamin Franklin took receipt, on Thursday 12 December, of an invitation to dine: 'Sir John Pringle requests Franklin's company at dinner next Sunday (15 December) to meet Mlle. Biheron and Dr. Ingenhousz before the latter's departure.'[78] Marie Marguerite Biheron would have been a fascinating dinner guest, especially for medical men. Born the daughter of an apothecary, she had studied life drawing as a student. She took to heart the point of her teacher that a comprehensive understanding of human anatomy, and not just of the superficial structures, was fundamental to success. She came to dedicate her life to preparing wax models of human dissections and, as Pringle remarked, these were so accurate that 'they want nothing but the smell.'[79] She had encountered hostility from most of the French medical establishment but nothing but encouragement from English physicians; and William Hunter, the gynaecologist and anatomist, was especially enamoured of her models of the female pelvis and its organs. This point might well have been raised at Pringle's dinner table. If so, it was an unwelcome reminder for Ingen Housz that waiting for him, at Versailles, was a real female pelvis that was still not performing as expected.

11
Hunkering Down

His presence is required at eight o'clock in the morning. Tuesday 24 March 1772 dawns gently but with a sun strong enough to burn off the mist over the Danube. It's maturing into a perfect spring day, a day good enough to lift the heart. As he walks across the inner courtyard of the Hofburg the sun is glinting on the gold finials of the rooflines. It's all counter to his anxiety. Having just returned from a year in Italy, a year in which he has almost killed one of the Empress' grandchildren through a complication of inoculation, he wonders how she's going to greet him. Kaunitz had taken the trouble to call in to reassure him last evening but he's renowned for trying, always, to mollify and trouble-shoot. On entering the large antechamber that leads to Maria Theresa's private apartments, Ingen Housz looks, immediately, to the huge double doors at the far end. The guards are standing either side and remain quite impassive. He can relax a little. He isn't late. Beyond the doors is where the Empress is generally found waiting to greet those allowed in to see her personally.

In fact, as the doors are opened to admit him to her presence, Ingen Housz knows, immediately, from the wide smile, that he has nothing to fear. He's walking on air when he returns to the antechamber and doesn't hesitate to approach when he sees a group of three other courtiers he knows, deep in conversation, at the far end of the room. He's within a few yards when one of them looks up and sees him. There's an abrupt change of expression and he sees his name mouthed silently. Two of them turn, abruptly, and head away out down the corridor as if called by a sudden, silent alarm. Brukenthal, effectively marooned, resorts to a clipped 'good day' and backs away, eyes averted, in the direction of the Empress' suite although, as Ingen Housz watches him, it's quite clear that he has no appointment and he is forced to take a pointless walk around the very margins of the chamber before scuttling off through a side door. Quite what he's done to upset them, to cause this crass avoidance, is a mystery to Ingen Housz. But then, there remain many mysteries about the court, one of them being how long it's going to take before he's accepted into the circle. In spite of everything, particularly his medical successes – or is it because of them – he's still an outsider. It will take yet more courage not to become a recluse.

INGEN HOUSZ AND Antoni left London for Brussels around Christmas 1771.[1] It was a departure more welcomed, perhaps, by servant than by master. Nearly four years before, Ingen Housz had strained every sinew, his own and those of many a horse, to reach Vienna as quickly as possible. He had been anxious, impetuous and excited. Not this time. He didn't purposely delay his progress but after nearly nine joyous months free to move at will among all his London friends and follow intellectual trails on a whim, one senses only a looming awareness of loneliness and a temptation, even, to stay put. It must also have played on his mind, during the long, cold journey, that both Pringle and Franklin, now his most established friends and advisers, were both well into their seventh decades of life. It was more than possible that he would never see either of them again. It took him a month to reach Versailles, a journey that need only have taken a week at most. His snail-like progress was, perhaps, symptomatic. He needed good news at the French court to lift his mood. He was out of luck. At the palace in February 1772,[2] roughly a year since his last visit, he found that Marie Antoinette had still not conceived. Even so his second attempt to examine her was also barred, again by dint of protocol.[3] The unyielding etiquette at Versailles was just as inflexible as before. Even five years later, when Joseph II himself visited France, in 1777, he too was struck by its stranglehold, reporting back to his mother that 'an aristocratic despotism reigns . . . everyone is absolute in his own department, but continually afraid – not of being controlled by the sovereign, but replaced.'[4] Count Mercy-Argenteau tried, as usual, to obfuscate his powerlessness when he wrote his next letter to Maria Theresa.[5]

> When Dr. Ingenhouse passed through here again, I took the opportunity for him to pay court to Madame La Dauphine and he obtained an audience of which he will give an account to Your Majesty, just as of the many observations which I have communicated to him in order to allow you to better appreciate the state of things.

Though the most junior royal physician at Vienna was ill suited to the mission by lack of rank and chosen only by the expediency that he was going to England, Ingen Housz must have felt responsible for his failure to glean, twice, any useful diagnostic indicators why Marie Antoinette was not getting pregnant. This time the dismal news could not be diffused in a letter and the audience with the Empress that awaited him at Vienna was likely to excite yet more parental agitation, putting him, the messenger, in a poor light. There may or may not have been impotence at Versailles but there was certainly impotence – political impotence – at Vienna.[6]

IF THE LEG of the journey between London and Paris had been a drag in the doldrums, the next leg, Paris to Vienna, was going to be one of wretchedness and, as his carriage eventually neared Austria, Ingen Housz's despair was to be complicated by physical distress.[7] There comes a time in spring, at each of the European latitudes, and at every height and location, when patches of frozen ground start to thaw and quicken. In the eighteenth century when roads were un-metalled, travelling at these times could be particularly treacherous and accidental injury more likely. Horses could stumble and carriages, suddenly losing part of their traction, might slew and tip. Even if actual precipitation onto the road and gruesome injuries were avoided, passengers suffered many a jolting insult. Perhaps, then, it should be no surprise that Ingen Housz, having arrived back at Vienna on 14 March,[8] was, within days, disabled with 'a violent *malum ischiadicum*' – severe low back pain.[9] The lumbago deteriorated and he was eventually confined to his bed for six weeks, the severe pains extending down his left leg as sciatica.[10] At first, however, he was just mobile enough to be debriefed at court and catch up with his few trusted friends, particularly Nikolaus Jacquin and Johann de Fries. To his dismay he was told that Gerard van Swieten was very ill. The senior royal physician was now himself a patient, at his summer residence at Hietzing, just to the west of the Schönbrunn.[11] Ironically, he, too, was struggling with severe leg pains but his were from a sinister recurrence of previous episodes. In a man of 72 who was developing gangrene it is almost certain that he had blocked arteries. Whether Ingen Housz was able to travel to Hietzing to visit him is not known, for within days of his return to the court Ingen Housz, too, was confined by illness.

It is not explicit whether Ingen Housz still had 'grace and favour' court apartments in Vienna on his return in spring 1772 but, by the summer of 1773, he was noting the levy of a quarterly *schuldernsteuern* (household tax),[12] paying a regular wage to a cook,[13] and commissioning his tailor to make livery items for his servants,[14] all suggesting a cutting of the royal umbilical cord even if the court was still paying his rent. By May of 1774 he was certainly running his own household, living very near St. Stephen's Cathedral.[15] Louis Mathis and Franz Mochet must have left his employ during mid 1770, their names thereafter disappearing from the household accounts[16] and, perhaps, as servants originally appointed by the court, their leaving had marked the loss of his privileged status – of being gifted with serviced accommodation in or around the Hofburg. Louis and Franz had been replaced by April 1772, by a hugely important appointment – of someone who would remain, faithfully and honourably, in his service for the rest of his master's life. Dominique Tede was an Italian, a native of Parma.[17] Dominique and Ingen Housz appear to have developed, and quite quickly, one of those master/servant relationships that cemented itself by quick understanding and mutual respect. For what would be more than a quarter of a century they were virtually never again apart. Dominique was even to become, it appears, a guinea pig in some of his master's future clinical experiments. All the good fortune in this personal circumstance may, however, have arisen by fluke. It suited Ingen Housz's immediate purpose to employ a native Italian speaker for he now knew that he was to return to Italy.[18]

Leopold, Grand Duke of Tuscany, and his wife, Maria Luisa, now had five children and number six was imminent. Ingen Housz was asked to go to Florence again in order to inoculate two of them – Francis, heir to the Habsburg throne after his father, now four months beyond his fourth birthday and his sister, the two-year-old Archduchess Maria Anna.[19] He was also asked to inoculate other individuals at the Tuscan court.[20] Josepha Deblands and Elizabeth Francois, two young women who were the nursemaids of Maria Anna, and twelve others, all pre-teen youngsters who were, presumably, courtiers' children. He may well have been asked to go to Tuscany immediately and inoculate before the extreme heat of summer, as he had done three years before, but illness clearly prevented him. By 18 April 1772 Ingen Housz had been suffering pain sufficient to resort to desperate measures; 'I have put upon it today a blister'. This description of his predicament comes from a letter that he struggled to write whilst supine in bed.[21] It was to a new friend in London; William Hunter, royal physician and gynaecologist. Hunter had commissioned Ingen Housz to play scout and to obtain for him, in central Europe, gem stones, mineral samples and other unique objects for his growing private 'cabinet' of natural artefacts at his home in Windmill Street, Westminster. Ingen Housz had not needed to exert himself too much, not that he was able, having the facility of being a close friend of Nikolaus Jacquin. It would have been Jacquin, almost certainly, who put him in touch with a priest/philosopher, called Francois Arnold,[22] who had a sizeable private collection of 'mineralia' – ores, crystals and fossils – which Jacquin approved as authentic and valuable.[23] Arnold, in financial straits, was anxious to sell and Ingen Housz, having been given £100 sterling, (£6,400) to spend on his behalf, by Hunter, made an offer of

Dr. William Hunter.
Courtesy of the
Welcome Library,
London.

100 ducats (£2,600). Curiously, he then wrote to Hunter that the collection would cost him 150 ducats. Perhaps he was including the costs of sending everything to England but, equally, he may have been in a bemused state, trying to organise 'business' from his sickbed while pain and drugs broke up his attention span. Perhaps he had resorted to laudanum, for he was certainly aware that he was not at his efficient best.[24] 'My present situation will, I hope, excuse this ill-wrote letter, which I had much trouble to finish as it is.' Nonetheless, the 'ill-wrote' details must have been to Hunter's liking for by 15 July a recuperating Ingen Housz had clinched the deal and was supervising the careful stowing of the collection into a packing case and arranging its transfer to London, via Amsterdam.[25] Unfortunately, despite the care taken, the destination markings were rubbed off the box during its long journey and it mouldered at Nijmegen for over six months. Pestered by Ingen Housz, the Vienna post office eventually traced its resting place.

When it did, finally, arrive in London it was a pleasant surprise for Hunter.[26] Subsequently Ingen Housz always insisted that addresses on crates put in the post were etched into the wood.[27]

Ingen Housz was mobile from 25 May[28] – 'today for the first time I have come out again in the sun although I am still limping' – and would have been relieved to find his physical problems remitting. He was able to visit Rudolf Gräffer, who ran 'the most visited bookshop in Vienna' and who still had Hoogeveen's *Particles* on sale. More copies had been sold and there was some further profit to send to his old teacher. 'In order to make up the deficit of your money chest, which may be covered with cobwebs, I am sending you a bill of exchange for 100 ducats' (£2,600).[29] This was witty and cheerful enough but much of the letter reveals how Ingen Housz was dejected again. His dislike of the Viennese, at least of those at court, was now spiced by a growing aversion to Vienna itself.[30]

> Because I am hardly used to the stormy and rather rough climate of Austria I cannot endure these regions unscathed . . . either I shall be doomed to die prematurely or I must leave the country in search of a climate more fitting to my constitution.

This was hardly the optimal frame of mind for someone now having to cope with the introspection of the funeral of a colleague and friend.

During the first days of June it was clear that van Swieten was terminally ill. His agonising struggle with gangrene was coming to its inevitable end and, bar the distressed Empress herself, '*toute la cour*' attended the administration of the extreme unction.[31] Van Swieten's wife, pitifully exhausted after nearly two months of shared suffering and sleeplessness, was persuaded to fall back to the family apartment in the city and her husband died on the 18th. The great man was buried, two days later, in the *Augustinerkirche*, the 14th century Gothic edifice that served as parish church for the Hofburg. His corpse was to lie only yards from the famous library he had founded. As the cortège passed the palace Maria Theresa watched from a balcony, paying her last respects until the hearse was out of sight. Ingen Housz was, we imagine, among the large throng of mourners: the whole of the university assembling in honour. Ingen Housz's personal wreath was placed among the many others.[32] It had been during van Swieten's sad decline, on Friday 5 June, that an affected Ingen Housz had written to Pringle, the floundering mood of his lost script revealed by the itemised response.[33] However, this sympathetic reply didn't arrive for six months; Sir

John had gone to Scotland. But it didn't seem to matter; once again Ingen Housz appears to have laid his ghosts by addressing them with quill and ink and sealing them up in letters to close friends. As four years before, the ploy seemed to give him courage and steady his determination. By the end of July he had rallied, had been issued with a new Imperial passport, had organised the refurbishment of a suitable travelling coach, no doubt checking the padding very meticulously, and was ready to leave Vienna for Italy once again. Both Antoni and Dominique travelled with him[34] but this time he was not accompanied by Alexandre Laugier[35] or any other medical colleague. On sound summer roads the party made good time to the Tyrol and crossed to the south of the Alps. By 21 August they were in Como,[36] a day's carriage ride north of Milan. Whether Ingen Housz had known that he should expect to find Grand Duke Leopold there during August, or it was by coincidence, is not clear. But the ever-diligent doctor, acting on behalf of Jacquin, took the opportunity to sell further sample pages of *Hortus Botanicus Vindobonensis*. And on this trip he also carried sample pages of Jacquin's next publication, his 1773 *Flora Austriacae (Flora of Austria)*, again heavily illustrated in colour.[37] Count Firmian, governor of Lombardy, was one customer and Ingen Housz might have sold another copy of the *Hortus*, he reported ruefully to Jacquin, but the potential buyer had noticed that page 75 was missing: perhaps it could be posted on, 'well rolled up in a tube?'[38]

Como was left behind on 22 August[39] and Ingen Housz arrived at Florence, presumably in convoy with the Grand Duke and his dusty entourage, eight days later.[40] There can be no doubt that he would have been looking forward to meeting Felice Fontana again. And it was probably Fontana who, again, using his local knowledge, helped Ingen Housz to track down a suitable donor case of smallpox from the Florentine population. By 8 September they had managed to find a source of smallpox matter: 'a girl of six who had the natural but the benign small pox'[41] and Ingen Housz had examined all the candidates for inoculation. He was therefore ready to inoculate the royal children and the other patients, once they had all been quarantined at the Palazzo Pitti. Uniquely, a very detailed case history survives in the Austrian National Library.[42] In Ingen Housz's own hand and in French, it details, day by day, the clinical progress of his most important patient, the four-year-old Archduke Francis. Ironically, it records Ingen Housz's one royal inoculation that went wrong, almost fatally. Medicine is so much easier in retrospect but in real time Ingen Housz allowed an impatience to proceed to dominate his better judgement from the very outset. At four o'clock on the afternoon of Wednesday 9 September 1772:

The inoculation was made by a small prick on the outer part of the midpoint of each arm between the shoulder and the elbow. His royal highness had had, for two days, a slight head cold and had all the inconvenience of a watery nasal discharge, but he is a strong subject, and in this case the understanding is that the malaise will always be finished in 24 hours and certainly within two days. And so, for this reason, it was decided not to postpone the operation.

Over the next 48 hours Francis, though still troubled by his cold, was well enough. However, by late evening of the 11th he had a 'high temperature' and a 'fast pulse'. Ingen Housz sat with his patient until four in the morning, until he was convinced that the signs had abated. However, the little boy spiked a fever again during the evening of the 12th although he had been lively in the day and had eaten normally. By day four the inoculation sites were becoming inflamed as expected. But intermittent fever and racing pulse then became more pronounced than was usual until, on the 16th, a week after inoculation, Francis also began to cough, just as the first pocks appeared. Next day there were many more pocks than Ingen Housz liked and, worryingly, the child was drowsy and complaining of back pain. The following morning he was very floppy, 'had a strong desire to sleep,' and was uninterested in food. By now he was not the only person in the palace who had lost his appetite. All this was far from usual and very worrying. On Sunday the 20th Francis was stronger despite a very restless night and no doubt many prayers were said. But more and more pocks appeared during that sad Sabbath, particularly on the boy's face.

Ingen Housz must have been somewhat relieved to be able to record that, by the evening of the next day, the 21st, there had been no new facial pocks for some hours and that the disappointingly large number already present had remained discrete. But by the 23rd there were new, unexpected, symptoms. Ingen Housz must now have sensed what he had always dreaded – that one of his royal patients was harbouring some serious form of complication. The little boy was very irritable from five o'clock that morning, burning hot and inconsolable. His nose was still blocked, his face and hands increasingly swollen and his florid pocks all suppurating painfully. On the 24th, overnight and into the 25th a crisis developed. Ingen Housz must have been frantic with regret and anxiety. In wondering how his friend Dimsdale might have treated the boy differently or coped with this emergency, he must have shuddered as it brought back a vivid memory of George Hodges in that snowed-in cottage in Berkhamsted. Then things took an unexpected turn.[43]

On getting up towards eight o'clock, (Archduke Francis) was more feeble than in the preceding days, he complained of a pain in the right shoulder and was unable to move his head or move his arm without difficulty. The skin of the shoulder was found to be tender, raised and inflamed . . . In the evening the pain and the fever were as at midday. The redness and the swelling of the skin over the shoulder very great. It all suggested that an abscess would develop. A soothing poultice was applied.

The journal ends abruptly at this point. Perhaps part was lost but we can infer that although Francis survived, his recovery took many weeks for Ingen Housz wrote to Pringle again on 20 October,[44] imploring a clinical opinion on best further management.

Ingen Housz's other inoculation patients came through the process without difficulties, including little Archduchess Maria Anna. But even in this group, the remainder of the cohort, the treatment was not wholly successful. The two nursemaids of the Archduchess failed to respond to puncture leaving Ingen Housz to advise in the face of a dilemma. Either his technique had been remiss or both women had, perhaps at a very young age and outside their awareness, suffered smallpox naturally and free of any telltale scars. In whatever case, the niggling uncertainty of whether the Grand Duke's household was now protected or not, can only have added to Ingen Housz's nightmarish experience, furthered his exhaustion, deepened his disappointment and loneliness. We have no evidence of the response of the Grand Duke and Duchess as parents and perhaps the silence is significant. Ingen Housz's popularity likely plummeted. On this occasion there appear to have been no gifts presented and no warm personal letters arriving from the Empress; in fact the very reverse. It may be that the very detailed and forensic report on the knotty clinical progress made by poor young Francis[45] was submitted to the Hofburg by Imperial ordinance, a closure on an unfortunate and very difficult near-tragedy.

Autumn was nearly over before Archduke Francis was fully recovered. Ingen Housz, his back pains still a fresh memory trace, had no intention of undertaking another long journey until the spring. 'I shudder to think of travel in winter, and of the inconveniences and dangers that I have experienced.'[46] So the royal inoculator remained in Italy, at what threatened to be something of a dismal loose end, between November 1772 and March 1773. After what could only be seen, by supporters, let alone antagonists, as a bungled inoculation of the young Archduke, Ingen Housz's future seemed uncertain. He had no imminent commitments. He had fought off waves of depression during early summer, eventually buoyed up by the euphoria of his physical recovery and stimulated by the challenge of having to perform yet more royal inoculations. But he had come close to disaster and the future was a fug. For the present, he had no material wants, indeed had now amassed a fortune. Not that money, of itself, brought happiness, a point given precedence by Pringle whose long-expected return correspondence (he had been away in Scotland) eventually arrived at Florence around the end of November.[47]

> I am sorry to see by your mournful sheen in your last (of 20 October), that instead of being happy . . . you are less satisfied with your lot than ever . . . an easy fortune, such as you now enjoy would have made most people happy, if human nature were capable of being so.

In fact Pringle's long letter, obviously constructed with much care and deliberation, went on to show much greater sensitivity than this. It contained a whole banquet of thought for Ingen Housz including a consensus 'of all your friends here' that Ingen Housz ought really to arrange his return to London forthwith.[48]

> . . . (having) never been happy since you first left England, I should therefore propose two schemes, of which you may take your choice. Either stay abroad at Vienna and abstain from all practice whatever, excepting inoculating any of the two (royal) families; or pluck up a resolution to tell the Empress that you find your constitution does not bear the air of Austria, as being too dry, or whatever other quality you should be pleased to give it, & that you find you can enjoy no health unless in your own country, or in England. If you choose the first plan, you will nevertheless have it in your power to make us a visit now and then; but I can only see that tho' this may be a relief for the time, your sufferings are likely to recur & in a double portion, after your return to your confinement. If you ask leave for the greater retreat, you may still show your readiness to obey in the quickest manner any summons Her Imperial Majesty may send you, when wanted for the service of any of Her grandchildren, in the way of inoculation. If you take any other measures than these . . . I shall have the mortification to think that we shall never meet again or see you here; you will run the risk of falling into the malum hypochondriacum (depression) not to be terminated but with your life. . . .

These were shock tactics of a sort and clearly an attempt to provoke Ingen Housz into taking one firm decision or another. To make amends, Pringle tried to bolster his maudlin protégé by pointing out that clinical calamities will always happen, and to the best practitioners. He would not have known himself, he confessed, how best to deal with the moribund little prince and even Dimsdale had had a disaster. 'Your friend the Baron has lately had dying under his hands the son of a family of distinction. I don't know how he bears it . . .' It was a reminder, verbatim, if one were needed, that eighteenth-century doctors offering inoculation were certainly embarking on a new kind of doctor/patient relationship. Although the practice was nearly always beneficial, should it go wrong there was nowhere for the practitioner to hide. The circumstances could be construed with brutal objectivity – that a perfectly fit individual, on a promise, had been subjected to an intervention that had, finally, taken away their looks, their health or even their life. This was fundamentally different, philosophically, to a skilled but humble doctor doing his level best in contesting a power of another dimension. Patients saw natural illness as a spontaneous predicament in which a physician's successes were to be applauded but failures expected and tolerated.

Pringle's next ploy was a masterstroke. Here one sees why, in the very week when Ingen Housz was reading the letter at Florence, his patron was being elected as the next President of the Royal Society despite what were generally seen, by some fellows, as unconventional background and experience.[49] He set Ingen Housz an entirely different set of challenges.[50]

> Mr. Walsh, a member of the R. Soc., has been making some experiments upon the fish torpedo at Rochelles by which is confirmed the late opinion concerning it viz. that its strange quality in giving a shock is by electrical power. Pray might you not amuse yourself for some days by repeating the same expts. at Leghorn; or somewhere on the Tuscan coast, where, I presume you will find the same fish? It certainly abounds at Naples.

Here was an ideal distraction, a trigger for sanative activity. Pringle was setting up a physical and intellectual challenge that would take the royal physician away from the introspection and morbid atmosphere of the Tuscan court and Ingen Housz took the bait. We should assume that he took Dominique with him for it was going to be necessary to ask the help of the local fishermen to locate specimens. He also invited along a new medical friend that he had met at Florence – Doctor Alexander Drummond.[51]

IT CAN BE difficult, in the twenty first century, to conceive a torpedo as anything but an underwater weapon of war. But a fish that can deliver a considerable shock – often fatal to its natural prey – was the obvious inspiration for the marine armaments inventor looking for a name. *Torpedo marmorata*, the European or marbled electric ray, sometimes called the 'cramp fish', is common in the Mediterranean and often found in Atlantic coastal waters off France and southern Britain. Up to two feet long and banjo-shaped, its dorsal marbling makes ideal camouflage for its natural habitat –

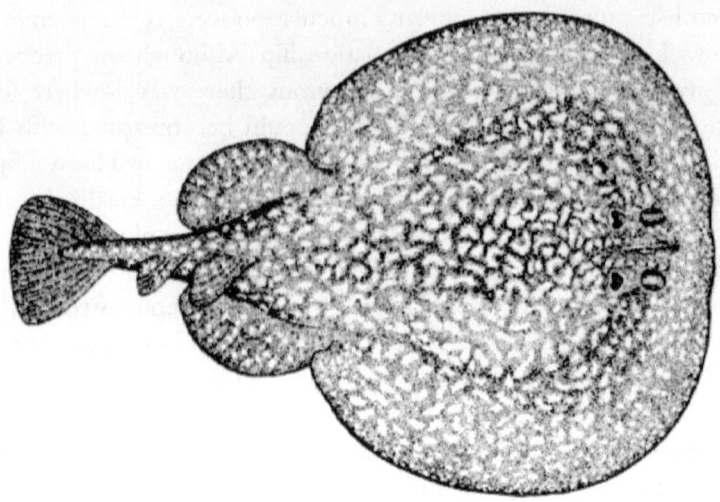

The European Torpedo fish.

the seabed a fathom or two down. It catches other bottom-living fish by stunning them with an electric shock delivered by its pectoral fins. This was not understood 250 years ago and the mechanism involved was a question of some intrigue among natural philosophers. The first formal reports suggesting that the numbing sensation, known to Mediterranean fisherman since antiquity, came from uniquely specialised tissue in the fish – the *musculi falcati* – had appeared in the early eighteenth century. The consensus, then, was that the fish secreted a numbing poison and this remained the accepted dogma until the early 1770s.[52] The invention of the Leyden jar and the widespread novel experience of electrical shocks gave a fresh insight and the toxin theory began to be questioned. John Walsh had returned, with a fortune, from India in 1759 having been secretary to Lord Clive. He was able to indulge his scientific interests and, in the summer of 1772, had travelled to the French Atlantic coast, to the Île de Ré.[53] The torpedo fish was common in the offshore waters there and he was able to experiment on freshly caught living specimens. He showed that the 'shocks' they delivered were of an 'electrical' nature for they were consistently transmitted through materials known to conduct electricity such as metal and water but never through insulating materials such as dry glass or sealing wax. He wrote, excitedly, to Benjamin Franklin on 12 July. Back on the mainland he demonstrated his findings to members of the Academy of La Rochelle invited to his lodgings on 22 July and sent more details to Franklin, on 24 August 1772. Another letter, from Monsieur Seignette, town mayor and secretary of their academy, corroborated the observations made. This correspondence was then used as the basis of a paper read to the Royal Society at their meeting on Thursday 1 July 1773.[54] Pringle, shown the original letters in early autumn 1772, had the perspicacity to see that this was an extremely significant discovery. Franklin had earned his justifiable fame by showing that lightning was atmospheric electricity and, now, here was evidence of electricity generated within living tissues. The force was turning out to be universal and fundamentally important and Walsh's observations would earn him the Copley medal of the Society for 1773.[55] Pringle knew that the topic would be irresistible to his protégé, Ingen Housz, and take him out of himself: the further verification of findings was probably superfluous.

Ingen Housz and Alexander Drummond travelled to Livorgno[56] (Leghorn) where they hired a fishing boat having a crew of eighteen and directed that they should sail out to where the torpedo fish were known to be common. It was on New Year's Day 1773 that the sailors took them some twenty miles from shore and soon proved their angling knowledge to be sound by netting five specimens, all more or less a foot in length. Ingen Housz must have generated some raised eyebrows by bringing an electrical machine up on deck and proceeding to charge up a Leyden jar. The sailors soon saw the point, however, when they had to agree that the sensation they felt when touching the charged jar was exactly the familiar one they often got from handling the caught torpedoes. The day's catch was put in a tub of seawater and Ingen Housz rolled up his sleeves.[57]

> I took one of the torpedos in my hand so that my thumbs pressed gently the upper side . . . of the side of the head whilst my forefingers pressed the opposite side. About one or two minutes after, I felt a sudden trembling in my thumbs. After some seconds more the same trembling was felt again . . . the same sensation as if a great number of very small electrical bottles were discharged through my hand very quickly one after the other. The fish occasioned the shock, or trembling, as well out of the water as in it. Sometimes it was very weak; at other times so strong that I was very near being obliged to quit my hold of the animal. (However) having insulated myself on an electrical stand, and keeping the torpedo in my hand, in the manner abovementioned, I gave not the least sign of being electrified, whether I received a shock from the fish or not.

Sooner or later every man who goes fishing has to collect up his tackle and return from bank or shore. Ingen Housz knew, by messages from Vienna, that many of the nobility had kept their faith in him as an inoculator despite negative feedback from Florence and were anxiously waiting for him to return and treat their children.⁵⁸ Not totally happy, though, with his torpedo experiments at Livorgno, he organised a departure from Italy that took him, first, to Venice⁵⁹ to make more observations that he could better report to Pringle. He had even toyed with the idea of going down to Naples and probably regretted not doing so for it was still very wintry around the Venice lagoon in February and not a single torpedo was brought ashore while he was there.⁵⁵ He also found that he had a disappointing 'catch' when he finally arrived back at Vienna, via Milan and the Brenner Pass⁶⁰ in late March 1773.⁶¹

> I found a great deal of the children of the Nobility inoculated at my arrival, who had waited for me a long while in vain. The natural disease, having made some havoc in some families of distinction, had roused the spirits.⁶³

The actual number of Viennese children who were still presented to Ingen Housz for inoculation that spring is not recorded but it could not have been very many. He was relieved: more than that, he should have been pleased. The fact that other doctors were at last inoculating widely, effectively and safely was a testament to his original mission although de Haen was still a looming presence. He had time, therefore, to sit down and send Pringle a detailed report⁶⁴ on his experiences with the torpedo fish, no doubt taking the opportunity to congratulate his great friend and patron on his election as twentieth President of the Royal Society.⁶⁵ There was also profound pleasure in making an important scientific contribution. It was Ingen Housz's first, even if it was not original work, and Pringle incorporated his protégé's findings into his 1774 Annual Discourse to the Royal Society that was dedicated to the recent discoveries on the torpedo fish. '. . .it is with pleasure that I can join the testimony of our learned and candid brother, Dr. Ingenhousz . . .'⁶⁶ Despite his frustration at Venice, Ingen Housz's return to electrical experiments seems to have triggered a seismic shift in his drive and there was a new enthusiasm for science. This had to wait, however, for around 22 May he left Vienna for Prague,⁶⁷ summoned by Count Nostitz-Rieneck, Czech nobleman and patron of the arts,⁶⁸ to inoculate his children. Ingen Housz arrived a week later: it was just over three years since his last descent on the capital of Bohemia and he was there just long enough to visit a few friends between his medical commitments.⁶⁹ The inoculations completed, we assume safely, he was back at Vienna in July.

INGEN HOUSZ'S LATER summer agenda for 1773 was dominated by personal and domestic decision-making. Although now perfectly fit in physical terms he was still sometimes at a low ebb emotionally. A letter to Hoogeveen in August was dominated by downbeat introspection and homesickness. He was suffering, he said,[70] from

> fits of depression as a result of frustrating cares bound up with a type of life, with its rigid manners, that is alien to the customs of our mother country. Experience shows me more and more how much the character of the Germans lags behind the sincerity of the Batavians . .

He seems to have been resigned to remaining as an isolated figure at Vienna, finding it difficult to make new friends, particularly at court.[71]

> I have tried in vain to strike up mutually sincere friendships with courtiers . . .
> I have found out how men held in high esteem by Princes feign friendship under a guise of sincerity but that they are actually mendacious and sly, capable of such intense malice and cunning that they are scheming the greatest humiliation and the worst disaster for a 'friend' the moment they embrace him.

Although he took the chance to take stock of all his finances, pay debts and settle with his servants,[72] his future was still enigmatic and he felt himself to be at a loose end.

> I do not know where in the world I will spend the rest of my life. I have hardly any idea of what is in store for me and even less can I surmise what the royal family think of me. I believed them to be well disposed toward me as long as I kept the Imperial Court in good health.

It was around this time, however, that he was summoned to the Schönbrunn for a private audience with Maria Theresa. He had reasons to dread the meeting having nearly killed one of her grandchildren, indeed the third in line to the Habsburg throne. Somehow he knew that he would never again be asked to inoculate any member of the royal family and, in the event, his intuition was correct. The Empress was far from recriminatory, however. As a reward for all his agonising struggles at Florence, she presented him with an extremely valuable jewel of diamond-encrusted silver that encased ivory reliefs. It has come to be known as the 'Maria-Theresa Brooch'.[73] The three small tablets of ivory, joined together in the silver frames, had been exquisitely carved into so-called 'micro-pictures' surmounting cobalt blue backgrounds. They were considered to be the most delicate sculptured pictures in the world. But the proposal she then made was an even more astounding token of her faith in Ingen Housz. He was asked to succeed the late van Swieten as senior body physician and head of the Vienna University.[74] It is amazing that Ingen Housz found the courage to turn down these proposals but his decision was based on a steely resolve. Somehow he knew that the cold insouciance he had met among the Viennese would only be worse if he was in power over them. He would also have had in mind that such high rank opened the floodgates to people begging favours and expecting him to lobby on their behalf. He had some experience of this sort of activity and had found it most unwelcome. He was already in the middle of some rather annoying correspondence with Father John Needham, director of the Brussels Literary Academy, who was trying to ensure the advance of a protégé at the court there.[75] And there had already been very uncomfortable friction between him and his best friend, Deckers, who had also expected him to pull strings on behalf of an acquaintance,[76] a request Ingen Housz had felt obliged to rebut, on the advice of van Swieten. That Ingen Housz declined the Empress's offer was a decision he never regretted. As he wrote to a friend later in life.[77]

> It was within my hands to be the successor to Baron van Swieten and to have, consequently, the ultimate say over all the universities and scientists of a mighty and widespread Empire. The Empress not only offered me a better income but also noble titles and public decorations. I turned all this down saying that I didn't wish for anything more than the continuation of what I had already received. From about 20 years experience I had found that I was able to relax and live far more happily than I would have when tempted by ambition and conceit . . . in this way I could earn a certain reputation for myself by doing useful, and at the same time agreeable, things without inducing envy in others.

It says much for the generosity of spirit of Maria Theresa that she did not take umbrage. Perhaps she was grateful that her royal physician had the grit to be so forthright and honest, knowing that a man who is unhappy in a post is unlikely to be a success. Ingen Housz was granted his request; that he simply remain one of the team of royal physicians and privy advisers. In fact, the Empress looked kindly on these modest ambitions and actually minimised his commitments without any amendments to Ingen Housz's very generous 'pension'. 'The Empress Queen promised me never to change anything in my situation, and declared me quite a free man. Also that I could reasonably expect to get leave to do what I will.'[78]

Ingen Housz's future may now have been like looking into a void but it was not his nature to be idle. Social life and financial dealings aside, late 1773, the closing months of his 42nd year, were to be the point in his life from which 'natural philosophy' took centre stage in his career. He was bursting with ideas and his apartment was soon part laboratory. Experiments seem to have occupied him wholly and contentedly. He certainly had few distractions. With no regular responsibilities at court and with two live-in manservants, a cook, a kitchen maid and a housemaid[79] he had the invaluable luxury of undisturbed time. We can read between the lines that, in having new livery made for his servants,[80] he had decided to stay in Vienna, at least for the immediate future. And if he had intended to remove to England or go home to the Netherlands he probably would not have taken the trouble to pack and send several boxes of gifts for his brother and family at Breda,[81] or write to Deckers pleading that they end the estrangement and restore their correspondence.[82] Indeed, having come to terms with new circumstances he even turned down the offer, made by Pringle in the spring of 1774, that he consider nomination to the expected vacancy of secretary to the Royal Society.[83] It was the second surprising shrinkage from the power and honours of a high-ranking position within nine months. It allowed him, though, to withdraw into a world where he, only, was in charge of his own drive, able to pose philosophical questions to himself and dedicate the necessary time to answering them without intrigue or interference. He took up with electricity again, writing to Franklin[84] that he had constructed a new type of disc electrical machine where velvet was rubbed against fur. And in making himself at home in Vienna he began to find himself hosting visitors to the city. The first appears to have been Dr. Francis Milman,[85] a young Oxford-trained physician who was a close personal friend of George Baker and who would rise through the profession to become, in 1811, president of the Royal College of Physicians of London.

Some time in 1774 Ingen Housz was asked, since he was still in Vienna, to wait on the Emperor.[86] Joseph II had become aware of the dangers of lightning strikes on gunpowder magazines. He would have heard of the disaster at Brescia and Ingen Housz would have been able to provide him with a first-hand account. He may also have heard, again from Ingen Housz himself perhaps, that members of London's Royal Society were advising the English government on installing lightning conductors on their arsenals.[87] Ingen Housz was asked to visit the Austrian army's munitions stores and advise similarly. There were few other matters taking Ingen Housz away from his laboratory bench for long, however. To vary the agenda, there were experiments on air. Once it had become clear that atmospheric air was not one substance, however counter-intuitive that was, philosophers began to wonder whether there might be many natural phenomena, diseases among them, that resulted from the constituents of the air varying relative to one another. Air analysis came into vogue. The first real possibility of investigating what was now known to be a mixture followed one of the discoveries of Joseph Priestley as published in 1772.[88] Sometime around 1770 he had begun to examine the properties of the gas that fizzed up when metal filings were dropped into strong nitric acid. He was particularly struck when he observed that, when mixed with ordinary air, this gas became very reactive and reddened and that there was a distinct pressure drop – the water level would rise if the mixing gases were trapped in a glass bell-jar over a water bath. Priestley also spotted something of great significance – that the better the 'salubrity' of the 'test' air (the purer its quality), the stronger the reaction of the gases. He might even have demonstrated this to Ingen Housz, Franklin and company in May 1771 when they called on him at Leeds. Developing this qualitative observation into a quantitative appliance – devising a reliable measuring instrument of air quality seemed to hold great potential and many an excited inventor took up the challenge. They were all looking for an instrument that would be as reliable and user-friendly, not to say as important, as a thermometer or a barometer. Fame and fortune tantalised the entrepreneurs and scientific instrument retailers made space in their shops. It was to be the story of the cactus flower, however. The vogue for testing the atmosphere was very fleeting. No single instrument came to dominate the market, no accepted standards were devised and as the massive liability to error of all the machines came to light, sales plummeted. Fundamentals were learned the hard way.

Gases, unlike liquids, were difficult to manipulate; they reacted differently if mixed vigorously as opposed to being left to mingle passively. There were massive pressure, volume and temperature considerations and no easy means of knowing when all possible interactions had occurred. And all this was with invisible substances liable to leak into the general atmosphere on the slightest bungle by the operator. Nonetheless, it was thought that it would only be a matter of time before there was proof that air was of significantly different quality in different places and circumstances. This gelled with the idea that health was a reflection of atmospheric purity and that some diseases, at least, resulted from breathing bad air – *miasma*. It was no coincidence that air analysis was a strongly supported research theme in England while Sir John Pringle was President of the Royal Society. His main claim to fame had been, after all, his books on the diseases of soldiers and prisoners alike – all conceptually based on epidemics resulting from fouled air. Air quality even became a fashionable topic of small talk. Franklin, very close to Pringle, even debated with another friend whether it was healthier to walk in the open air as opposed to riding in a crowded carriage.[89] And showing his usual wit, he even lampooned Priestley who was, he said, 'apt to give himself Airs'.[90]

Another country where there was now very active research into air analysis was Italy, specifically the northern states governed by the Habsburgs. Among the inventors was Marsilio Landriani, professor of physics at Milan. His particular machines never caught on but the name he gave them did – the *eudiometer* – difficult to translate, but basically an apparatus for measuring 'weather goodness' i.e. air quality. Meanwhile, Felice Fontana, at Florence, was designing his own eudiometers, work proceeding while Ingen Housz was with him in Tuscany in 1772. In fact, if Ingen Housz hadn't been party to the conception of eudiometry (revealed by Priestley in 1771) he was certainly present at the birth (with Fontana in 1772). It was all food for thought, especially for a disciple of John Pringle. Fontana's genius did eventually lead to some glass instruments that were reasonably useful; reliable enough to support, for a few critical years, some successful air investigations. And for those few years it became fashionable to go on 'eudiometric' tours. Philosophers took their instruments to widely differing locations to try to measure differences in air quality. They compared mountains with marshes, rural hamlets with city centres, beaches with heaths. Fontana himself, visiting London in the late 1770s, even compared air at street level with that swirling around the external stone gallery of the dome of St. Paul's Cathedral, 173 feet above the ground and, even, over 300 feet above the pavements below, on the 'golden' gallery near the very top.[91] Ingen Housz, too, took a eudiometer with him on a later sea voyage. The results of all the investigations proved to be conflicting, however, and the differences in air quality, even when they were vaguely consistent, were subtle at best. Eudiometry, as a supposed expertise, began to come under huge suspicion. Fontana's own sustained efforts to produce and promote a standard piece of equipment that was superior and dominant, failed. By the early 1780s there were as many different kinds of eudiometer as there were sceptics in their value. Ingen Housz became an increasingly important player in these events and was, perhaps, the only real beneficiary of the instrument during its very short popularity; the one instance where eudiometric assay was integral in a seminal scientific discovery.

In May 1775 Fontana published the results of all his efforts to design a practical eudiometer. The pamphlet,[92] printed in Florence, was no flimsy pot-boiler and must have been a long time in preparation. But as early as 1773, back at Vienna, Ingen Housz had convinced himself that all of Fontana's devices suffered from inherent flaws that were virtually fatal 'because in his (apparatus) the quantity of nitrous air forced into the vessel can't be ascertained exactly.'[93] Knowing when all possible chemical reaction had finished was also uncertain in Fontana's apparatus and the results were subject to such variation that the readings were virtually meaningless. In other words, among the first eudiometers invented, the early ones devised by Fontana had more potential than immediate utility.

Ingen Housz saw the need for improvements and, between late 1773 and early 1775, came up with ingenious modifications, designing not just one new eudiometer, but two. He had now lived in Vienna long enough to know where he could find an innovative glassblower, a foundry that could make him gas-tight brass taps, and suppliers of mercury, of pure iron filings, and of *aqua fortis* – concentrated nitric acid. The fundamentals of his designs might have been 'Fontanic' but the improvements he was to make were entirely his own. Vienna, too, was going to be a birthplace of types of eudiometer. The manual dexterity needed to assemble and manipulate gas-tight components effectively was also entirely his own asset, a driving skill that was highly complementary to his cerebral agility. Here was a further demonstration of the combination of intellectual grasp, logical rigour, systematic application and manipulatory finesse that had made him such a successful inoculator.

In the shadow of St. Stephen's Cathedral, from mid 1773, Ingen Housz served an apprenticeship to trial and error, and honed skills that would lead to great things. In the first of his own eudiometers he substituted a soft rubber bag, tightened around tubing at its neck by a drawstring, for one of Fontana's usual glass chambers. In it he put the air to be tested. Rubber was then a novelty: called *caoutchouc*, it came from central America, prepared from the sticky latex of several tropical trees (*cahucu* being Indian for weeping tree). This 'balloon' was then joined, via one arm of a glass T-tube, to a sealed glass vial, replacing the other chamber of Fontana, in which the nitrous air was produced. Even here Ingen Housz was inventive. The violence of the chemical reaction between the acid and the metal filings was held in abeyance, giving him time to seal all the joints, by wrapping up the filings in a paper parcel before immersing them in the acid. The third arm of the T-tube led to a long straight tube that was marked off in equal divisions and held vertically, its downward open end submerged in water. Until the last, crucial, moment, brass stopcocks isolated the three components. Then, as the acid finally ate through the paper, and the glass vial filled with the fumes of 'nitrous air' (nitric oxide), the valves between vial and rubber balloon were opened. The balloon was then squeezed repeatedly so that its air was propelled backwards and forwards along the tubing and mixed thoroughly with the 'nitrous air.' The usual volume and pressure drop could now be seen by the shrinking and puckering of the balloon. When this had stopped, taken to be a signal that the gases were no longer reacting, the stopcock on the side arm was opened and water was drawn up by the negative pressure within. The higher the water column came up the side arm, the more notches passed, the 'purer' must have been the test air (or, what we now know and couldn't have been known then, the higher the concentration of oxygen). Ingen Housz had the self-discipline to repeat the experiment many, many times and the probity to admit that the results were still unacceptably variable. An extra dose of ingenuity had still not produced a reliable instrument and Ingen Housz was honest enough to record his verdict.[94]

> By repeating several times the experiment in the same place, I found the rise of the water nearly the same, though not so exactly as I could have wished: the variation I ascribed partly to the elastic bottle not always being of the same firmness or elasticity, which it loses more or less by squeezing.

Disappointed but not exhausted he made a new approach to the problem and it was revolutionary enough to make it entirely his own. It was also pleasingly simple, if rather demanding on dexterity. He asked his glassblower for a long straight tube only an eighth of an inch in diameter and had brass rings glued on either end to protect his fingers from any sharp edges. 100 equal divisions were then etched along the length. He now had a long transparent pipe of such dimensions that a 'column of quicksilver might slide through the whole without dispersing itself, filling always the whole cavity.'[95] After some practice, he was then able to plunge one open end of the tube into the neck of a tilted vial where nitric oxide was being produced as per usual, all the time holding a short column of mercury just inside that open end by pressing a finger over the proximal end – by using what we now would call pipette technique. By lowering the pipe into a near horizontal position and intermittently releasing his finger he could let the mercury column advance slowly along the lumen of the pipe towards him and, in so doing, draw 'nitrous air' (nitric oxide) along the lumen of the pipe. Then, with the mercury column at the halfway mark he rapidly removed the tube from the vial and jammed a finger of his other hand over the far end, trapping the nitric oxide. He then turned the pipe, up one end and down the other and vice versa so that the mercury column flew from one end to the other so mixing the 'nitrous air' and the 'test air' very 'intimately'. One end of the tube was then lowered into a tub of mercury and the lower finger removed. The height of the rising column of mercury then represented the 'purity' of the air. Ingen Housz gave no verdict on the efficiency of this 'micro-technique' – but, sad to report, this appealing and highly portable method of gas analysis never caught on anyway. Perhaps it held no attraction for instrument makers: it certainly would have had minimal profit margins.

On Friday 3 November 1775 Ingen Housz wrote a long letter to Sir John Pringle that reported his thoughts and experiments with 'nitrous air' (nitric oxide). It was intended as a formal submission to the Royal Society but more than half the letter concerned 'experiments on platina'. Despite the diversity of the two topics, a full transcript was read to attending fellows on 15 February 1776 and published in the *Philosophical Transactions* of the Society.[96] Thursday 15 February 1776 was therefore a very important day for Ingen Housz. He was probably unaware of the actual event at the time, being so far away from London but it was the occasion of his first real contribution to the Royal Society and would be the first of many. From here onwards there was never a danger of Ingen Housz being accused, justly, of being, among all the members, just one of the passengers. It is no surprise that his article was welcomed and put before fellows at an early date. New designs for eudiometers were highly topical and certainly of deep interest to Pringle. The information on 'platina' would have been seen as a bonus. Then an exotic substance in Europe, it always caught the attention and William Watson had been prominent in stirring up interest.[97]

Platinum, as it is now called, had first been 'discovered' in South America by 16th century invaders from Spain and Portugal. Small quantities found their way back to Europe in the early 1740s, by which time other explorers of 'New Granada' had found evidence that some tribes of native Indians, seemingly extinct, had been able to work the substance into primitive jewellery. Since it appeared to be a metal, the question of its nature and classification was important. Was it a pure substance and, therefore, potentially the eighth known metal, or was it just a naturally occurring alloy? Many suspected it to be simply gold adulterated by iron. The time-honoured way to free ores from impurities and analyse their constituents was to use intense heat and run off the molten substance, several times if necessary. From the outset platinum proved to be unique – it was virtually impossible to achieve and maintain a temperature that would melt it: philosophers were going to have to be more inventive. William Lewis, a fellow of the Royal Society had been encouraged by Watson to investigate platinum when a small quantity had become available, fortuitously, in London in the early 1750s.[98] Lewis used it to very good effect, doing such rationally-organised and pleasing experiments that the work won him the 1754 Copley Medal of the Royal Society.[99] He continued making observations and published a well-received book, in 1763.[100] This was only a year before Ingen Housz had first arrived in London and his interest in platinum must surely have been inspired by William Watson between 1764 and 1768.

Ingen Housz may have been very interested in platinum from the mid 1760s but it was not something that could be easily obtained. So, when given a small quantity by Count Dietrichstein,[101] an equerry to the Emperor, Ingen Housz would have had pent-up ideas to test. From his description, that it was 'fine', we can assume that it was a reasonably pure sample, albeit in many small fragments. Contrary to some reports, Ingen Housz found that the 'flat, smooth and shiny bright' particles he had been given were attracted to one of his magnets.[102] The magnetic force was certainly weaker than with iron but definitely always present, even if to demonstrate it one sometimes had to overcome inertia by floating the tiniest fragments on the surface of some water. Actually, the weak magnetism was solely due to the presence of some contaminating iron, a fact that would become clear only in later years.

Ingen Housz was yet another investigator of platinum, however, who found it to be stubbornly heat resistant. When he generated extreme heat in a flame by using a blowpipe none of the smallest fragments, even, would melt. They did, however, lose their sheen and their magnetic sensitivity, some of the iron presumably having been burned off. With remarkable ingenuity, however, he did find a way of fusing his pieces of platinum and making a tiny sample of pure metal, making the paper to the Royal Society particularly valuable. He lined up many of the smallest fragments along the bore of one of the tiny glass tubes that he had used in his second eudiometer design, packing them close together and sealing them in with a pin/wax plug at either end. He then used a battery of three large Leyden jars, ready charged, to drive an electric current through the assembly. He could see, down his microscope, that the platinum pieces had fused together into an 'uninterrupted cylinder of glistening metal'.[103] Accessing this short length of platinum 'wire' proved to be most difficult, however. He had to resort to smashing the tube with a hammer and extricating the metal from amongst all the shards of broken glass. The platinum had significantly changed in texture and appearance but he had, at least, managed to melt a sample. As with his eudiometers, this was far from the last word on the subject but his originality and ingenious methodology alone convinced the Royal Society to approve the work. Ingen Housz's first real contribution to their annals was a promising start to a new career.

A VEXATION THAT Ingen Housz suffered, increasingly from late 1773, reverberated from London. It arose from his deep friendship with Benjamin Franklin and the fact that the American was deeply involved in affairs of state that were becoming intractably difficult. Being so remote, Ingen Housz was impotent to support his friend and was to write many pleading letters trying to find out exactly what was happening and why Franklin was becoming so unpopular in England. Except for a brief return to Philadelphia in the mid 1760s, Franklin had lived in England from 1757. In effect, he had migrated to London at the age of 51 and was now approaching 70. He had been 'politicking' for two decades. Pennsylvania, his real home, was just one of a number of isolated British colonies across the Atlantic. There was no such country as 'America' but Franklin had served, effectively, as its 'ambassador' in London. More properly he was 'agent' for Pennsylvania and, from 1770, also for Massachusetts. Although easily the most famous American colonist in England he had always been against their separation from the mother country. He had been a most diligent moderator in all the many 'family' squabbles. And it is clear that if Debbie, his wife, would have consented to leave Philadelphia, he would have settled permanently in London.[104] He valued the intellectual life of the city, particularly the stimulus of the Royal Society and, by the early 1770s, virtually all his close friends were English. It is not surprising that his countrymen of birth had come to view Franklin, until 1774, as having a suspect affiliation, especially when they remembered that his son, William, was governor – the King's personal representative – of the colony of New Jersey.

In 1772 Franklin did something for which his motives have long been debated. It would bring his precarious political handstand crashing down. Some half-dozen letters had somehow come into his possession.[105] They were from Hutchinson, the Massachusetts governor and his assistant and brother-in-law, Andrew Oliver, to contacts in England and dated between 1767 – 1769. The correspondence revealed, very convincingly, that Hutchinson and Oliver had been serving two masters at once and that their dominant allegiance was to the British throne, seeking to protect their careers and fortunes. Seeing their political significance, Franklin bundled them up and sent them to Thomas Cushing, leader of the Massachusetts Assembly, pleading that he read them in private, digest them confidentially and return them post-haste. There is a strong consensus that Franklin knew that he was tempting fate and, predictably, the gist of the correspondence rapidly surfaced in a pamphlet at Boston. Copies were soon crossing the Atlantic. A hailstorm of retribution blew up and Franklin was forced, after only a few months, to admit to his culpability when a more likely, but totally innocent, suspect was having to defend his reputation in duels and had already sustained one nasty injury. Franklin was instantly a very unpopular figure in London, the frostiness making his diplomatic function very uncomfortable indeed. He was still having to represent increasingly aggressive demands from the colonial assemblies to a very distrustful Lord Dartmouth, colonial secretary in Whitehall. Only some unresolved and volatile issues forced Franklin to stay, reluctantly, in England. He confessed to Ingen Housz, in September 1773, that

> I had purposed to return to America this last Summer but some Events in our Colony Affairs induc'd me to stay here another Winter. Sometime in May or June next I believe I shall leave England. May I hope to see you here once more.[106]

Later in the autumn, with several colonial demands still being chewed over by the Westminster politicians, there was an event in Massachusetts that was more symbolic than significant, but it heightened the tension between England and the colonies. Three ships of the East India Company were trying to dock in Boston harbour and unload their cargos of tea. This, thanks to yet more inept legislation in London, was a provocation too far for the Bostonians: East India cargoes were now being exempted from colonial import duties, the lower prices of the goods undercutting the businesses of other local merchants. In the stalemate, the Boston stevedores

The Boston 'Tea Party'.

left the fleet rocking at anchor and the crews rowed themselves ashore. On the night of 16 December 1773 men dressed as American Indians stole out to the ships and ten thousand pounds-worth of tea was tipped into the harbour during the so-called 'Boston Tea Party'. Six weeks later, in Whitehall, Franklin made his way to the Cockpit, the assembly room of the Privy Council.[107] Expecting to contribute to a debate on a Massachusetts petition for a more independent governor, he found himself harangued by Alexander Wedderburn, a notoriously belligerent solicitor-general, for the putative crimes of stealing and publishing the correspondence of high public officials; the letters of Hutchinson and Oliver.[108] Totally isolated and unprepared, Franklin accepted the offer of a three-week adjournment to recruit and brief a defence team. Unhappily, news of the 'Boston Tea Party' now arrived in London and, when he next entered the Cockpit, Franklin found that he was in a show trial. In reality all America was in the dock and he was a convenient scapegoat. Privy council meetings were usually routine, low-key, tedious and poorly attended. But this one, Franklin couldn't help noticing, had even attracted the Prime Minister and the Archbishop of Canterbury among other celebrities; and a heaving throng of the general public. Standing imperviously for two hours, the 68-year-old Franklin was verbally lacerated by Wedderburn. Overwhelmed and outclassed, his defence team muttered ineffectively and he felt himself fortunate to leave as a free man. Wedderburn's speech was soon circulating in print, an alarming document for Ingen Housz, who immediately wrote to London. Franklin replied reassuringly and his letter is probably one of the best records of his prevailing mood and attitude at this personally difficult time.[109]

> I am very sensible of your kindness in the concern you express on account of the late attack on my character before the Privy Council and in the papers. Be assured, my good friend, that I have done nothing unjustifiable, nothing but what is consistent with the man of honour and with my duty to my king and country, and this will soon be apparent to the public as it is now to all here who know me . . . You know that in England there is every day, in almost every paper, some abuse on public persons of all parties, the king himself does not always escape, and the populace who are used to it love to have a good character cut up now and then for their entertainment.

Franklin knew, though, that it was time to go home. Conflict was inevitable and it could well lead to a struggle for complete independence of the colonies; effectively a civil war. It must have been painful, if not ironically amusing, to learn that the ammunition at Purfleet, the very stores that he had helped to preserve against lightning, was being loaded onto ships and sent, with large numbers of troops under General Thomas Gage, to America. Even so, Franklin did not leave London for another year. A cabal of moderate peers had contacted him secretly. They involved him in a conciliatory, conflict-avoidance plan but this only fizzled out in mockery when it was unwrapped in the House of Lords early in 1775. So Franklin finally left London, for Philadelphia, on 20 March 1775; there was nothing more he could do to pull the British government back from the brink. He was still on the Atlantic when the opening shots of the American War of Independence were fired at the battles of Lexington and Concord on 18 April.[110] Ingen Housz, still remote from events, was again writing to Huck[111] and to Collard[112] begging them for news and still expecting that Franklin would 'accommodate matters between the colonies and England.' He was wrong and it must have come as a personal hammer blow to learn that his great friend was back in Philadelphia. This put him quite beyond reach. One of the main attractions for returning to live in London had evaporated. Fortunately, Ingen Housz had become more attuned to settling in Vienna by 1775. His apartment-cum-laboratory was proving productive, he had a stable and happy household and he was being treated with far less 'suspicion' by Viennese society. Strangely, he may have benefited from Maria Theresa's remarkable job offers despite his refusal. There were few secrets in Vienna and the royal admiration had boosted his personal prestige. But, more than that, his own surprising resolution, to eschew the actuality of power and influence, could even have made him more popular; the daring rebel is often more fêted than the dogged retainer.

DURING 1775 INGEN Housz had been making further investments in Eastern Europe, buying, on the advice of Nikolaus Jacquin, more shares in mines, in Saxony near Chemnitz and at Kapnick in Hungary.[113] There were good prospects of high dividends from the subterranean explorations at such sites as metals became increasingly important in the beginnings of the technological explosion that would come to characterise the next century. Ingen Housz had actually become very friendly with the whole of Jacquin's family and was a regular visitor to their home. The professor of botany and his wife now had three bubbly children. In 1774 Josef was eight years old, Gottfried was seven and daughter Franzisca five. Ingen Housz had also met Agatha, Nikolaus Jacquin's unmarried sister, who had moved to Vienna to live with her brother's family. But it was the children who would have, immediately, given Ingen Housz a surrogate family life, reminding him of all his nephews and nieces at Breda. He was certainly mindful of Louis's children. He had arranged annuities totalling 24,000 French Livres (£64,000), a sum in which they would all share on his death.[114] Even so, he still had some £2000 sterling (£127,000) that he could afford to invest in stocks in London.[115] He had also spent money having medallions cut and forged, this time with his name aboard and, indeed, a new head and shoulders portrait. This time they bore the facts, in Latin, that he was both privy counsellor and doctor to the royal family. Inoculation was not featured. Ten were sent to Louis at Breda and two to Deckers at 's-Hertogenbosch (see p. 242). There may have been an element of vanity but more significant, perhaps, was their symbolism – that his time as a royal inoculator was indeed completed.[116] Dr. Angelo Gatti, professor of medicine at Pisa and a keen inoculator, had just travelled from Paris to Florence to inoculate some more of Leopold's children.[117] In 1775 Ingen Housz also wrote, post-haste, to his brother when he heard that Marianne de Kanter, their childhood nurse and surrogate mother, was ill.[118] Louis was told to be sure to charge half the cost of treating her illness to Jan's account. Marianne's name fails to appear in any later correspondence and, sadly, she may have died that year.

In May 1775 Ingen Housz had been invited to join a royal cavalcade that was intending to visit Constantinople and then to sail across the eastern Mediterranean to Egypt. Ultimately, he never joined the group, possibly because he was ill with nausea and was not fit to travel.[119] His brief illness was a herald of future trouble, perhaps, but it was soon to be forgotten in a looming domestic imperative.

12
Mesmer & More

Philadelphia in the British colony of Pennsylvania. Monday June the 10th 1776. By two o'clock in the afternoon it's unbearably hot and stuffy in the assembly room of the Pennsylvania State House. It could be cooler but opening the windows admits swarms of summer insects. It could be cooler but the room is very overcrowded: none of the colonial delegates can afford to be elsewhere today. It could be cooler but the debate is aggressive, bordering on ill-tempered, the topic urgent, a decision vital. The motion – that 'Americans be absolved from allegiance to the British crown' is still not acceptable to some delegates: the representatives from New York, especially, are torn. However, John Hancock, president, on the dais by the empty fireplace, has watched how the number of delegates remaining opposed to full separation from England has dropped to what is now a minority. In the midday break other members of the Continental Congress had agreed with him that separation now held sway: that Ben Franklin had been right all along. There could be no going back after the blood loss of last summer. But no-one is sure if they should sit it out for unanimity or should there be a show of hands? Can a mere majority decide something so important? All this is spinning in his head when Hancock retakes the chair and calls the meeting to order for the afternoon session.

After another hour there's been no movement; plenty of words, many expressed with feeling, but they're no further on. The temperature has racked up a further five degrees – in both senses. Then Ben Franklin lifts himself out of his chair and motions to Hancock that he would like the floor. A silence descends. The other delegates acknowledge his wisdom. He is the oldest delegate and knows the ways of the Westminster politicians from painful personal experience.

'Is it not time, my friends, for us to see that we have the possibility, at least, of a decision to separate from England and to claim our freedom? In fact, is it not time that we should accept, in reality, that this has already happened – by the shedding of blood; blood of friends of ours that we should not forget? So can we not simply accept our situation and our common fate on this head, as foregone?

The New York delegates rapidly lean towards each other, and a loud and staccato whispering begins. John Jay moves to stand up, loudly scraping the Windsor armchair on the wooden floor. Franklin has stopped speaking but stays standing without looking round. Hancock flips his hand regally for the interrupters to be quiet and listen out the speech. They look very unhappy – everyone senses that they fear an ambush, that their moderation is about to be subsumed by a power play. It's some seconds before Jay slowly lowers himself into his seat and Franklin resumes.

'. . . as foregone. No. I cannot think that we should proceed in this way; we need to show that we are civilised men controlling our destiny. We need to move in a way that will be seen to be both honourable and esteemed by lawyer friends when we come to a time that we can all feel resolved. We must make a bid and reveal our hand. I dare to suggest that, even though we have not yet a decision amongst us, we should see that one may be close and should, therefore, have prepared, for it best to lie in readiness, a full declaration of our intent. Perhaps at this time, on such a hot and uncomfortable day, we could make some progress, at least, to this purpose, if such notion be in acceptance. Mr. President, I make so bold to move.'

There is huge relief in the room, almost an audible sigh. A decision of some sort can be taken and rescue them from the stalemate and the stifle. It happens so quickly that many can barely remember it, even a few hours later. The deadlock broken by the wily Franklin releases a common purpose – a 'declaration of independence' should be drafted by a sub-committee and debated, at a later date, by the Congress. The committee members are soon chosen and Thomas Jefferson, from among them, to be the draughtsman. As the delegates pick up their belongings and begin to leave the chamber, John Hancock comes down the room to thank Franklin for his intervention and make the obvious comment.

'Whatever happens, we must all hang together.'

Franklin may be hot and tired but his wit is still razor sharp.

'Yes, we must indeed all hang together, or most assuredly we shall all hang separately.'

BY 1774 THE ill-tempered debate on smallpox inoculation at Vienna appeared to have died down although de Haen was still a brooding presence. The determination of the Empress had forced the issue and its introduction had been a success. Ingen Housz was no longer the unwelcome intruder of the medical fraternity. Although he had turned down Maria Theresa's offer that he succeed van Swieten as her senior body physician, as the head of the medical faculty of the university, and as court librarian,[1] the very fact that he had been seen as her first choice for these responsibilities seemed to secure his social position. He sought only a quiet life, free of further controversy and certainly free of spiteful, competitive, professional backbiting; he hadn't left his practice in Breda in order to suffer these stresses higher up the medical hierarchy. He was very glad to find that he could enjoy his increasingly friendly integration into the professional life of the city. But there were always fractious disputes blowing up in intellectually claustrophobic Vienna. Ingen Housz would have been aware, late in 1774, of a growing and very public row between the professor of astronomy and a doctor popular with the aristocracy. It must have been to his horror that he saw this spat come in his direction.

The doctor in question was the 40-year-old Franz Anton Mesmer. Born in Swabia in 1734, Mesmer had come to Vienna to read for a law degree in 1759.[2] He had already studied philosophy and theology at a Bavarian university. By 1760, however, he had become a medical student following the 'modernised' courses pioneered by van Swieten and de Haen. Then nearing 30, Mesmer was older than most of his fellow students; the same age, indeed, as some of the younger teachers and he became a very close friend of a near-contemporary fellow Swabian, Anton Störck, already professor of pharmacology. Mesmer qualified medically in 1767 having successfully defended his seemingly esoteric and oblique dissertation – on the influence of the planets on human health.[3] The new doctor was now able to consider marriage and chose the daughter of an aristocratic physician. Maria Anna von Bosch was a 43-year-old widow with a teenage son. She had considerable wealth inherited from her father and from her first husband. The ostentatious wedding was at St. Stephen's Cathedral on 10 January 1768, the archbishop of Vienna being the celebrant and one of the witnesses was Störck. Mesmer was known to be the son of a gamekeeper and the Vienna gossip shop inevitably traded in the assumption that he had married for money. It is certainly true that his union gave him a large house in the fashionable Landstrasse district next to the Prater, the famous Viennese public park, and it was in these salubrious surroundings that he set up a medical practice.[4] The newly-weds moved in high social circles including, even, Kaunitz, the Chancellor; and the Mozart family were also among those who participated at their frequent musical evenings.

For some five to six years Mesmer successfully built a reputation based on conventional medical practice, acquiring a growing popularity among the Viennese 'upper crust'. Sometime, though, in 1774, a cousin of Frau Mesmer came to stay. Franzisca Oesterlin was in her twenties. She soon proved to be not so much a 'houseguest' as a guest who was 'housebound'. She suffered from a gamut of hysterical symptoms that climaxed in dramatic episodes provoked by over-breathing. She could bring on a wide variety of pains, hallucinations and even convulsions. Her illness had stubbornly resisted all conventional treatments until the repeated bleedings and blisterings had added profound physical weakness to her symptom-tree. Mesmer became fascinated by her clinical complexity and decided to try an experimental therapy that had just been conceived by Maximilian Hell, the Vienna professor of astronomy. The magnetic pull of a 'lodestone' had been known for many centuries and occasionally tested for therapeutic gain but, from 1750,[5] it had been possible to obtain, from England, much stronger magnets made of steel. With genuine motives – he himself was a sufferer – Hell had shown, to his own satisfaction at least, that by laying the new type of magnets along body axes, he could relieve the chronic pains of rheumatism. This was less bizarre than it now seems, perhaps, taking into account the then state of knowledge and that the professor of medicine himself, de Haen, was an active proponent of electrical treatments. If it was generally accepted that electricity involved the movements of a helpful universal 'fluid' there was no reason to discredit a parallel theory involving a magnetic 'fluid' that might also influence body function.

Possibly Hell would have forgiven other doctors for stealing his ideas but Mesmer, at his Landstrasse clinic, rapidly took 'magnetic medicine' into territory having no theoretical basis. Having tried magnets on his wife's cousin with great 'success', he somehow jumped to the conclusion that, for him, placing magnets on the patient was actually superfluous. He began to claim that he himself could radiate magnetic forces into his patients. Largely on the basis, then, of his experience with just one, most unusual, patient, he began to advertise his powers of 'animal magnetism.'[6] He also claimed that he could magnetise inanimate objects such as cloth or wood, even through the intervening walls separating rooms. Sensing the need for authoritative approval, he invited his friend Störck to his clinic to be a witness to his new therapy. Despite their close personal relationship, or perhaps because of it, Störck refused to become involved. He must have sensed an embryonic controversy but it did not stop him suggesting that Mesmer might like to invite Ingen Housz.

Ingen Housz was no virgin with magnets. One of his friends at the Royal Society in London – and one of the sponsors for his fellowship election – had been Gowin Knight . As one of the most successful makers of steel magnets,[7] Knight's work was known to Ingen Housz, who certainly already owned some powerful magnets.[8] There is also some evidence that Ingen Housz, too, was experimenting to see if magnets could treat disease and perhaps even working jointly with Hell.[9] According to the same source,

Dr. Anton Mesmer.
Courtesy of the
Wellcome Library,
London.

Ingen Housz rapidly lost any enthusiasm for magnetism as a therapy but his known scepticism made him an even better witness for Mesmer's purpose: if Ingen Housz could be converted to the cause, it could be a breakthrough. Against the grain, perhaps, Ingen Housz did allow himself to be involved, accepting an invitation to the Landstrasse, to Mesmer's clinic. In January 1775 he reported to Sir John Pringle[10] that 'many people' in Vienna – 'even physicians' – believed in 'magnetical cures'. But he was to add, finally, a revealing facetious remark viz. '. . . as everything is believed here'. He had obviously gone to judge Mesmer's 'performance' with a secret plan in mind. The visit was timed to coincide with one of Franzisca Oesterlin's treatment sessions. The two versions of what happened in the next half-hour and the conclusions drawn are so disparate that at least one of the reports must be badly distorted. The stronger motive for selective recording belongs to Mesmer for it was he who was playing for high stakes – fame and riches on the one hand, humiliation on the other. According to his *Memoir on the discovery of animal magnetism,* published four years later,[11] Franzisca was in a state of 'syncope with convulsions' when Ingen Housz arrived.[12] Mesmer immediately invited his guest to touch the patient. Nothing happened. He then took the royal physician's hands in his own 'in order to transmit the magnetism'. When Ingen Housz then touched the patient again she responded with a convulsive jerk. And when he then moved his hand her tremors followed its trace over her body. Mesmer then demonstrated his unique powers by creating convulsive muscular movements in his patient by just pointing at her, even if Ingen Housz was placed between them. As Mesmer remembered the ensuing discussions, Ingen Housz had to admit that he had 'changed his mind' about animal magnetism but the quotation of his actual words reveals masterful diplomatic evasion: '. . . he required nothing more, being now convinced'. Mesmer seemed not to spot the clever ambiguity of the testimonial. On the face of it Ingen Housz never said in which way he was convinced but his consequent advice to Mesmer that he 'should not publicise his work for fear of ridicule' was far less ambivalent. Ingen Housz, in his immediate letter to Pringle on 13 January 1775,[13] gives the other side of the story. He wrote in his faltering English so we must allow for eccentric idioms: nevertheless the meaning is clear.

> Père Hell and Dr. Mesmer pretend to have found out a magnetical cure for epilepsies, rheumatisms, haemorrhoids, haemorrhagicals &c. They apply little magnets at various parts of the body. They pretend to magnetise bread, wood, stones, dogs, mens and every other thing. They fill a bottle with magnetisme just as well as with electricity. I was called to see the experiments but found them manifestly mistakes. The woman, who was the chief subject of these experiments, had got convulsions as soon as touched by a man having a magnet in his pocket or who has been touched by one who wears the magnet, got no convulsions when I touched her though I had strong magnets in my own pocket, which she did not suspect.

Ingen Housz elaborated further in another letter, this one to Richard Huck, a week later.[14]

> Since I wrote to Sir John Pringle, Dr Mesmer published a pamphlet upon the power of the loadstone in diseases but alledges only the experiments made upon the choking woman of which I was an eyewitness, but found them ridiculous and imposition of the part of the woman. Abbé Hell, being published as a witness, is much angry at it and gave the lie to the Dr. in the Vienne Newspaper of the 14 Jan, saying openly he never was present at the experiments and he thought them even incredible.

Mesmer did not give up on animal magnetism despite the growing professional cynicism and ostracism. Patients began to flock to his clinic, many later to report extraordinary 'cures' for, as still the case today, over a third of symptomatic illnesses are psychosomatic. With hindsight it is easy to see how Mesmer must have possessed, in fact, an innate ability to put people into trances and help them by autosuggestion, irrespective of magnetism or any of the other physical paraphernalia he gradually collected. However, he himself did not acquire this perspective until, perhaps, his final years, in the next century. His persistent belief in 'animal magnetism' cost him his relationship with his wife and, finally, all hope of remaining a reputable Vienna physician. He therefore left the city to respond to requests for treatment from notable patients elsewhere in the Empire and, in the very final days of 1777, estranged from his wife and stepson, moved to Paris.[15]

THE FACT THAT Mesmer was hounded out of Vienna by a tide of negative opinion led by Ingen Housz signified a sea change in the standing of the once-shunned royal inoculator. And if his professional standing had blossomed so, apparently, had his social status. Ingen Housz was, more and more, the welcome guest at soirées and balls, concerts and dinners. He was beginning to feel at home in Vienna, a feeling that was reassuring, at least, when one of the possibilities for his long-term future closed down most emphatically. He had often thought of immigrating to America but now the new continent was a war zone. By the autumn of 1775 there had been three serious conflicts; at Concord and Lexington in rural Massachusetts, and also at Bunker Hill on the very outskirts of Boston. Over three hundred British soldiers had lost their lives as well as about 170 rebel colonists, not counting those who had died of their wounds after the battles.[16] With fighting on this scale, the colonies saw the need for a unified and organised defence of their rights and freedoms and appointed a commander for their 'army', a young colonel called George Washington. It was symbolic of the rift that had so opened up between two communities that there was now, de facto, a civil war between them; between the population living on the European land mass and the one on the eastern seaboard of North America – all supposedly English. There could be no turning back and bloody skirmishing would escalate into a long and significant conflict that would win the Americans their independence. Ingen Housz voiced his disappointment in a letter to Franklin,[17] a rather selfish emotion in the circumstances, but he was as confused as many others.

Since I received your kind lettre dated March 18 1774, great things have happened of which your country is the theatre. That country is become the seat of horror and bloodshed which I took to be the seat of tranquillity and happiness and which I was formerly much inclined to chuse as a quiet retreat in an approaching old age.

His phrases held much more meaning and portent of his future than Ingen Housz could have realised at the time. The rest of his life would be acted out, unavoidably, against a background of revolution and violence, anxiety and financial instability. But as a life in America was forsaken, a foreseeable future in Vienna was accepted; on the rebound perhaps. His servants must have noticed that their master was enjoying a much busier social life and setting off, more and more, to one or other of the homes of the Jacquin family – in the Rennweg, the house for the curator of the royal botanical gardens or in the Bäckerstrasse very close to St. Stephen's Cathedral.[18] Perhaps Ingen Housz's household sensed that something profound was imminent and the proof of their suspicions came in the autumn of 1775. The officious pressure exerted on Ingen Housz, when he had first arrived in Vienna – that, as a courtier, he should urgently consider marriage – was a dim, distant distress

Professor Nikolaus Jacquin and his son and successor, Josef Jacquin. Courtesy of the Wellcome Library, London.

but now he fell under a parallel obligation that was well outside any official umbrella. In fact it was both intensely personal and easier to agree to; a suggestion from someone highly adept at pushing on open doors.

It was Maria Theresa herself who confronted Ingen Housz, sometime during the summer of 1775. She was obviously still very fond of her family's 'saviour' and she had watched him integrate into the Jacquin household. Nikolaus Jacquin was also among her most favourite courtiers; she seemed to have an innate attraction to Dutchmen, and she saw it as her responsibility to ensure that his family was happy and flourishing. In taking a personal interest she had come to know that Jacquin's younger, unmarried sister, Agatha, had come to live with them at Vienna. The story that has come down to us[19] is that the Empress saw no good reason why she should not make a bold, but pragmatic, proposition; putting it to her favourite physician that he 'should marry the sister of Professor Jacquin.' Maria Theresa was, after all, no novice in arranging marriages; she had introduced every one of her married children to their spouses, the unions always motivated by international political manoeuvrings. This was not the drive for marrying off Ingen Housz but there were some very obvious local considerations, and in the wake of his previous refusals to her, Ingen Housz appears to have agreed without daring any discernible quibble. There were certainly, now, societal pressures on him with a growing Vienna social life. It was difficult, as a bachelor, to return any of the numerous domestic invitations he was receiving and be a satisfactory host on his own account. A wife would be able to help him keep up a socially desirable home; a welcoming oasis rather than a mere bastion of masculine austerity. In late November, Ingen Housz sat down to write another letter, a very surprising one, to 'my dear friend Mr. H. Hoogeveen.'[20]

> A lengthy silence has at last produced something momentous . . . something important, something unusual, something you would not have expected. After long and careful consideration I have actually made a decision of which you may not wholly approve. For a long time I have been in doubt as to what I should do: return to my native country or spend the rest of my life in these regions . . . To marry a girl too different in age from myself would not suit me for various reasons: I like a quiet life but it would change into a restless existence, I think, in the midst of lots of young people and this is not even without danger, especially in circles where conjugal fidelity is a rare thing and chastity among women hard to find. Some years ago I made the acquaintance of a young Dutch woman, born at Leyden, Agatha Maria Jacquin by name, who lived in the house of her brother, an adviser in the

mining industry, Professor of Chemistry and Botany; well-known in the scientific world and a friend of mine. She looked after his household and led a quiet life amid the day-to-day chores; she was used to spending her spare time in reading books and is now 40 years of age, I myself having reached my 45th year. She is now my wife.

Agatha Jacquin had been born, like her brother, in Leiden. She was some eight years younger than him, born on 1 July 1735.[21] It was no surprise that she had joined him in Vienna when, after the early death of their father, their mother had also died. From her perspective, marriage would give her a household of her own to manage, the opportunity to express her own tastes, and make her a much more acceptable social invitee.

Many letters were penned the day following the wedding, each giving the 'surprising' news to family and friends. The signing of the actual marriage contract,[22] followed by the public declaration of vows and religious blessing, had all occurred on Friday 24 November. Abbé Sebastian Jacquet, a clerical friend[23] of the Jacquins, had been the witness to the bride's signature and Nikolaus Jacquin himself witness for the groom. Ingen Housz was obliged, by the contract, to make arrangements to support his widow in the event of him dying first. He was required to leave a capital sum of at least 20,000 Florins (£116,000) suitably invested at Vienna so that Agatha would be comfortable on the annual interest payments. Only then, on her death, would the capital pass back into the Ingen Housz family. And in a separate fund, perhaps with funeral expenses in mind, Ingen Housz began to invest 200 florins (£1100) annually in the Vienna physicians' pension club.[24]

It can only be an impression, but it seems more than likely that the marriage was entirely one of social expediency and mutual comfort not involving sexual intimacy: that it was a *mariage de raison*, a marriage of convenience. We do not have any of the letters that Ingen Housz sent to his brother and all his close friends, except for the scribbled and heavily corrected draft, in Latin, of that sent to Hoogeveen.[25] Ingen Housz did record, though, in his letters diary, the emphasis he had placed, in his outgoing correspondence,[26] on the fact that he and his bride were both well into middle age. And Richard Huck, writing back from London in mid-December, seemed to pick up on these platonic but agreeable circumstances among his congratulatory, and envious, even, remarks.[27]

> Permit me to congratulate you on the change in your condition. May you enjoy all the comforts and felicity which I wish you, and you will be a very happy man. I beg likewise that you would assure Mrs. Ingenhousz that she has added to the number of her friends by her marriage and that she may

reckon me in the number. Sir John approves very much the step you have taken. He feels the inconvenience and melancholy situation of a solitary old man: no comfort at home and nobody caring for him abroad. I wish that both he and I had been comfortably married some years ago. Repentance avails little.

And in a letter written almost a year after his wedding, to Benjamin Franklin, Ingen Housz was still drawing similar implications as well as stressing that he was still free to travel.[28]

You will be surprised to hear that I am married with a Dutch lady, the sister of Mr. Jacquin, professor of Botany and Chemistry at this University... but you will, perhaps, not approve that I took one who is but 5 years younger than I. However, after a great deal of reflexions, I thaught that such a wife did suit me better at my age than a young girl. This changement in my life will not prevent me from visiting my friends in England.

Franklin, moving about the globe once more, never received the letter but heard the news of Ingen Housz's marriage on the Paris grapevine shortly after his arrival there in December 1776.[29] He himself was now a widower and just as positive about marriage as the groom's other friends.[30]

They tell me here that you are married. I congratulate you on that happy Change in your Situation. It is the most natural State of Man. I have lately lost my old and faithful Companion; and I every day become more sensible of the greatness of that Loss; which cannot now be repair'd. Present my respectful Compliments to your Spouse, and believe me ever, with sincere and great Esteem, My dear Friend, Yours most affectionately...

During the immediate months after his marriage, Ingen Housz wrote few letters and perhaps this is not surprising. Agatha would have had many suggestions on improving their home, on applying her 'woman's touch' in her newly found power-base.[31]

The leading, perhaps even decisive, say in the look, arrangement and management of domestic life was something that honourable and happily married men were prepared to cede to their wives. These were due female privileges authorised by custom. That men acknowledged this authority ... must speak of the importance to men of the benefits of successful domesticity.

There would have been many homely matters that took up worthwhile time and there would also have been, almost certainly, an initial flood of invitations to the newly-married couple. Agatha was also enfolded further into the Ingen Housz family in accepting an invitation to be Godmother to the latest addition at Breda.[32] A fifth daughter to Louis and Maria was born on 21 September 1776 and was baptised Josina Agatha.[33] A hundred ducats (£2,600) were sent as a baptism gift.[34]

There was certainly no drop in the number of contacts with Ingen Housz's new in-laws. Nikolaus Jacquin, continued to advise him on acquiring mining shares and his letters diary refers to many new purchases in Saxony and Bohemia, some of which he made on behalf of his brother.[35] And it was from Bohemia that a letter arrived that would result in a temporary interruption of the new domestic bliss. Ingen Housz was asked to return to Prague once more, again to inoculate some noble children at the castle, those of the Supreme Burgrave (governor) of Bohemia, Joseph Nepomuk, the Prince of Fürstenberg, and he agreed to travel in late May 1776.[36] By this time he had also decided to travel on to Berlin[37] and he was then away from home until mid-September. Agatha does not appear to have travelled with him, a pattern that set a strong precedent for their future. His activities in Bohemia and Prussia, other than inoculating at Prague, are not recorded: there is a striking absence of surviving correspondence. In the first flush of marriage, he wrote, perhaps, only to Agatha during these months and such private communications have probably been destroyed.

INGEN HOUSZ'S LABORATORY bench seemed to gather some dust during the first year of marriage, and perhaps this is no surprise. However, he continued, if distractedly, trying to improve his new electrical machine, a large one he had constructed from pasteboard with fur pads; and he had a new idea that required a very small insulated Leyden jar. For the large apparatus, he wondered what would happen if he added further discs on the same central axis, alongside the original one, and found, as he reported excitedly to Franklin,[38] that he could generate even more electricity. We do discover, though, from a lecture he gave to the Royal Society, a couple of years later,[39] that he had just received a new electrical 'toy' and that he had taken it with him to Prague that summer, using the opportunity to discuss its operation with the Bohemian professor of anatomy, Josef Klinkoch, to whom he refers as 'my learned friend.'[40] The apparatus was an example of Volta's 'electrophore'. It had been sent to Ingen Housz by Grand Duke Leopold, probably on the prompting of Fontana.[41] Whether it had actually been made by Volta himself is not clear but it is at least possible. Although not strictly the originator of this intriguing piece of apparatus, the invention is usually attributed to Alessandro Volta[42] for he significantly improved its design and power, reporting his findings in early 1775. This was the first

The Electrophorus; electrical capacitator invented by Volta.

alternative means of storing static electricity since the Leyden Jar of 1744. It was also Volta who coined the word 'electrophore' – from two Greek words, *electros*, and *phoros* implying storage. It consists, traditionally, of two circular plates of similar diameter although there is nothing magic in its being circular. The bottom plate, which rests on a working surface such as a wooden table, is made with, or at least coated with, non-conducting material. Originally this was a resin coating over wood or metal but today it would be plastic. The upper plate is of a conducting metal but it has a central handle made of an insulator. To use it one charges the base plate by friction with, say, fur or cloth. The top, metal, plate is then placed on the lower one using the handle and its top surface is touched, momentarily, with a finger, conducting the charge there to earth. The top, metal, plate can then be separated again, lifting by the insulated handle, and is found to be highly charged. Having used this electrical force for some purpose, the plate can then be replaced on the resin base and recharged in the same way. Even Volta was surprised how many times this could be done without loss of power and just what was happening in these electrical interchanges took some years to understand. It was seen, however, as an improvement on the Leyden jar and Ingen Housz was among the first to notice, with some pleasure, that it was far less prone to loss of function in damp weather. All this is less arcane than might be thought. The phenomenon of electrical storage, now known as 'capacitance' and the ability to repeatedly drain a small amount of electricity from the store by finger contact, so reversing the charge, is the basis of one type of a very modern device which is finding universal usage – the touch screen.

After returning to Vienna, sometime over the winter of 1776/7, Ingen Housz was approached by the personal physician of Karl Anselm, 4th Prince of Thurn and Taxis,[43] one of the richest and most powerful nobles in Vienna. The Thurn and Taxis family had become the proprietors of Europe's leading international postal service since humble beginnings in the Middle Ages. The prince, born 1733, was postmaster general at the Imperial court. After 23 years of marriage he finally had two sons, now approaching seven and five years respectively; the first of his children to have survived beyond infancy. These boys were obviously extremely precious to him and his wife and they were most anxious for them to be inoculated against smallpox. Ingen Housz was asked if he could travel to Germany sometime in the spring in order to treat the children.[44] He accepted, even though it meant a journey of over 350 miles to the Thurn and Taxis family home at Dischingen, some 25 miles northeast of Ulm in Baden-Württemberg. In preparation, Ingen Housz wrote to Linz, to a friend[45] who had borrowed his carriage, and asked for it to be returned to Vienna. Then, in the very first few days of the New Year, 1777, he received startling news that would make him regret another recent decision. He had declined an invitation to join the Emperor on a royal visit to France, scheduled for spring 1777. This adventure far from attracted him; it was bound to be full of tiresome pomp and ceremony even if the Emperor was intending, as usual, to minimise formalities by travelling incognito.[46] Now Ingen Housz came to lament his negative decision: amazingly, Franklin was back on the European mainland and, tantalisingly, in Paris. Ingen Housz's astonishment is apparent in an immediate letter to his old friend.[47]

> . . . a few days ago I was informed by the newspapers and private lettres that you are arrived at Paris. This piece of news, as astonishing as unexpected was very agreeable to me, being happy to understand that you are safe and in good health and that I have some hopes to meet you once more upon this world . . . I am highly rejoyced to be informed of your arrival and I have the greatest reason to regret my not accompanying my Imperial Master to France.

Ingen Housz reacted to the new circumstance not by asking to relent on his earlier decision, but by applying for more extensive leave of absence on the grounds that he needed to go to London. This was not totally disingenuous – he was worried about his investments there 'for I have still my whole inheritance of my father in the stocks and bankruptcies become frequent; I dare scarce to trust my money in their hands.'[48] His estate was languishing in value under the economic cloud of the civil war between England and its American colonies. It was time to go and rescue his savings. He wanted, also, to introduce around London a very talented young physician friend who had become one of his closest companions in Vienna in recent months. This was Franz Schwediauer,[49] a graduate of the Vienna medical school. Ingen Housz may also have been in receipt of a letter from Sir John Pringle who, as President, was seeking a candidate to give the 1778 'Bakerian' invitation lecture of the Royal Society; an annual celebration of the life and bequest, since 1775, of the late fellow of the Royal Society, Henry Baker. We have no evidence that Pringle's invitation came in early 1777 but it certainly surfaced at some point during the year; perhaps more as an expedient decision when his patron discovered that Ingen Housz was soon to be in London anyway. In any case, it was natural enough, was it not, for Ingen Housz to travel to London via Paris? And was he not, anyway, about to rumble off to southern Germany, almost half way to the French capital, with his inoculation lancets? The tempting tessellation was too much. However, he was in no position to leave Vienna immediately for he was still supervising the finishing touches of the erection of lightning conductors on all the gunpowder magazines around Vienna. And he also received a new commission from the court. The Emperor had been asked, by the republic of Venice, for specific instructions for so protecting their own particular arsenals and, for quite obvious reasons, the task was offloaded to Ingen Housz.[50] Meanwhile, when he was not so engaged nor preparing for his journey, Ingen Housz went on tinkering with his new electrical machine and, characteristically, conceived a useful purpose for it; a new invention that, being portable, he would take with him on his travels and go on improving. He appears to have brought it to perfection by the time that he was in London in early 1778[51] but the inception of his 'new way of lighting a candle' appears to have been in Vienna about a year before.[52]

Lighting candles was a regular chore in the eighteenth century and, in the absence of another flame, a spark had to be generated by striking a flint and igniting some sort of kindling material. Here was an opportunity for innovation but a credible tool for lighting a candle would need to remain functional for many weeks and work instantly. Igniting ether gas with a spark from an electrical machine was a popular trick that many a philosopher would use to instruct students and entertain friends but ether was very expensive and too easily lost to evaporation. Leyden jars, on the other hand, held a charge for a considerable time and Ingen Housz knew from his friend, Tiberius Cavallo, that even one small enough to fit in a pocket would deliver a spark for up to a month if exposed surfaces were insulated with wax. Ingen Housz's own inventiveness centred on the preparation of a readily inflammable material to be fired by the spark, for a kind of 'match' to be ignited by an electrical discharge. His suggestion was to wrap some cotton wool around the extremity of a long pin and roll the padded tip in some very dry, powdered gum resin. Then, by applying the bald end of the pin to the outside of the charged jar and turning the coated end to the discharge knob of the jar, a spark would jump the gap and ignite the resin. The cotton would also burn and continue just long enough for him to be able to light his candle. Soaking the small cotton plug in turpentine oil was an alternative method but had the disadvantage of filling the room with smoke.

This little invention must have given him much amusement and it was not entirely flippant. The Royal Society of London, no less, would, in July 1778, invite him to read a paper describing his apparatus and it was written up in their journal, the *Philosophical Transactions*[53] – his second paper to be published in these esteemed volumes. Nonetheless, it appears to have been seen as something of a toy and did not catch on. Perhaps his audience – aristocrats, gentry and affluent professionals – did not see the point when it was their normal habit to expect a servant to light their candles for them!

JOSEPH II LEFT Vienna for France on 1 April 1777[54] and Ingen Housz was still wondering, the next day, what had happened to his warrant from Maria Theresa, to go on from Germany to France, The Netherlands and to England. A letter to Franklin[55] was written in terms that were equivocal but his American friend would have read between the lines without the slightest difficulty: Ingen Housz would soon be on his way to Paris. And, indeed, Ingen Housz was writing to Sir Robert Murray Keith, the new English ambassador at Vienna, from Ratisbon (Regensburg) before the end of May.[56] It had been around the 20th of the month that Ingen Housz finally set off for southern Germany, armed with some lancets and a letter of credit from Stametz bank for 5000 Florins (£29,000) to be available at Paris, Amsterdam and London.[57] He was accompanied by his young acolyte and friend, Franz Schwediaur, and by Dominique Tede.[58] Murray Keith was given the privilege of an extract of a letter to the Empress,[59] the contents of which were a transcription of Franklin's personal opinions on the progress of the American War of Independence; the view from the rebel camp, so to speak. Being such a close and high level American contact, Ingen Housz had asked Franklin for his views on the present state of the conflict, information that he might then render to the court. The royal physician obviously thought that he could be of political service to his Imperial mistress. It was an ingratiating exercise that he would repeat some six months later.[60]

The two Thurn and Taxis boys had been 'safely inoculated' at Burg Trugenhofen, the Thurn and Taxis castle near Dischingen, by the beginning of June[61] but Ingen Housz lingered there for a further month, somewhat longer than it needed to supervise the boys' recovery. Perhaps he was still waiting for the actual written permission to proceed westward or just relaxing in pleasant company and surroundings. Certainly, when he and Schwediaur left Dischingen on Wednesday 2 July it was as though they had been let out of school: the mood was obviously upbeat. Schwediaur was anxious to reach London so they agreed to make fairly rapid progress through Mannheim, Cologne and Achen to Brussels and Ostende. At the same time, Ingen Housz was not to be deterred from making many a small detour and the expected rate of progress was soon lost to all sorts of side interests.

Ingen Housz kept a very detailed *Journal de Voyage* that has survived, seemingly intact.[62] We can therefore follow his movements, log his new, and old, friends, read his thoughts on places, watch him collecting new prescriptions, shrink slightly at some of the gruesome post-mortem material he saw and smelt, and wander voyeuristically through his shopping lists. But what is striking, and probably almost unique, is the endless range of other topics that interested him as he zigzagged across northern Europe from one expert-enthusiast to the next. His travel diary is a manifesto of his polymathic and enlightened curiosity and it must all have been very instructive for his young companion. We can also sense Ingen Housz's attitude and mood. He was on vacation. The *leitmotif* was pleasure, well personified by the band of comic actors the party bumped into at their lodgings in Cologne. Ingen Housz was buoyant, if not boyish. His brief notes convey a spirit enjoying refreshment: it is as if he was being transfused, colour is returning and energy building up. The secret was probably that he knew, all the time, that he was getting closer to many old friends in England and especially to Franklin at Paris. If the purpose had been a direct journey to London, their final destination, they could have been there within a month but the latter half of 1777 was to be a time of free-wheeling exploration.

From Dischingen their coach took the descent through Eysbach and Geisingen to Stuttgart where they stayed for a couple of days. After visiting the royal palace at Ludwigsberg on the west side of the Neckar, they descended on the village of Kornwestheim where mathematician and astronomer, Friedrich Hahn was the priest. Hahn had invented an early kind of calculating machine that would perform all four arithmetic functions 'in an instant' and for all the numbers between 1 and 900 million. He was delighted to demonstrate it and was thinking of selling it - but not at that moment, not before he had made himself a replica. Back in Stuttgart they turned north and made their way to Mannheim, the next 'base camp'. The new opera house was very impressive but Ingen Housz was more interested in the observatory in the grounds of the Elector Palatine's palace. Here he found Father Christian Mayer, the court astronomer. Having been to St. Petersburg to observe the 1769 transit of Venus across the sun, Mayer had remained very interested in Earth's nearest planetary neighbour. He was now convinced that a satellite orbited once a year. There was also a new observatory at Schwetzingen, at the summer residence of the Elector, Karl Theodor, which Ingen Housz also had the opportunity to see for 'I was invited by the Elector to meet, for two to three hours, with he and all his Ministers . . .'[63] Perhaps the elector was anxious to meet an intimate of the court at Vienna, knowing that he was destined to become Elector of Bavaria, a significant step nearer to the pull of Habsburg power. Ingen Housz, for his part, was obviously just as pleased to have met the elector's nephew, Maximilian Joseph: that this young officer was already a colonel in the French army was less impressive than the fact that he managed all the local mines. Although his interest was not totally mercenary, Ingen Housz now had considerable sums invested in mineral extraction and information is power, after all.

A few hours after leaving Mannheim the party joined the Rhine to follow it downstream. At Oppenheim their attention was drawn to the wine industry. The precious vintages made in the adjacent village of Nierstein were of particular interest, Ingen Housz carefully noting their alarming prices. Arriving in Frankfurt – 'a wooden town' was the descriptive shorthand – Ingen Housz immediately decamped to the porcelain works at Höchst. Factories of all sorts seem to have had a particular fascination just as they had done when touring England with Franklin in 1771. He was impressed by the fine painting in the finishing shop but process was not totally to the exclusion of product. Ingen Housz turned from visitor to shopper: he could not resist a statuette of Venus, in biscuit, which he pocketed for five Livres (£13). A few hours further along the Rhine valley was Mainz. The professor of medicine was enthusing over a new prescription for whooping cough but Ingen Housz was just as interested in the cathedral library. Here there were many books surviving from the very beginnings of European printing, something of a personal interest after the effort he had made for his compatriot, Gerhard Meerman, when trying to promote Meerman's book on the history of printing, in London, in 1768. The next subject that seemed to dominate his interest was water. That the waters of the Main and the Rhine were distinctly different in colour was the first noteworthy observation. There are many springs in this part of Germany and, at Shalbach, a local professor told him of the efficacious powers of their local water; that it was well aerated and slightly salty. On the other hand the waters at Slagenbad were reputed to be the purest in Europe. At Wiesbaden the water was different yet again: there was a sulphurous fug.

From Bonn the travelling coach was hauled up the west bank of the Rhine to Aachen on the German-Belgian border.[64] Here Ingen Housz mothballed his coach and lightened his luggage: some of the remaining journey would be on water. He paid his respects at the tomb of Charlemagne but it was obvious, that for the living residents, much of the town's life turned on the fact that it was also a spa. But the town of Spa was, as its name implied, the centre of the thriving health restorative industry based on 'taking the waters'. Among the florid bathers hoping to flush out their livers and bladders Ingen Housz recognised several benighted and aristocratic acquaintances from London. At Liège he talked at length to a local amateur geologist who had found that the town sat on huge seams of 'Glauber's salt' (sodium sulphate) that were just waiting to be exploited. Equally fascinating, to the son of a leather merchant, was a tannery in the town that specialised in making insoles for shoes. The pickling process for these really tough accessories took seven years and Ingen Housz bought some – no doubt to show to Louis rather than to wear. Four hours of pleasant cruising down the Meuse took Ingen Housz, Schwediaur and Dominique to Maastricht where the Dutch architecture made one of them, at least, feel at home. Ingen Housz was amazed by the degree of religious tolerance in the town. Its council was half Catholic and half Protestant Reformed church, a standing arrangement even though the former faith was that of a substantial majority of the population. But the real surprise was the St. Pietersberg mountain, another geological idiosyncrasy. This hill, unremarkable on the surface, was merely the exposed 'incline' of a vast deposit of a very curious sandstone that, though soft enough to be scraped away with one's fingernails, was incompressible and formed very strong blocks. Over many centuries it had been extracted as building material so that a vast area of land around the town was undermined by a maze of substantial tunnels.[65]

> The roads into the interior are innumerable and there has never been a map of the labyrinth. There are over 6000 different vaults . . . some higher than a church. The man who took me there and who conducts almost everybody has done nothing else all his life . . . knows the way extremely well. He took me,

twice, by the subterranean road, as far as the Convent de Recollet on the banks of the Meuse... A man who dares to go there without a guide will inevitably be found dead, his dog the same. An army of 100,000 men could be lost there.

The coach from Maastricht to Brussels ran three times weekly, they discovered; a journey they were to find themselves sharing with a tax collector and his daughter. In the Belgian capital again Ingen Housz was pleased to see how the 'Place de Lorraine', the new city square, had progressed. He was also intrigued by the fact that he was now a mere three-day carriage ride from Paris, a short hop from being with Franklin. Even though Schwediaur was still anxious to be in London, the temptation of a joyous and unexpected reunion with Franklin proved to be too strong and Ingen Housz was in the French capital by the beginning of the second week of August, arriving on Saturday the 9th.[66] He was to stay some ten weeks despite bouncing contrary news back to Vienna on the 6th of August.[67]

> ... according to lettres I received from Franklin he could easily escape me to America if I waited longer to visit him and as I have some business of my own to perform there, I intend to take a trip for some days to Paris, where I will be till the 20 or 23 of this month.

FRANKLIN HAD BEEN despatched, yet again, to cross a winter Atlantic. This time it was to lobby the French to support the colonists' lunge for independence not, as Ingen Housz had early imagined, to lead a parley for peace. 'In the annals of diplomacy his was an original one: Franklin was charged with appealing to a monarchy for assistance in establishing a republic.'[68] The American was ensconced at Passy, a village conveniently sited between Paris itself and Versailles. Ingen Housz must have jumped eagerly into the carriage he hired to take him on there from central Paris. Half an hour later he had crossed the Seine and climbed what was then a verdant hill and was dismounting in the grounds of the so-called 'Hotel de Valentinois'. This huge estate, with all the buildings, had only recently come into the ownership of a very successful and well-connected Parisian *nouveau riche* entrepreneur from Nantes called le Ray de Chaumont. Anxious to ingratiate himself with the new American delegate so that he could become the operator of a highly profitable supply line of munitions to the rebels (he had long hated the English anyway), he had persuaded Franklin to become his non-paying guest at Passy. Franklin's proximity guaranteed Chaumont privileged status among those jockeying to supply the needy Americans; it put his market in his backyard.[69] Franklin, for his part, had been glad to leave the fetid alleys of the city. He had brought two of his grandsons with him, the sixteen-year-old William Temple Franklin to act as companion and amanuensis; the younger boy, Benny Bache, to be installed in a school. Rural Passy was certainly a healthier prospect than lodging in the city itself and a reassuring remove from the corrupting debauchery that might have tempted young 'Temple'.[70] Franklin had moved into Chaumont's neoclassical garden pavilion that spring and to make the best of his unexpected reunion with his great friend, Ingen Housz himself also moved in. Chaumont appears to have been only too pleased to buttress his personal popularity with Franklin by having this further guest.

There must have been an explosion of news and nostalgia between the two old friends. Now, at last, the true story could be told of how the American agent had come to be such a disreputable figure in London. So, too, could the story of the opening skirmishes of the American War of Independence – from the colonial perspective. No doubt the two friends exchanged their latest reports on common friends, Ingen Housz tripping over his rusty English. He was able to bring Franklin up to date with the progress of smallpox inoculation in Europe and how lightning rods had now appeared on the munitions stores at Vienna and were being constructed in Venice and Brussels. More amusing was the new invention to combat electric storms; the story a perfect vehicle for the delightful wit of Franklin. One of his new French friends, an otherwise reputable physician and renowned botanist called Jacques-Barbeu Du Bourg, had designed an umbrella fitted with a lightning rod.[71] Walking in a storm could, now, perhaps be safer but Du Bourg's sanity seemed to be as much on trial as his invention. It was not wholly a laughing matter, though, and perhaps reminded Ingen Housz of Mesmer and human gullibility. The story of animal magnetism was naturally interesting to Franklin – what topic wasn't – but at this time neither man could have known of Mesmer's intention to come to Paris and that the 'trial' of magnetic therapy held at Vienna was to be repeated, this time with Franklin on the jury. Remembering with such pleasure their trip to the north of England six summers before, the two friends began to plan a similar journey through part of France and Ingen Housz went looking for a suitable coach in Paris[72] but, in the end, it all came to naught. Perhaps politicking was keeping Franklin busier than he liked to admit. Nothing could stop the story-telling however; everyday life in America being a rich source. Ingen Housz recorded, in his travelogue, the peculiar fact that American indians would put lice in their soup and Franklin, despite having been a city resident all his life, delighted in a ploy of pig farmers: that the colonial method for

Jan Ingen Housz; medallion of about 1774. Counsellor and doctor to the royal family.

convincing a hog to walk in one direction was to tie a string to a back leg and pull the opposite way! Was this not a superb tactic in politics? Not many days passed, however, without the two old friends dedicating time to discussing science. They debated the workings of Volta's electrophore, the latest means of storing electric charge, and tried to explain its surprising properties in the context of Franklin's theory of positive and negative charge. These were very important conversations for Ingen Housz – he was to call on them when writing a lecture given to the Royal Society in London the following summer. Ingen Housz must have learned, this time with immediate interest, that the French royal scientific laboratory was also at Passy. The philosopher in charge was Jean-Baptiste Le Roy. Here, most certainly, was an institution to be visited and a fellow enthusiast to be turned into a friend. And it was from Le Roy or another amicable contact at the Paris Royal Academy of Sciences that Ingen Housz learned of the presence of another good friend in the city – Felice Fontana was in residence. It was a golden opportunity to explore eudiometry again and we know that the two exponents spent many hours together perfecting the design and use of Fontana's latest apparatus.[73] It was a chance occasion but an invaluable investment of their time.

Ingen Housz was unable to spend all his time at Paris in recreation: he had business to transact. Louis had given him the power to speculate with a considerable sum of money and Franklin had pointed him in the direction of his grand nephew, John Williams, whom Ingen Housz already knew from their tour of England in 1771. Williams had recently taken the post of American agent at Nantes and would be in a good position to know of bright investment opportunities in Pennsylvania and the other colonies and to make the purchases. Ingen Housz asked him to find a suitable venture in America for 13,660 French Livres (£38,000).[74] In fact, Williams long hesitated for he was unable to find an agency willing to insure the money against loss at sea[75] and over a year later he was still seeking advice, from Ingen Housz, as to how he wanted it subdivided. There is then a strong hint, from correspondence, that Ingen Housz withdrew from the American plan and diverted the money into buying shares in a canal project in Murcia, in Spain.[76]

By spending the more everyday sum of one Ecu, Ingen Housz was able to admire some of the tricks of Monsieur Comus, one of the more famous 'pavement conjurors' of Paris. This enterprising breed would quickly adapt to their purpose – to amuse the paying public on the streets – any new development in chemistry or electricity. Half magician, half scientist, Comus had even been patronised by the king. For his fee Ingen Housz was entertained by a trick in which a sheet of paper impregnated with a secret red compound of gold was turned, by pressure, into gold leaf.[77] Comus was also able to amuse his audience by making a paper figure dance between two charged metal electrodes in a sealed bell jar. Another scientific phenomenon was invoked when pumping out the air stopped the jig. In fact, these are only two of Comus's little tricks that Ingen Housz took the trouble to record and there is no tangible sign that he was affronted by the rough and ready circumstances or by using experiments as entertainment. Indeed, it was a precedent for some of his own personal popularity later in his life.

On another day Ingen Housz visited the suburban home and workshops of Louis-François de la Tour, the celebrated printer. Ingen Housz was particularly intrigued by the exquisite quality of the paper de la Tour used for his exclusive and demanding clientele, taking away with him an unbound but annotated copy of a Tacitus Histories for four Louis D'Or (£250). The Latin would have been impenetrable to Franklin but there's no doubt that he, a retired printer after all, would have run his knowledgeable fingers over the leaves and confirmed the physical value of the book. But before returning to Passy, Ingen Housz was shown de la Tour's wonderful collection of Chinese artefacts, acquired, during twenty years, from missionaries returning from Peking. Among them, naturally enough, was an instructive set of prints that told the story of ancient Chinese papermaking. Then, on 29 August Ingen Housz returned to the classroom. One Abbé L'Epée was teaching a lesson at the school he had founded, and funded, in 1754, for deaf-mutes. L'Epée deserves to be more famous now although he was well known by his contemporaries: the Emperor himself had paid a visit to the school during his Paris soujourn that spring.[78] Once, on a pastoral visit, Father L'Epée had come across two little girls who were being punished for their antisocial behaviour and mutism. He realised, however, that they were not so much wantonly evil but profoundly deaf, a humbling and inspiring experience that came to dictate the rest of his life. By adopting and adapting the rudimentary gestures that the deaf then used for communication, he developed the first ever true sign language, the direct antecedent of those in use today.

Ingen Housz found another Parisian pioneer at work in his unique garden, one he had given over wholly to the cultivation of watercress. Aptly named Sage,[79] he had developed two processes that are commonplace today and he might well have been their originator. Sage grew his cress in water-filled ditches but without burying the roots – in other words he was practising hydroponics. He had also shown that cress responded well to what we now call 'cut and come again'. As long as flowering was prevented, the cress soon regenerated any leaves cut away and the never-ending harvest could be considerable. Ingen Housz duly noted the details, as always. At the home of the Duke de Chaulnes he was on more familiar territory. In a hands-on tour of all the scientific apparatus of this keen amateur, Ingen Housz must have found pleasure and pride as he spotted a disc electrical machine of his very own design. This one, the largest he had ever seen, had a disc over five feet in diameter and the sparks generated would, often, leap a gap of two feet. The ingenious duke had also been impregnating water with carbon dioxide, etching glass with vitriolic acid, and had even taught himself to tranfer gases from one submerged jar to another, without getting a soaking, by using a pair of obstetric forceps. The damp back-alley smell pervading his house was explained by the fact that he was trying to repeat Carl Scheele's method for extracting phosphorus; by evaporating vats of pooled urine. When Ingen Housz called back a few days later – on the evening of 3 September – the stench was worse, the experiment still not finished. Before leaving Paris, Ingen Housz made one last foray into the galaxy of skills of this remarkable city. He went to watch the process of silver plating and brought away some kitchen utensils and a table service to send, we assume, to Agatha in Vienna. Altogether, he spent 1750 livres (£4,550) in a day: it was serious home-making, if at a distance.

On Friday 17 October Ingen Housz made his farewells to everyone at Passy. Franklin provided him with a letter that was designed to ensure him a safer sea crossing to England.[80]

To all commanders of armed vessels appertaining to the United States of America. This may certify that . . . the bearer, Dr. Ingenhauss is not an enemy of the said states, nor a subject of Great Britain (and is) going to England on his own private affairs. I do therefore request of you that you will not treat him as an enemy . . . as he is a person of distinguish'd merit, long known to me, and my intimate and dear friend. I recommend him particularly to your civilities, assuring you that I shall esteem and acknowledge them as done to myself. This 17th day of October, 1777. B. Franklin. Commissioner in France from the Congress.

Franklin must have known of the presence of armed American ships hunting commercial vessels in the English Channel and gives new meaning to the phrase 'gunboat diplomacy.' At the very least, it serves, still, as a moving testimonial to the depth and sincerity of the relationship between the two men.

In return, Ingen Housz took with him, to London, another Franklin manuscript, a comprehensive summary of the current medical problems being suffered by his 71-year-old friend,[81] so that they could be placed, surreptitiously, under the eyes of Sir John Pringle. The hope was that Pringle would give, in confidence, his valued medical opinion for a long-standing companion even though Franklin was now, technically, an enemy. Ingen Housz had obviously recommended Sir John's opinion, as being better than his own, for what were, in the main, skin disorders. He promised to send the recommendations to Passy, a pledge he was able to honour in March 1778.[82] Ingen Housz then spent the night at Paris so that he was on time for the early morning coach back across Picardy to Brussels. It was necessary to resume his broken journey to London.[83]

T HE JOURNEY BACK to Brussels took the usual three days with overnight stops at Péronne and Mons. Ingen Housz must have found it extremely frustrating to find news[84] at the Belgian capital that he would so have loved to discuss with Franklin. In the latest *Courier de Europe*, of 17 October 1777, ironically a London publication, it was reported that Benjamin Wilson, disgruntled dissident on the Purfleet arsenal committee of the Royal Society, had engineered the changing of Franklin's pointed lightning rods on the royal palace at Kew to the rounded ones that he favoured. It appeared that the King had been swayed by politics, preferring not to live in a building that had antennae symbolising America. Ingen Housz had not spent over two months in the company of Franklin without having his satirical teeth sharpened and commented, to himself, 'that it was an expression of Majesty in a King to make a point of baring his 'head' at the lightning when it had been a rebellious hand that had tightened up the buckle on his 'hat'.[85] However, there was more ominous news at Brussels. He was summoned, urgently, to Breda. He arrived in the last week of October 1777 to find that his sister-in-law, Maria, was seriously ill.[86] He saw it as his duty to stay and support his brother and the children through this crisis and it was approaching December before Maria was definitely recovering and he felt able to continue his journey. Louis and Maria now had nine live children, five boys and four girls, ranging from 13 years down to a toddler of two.[87] The older ones now remembered their popular 'Uncle Jan' from previous visits. It was six years since he had last visited and now thirteen years since he had closed his practice and left Breda. However, he still knew some of the town's family doctors, one of whom, a Dr. Kremers, gave him another prescription for whooping cough.[88] Perhaps to distract themselves from the anxieties of the sickroom, the two brothers gravitated, one nice afternoon, to the garden and Louis described to Jan the methods he had invented to reduce insect damage to his fruit crops. One ploy for protecting his peaches was to provide earwigs with little bamboo 'drawbridges' so that they were tempted up onto the fruit by the shelter and warmth. Then, before they could do any harm they could be shaken out 'en masse' and killed underfoot.[89] Whether playing 'agent provocateur' was effective horticulture or not, Ingen Housz duly noted the details on returning indoors. Perhaps he promised to discuss Louis's ideas with Nikolaus Jacquin and make a point of seeing how the orchards were managed at the Schönbrunn.

After leaving a relieved household at Breda in late November, Ingen Housz embarked on a frantic burst of meetings and activity: he was probably conscious of the lost timetable and had, perhaps, arranged to meet Schwediaur at the Channel coast before the end of the year. His rate of travel certainly accelerated in the next few weeks. At Rotterdam he stayed at the home of his friend, Salomon de Monchy. The *Bataafsch Genootschap der Proefondervindelyke Wysbegeerte te Rotterdam* (The Batavian Society of Experimental Philosophy), of which de Monchy had been the first director,[90] had been founded at Rotterdam in 1769. Ingen Housz and Franklin had both been elected corresponding members in 1771.[91] The second half of the eighteenth century was a time of huge growth in the number of such societies all over Europe and the one at Rotterdam had, typically, raised the profile of science and medicine in the city. Ingen Housz was therefore able to find several interesting volumes in the bookshops, one of them rekindling his interest in the plague and its history. He was also able to witness local research: a doctor in the town was analysing human bile, presumably obtained from post-mortem gall bladders. And he also met Doctor Lambertus Bicker, the secretary of the society, who was adamant that he and some colleagues had shown, by experiments, that the life of the active agent of smallpox matter was confined to two or three days unless it was kept in very humid conditions. One can well imagine the royal inoculator discussing these findings in endless detail. Light relief came in the politely reluctant tasting of the gin itself after touring the distillery at nearby Schiedam and one last call before heading north was at Corper's, the best baker's shop in Rotterdam, to buy gingerbread. At Gouda, one of the town's doctors gave him a conducted tour of the sewerage system.[92] This was innovative in that it was wholly subterranean and out of sight but its presence was soon obvious if there were any blockages. If the stench didn't subside it meant digging holes, sometimes even under houses and, sadly, the drainage maps held by the town council often proved to be misleading.

On arrival at Amsterdam, Ingen Housz was soon reminded[93] that the open drains of this city were always offensive, but somehow they were less sinister. Here his time pressures seemed to impose most heavily. Nonetheless, he was able to collect some popular local prescriptions, to note the details of a new treatment for anal fistula – allowing a weighted length of silk to slowly cut its way through the buttock like a cheese wire, enticing the infection nearer and nearer the surface (rather than laying open a huge chasm with a knife) - and debate a new theory of colour vision. Still among the medical fraternity, Ingen Housz came across Andreas Bonn, the professor of anatomy. Together they turned over the skeleton of a poor man whose bones had turned cancerous; the hideous distortions to his frame and dislocation of his organs had been responsible for his early death. Another meeting in Amsterdam, a very important one for Ingen Housz's immediate future, was with the English instrument maker John Cuthbertson, who had settled in the city. He and his friend, Hendrick Aeneae, teacher of mathematics, were both keen amateur scientists and had developed a way of producing volatile ether by heating together 'oil of vitriol' (sulphuric acid) and 'spirit of wine' (concentrated alcohol.) They had gone on to show that a small quantity of the gas, mixed with air, could be ignited with explosive consequences, a force they had turned to good use in designing a type of air pistol. Ingen Housz was clearly impressed and inspired.

Heading south again, to The Hague, Ingen Housz paid homage to Pierre Lyonet.[94] Though nearly seventy and long retired as a translator in the United Provinces court, this amateur naturalist was still hard at work. In the view of many, Lyonet was the father of entomology and was certainly, then, the world expert on the life cycle of butterflies and moths. His anatomical drawings of caterpillars and chrysalids were superlative and the artwork inspiring. Ingen Housz was sad to find that there were no plans to make prints but Lyonet was adamant that there was 'no engraver in the whole world' who could do justice to his draughtsmanship.

Ingen Housz browsed the familiar bookshops in Leiden; perusing his purchases, perhaps, on the tow-boat that took him on by canal to Delft, where he arrived around 13 December.[95] Here, of course, he headed straight for the Latin school and his old teacher, Hendrik Hoogeveen. The two friends, author and informal 'literary agent', presumably made a final assessment of the success of *Particles*. Hoogeveen had embarked on a new project, an analogical dictionary of Greek, but he was so busy as head teacher of the school that he thought it very likely the work would be unfinished at his death. Perhaps he could interest one of his sons before it was too late. Maybe Ingen Housz was meant to take this as a hint but appears to have made no promises and had to hasten on to Rotterdam: he was to be a guest at a meeting of de Monchy's experimental philosophy society. He was to travel on to Brussels from where he wrote to Franklin, 'I am to visit the gunpowder magazines on purpose to direct the officer appointed to erect proper conductors upon them.'[96] Even so, there was still just time for one more appealing detour before having to leave his homeland. It was to Middelburg, the westernmost part of the Netherlands, on what is virtually an island in the North Sea. De Monchy recommended Ingen Housz to his son-in-law there, who seems to have acted as local guide. There was much to admire in the coastal defences and one can savour the justifiable pride, in the notes he made, when Ingen Housz saw the efforts made by his countrymen to defy the sea.[97]

> The Isle of Walcheren has a circumference of about 12 hours, it has three cities, three smallish towns, 32 villages, and a considerable number of beautiful country estates. The ground is very fertile, being clay mixed with sand. Nearly everywhere it is surrounded by dunes, except at the west side where the sea has penetrated in former times and once destroyed the town of Westkapelle. Instead of dunes they have made a dyke, half an hour in length and of considerable width, gradually sloping into the sea; the sloping surface covered with plaited straw that has to be renewed every year. . . They are always busy, in one place or another, to protect this dyke against the violence of the sea. . . The old inhabitants of Walcheren must have prayed to a certain goddess called Nehalennia for in the submerged ruins of the lost town of

Domburg . . . many altar stones were found, always with a statue of or an inscription to Nehalennia.

Perhaps Ingen Housz mused wistfully that it could have been this goddess, 'the spirit of boats' and 'the protector of seamen' who had saved his father from drowning in these waters in 1714; a sobering thought for someone who was also about to cross them to England.

13
Apogee

As the light gathers itself and the thin dawn chorus breaks up in search of breakfast, a well-dressed middle-aged man who's become fondly familiar, now, to the villagers, is already returning to the grounds of the manor; to the steward's house that's become his summer home and laboratory. He's been out very early, to the horse chestnut tree that overhangs the duck house. His feet are wet, though more from the dew than slipping in the black slime at the pond's edge, and he's managed to drag down one of the of the lower branches and relieve it of some of their louche palmate leaves. Back indoors, natural needs attended to, he puts two handfuls of his booty into a dry gallon jar and holds them up inside with splayed fingers and thumb as he inverts the glass vessel and drops it, squarely, into a large dish of water, sliding his hand away at the very last second. The jar's contents, the leaves and the air around them, are now fellow prisoners above a water seal, cut off from the atmosphere: how they interact will become a matter of scrutiny later in the day. Ingen Housz shakes his hands to dryish and completes the job with a thick cloth that he then shakes fully open and drops over the apparatus, shutting out, for now, all the light. While his hands are dry he records this experiment so far, adding it to the written list of all the others - number 180 – and then remembers to note, in the margin, that today is August the 24th – Tuesday, he thinks, he'd best check with Dominique. Deliberately cut off from all social interaction or commitments other than his experiments, each week is rolling seamlessly into the next. All that matters is that, soon, the days will be shortening and cooling: with the sun now known to be the critical factor in his trials there's no time to lose.

He takes two more bell jars off the shelf and incarcerates some more fresh vegetation, some henbane leaves - experiment 181 - and then a whole plant of water figwort - experiment 182. The three shrouded jars stand in a line at the back of the kitchen and draw Dominique's attention when he returns, armed with breakfast. He puts a jug of fresh cream and some still warm milk on the scrubbed wooden table. He'll collect some bread from the farm later – it's always so fresh it lasts them two days and they've plenty for the moment. He prises a lid out of its dock on the top of the range and slides a heavy blackened kettle over the glowing porthole. All the while his master has been working at the other end of the table, conscious, now, of his hunger. Four smaller bell jars that have been standing there overnight are now ready for examination. Each contains a handful of water pepper leaves trapped in water though each has a large bubble of air above them. The findings will be written up as experiments 183, 184, 185 and 186. To analyse their contents, Ingen Housz needs some fresh nitrous air and there is just time to make some before breakfast is ready. He goes out through the scullery and, under the overhang where the unwanted fumes can escape – a lesson they learned soon after first arriving – he drops several coils of fine copper wire into some spirit of nitre in a flask. The bung, pushed home firmly, is pierced by a short tube and with his thumb over the end of it, Ingen Housz takes a deep breath and plunges the vessel deep under the water in a wooden tub standing on the lead drainer. He slides it sideways under a split shelf: the bung is wider than the groove and the flask is pinioned, upright, by its buoyancy. The water is bitingly cold at this time in the morning but he clenches his teeth and holds a succession of small vials, each turned upside down under the water, over the end of the pipe until the fumes, bubbling up, have displaced all their water and he can slide the tightly fitting trap-door across each of their necks. He expects to need twenty or thirty such identical 'measures' during the day and will need to prepare several more batches but for the moment the bubbles are slowing and his fingers are numb anyway. He'll be glad, on his next trip to London, to buy some more curved glass piping so that he can make his nitrous air in a small trough of water – so much less painful. The kettle has boiled and Dominique has mixed some chocolate in a mug, adding the cream drop by drop. It's time for breakfast, which, once he has convincing circulation back in his arms, Ingen Housz can sit down to enjoy. The hot drink is very welcome; sedating almost, and it would be easy to sit and muse but after some bread with butter and jam, and some cheese, all washed down with a hearty quantity of milk, he is soon back to his experiments.

It's time to make an assessment of today's 'common air' – of the atmosphere; as if to set a 'gold standard.' Standing a few yards away from the house where there's a slight breeze, he inverts a clean quart jar over a dish of water and brings it in to the wooden tub in the overhang. Making sure everything he needs is within reach, he rolls his sleeves up very high once more and, knowing what to expect, holds his breath as he lowers the whole assembly to the bottom of the tub of water. He slides a funnel, bell down, along the split shelf and then tilts the jar until, nearly horizontal and positioned under the funnel, the air belches up to displace the water in a glass column that he is now also holding in position above. When the column is full he stands the jar back on the bottom of the tub and uses his free hand to reach for his eudiometer tube; for the first of what will be many times that day. Holding the main stem of the water-filled pipework under the surface and making sure that it's full of water, he lets the measure of 'common' burst up into it and then, moving quickly but with control, also lets in two of his fresh measures of his nitrous air. He lifts the tube clear of the water, his palm pressed to the open end and rocks it vigorously, ice-cold water running down his arms and chest all the while. The fumes within react violently for a few seconds and a red colouration appears. He then clips the tube, open bottom end under water again, into the wooden frame that is the superstructure of his eudiometer and perches the whole assembly on the submerged shelf. A column of the water rises up the tube. So far his eudiometer results for the air at Southall Green have always been between 170 and 190 and today is no exception, reading 182, and there seems little point in taking another reading. The Fontana eudiometer is very satisfying to use for the higher the final reading the purer the sample of air — nothing difficult to understand there — and readings less than, say, 150 must mean that that air is very impure, 'fouled'. But, he thinks idly, the whole new idea that 'common' air, the invisible fluid we all live in, is actually a mixture of other airs, and that the proportions can differ, that its quality can vary, is very difficult to accept. Thanks to men of genius, though, some of them his friends like Mr. Cavendish and Mr. Priestley, it's now known to be the case. He laughs to himself — if it isn't true he's wasting a whole summer of his life. Dominique, bringing through a tray of hot ashes from under the fire in the range, breaks his reverie and Ingen Housz gets back to work.

The label that he'd numbered 183 and tied to the first two-pint bell-jar the previous evening reminds him that he had submerged a handful of water pepper leaves and trapped them over a dish of water in the usual way. He had then let up two ounces of some high quality dephlogisticated air — a sample that had given a eudiometer reading of 220. He notes that the volume of air at the top of the jar has halved overnight before taking it outside, keeping it carefully seated in its dish. Once again taking the plunge into the cold water, he takes his next eudiometer reading using a sample of this air and two of his nitrous air vials. Water rises up the tube but only slowly this time and the end result is 15 — down by a dramatic and convincing 205 units - the water pepper has obviously contaminated the remaining air around it during the night. Ingen Housz rubs his arms and hands dry and takes up the quill. 'Experiment 183: in the morning I found the air diminished to one half and was almost not reduced by nitrous air.' The same methodical manipulations are made for his remaining overnight specimens, experiments 184, 185 and 186, and the results are all very similar. Whether the plants were trapped in purified air like the water pepper, in ordinary 'common' air or in air vitiated by his breathing, it is always spoiled, or spoiled further, overnight. This is the result he has always found and he nods in satisfaction. Yet again his observations show that his discovery of what he has taken to calling the 'influxus nocturnum' — a drawing into plants of good air in the hours of darkness while they poison what remains — is universal.

The morning is now well on and Ingen Housz lifts his jar of chestnut leaves, experiment 180, onto the table next to a short stub of candle, which he lights. He tilts the jar to one side, just enough to squeeze the candle under the rim and he watches the flame dwindle and go out. The air in the jar is already fouled but the 'candle test' is rather rough and ready and he thinks he must double-check the others. He now takes the jar out into the garden and fully exposes it to the sun. He sniffs the air and scans the sky, recording, back indoors, the weather to be 'sunshine, windy, temperate.' But while in the yard he crouches low into the henhouse and opens one of the deep drawers on his left. He takes a three-day old chick from among the straw noting, mentally, that if the two eggs at the back haven't hatched by next morning, they can be thrown away as sterile. Back in the house he puts the chick at the bottom of a wooden pail and turns to the henbane experiment also set up in that morning, number 181. He takes the jar out to the tub and bubbles some of the trapped air up into a quart jar. Back at the table he picks the chick out of the pail and releases it under the glass lip of the jar. He peers intently, one eye on his watch. In less than thirty seconds the perky little bird shows obvious signs of distress and, in under two minutes, has flopped onto its side, still and beak agape. Ingen Housz quickly pops it back into the pail where it soon recovers. He then takes the gallon jar of henbane leaves with their remaining poisonous atmosphere, still experiment 181, out into the garden and the sun; once again a companion for the jar of chestnut leaves. The last large jar standing under a cloth has lost its label but he finds it on the floor, still clearly marked 182. Here, too, he takes a small sample into a quart jar and pops in the recovered chick. This time it collapses very quickly and dies. He is therefore not surprised to see a lighted candle go out almost immediately when slipped into the jar and, as a further check, that the eudiometer reading is virtually nil. The air in the gallon jar of water figwort leaves is obviously 'completely spoiled' and he looks forward to seeing how dramatic the effect of the sun may be as he places it out on the terrace, alongside its earlier companions, the chestnut and henbane leaves. He whistles quietly to himself as he realises that, for once, nearly a whole morning has passed without a slip, without air escaping from any of the jars and without breaking a eudiometer tube, something that costs him two shillings every time.

 It's well after eleven and Dominique appears with a pint of small beer and nods as Ingen Housz reminds him to fetch another carboy of spirit of nitre from the cellar and gives him some indications of which glassware can now be washed, dried and put away. Ingen Housz welcomes the liquid distraction but in his mind he's already organising a new comparative experiment – number 187. He takes a little time over his beer to write in his notes the aim of his new venture.

 'As I had found now that plants yield bad air in the night and in a dark place in the daytime, I was wanting to see the difference of the air affected by a plant standing in a room against the window towards the sun and one standing in the back part of a room which was very light but in which the sun did not at that time shine.'

 He finds two peppermint plants outside in the herb garden that are about the same size and takes a fork to uproot them. Back in the kitchen he traps each in a quart jar inverted over water just as he had done nearly five hours before for the horse chestnut leaves. He places one jar on a well-lit windowsill and the other on the floor at the back of the room.

Between twelve and one o'clock he returns to experiment 180, the chestnut leaves soaking up the sun. After only two hours or so in the garden they look no different but the air around them is much more reactive in the eudiometer than the 'common' air, reading 215. By one o'clock he's also retrieved and re-tested the air in the jar of henbane leaves, experiment 181. This, too, is already 'more salubrious' than the 'common' air although the eudiometer reading is less spectacular than the horse chestnut example. Anxious as ever, to move on, he clears everything out of the water tub, shakes the eudiometer tube dry once more and puts on his hat and outdoor coat, forgetting the third gallon jar standing out in the sun and experiment 182 goes wanting. He takes the garden gate into the meadow beyond and crosses to the bank of the stream opposite the forge. It now looks like rain is coming: at least poor Dominique may be relieved of some of his trips to the village pump if they have enough of a downpour to fill the water butts. For the second time today Ingen Housz risks a wet foot in reaching out to pilfer some leaves – this time from a pollard willow. He returns to the house by a different route for he's spotted, a few days ago, a nightshade plant among the ivy on the shady north side of the church wall. And from the hothouse next to the manor he picks a couple of ripe lemons to put in his pocket and some small but ripe peaches that he places carefully on the deck of a small trug. He now has the materials for a new batch of experiments. The willow leaves soon find themselves in the locked-in atmosphere of a gallon glass jar, experiment 188, and so do those of the woody nightshade, experiment 189. He takes the two jars out into the sun, finding with annoyance, the forgotten jar of water figwort leaves. The small peaches are then placed in a quart jars, six in one, experiment 190, and two in the other, experiment 191, and stood, dry, on the kitchen table, over plates covered with just enough water to trap in the air in them. He then sets up a parallel experiment, number 192, with the riper of the lemons, though in a smaller jar.

At a quarter to three Dominique announces that dinner is almost ready and places a small glass of red wine on the 'dining' end of the table. Ingen Housz needs no second bidding. He dries his hands yet again, rolls down his sleeves, and starts to record the details of the experiments he has just set up. He's a little anxious about the exact numbering and ducks back outside to check the labels. On returning, he puts his coat back on; it has turned rather cool and dark and Dominique won't be banking up the range until evening. Fresh bread, more cheese, and some cold mutton appear in front of him; a plate and a sharp knife and, sure enough, having become aware of the appetising smell in the last half hour, Dominique puts a dish of gruel in front of him with a spoon already bathing in it.

His work resumes soon after four when he sets about preparing some fresh nitrous air. He manages to fill seven measures this time and uses the first two of them to retest the air around the Henbane leaves that he'd put back in the daylight of the garden at one o'clock, experiment 181. He's so engrossed that he doesn't hear the church clock striking five just as the level settles in the eudiometer tube. The reading is rewarding – significantly better than today's atmospheric level and, sure enough, a trapped candle now burns happily. Either side of six o'clock a brief storm passes and the daylight seems almost to be brought to an early end. Ingen Housz sees no point in leaving his comparative experiment with his mint plants, 187, to go any further; both are now, effectively, in the gloom. Carefully marking, with label marked 'dark', the jar from the back of the room, he also collects the one on the windowsill and takes both out to the scullery to test the air in them. They'd been locked in some five hours ago and the afternoon has been mostly overcast. He thinks there may be little or no measurable difference between them and knows that he may have to use many of his remaining stock of nitrous air measures to get some very accurate and consistent readings. After letting up one measure of the 'windowsill air' and shaking it with two of the nitrous air, the reading from the top of the eudiometer tube down to the water level is 194 units. After another measure of nitrous air it's 227 and, after another, 321. After one more it's 421. The last 100-unit measure of added nitrous air has not been reduced; it just increases the whole volume by exactly 100, so saturation has obviously occurred. The final reading must therefore be 600 units – the six measures in all – minus 421, the current level of water in the main tube, making 179. Air from the other jar is also reactive but not quite so – the sequence is 196, 237, 337 giving a final reading of 500 minus 337 – 163. In other words, after a very busy, and again wet, half an hour, he's been able to show that the purity of the air in the jar on the windowsill is higher than its counterpart at the back of the room. He makes a mental note: he must always be very careful never to assume the light intensity in a room to be universally the same.

He has found, on many days, that his hands and arms somehow acclimatise to being dunked in cold water much of the time and that he usually finds working with the eudiometer less painful by early evening. Today is no exception as he ploughs through assessments of the air now surrounding the willow leaves, experiment 188, and then that around the nightshade foliage, experiment 189. Since two o'clock when he had taken them out into the sun, the leaves have 'cleansed' the 'common' air surrounding them and his readings, both above 400, would suggest that the proportion of dephlogisticated air has doubled, a very surprising result in such a small time but he writes up the findings well remembering his absolute surprise and amazement the first time he had obtained such spectacular readings back in June. All discomfort and tiredness forgotten, this calls for a pipe of tobacco in the armchair. He thinks through, yet again, all the possible factors he should test. The willow and the nightshade clearly behave, in the sun, just as all the other green vegetable material he had tested. Results have been most spectacular when the sun was really bright but even on rather lacklustre days like today the leaves of all the plants he'd ever tried always purify the air around them if they're in the light, even for just a few hours. Whether they're sealed in dry, or under water, seems not to matter, nor does the source of the water – rain water, river water and pump water all give the same findings. Using Fontana's eudiometer properly gives very accurate readings and confirms that the gas reactions have completed. No, there's no doubt that he should now try to find some time, every day, to begin to organise and write up his manuscript: Sir John had been adamant – there was already much more to 'put before' the public than could be done in a paper printed in the Philosophical Transactions and there would be much less delay with a book.

Reinvigorated, Ingen Housz springs back to his current experiments on the ripe fruit that he had imprisoned around two thirty. The findings are very similar in all three experiments – the peaches, experiment 190, even just two of them, experiment 191, and just a single lemon for that matter, experiment 192, have, over three to four hours in a poorly-lit room, all 'spoiled' the air around them so much that a lighted candle dims immediately or actually goes out and the eudiometer readings are very low, indeed barely measurable. Night is drawing in now but Ingen Housz's working day isn't over. At nine o'clock it's dark but he uses the three remaining measures of his nitrous air to test the night air, working in candlelight and getting wet yet once more. He's glad to dry his hands and arms for the last time of the day and to see Dominique, with his gift for timing, reappear with firewood and to put the kettle on the range. Having put all the used glassware in the stone sink for cleaning and drying, Ingen Housz then sits with his notes to record his evening's work. There are, however, a couple of tomorrow's experiments to prepare for. He seals the other lemon he picked into a jar to be left overnight and, in another one, some fresh henbane leaves. These preparations will become experiments 193 and 194 and he makes the appropriate notes in the experimental log. He scans back up the page – in an eighteen hour day that has been entirely typical in duration and intensity, he's completed 13 experiments, many of them involving repeated gas analyses, established two others, written up all the findings and prepared for tomorrow. He has also found time, somehow, to let his mind wander through all the implications of his lonely discoveries. Is he really still the first and only explorer of nature to have found out the effect of sunlight on leaves? Can he get his discovery of its sum and substance into print before anyone else rushes out the same breakthrough? Might his findings have already been stolen by someone unscrupulous enough to claim them as their own? He's glad to be exhausted – he knows he needs to be to have any hope of sleep.

After all the digressions and delays, Ingen Housz finally crossed the English Channel to take the London coach in the second week of 1778, arriving in the city on 12 January.[1] He probably preferred the short crossing because there had been yet more reports of armed American boats hijacking shipping, even neutral commercial vessels. His letter from Franklin wouldn't protect him against bullets or shrapnel and poor Schwediaur, rejoined as travelling companion at Calais,[2] had no such evidence of neutrality in any case. In London Ingen Housz found that his old lodgings at Northumberland Court were not available but he did manage to find rooms, the next day, on the south side of Pall Mall,[3] the very street where Sir John Pringle had long lived. Ingen Housz's rooms, for which he paid a pound (£63) a week,[4] were at the back, for he refers to overlooking the elm trees in the gardens of Carlton House, the royal residence.[5] Ironically, he was living on, or very near to, the present location of later buildings that now house the Royal Society. Schwediaur presumably made his own arrangements but, whatever they were, from now on the two visiting physicians were always seen out together.[6] Ingen Housz was very assiduous in introducing his young friend all over London; it was a good time to do so. In the middle of the winter 'season' everyone of note was in residence. More than a decade after having been a disorientated novice in the city, Ingen Housz was now patron and guide in his turn. Among many others, they dined with the politician, Lord Shelburne.[7] For Ingen Housz this was to be a very important rekindling of not one, but two, previous friendships. He had probably known the second Lord Shelburne from his London visit of 1771[8] but since 1773 his lordship had been, by a happy coincidence, employing as personal adviser and librarian, another friend of Ingen Housz. This was Joseph Priestley,[9] tempted away from his Unitarian ministry at Leeds to be an intellectual companion to Shelburne in London and, during summer months, at Bowood House in Wiltshire. Ingen Housz and Priestley began to meet regularly during the spring of 1778. In scientific terms they were happy and profitable sessions although Priestley would likely have been very envious of the recent intimacy between Ingen Housz and Franklin. Some of the meetings were in Ingen Housz's lodgings[10] and others, perhaps, at Shelburne House in Berkeley Square, where Priestley had rooms and a laboratory.[11] But their contacts were cut short when Priestley was obliged to return to Wiltshire in April,[12] to be with his family near Shelburne's country seat, Bowood House.

The early and long anticipated reunion with Sir John Pringle was not so auspicious. Pringle had had a midwinter slip on some ice[13] and had hit his head. In his 70th year, his powers of recovery were overtaxed and he was below par for many weeks. Nonetheless, he was able to function, if without the previous head of steam that was familiar to all his friends, and Ingen Housz was able to give him regular support and companionship being, now, a close neighbour.[14] 'I can scarce avoid to pay allmost every day a visit to our common friend who seems to find a kind of confort in my company and who I respect too much for not to oblige him.' Only in May did Pringle feel well enough to rejoin his friends at the Royal Society dining club: Ingen Housz also attended, as his guest,[15] on Thursday the 21st. All this must have been very poignant for Ingen Housz and he had the delicacy to wait before pressing his old patron for an opinion on Franklin's skin problems, withholding the written case history until March. Pringle did have, however, the grace and the stamina to reply, with considerable attention to detail, and Ingen Housz promptly sent the recommendations over to Paris.[16]

Pringle was still President of the Royal Society and had given Ingen Housz a huge honour by promoting his name for the annual invitation lecture – the 'Bakerian' - to be delivered in early June. Ingen Housz was approved by the council and spent some of the late spring preparing his talk.[17] Also on arriving in London, and almost by tradition, Ingen Housz immediately became engaged in a translation exercise. It must have been within days of his arrival in early January that he was shown a manuscript by Nathaniel Hulme, a London physician. It was a potential book promoting the drinking of carbonated water, after dosing the patient with an alkali/acid regime, in the hope of dissolving and washing out bladder stones.[18] It is very possible that Ingen Housz was introduced to Hulme by Priestley – the inventor, after all, of the apparatus that produced the carbonated water.[19] Ingen Housz saw the importance of this new therapy immediately but he also saw that, written in English, it would not be appreciated on the Continent. Hulme appears to have given him his willing permission to translate his text into Latin and for this *lingua franca* version to be published, by Ingen Housz, at Leiden This would make it accessible to physicians all over Europe, a much wider distribution than Hulme could otherwise ever have hoped. The Latin version was rapidly produced by an enthusiastic and industrious Ingen Housz.[20] It ran to fifty-two pages after he had written, himself, an introductory preface and by late February Luzac and van Damme, publishers and booksellers at Leiden, were waiting only to press the final pages.[21] This worthy exercise, performed at some speed, was another success for Ingen Housz as translator and literary agent, and it would not be the last. His ambition that Hulme's work be available across Europe was certainly fulfilled: by 1782 there had been translations into German and into Dutch.

For his Bakerian lecture, on Thursday 4 June,[22] Ingen Housz chose to discuss the probable mechanism of the newly invented 'Volta' electrophore using the concepts of positive and negative electricity long proposed by Franklin. Discussions that he had enjoyed with Franklin at Passy were turned to good advantage and the lecture was well received. Ingen Housz was able to convince his audience that he had a depth of understanding and that it was fitting to apply Franklin's concepts of positive and negative electricity. Today, the process is perfectly clear as an example of 'electrostatic induction' – that when an electrically neutral body (its positive and negative charges being equal in number and randomly distributed) is approached by another, charged, body (for example, positively charged), all the positive electrical elements are repelled to the far end of the receptive body and the negative counterparts attracted to the near end. Ingen Housz had been able to construct a model using a plate of glass with metal foil on either side, to represent the top disc of an electophore.[23] After being placed, flat, on some charged resin, the upper layer of foil over the glass became negatively charged and the lower foil positively charged. Contact with the charged resin plate (which had taken up negative charge only) was 'inducing' or persuading the negative charge in the glass to move away to the top and 'inducing' the positive charge to collect underneath. So, when the top foil was earthed it left the glass with a predominant positive charge that could be replenished at will since none of the charge originally given to the base plate had been utilised except to 'induce' charge separation within the glass/foil sandwich. It is not clear whether Ingen Housz was the first person to devise this conceptual explanation of function but it is certainly the case that he had had a unique opportunity to discuss all these issues with the then doyen of electricity, Benjamin Franklin, when the two men had been together at Passy in summer 1777. In fact, Ingen Housz's lecture may have been more provocative in the political domain than for its intellectual demands, and he may have misinterpreted any unease in his audience for he made no apparent concessions to the fact that Franklin was currently a very active and powerful enemy of the state.

Despite his recent poor health and prolonged recovery, Pringle managed to keep up his tradition of writing and delivering an 'Annual Discourse' related to the prize-winning paper presented by the winner of the Copley medal at each anniversary meeting of the Royal Society every autumn. His sixth such discourse, on the theory of gunnery and gunpowder, was delivered on Thursday 30 November 1778.[24] It was to be his swan song. He had come to a realisation that his former vigour and intellectual tolerance were unlikely to return and he had become rather introverted and depressed. Ingen Housz was still trying to support and encourage him but with little success, as he bemoaned to Franklin in early October.[25]

> Our friend begins to linger sometimes and seems to be tired . . . he is not able to bear some vexations inseparable from an exalted position . . . (and) looks with an eye of indignation upon those who obstruct him in what he is conscious of being right.

There was dismay when Pringle also chose to use the meeting to tender his resignation as their President. He had the self-understanding to know that he was rapidly approaching a time when he would no longer be able to do the position justice and there was, also, the long-running controversy with the King regarding the best shape for lightning conductors. Pringle belonged to the 'pointed' majority whereas his royal patron and patient had been persuaded into the 'ball' camp partly because 'points' were favoured by Americans. Pringle was succeeded by the 35-year-old Joseph Banks, who was elected unanimously by the assembled fellows.[26]

EXHILARATED BY THE reception of his Bakerian lecture and with London society departing for their summer residences, Ingen Housz turned his sights back towards Paris. His commitment to Schwediaur was now fulfilled – the young man was already well ensconced among the English medical fraternity. Ingen Housz would also have read, or heard, that the French queen was now heavily pregnant. As an Imperial counsellor it was a relief to know that the Bourbon succession might now share Austrian blood. He was in a position to suspect, quite correctly, that Joseph's visit to Versailles the previous summer had been a 'trigger.' Only the Emperor would have been able to pierce the shell of protocol and unearth the nature of the problem. But as the 1778 summer drifted by, Ingen Housz remained in Pall Mall, hunched over his desk, despite having written to Franklin to invite himself back to Passy on the pretext of 'bringing over' the ingredients for the skin therapies recommended for Franklin by Sir John Pringle.[27] Perhaps he came under political pressure via whispered words of advice: that he should not infuriate his English hosts by visiting the enemy camp. Or, equally credible, is simple inertia. He had become engrossed in a new literary exercise; a book on smallpox and inoculation.[28] But living in the metropolis generated too many diversions; distractions that he found irresistible. He continued to dine weekly with Royal Society fellows and, throughout his time in London, many mornings were spent in the shops. Besides the never-ending list of chemicals, glassware and scientific instrument parts such as hollowed-out glass slides for his microscope, there were more commonplace purchases[29] – new shoes, shoe-buckles, black silk stockings, some 'shaggy' velvet, several yards of silk for a coat lining and seven shillings-worth of 'bawbels.' He was trying to find good quality English cutlery for his brother but Louis's specifications were proving difficult to meet.[30] However he was able to find a suitable longcase clock mechanism for the Breda household and arranged for its shipping to Amsterdam where a case was constructed.[31] He visited the studio of the famous James Tassie, the miniaturist, and ordered a head and shoulders profile. He spent £3 – 6s – 0d (£207) for 22 copies[32] in jasperware at Bentley and Wedgwood, the London retail outlet for Josiah Wedgwood and his partner. But, increasingly, the items that Ingen Housz was taking back to his Pall Mall lodging were clothing. Or, sometimes, he was just enhancing his wardrobe. He was so impressed by some mother of pearl buttons stitched onto a coat, that he bought some spares for a waistcoat in Vienna. Was he more fashion-conscious than before or is it that we just happen to have a period where his shopping lists have survived? He bought a ticket for a masked ball at the Pantheon, the London assembly room[33] and chose for himself a black silk mask with a 'goze' (goat's) beard and a pair of white gloves,[34] spending a total of £2 – 9s – 0d (£155). And even when he tried to settle to his manuscript, his popularity meant frequent interruptions from a constant stream of 'friends and acquaintances' calling at his rooms.[35] By September the inert smallpox manuscript was causing him such heartache, that when Felice Fontana arrived from Paris,[36] he was only too glad to put it aside and help to meld his old friend into the life of the city. Fontana even moved into Northumberland Court where, they found, Ingen Housz's old rooms with the Pitters family were now available.[37]

The two Imperial physicians immediately returned to air experiments, testing the many ideas they had conceived since the previous October. They returned to their quest for perfection in the use of Fontana's eudiometer[38] and they probably conceived, at this time, a critical sophistication. Although the core of the apparatus was still a cylindrical glass tube sealed at the top, as long conceived by Priestley, they had the insight to see that three mandatory improvements were needed. One was to ensure that the starting volumes of the gases used were very accurately measured and reliably identical.

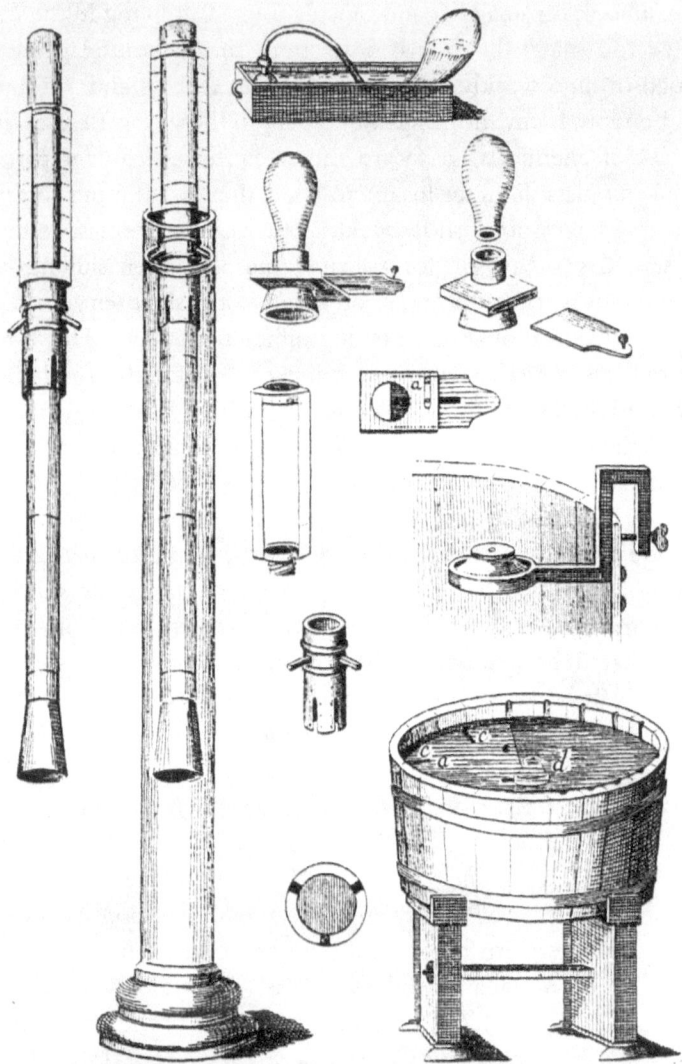
Essential components of the Fontana Eudiometer.

Secondly, the actual mixing needed to be regularised, that it be immediate and thorough; by controlled oscillation. And thirdly, a protocol ensuring that the chemical processes were totally completed – that the test air had been totally 'saturated' by the 'nitrous air' (nitric oxide). To prove saturation they introduced a refinement in which repeated increments of 'nitrous air' were bubbled up into the test air sample until it was certain that no further shrinkage was occurring. More subtle arithmetic was then needed to calculate the result but at least this was found consistent on repeat testing.

This advance obviated any worries about the quality of the 'nitrous air' – it required just as many volumes as were necessary to achieve and prove 'saturation' and as long as the final calculations accounted for the varying number of increments, the readings were meaningful. This was a point of function that Priestley was never to understand, indeed quoting, at a later time, the 'Fontana/Ingen Housz' tolerance of quality in the nitrous air as a fatal flaw in the use of the apparatus.[39] Italian design flair solved other problems. The flasks or 'measures' of 'nitric oxide' that they used were each identical but also fitted with a machined metal trapdoor that was spring-loaded so that their gaseous contents were guaranteed to be of equal volume. And since it was difficult to etch glass into small enough divisions – up to a hundred per inch – Fontana invented a brass collar that was spring-loaded so that it could be slid up and down the eighteen inch glass tube to superimpose major graduations, allowing very accurate readings of fluid levels. Fontana never published the details of 'his' eudiometer. Permission granted, that fell to Ingen Housz.[40] After some four years of trying, Fontana and Ingen Housz now had a unique eudiometer and a practised technique that put them far ahead of competitors. It is apparent that Ingen Housz's attention was now almost exclusively given to experiments with gases. It is equally clear that, with some cynicism, he was using his neglected smallpox manuscript as a diplomatic cover for remaining in London. He even reinforced his reluctance to return to Vienna with another excuse. This was a summons to Brussels, in October, to inoculate the four-year-old daughter of the sixth Duke of Arenberg which, in fact, he rebuffed – the 'little princess' would be best inoculated in spring – 'the season that people prefer for the process to begin.'[41]

  Some of the joint experiments with Fontana were not without danger. They did, however, lead to two papers that were read to the Royal Society. The first was written by Fontana or, more correctly perhaps, by John Paradise, a friend and Royal Society fellow, who certainly read it on his behalf. It was featured on the agenda of the meeting on Thursday 11 March and described the results of animals and, on one occasion, Fontana himself, breathing 'inflammable air' (hydrogen).[42] It was a meeting where we can be sure that Ingen Housz was in attendance along with Fontana's Italian friend, Tiberius Cavallo. Both of them had been with Fontana as he had performed many of these experiments and Cavallo had been on hand, fortunately, to resuscitate Fontana when he almost asphyxiated from trying to breathe the hydrogen. After the third deep breath he had lost consciousness and, when he came to, suffered intense chest pains until the next morning. Various small birds were not so fortunate. Fontana found that they always convulsed and died within a few minutes when put in a chamber of the pure gas. The obvious conclusion of the paper was that 'inflammable air' was not respirable; indeed it was highly toxic unless massively diluted in the atmosphere. This may appear naive but, without our modern understanding of hydrogen and its properties, this research was ground-breaking and important. A new paper by Ingen Housz was read to fellows two weeks later.[43] It was an exposition of a simpler way of preparing volatile ether and using it as the explosive propellant in air pistols. This was applied science and Ingen Housz freely admitted that he was largely relating what he had first seen in Amsterdam in November 1777 when he had been entertained by John Cuthbertson and Hendrik Aeneae. The last third of the paper, however, discussed the theory behind gunpowder and his thoughts here were highly original and intuitive.

Ingen Housz spent much of spring 1778 trying to improve the ether air pistol and it was certainly a novelty to use this type of 'inflammable air' as an explosive. He had shown, by trial and error, that for maximum effect, the volatile ether had to be well diluted and mixed with a bulk of atmospheric air. Just one drop of liquid ether in about ten cubic centimetres of air, the internal volume of his pistol, was about right to create 'a very explosive force', as he formulated for Priestley in a letter of 1 March 1779. Priestley witnessed many of these trials[44] and later published the letter as an appendix in his 1779 book, *Experiments and Observations* of 1781.[45] It was certainly then possible to prime and fire one of these noisy weapons with such a force that the ball would hit a wooden board with an impact that left a deep impression, with any ricochet putting the marksman at risk. But if the 'ether pistol', or any kind of air pistol, was to supersede the powder musket, the loading and priming had to be as slick and the materials just as portable and ready to use. The 'Ingen Housz modification' was the use of a small rubber balloon with a draw-tie neck as the ether reservoir, pre-loaded by means of a glass pipette, and a charged Leyden jar small enough to fit in a pocket to provide the sparks as necessary. He had discovered that, by these means, it took only five or six seconds to prepare for a firing. However, he admitted that at its present state of development, the 'ether/air pistol' was more useful as a demonstration of scientific principles and as a source of 'pleasing amusement' to philosophers and their friends than a reliable weapon, even the one he had had manufactured in brass by Edward Nairne, the London scientific instrument maker.

A theoretical basis for the awesome force of gunpowder had been developed by Isaac Newton, among others. There was a strong consensus that a gas was released in such volume and so instantaneously that the expansive force was virtually incalculable. Now, following Priestley's isolation of 'dephlogisticated air' (oxygen), and having seen how much more powerful was the explosion when inflammable air was burned in this 'new' gas, Ingen Housz took his audience down a fairly obvious intellectual highway. It seemed highly likely, knowing now that gunpowder released, separately, both 'dephlogisticated air' and 'inflammable air', that there was an instantaneous reacting of these two powerful gases when gunpowder was ignited. From the massive increase in bulk, at least several hundredfold, it was easy to see how gunpowder could be so violently destructive. Ingen Housz's mind must, surely, have returned to the horrific scenes he had seen at Brescia in 1769. Then, almost as an afterthought and, presumably, following very late developments, he returned, in a substantial appendix,[46] to the matter of air pistols. He and Fontana had followed their instincts and tried igniting Ingen Housz's pistol with a mixture of ether and oxygen rather than with the earlier ether/air mix. After some tentative trials igniting the two gases in a strong glass vial and being deafened by the reports, the two madcaps tried the obvious experiment using Ingen Housz's brass air pistol, introducing the oxygen through a small hole drilled in the side of the barrel. It was Fontana who actually fired the gun but only after wrapping it up in a towel. A cork, substituting for a bullet, was shattered to pieces as it hit the wall of the room and Fontana felt such a shock through his hands and arms that he was not prepared to repeat the trial unless a stronger pistol could be obtained. It was only possible to communicate this reluctance, however, after some minutes; both men had been temporarily deafened. Nairne was called on to make an altogether stronger pistol using brass the 'thickness of a half-crown' to form a larger air chamber leading, only then, to a narrow barrel. Ingen Housz and Fontana believed that this weapon would fire safely and put a lead bullet, with some leather packing, in the barrel. They were wrong and they were fortunate: the gun was 'rent' into pieces by another deafening explosion. Quite where these experiments were taking place is not recorded but we have to assume some extremely tolerant neighbours or significant local consternation. Nairne's next attempt – making the gun yet stronger and the far end of the air chamber conical to temper the expansion – was more successful and it must have been a weapon of this design that Ingen Housz took back with him to Vienna for these risky experiments were to continue.

THE INVITATION GIVEN to Ingen Housz to deliver the next Bakerian Lecture to the Royal Society, scheduled for the evening meeting of

Thursday 3 June 1779, was a double honour. It was the second consecutive request and since Joseph Banks was now the President there was no element of long-standing personal patronage: the council had approved his selection on merit alone. Submitting his manuscript for Banks' approval on 29 April, Ingen Housz again chose to speak on electricity; the one universal diversion of all contemporary philosophers.[47] Even if electrical phenomena were not their current field of experimentation, nearly all Enlightenment 'savants' had experience of the friction-generated sparks. He took the opportunity to give a history of the flat-plate 'contrivance' that he had, himself, invented about 1763, when still a family doctor at Breda. Since then he had made successive improvements and the progress of these formed the bulk of his lecture. This time there was only a single, passing, mention of Franklin and Ingen Housz was also diplomatic in obfuscating the plagiarism of Jesse Ramsden.[48] On first arrival in London, Ingen Housz had 'shewed it (his plate glass invention) to several of my acquaintance, and in a short time I found such machines ready made at Mr. Ramsden's.'

Originally, Ingen Housz's machine was a circular glass plate, with a central wooden spindle, that could be cranked at high speed. It was spun against several fur-covered pads and the charges led away to a 'prime' conductor. It was immediately obvious that the electricity generated was more powerful than could be obtained from glass globes or cylinders. When Fontana had made a similar machine in Florence, but with two discs on a common spindle, they had found that it could deliver a nasty shock to the operator. Ingen Housz had long known that the composite size of the glass disc, or discs, governed the power of the charge but safety alone restricted the glass dimensions – very large plates could shatter and were especially dangerous if spinning at speed; no charge was worth sustaining a shard wound that might even be lethal. He had therefore moved on to making safer machines using large circular pieces of cardboard that were made 'electrically excitable' by soaking them in amber varnish and allowing it to dry thoroughly and harden.[49] His ingenuity was again at a premium: the apparatus was now much lighter and flying glass shrapnel not an anxiety. Nor was there a limit on disc size or number and he even made, on one occasion, a machine that had twelve such discs on a common spindle, each varnished board separated by an insulating disc of wood. Indeed, the only limitation now was the mounting inertia if all the plates were rubbed by hare fur pads. Once the discs were moving however, they could generate a charge of such force that it was common to see sparks that were thick, very bright and up to two feet long: very few of his friends, he had found, were keen on drawing them onto their knuckles more than once.

The search for perfection was far from over, however. Ingen Housz reported, with creditable honesty, that cardboard disc machines were very prone to loss of function in damp atmospheres – in unheated rooms in winter for instance. Currently, he told fellows, he was working on a machine that simply used a sheet of brown wrapping paper pinned to a circular wooden frame and intended to make one of stretched silk. That he was still trying to improve his invention some fifteen years from its inception and was modest enough to admit it, was typical of the man. Ingen Housz was given the usual honorarium for delivering the lecture – four pounds sterling[50] and presented with a bronze medallion - the one shown in the illustration.

Medallion presented by the Royal Society on the occasion of the Bakerian Lecture 1779.

Also in June 1779, at their meeting on Thursday the 17th, only two weeks after his second Bakerian delivery, yet another paper by Ingen Housz was read to the Royal Society.[51] He had pulled together his ideas and trials for tackling what was an everyday problem in navigation. Many an idle moment, during 1778, had been filled by trying various new ways of floating magnetised needles as used in marine compasses. Although the first, primitive, mariners' compasses had been invented at the beginning of the 14th century they were still notoriously subject to vibration and could be, even if they remained undamaged, dangerously inaccurate in rough seas. The basic principle was secure, however – a centrally suspended and magnetised pointer, usually with the north end marked, was allowed to swing freely and it would always align with the earth's magnetic field. A number of scientific academies had long offered prizes for successful innovators but Ingen Housz was motivated more by the intellectual challenge. The more powerful magnets now available, developed by his friend and fellowship sponsor, the late Gowin Knight, made it possible to suspend them within fluids, rather than just floating them on the surface, without imposing too much inertia and making them insensitive. One of the first suggestions for using fluids to damp unwanted movements in mariners' compasses had been put before the Royal Society by the famous Edward Halley in 1690.[52]

Ingen Housz's ideas were not, perhaps, wholly original but after four uninspired suggestions for contriving to hang actual needles just underwater, he described a more innovative idea and the one that made his paper worthwhile. The needle, he suggested, should be enveloped in a sealed glass tube of air such that when immersed in water its innate buoyancy was offset by a central point pivot hanging down from the glass lid of the compass. To aid stability, an opposing pivot was placed rising up from the bottom. By this means, he argued, the hermetically sealed needle lost none of its ability to find the magnetic meridian but all tendency for unwanted vibration or ship movement were offset by the inertia of the liquid medium – water or, if preferred, an oil such as linseed since this would prevent rusting and would not freeze in cold weather. Finally, he took this idea one stage further, a significant mark of his native ingenuity. He saw that if he made his hermetically sealed tube of thin steel and magnetised it, he could actually dispense with the needle within: the tube itself would act as the compass direction indicator. There appears to have been no immediate practical outcome from all this intellectual exertion: 1813 is recognised as the year in which the first commercially successful fluid-damped mariners' compass appeared on sale.[53] Nevertheless the paper, his sixth to be published in the *Philosophical Transactions*, helped to further establish Ingen Housz as a useful contributing member of the Royal Society, a man of eclectic talents and ability.

I T IS PROBABLY a safe assumption that Ingen Housz saw the end of the 1779 London 'season' in more abrupt terms than the rest of his friends. Beyond it was an extended period that he had been preserving free of commitments of any kind. On 9 May he wrote to the court chamberlain at Vienna, Count Orsini-Rosenberg,[54] to apply for permission to stay in England for another summer and winter and informed Agatha of his intentions a week later.[55] Ingen Housz appears to have assumed that his request would be granted but he did not really know for some six weeks. It was only on 29 June that he reneged on his promise to go to Brussels and inoculate, 'in spring', the Duke of Arenberg's daughter; writing that he would not now be doing so until October.[56] Officialdom at Brussels must have been stupefied by the apparent perversity but they could not have known Ingen Housz's personal priorities. Having successfully delivered his second consecutive Bakerian Lecture on 3 June, a further watershed date had been Thursday 17 June, when he read his paper on marine compasses to the Royal Society. Afterwards a seductive freedom beckoned. There is no indication when it was that his ambition for using this privacy changed but by June the plan to finally finish his book on smallpox inoculation had been substituted.

There had been a sea change in Ingen Housz's thinking. His mind had begun to focus on the intellectual sphere of gases. In a sequence that we can never know and which he, himself, did not perhaps discern consciously, a string of seemingly unconnected ideas had begun to jostle in his thoughts. In what may be the wrong order, then, we must read his mind. One can put the latest influence first. Ingen Housz now knew himself to be an adept and judicious operator of Fontana's eudiometer. His authority could not have been bettered. Having dabbled with his own, unsatisfactory, versions of the instrument at Vienna around 1775,[57] he had then spent hundreds of hours, in Paris in 1777 and, over the last winter and spring, in London, working with the inventor himself. Together they had come to recognise and eliminate the antics of the apparatus and so avoid inconsistencies in its performance. Between them the two friends had gone on fine-tuning the apparatus until they had recognised and knew how to overcome seventeen 'sources of error' which, if not avoided and allowed to be cumulative, could give wholly false readings of the integral 'nitrous test.'[58] He therefore knew that he had the ability to reliably assess the 'quality' of different air samples, even samples that varied only by hair-splitting differences. In other words Ingen Housz was confident that he could assess air 'salubrity' or quality (its oxygen content, to use modern parlance) to a very fine degree.[59] 'I even think with the Abbé, that . . . we may, with as much precision, judge of the degree of purity of common air, as we now are able to judge of its degree of heat and cold by a good thermometer.' He could see that Priestley's original ambition for his 'nitrous test' had, in fact, been fulfilled.

Ingen Housz was also aware that both Priestley and the Swedish chemist, Carl Scheele, had come up against an apparent conflict. It was

Priestley who had first shown[60] that plants somehow purify or revitalise the air, reporting that

> ... on the 17th of August 1771 I put a sprig of mint into a quantity of air in which a wax candle had burned out, and found that, on the 27th of the same month, another candle burned perfectly well in it.

He had also gone on to show that plants could restore fouled air such that animals could then function in it.[61]

> I put a sprig of mint into a glass jar (of fouled air), standing inverted in a vessel of water (and) when it had continued growing there for some months I found that the air would neither extinguish a candle, nor was it at all inconvenient to a mouse which I put into it.

Whatever was vital, in air, for plants was not the same as for animals and Priestley went on to show that plants, trapped in 'vitiated' or 'used' air, could even restore its 'vitality' – indeed they thrived on it. Ingen Housz was fascinated.[62]

> I was struck with admiration . . . (by) his important discovery, that plants wonderfully thrive in putrid air; and that the vegetation of a plant could correct air fouled by the burning of a candle, and restore it again to its former purity and fitness for supporting a flame, and for the respiration of animals . . .

The operative word was 'could'. Such experiments, in Priestley's hands, gave inconsistent results – sometimes the air was not improved and he had concluded that restoration of the air depended on the healthy growth of the plants.[63] 'This restoration of air I found depended upon the vegetating state of the plant . . .' And Carl Scheele, in his book of 1777,[64] had challenged Priestley's findings, for he had found no instances of plants 'making fouled air wholesome' again. 'I kept plants, in the dark as well as exposed to sunlight, in a flask which was filled with vitiated air and carefully secured . . . I tested a little of this air every two days, and always found it vitiated.'

Here, then, in early summer 1779, was a conundrum, an incomplete jigsaw. How was it that plants could purify the air and why did they only do it sometimes? It was Ingen Housz's particular brilliance that he found and fitted the missing pieces, although it was mostly a subconscious process. It is worth noting, however, that two very important factors had already materialized. Priestley seemed to have been unaware that environmental conditions might be important – were his experiments set up in the sun or in the shade? He did, however, use green plant material although this may have been by fluke or by instinct rather than because of any sentient strategy. Scheele, on the other hand, obviously did recognise that light or darkness could be influential but his experimental model involved the emerging roots of germinating peas rather than green stems or leaves. Neither philosopher could satisfactorily explain their results. Scheele, in fact, made no attempt at an explication but Priestley was wont to relate his findings to the growth processes of his plants, that they 'absorbed the phlogiston'[65] from the surrounding air to use as an essential ingredient in accrual. At some stage, probably when least expected, and perhaps during a restless, warm night in London, Ingen Housz must have seen that he could try to unravel this confusion using Fontana's eudiometer: that disciplined experimentation and dogged assiduity might unlock the mystery.

Sir John Pringle had always been impressed by Priestley's findings. He had been given a personal demonstration when he had visited Leeds with Benjamin Franklin in the summer of 1772[66] and the reported inconsistencies had not worried him. With the vision worthy of a President of the Royal Society he had seen, here, a supremely important pattern of nature. He saw the makings of a life-cycle linking plant and animal kingdoms and his grasp of its centrality had featured prominently in the first of his annual 'Discourses', read to the Royal Society in November 1773.[67] Pringle articulated how it must be that plants and animals survive interdependently – that plants thrive in fetid air, thereby cleansing it in favour of animals, whose needs are opposite, and whose respiration reverses the process.[68]

> From these discoveries we are assured, that no vegetable grows in vain, but that, from the oak of the forest to the grass in the field, every individual plant is serviceable to mankind; if not always distinguished by some private virtue, yet making a part of the whole, which cleanses and purifies our atmosphere.

Ingen Housz could want no more prestigious inspiration than this. Unlike Priestley or Scheele, as far as we know, Ingen Housz also wove into his thought processes some observations made and published, 25 years before,[69] by the Genevan philosopher, Charles Bonnet. In this book, in French and therefore fully accessible to Ingen Housz, Bonnet had described his observations on the bubbles that collect on the leaves of plants when they are plunged into fresh water. He believed that the leaves somehow 'drew' the air bubbles from the water in response to heat from the sun for they did not emerge from water which had been boiled nor were they seen to form at night. Here, too, was a scientific story left incomplete and which did not have that ring of truth that good scientists can always recognise.

Ingen Housz, Priestley, Scheele and others were all pioneers in what is now called 'pneumatic chemistry', the rational study of gases, but they were really second generation. It would be unjust to omit the pervading influence of the late Stephen Hales, an experimental philosopher fostered by Isaac

Newton. Hales, a curate at Teddington, had studied animal physiology for many years and took the Harveyan view of the circulation of the blood as the model for the circulation of sap and water through plants. He admitted, however, that he had found no evidence for such when he published his famous treatise in 1727,[70] but did make the inspired but hypothetical observation that plants draw 'through their leaves some part of their nourishment from the air.' Here was a concept that excited his successors, Ingen Housz among them, for it had come to be known that atmospheric air was a mixture of gases and that its constituents could fluctuate with respect to each other. All this, and more, must have been racing through Ingen Housz's mind as he tried to concentrate on preparing his 1779 Bakerian lecture.

Some successful scientists have a very practical bent, are good with their hands, and become persistent and diligent 'experimenters.' Others can never seem to master an apparatus and their laboratory bench will be littered with the crash sites of failed experiments. The latter group often compensate by being the more inspired thinkers, the more incisive questioners, by being more agile and aggressive with ideas. Occasionally, lateral thinking and manual dexterity – crossword and crosswire – combine in lucky individuals who, opportunity given, can forge a breakthrough. One such was Ingen Housz. The intellectual distance he travelled may have been modest and he was propped up, in the famous Newtonian phrase, 'on the shoulders of giants'. But by June of 1779 the revolutionary ideas of Hales, the conflicting observations of Priestley and Scheele, the mis-readings by Bonnet, the inspiration of Pringle, the ownership of a Fontana eudiometer and the skill to use it, all gave Ingen Housz a unique assembly of resources. In that month we find some evidence, tangential on the face of it, that his cerebral processes were moving on to practical considerations. A small volume survives, in the archives at Breda, that is entitled '*aanteken boekje*' (memo book).[71] It begins at his arrival in London in January 1778 and the next few pages record, effectively, his humdrum shopping in the city over the next 18 months. But the entries for June 1779[72] are worth reproducing verbatim. After paying £1 11s 0d (£97) for a pair of doe's skin breeches he went on to buy:

three deal tables for the country house	£ 0 12s 0d
half a dozen table knives and forks	£ 0 7s 6d
a cushion for the arm chair	£ 0 2s 0d
a mahogany dining table	£ 0 18s 0d
a brass socket of the invention of Abbé Fontana for containing the glass measure for air	£ 0 9s 0d
for another ditto – thicker and stronger	£ 0 10s 6d

a tub for experiments with air and a stand to put it upon	£0 18s 0d
different articles of linen for the housekeeping at Southall Green	£14 6s 8d

The outlay of £18 3s 8d (£1143) is probably not the total sum of Ingen Housz's purchases that month but their nature is a valuable set of clues that take us, in a very tangible way, into the next five months of his life, a time 'in which I never worked so much and with so much pleasure as I did this time in England'.[73] It was to be worth it: it was to be the apogee of his scientific career. Ingen Housz had long intended, for the purpose of writing a book and needing a peaceful retreat, to lease a country dwelling. It appears to have been only partly furnished and, from the fact that he knew of specific needs for their comfort there, he had obviously already visited, presumably with Dominique. He finds he has to spend a considerable sum on linen but most of his purchases relate to an otherwise undeclared intention – to spend his time performing clandestine 'pneumatic' experiments, the 'tub' or 'trough' being a classical piece of apparatus of this discipline, the brainchild of Hales. These revealing clues are readily available in his little notebook, a personal item where confidentiality would have been assured. Otherwise, it has been impossible to find any other documentary evidence of his plans and in particular to locate, definitively, the exact property that he rented.[74] His presentiment for secrecy may have been to avert the daily interruptions that had annoyed him so in London but it was to prove wise for other reasons. Southall Green was then a very small hamlet some two hours carriage ride to the west of Westminster. It enjoyed rural seclusion in the divarication of two arterial roads heading out of London – that to Bath passing south of the village and that to Oxford skirting the north of the parish. The Grand Union Canal and the railway, such strong features of Southall (as this densely populated London suburb is now called) were still in the future. It was here, then, in a house that must have been private and substantial that Ingen Housz spent that summer, between June and 22 October, when the first frosts hastened his return to his lodgings in Pall Mall.[75]

IN MID-JUNE 1779 a carriage bearing two middle-aged men chalked its way into the small, secluded community of Southall Green; and so did a wagon groaning under furniture, trunks and crates. Settling in would probably have tried his patience. He would have been longing to start his investigations. Perhaps the house had been leased with some staff since Ingen Housz's own 'household', in England, consisted only of his manservant Dominique. There may have been, at least, a cook/housekeeper but the two men may well have subsisted in a rather Spartan environment, relying, perhaps on food cooked at a neighbouring household and a few hours of service each week from a visiting maid or cleaner. There was probably a team of gardeners, for the grounds around the property accommodated select flowering plants, vegetables and herbs, soft fruit, top fruit and specimen trees such as a Cedar of Lebanon. Ingen Housz also records exotic fruits such as mulberries, peaches and lemons. There was certainly a well in the garden and nearby, in the middle of the hamlet – 'on the highway' – there was a pump atop a water-trough 'where horses could drink.' We know of these botanical and aquatic attributes of the property because they were to be among the raw material for Ingen Housz's activities there.[76]

Within hours of his arrival, we can imagine Ingen Housz carefully unpacking his eudiometer tubes and all the other glassware, checking it anxiously for damage. Then there was the pneumatic trough to set up and fill and all the chemicals to store away where they could not find their way into the soup. As the domestic unpacking continued, Ingen Housz was probably already preparing measures of 'nitrous air' and sampling the atmosphere at Southall Green. He seems not to have kept any record of his first month of 'retirement', a time in which he probably explored his ideas with no fixed plan. It was during this 'lost' time, however, that he very quickly stumbled across one of nature's most important processes for maintaining life on earth.[77]

> ... from Dr. Priestley's works, I had little doubt but there was some quality in plants proper for correcting bad air, and improving ordinary air. My curiosity led me to investigate in what manner this operation is carried on, whether the plants mend air by absorbing, as part of nourishment, the phlogistic matter, and leaving thus the remainder of the air pure (to which Mr. Priestley inclines the most); or whether perhaps the plants possess some particular virtue hitherto unknown, by which they change bad air into good air, and good into better, which I suspected to be the case. I was not long engaged in this enquiry before I saw a most important scene opened to my view; I observed, that plants not only have faculty to correct bad air in six or ten days, by growing in it, as the experiments of Dr. Priestley indicate, but that they perform this important office in a compleat manner in a few hours; that this wonderful operation is by no means owing to the vegetation of the plant, but to the influence of light of the sun upon the plant.

Seclusion brings disadvantages as well as benefits and having no one with whom he could debate his exciting findings must have been exasperating, especially for someone who had so many friends. He had the self-discipline, though, to see that he would need to repeat his early experiments many times if he was to defeat the inevitable scepticism he would find in the always overheated philosophical cockpit. But he knew, now, that he could explain the inconsistent findings that Priestley had found so frustrating, not understanding that he needed to allow for the factor of light. The sun must be the missing link. But Ingen Housz had the experience to know that other such environmental circumstances, if he was not extremely careful to think of them all and disprove any influence, could easily put him in the same unhappy embarrassment as Priestley. Inspiration, as always, was to be lost to perspiration, of which there was to be plenty.

The original and very detailed notes of Ingen Housz's 1779 experiments[78] have survived as an evocative testament to his subsequent diligence. They reveal a creditable degree of organisation and efficiency (see page 526). Whilst one set of experiments was 'cooking' in the sun, or in the shade, Ingen Housz was busy at the eudiometer, measuring the quality of air in the jars from previous trials or setting up the next batch of tests. He seemed never to rest, never to tire and worked from dawn to dusk seven days a week. His notes began on Saturday 17 July with what he chose to call experiment no. 1 but many dozens of experiments, perhaps more, must have preceded this. Then, on 1 August, he began to number his experiments afresh, anything from four to seventeen per day, depending on their complexity. Number one was that Saturday morning and by Monday 4 October, nine weeks later, he had reached experiment 353. Ingen Housz referred, later, to over 500 such experiments,[79] implying that he had been just as busy in June and early July. But the message was always the same, the evidence base for what would be his conclusions could probably not have been bettered, at least by any one sane man working on his own. He worked with such obsession, fanatically almost, during those summer months of 1779 that it is virtually impossible to summarise his activities coherently.

Ingen Housz was certainly totally absorbed at Southall Green from mid June until Friday 6 August, when he returned to London for two days.[80] It is noteworthy that he was not, therefore, in the city on a Sunday, usually the favoured day for one of Sir John Pringle's regular soirées, although he certainly called on his old patron. Although Pringle had retired as Royal Society President and was not so vigorous – he had been to Bath to take the waters in May and would visit again in late September[81] – he was still the pivotal figure of a very prominent group of philosophers. Ingen Housz appears to have outlined his findings on leaves and sunlight to his old friend and patron in private, assuming, wrongly as it turned out,[82] that they would be kept confidential. Pringle advised that it would be wise to perform yet more experiments for on returning to his retreat Ingen Housz performed 78 more experiments in the next nine days. There then follows a three-day gap in the experimental sequence and he probably returned to London for another brief visit. We may assume he was reporting to Pringle again and that he fully convinced him this time for there was a subsequent dropping off in the tempo of the experiments. Perhaps Pringle agreed, now, that it was time to start formally writing up results. But he appears, still, to have made one or two suggestions for further experiments. Ingen Housz made a significant detour on his way back to Southall – to the banks of the River Thames – from where he collected some rushes and water lilies. And, during the next few days, he made an excursion across to the south bank of the Thames[83] where he 'obtained some branches of foreign plants from Mr Eton (Aiton), the King's Gardener at Kew.'

Ingen Housz must have smiled wryly to himself as soon as he recognised the critical factor in the 'air cleansing' facility of plants. He had set himself up in a 'laboratory' for the investigation of the one environmental factor for which an English summer is notoriously lacking – sunlight. However, he was probably ideally placed. Intemperate glare and its concomitant heat would probably have ruined many of his experiments by damaging his plant specimens. Many of them were, after all, in artificial circumstances and he was obviously conscious of this.[84] At first he bent pliable land plants, 'the root remaining in its own earth', into inverted glass jars and sealed them in so that the air surrounding them was trapped. At the same time, he saw no reason not to study aquatic plants in their natural environment – under water. But when he saw that his findings were so clear-cut in only a few hours, he realised that he could take cuttings away from plants, terrestrial ones included, and safely submerge them so that he could watch and test accurately any air they released as bubbles. He was then able to repeat and much extend the observations made by Bonnet[85] concluding, correctly, that the process must be more complex than the leaves merely acting as a focal point for the small amount of air dissolved in any water.

> This air . . . appears soon upon the surface of the leaves in different forms, most generally in the form of round bubbles, which, increasing gradually in size, and detaching themselves from the leaves, rise up and settle at the inverted bottom of the jar . . . It is not very rare to see these bubbles so quickly succeeding one another, that they rise from the same spot almost in a continual stream . . .

Ingen Housz observed closely how and where the bubbles formed and recorded, with fascination, that this was as unique a species characteristic as plant silhouette or leaf shape. This was the end, however, of his scrutiny of the work of others. For the rest of that summer his work was unprecedented and his evidence groundbreaking. His ceaseless drive to test his theories repeatedly in different circumstances and with different plants is quite astounding as is his tangible fear of jumping to untenable conclusions. His diligence was a credit to Enlightenment attitudes and to disciplined scientific method. Many generations later his experiments were seen as 'a masterpiece of manipulation and self-criticism.'[86] He certainly used wide-angle vision. He tested plants in different kinds of air, at different temperatures, at all times of day and in all weathers. He used water from different sources, some in transparent glass, some in darkened glass. He prepared his own 'nitrous air' (nitric oxide) using different metals and he repeatedly crosschecked his 'Fontana' eudiometric readings by using, alongside, a 'Priestley' apparatus, lighted candles or, sometimes, unfortunate chicks. And the different plant components he used were legion – leaves, obviously, but also stems, roots, flowers and fruits. Fortunately, as the son of an apothecary and after medical training, he had a superior knowledge of plants. Herbs and medicinal plants were prominent but he also used culinary plants, poisonous plants and even exotic foreign plants. In some of his control experiments he also used dead plants and other inert materials.

Ingen Housz's early discovery that sunlight was the critical factor in enabling leaves to 'cleanse' the air around them governed all his subsequent work at Southall Green. It is as if he had immediately won a fortune on the French lottery, to which he regularly subscribed,[87] and then spent three months checking, and rechecking, his ticket number before being convinced of his success. But in scientific endeavour such behaviour is far from eccentric. Indeed it is mandatory and, as a consequence, the previously unsuspected power of sunlight was not his only discovery. There is an impressive catalogue. Using modern scientific concepts and language, since each of them has stood the test of time, we can list these as follows:
  i. that in sunlight the green parts of plants release oxygen;
  ii. that the oxygen is released, in the main, from the underside of leaves;
  iii. that the speed of the process is governed by light intensity;
  iv. but with a time delay after first exposure to the sun and a run-on after sunset;
  v. that the process is not dependent on heat;
  vi. that the reverse of (i) happens in darkness i.e. oxygen is absorbed by plants;

vii. but this reverse process is on a smaller scale than (i);
viii. that young leaves are less effective than mature ones although growth itself is not responsible for the processes;
ix. that roots, flowers and fruit absorb oxygen by day and night.

And by continuing in our modern idiom, we can summarise with the assertion that it was Ingen Housz who discovered the crux of photosynthesis, notably the vital 'heart' of the process – that green plant material absorbs carbon dioxide from the air and replaces it with oxygen <u>in sunlight</u>.[88] Pringle's seminal concept of a life cycle, that had somewhat preceded the evidence, had been proven; for animals do the opposite, continually absorbing oxygen and releasing carbon dioxide. Having surveyed the situation very much in the modern vernacular we owe it to Ingen Housz to present how he expressed his own summary of 'the secret operations of plants I discovered in my retirement'. Although it must be one of the longest sentences ever written,[89] even in the eighteenth century, it is remarkably eloquent for a man not using his native tongue, allowing that it was probably edited by a native speaker of English.

> I found that plants have, moreover, a most surprizing faculty of elaborating the air which they contain, and undoubtedly absorb continually from the common atmosphere, into real and fine dephlogisiticated air; that they pour down continually, if I may so express myself, a shower of this depurated air, which, diffusing itself through the common mass of the atmosphere contributes to render it more fit for animal life; that this operation is far from being carried on constantly, but begins only after the sun has for some time made his appearance above the horizon, and has, by his influence, prepared the plants to begin anew their beneficial operation upon the air, and thus upon the animal creation, which was stopt during the darkness of the night; that this operation of the plants is more or less brisk in proportion to the clearness of the day, and the exposition of the plants more or less adapted to receive the direct influence of that great luminary; that plants shaded by high buildings, or growing under a dark shade of other plants, do not perform this office, but, on the contrary, throw out an air hurtful to animals, and even contaminate the air which surrounds them; that this operation of plants diminishes towards the close of the day, and ceases entirely at sun-set, except in a few plants, which continue this duty somewhat longer than others; that this office is not performed by the whole plant, but only by the leaves and the green stalks that support them; that acrid, ill-scented, and even the most poisonous plants perform this office in common with the mildest and the most salutary; that the most part of leaves pour out the greatest quantity of this

dephlogisticated air from their under surface, principally those of lofty trees; that young leaves, not yet come to their full perfection, yield dephlogisticated air less in quantity, and of an inferior quality, than what is produced by full-grown and old leaves; that some plants elaborate dephlogisticated air better than others; that some of the aquatic plants seem to excell in this operation; that all plants contaminate the surrounding air by night, and even in the day-time in shaded places; that, however, some of those which are inferior to none in yielding beneficial air in the sun-shine, surpass others in the power of infecting the circumambient air in the dark, even to such a degree, that in a few hours they render a great body of good air so noxious, that an animal placed in it loses its life in a few seconds; that all flowers render the surrounding air highly noxious, equally by night and by day; that the roots removed from the ground do the same, some few, however, excepted; but that in general fruits have the same deleterious quality at all times, though principally in the dark, and many to such an astonishing degree, that even some of those fruits which are the most delicious, as, for instance, peaches, contaminate so much the common air as would endanger us to lose our lives, if we were shut up in a room in which a great deal of such fruits are stored up; that the sun by itself has no power to mend air without the concurrence of plants, but on the contrary is apt to contaminate it. These are some of the secret operations of plants I discovered in my retirement.

 Ingen Housz must have made arrangements to have his book published in early September 1779. He had hurried back from Southall Green to central London on Sunday 12 September, presumably clutching a first batch of manuscript pages, and stayed in the city for five days. He makes no reference to someone helping him with the script but the fact that he chose Peter Elmsly of Fleet Street as his publisher is no surprise. The two men already knew each other through the works of Hoogeveen and Jacquin and Elmsly was well known for publishing the books of foreigners. This book, though, was in English and it is a moot point how the very good quality script came about. The book is written with impeccable grammar and with a literary quality that was probably beyond the author himself but, surprisingly, no acknowledgement is given for any help from a native speaker. There is a clue, from later correspondence,[90] that the editor might have been Richard Huck, someone very close to Pringle and who might have been persuaded to help whilst preferring anonymity. About a third of the printed pages are, in reality, a detailed manual for using the latest Fontana eudiometer with accuracy, pages that could even have been written before June. Nonetheless, the remainder of the book is a detailed description of Ingen Housz's hundreds of experiments, re-sorted by 'investigatory facet'. The manuscript must have been written at an incredible pace: the occasional mis-copying of data to be found on very detailed comparisons between his experimental notes and printed text[91] is no surprise and does not detract from the whole. By long tradition in publishing, the dedication (to Sir John Pringle) and preface were written last; dated 12 October 1779 and presumably hurried into Elmsly with the ink barely dry. The book, 319 pages (including index) in folio, was then finished and stitched within hours and went on sale shortly thereafter at three shillings and eight pence (£11.50) per copy,[92] Elmsly taking a 5% commission on every sale. About 600 copies of *'Upon vegetables, discovering their great power in purifying the common air in the sunshine, and of injuring it in the shade and at night.'* were printed,[93] and Ingen Housz himself must have taken some twenty to thirty copies – perhaps from a pre-arranged complimentary overrun. *Vegetables* was not the first book that Ingen Housz had brought into the world but this time it was his very own, the documentation of his supreme scientific achievement.

ON FRIDAY 15 October Ingen Housz caught the coach to Bath.[94] Sir John Pringle had returned there to take the waters again and if his protégé wanted to present him, personally, with a copy of 'the' book before leaving England, it meant a trip to the west of England. Ingen Housz also took the opportunity to accept a long-standing invitation to visit Lord Shelburne at his country estate, Bowood House near Calne in Wiltshire. Calne is astride one of the London-Bath routes so the detour was an easy one and made on the way west, Ingen Housz arriving at Bowood, it appears, on 16 or 17 October.[95] Priestley was at Bowood so there was a double reunion and two book presentations. That Priestley was one of the very first recipients of a copy was to become a matter of very high significance in the near future. Ingen Housz was introduced to Christopher Allsup, a family doctor of Calne and who was the usual 'practitioner' called upon for medical problems at Bowood. Allsup was someone Ingen Housz would re-encounter later in life and emerge as a very significant medical colleague.

Bath is only a few hours journey from Calne and Ingen Housz was with Pringle for a few days around 20 October. Pringle was still not his former dynamic self and the meeting must have been a painful one, having all the appearances of being likely to be their last (as, indeed it was). Understandably, he was enchanted by the book and promptly nominated Ingen Housz as a member of the newly formed Bath Philosophical Society although the new

> EXPERIMENTS
>
> UPON
>
> VEGETABLES,
>
> DISCOVERING
>
> Their great Power of purifying the Common Air in the Sun-fhine,
>
> AND OF
>
> Injuring it in the Shade and at Night.
>
> TO WHICH IS JOINED,
>
> A new Method of examining the accurate Degree of Salubrity of the Atmofphere.
>
> By JOHN INGEN-HOUSZ,
> Counfellor of the Court and Body Phyfician to their IMPERIAL and ROYAL MAJESTIES, F. R. S. &c. &c.
>
> ---
>
> LONDON:
> Printed for P. ELMSLY, in the Strand; and H. PAYNE, in Pall Mall. 1779.

Title page of Jan Ingen Housz's book on plant experiments 1779.

member did not learn of his nomination and subscription-free election for some years.[96] The lonely, introspective journey back to London was accomplished by about 26 October giving Ingen Housz only a few days to pack and prepare for his departure, to Belgium in the first instance.

Ingen Housz left London over the last weekend of October 1779. He left Schwediaur, who had been ill,[97] in England. He made his way, with Dominique, to Gravesend where they boarded a packet boat for the continent. The ship lay bobbing at anchor for two days because of unfavourable winds in the Thames estuary but Ingen Housz was far from idle. Much to the intrigue, and amazement perhaps, of the other passengers he filled, with harbour water, one of the ship's buckets on deck and proceeded to test the air with a portable eudiometer acquired from Benjamin Martin, the famous spectacle and glass instrument maker of Fleet Street. Differing locations were now the point of the trials and, as he reported to Sir John.[98] 'I experimented as I proceeded on my journey; in noisy inns, on shipboard and such like places little adapted to philosophical application. I hope you will make some allowance for the inaccuracies . . .' Sir John Pringle had encouraged him to try to obtain good evidence for what he, himself, had long believed – that the 'salubrity' (quality) of the air around coasts, and certainly out at sea, must be exceptionally high. Ingen Housz was equally convinced

> that great bodies of water, such as seas and lakes, are inducive to the breath of animals by purifying and cleaning the air contaminated by their breathing it; so that the salutary gales by which this infected air is conveyed to the water, and by them returned again to the land . . .

At Gravesend, however, he found the air quality exactly as it had been in central London. This was disappointing, but no reason not to persevere and on 3 November, eventually under way, he tested the air out in the Thames estuary. They were between Sheerness and Margate and the weather was 'agreeable and warm and the sun shone very bright'. He was much encouraged by the readings – they suggested that sea air was certainly more 'salubrious'. His next attempted trials, in the middle of the English Channel, had to be abandoned. The ship's motion – it was 'tossed by the wind and waves' – made it impossible to use his apparatus. Unbowed, he filled three glass phials with the ambient air to test, later, ashore. Further trials were made on the beach at Ostende, and then in Bruges, Ghent and, between 12 and 21 November, in Brussels. His readings began to suggest some sort of pattern: nowhere gave him readings as high as those he had obtained off Margate but all the differences were subtle. He was frustrated that the samples he had taken mid-channel had been spoiled by their corks working loose. He certainly found no evidence that the citizens of the lower parts of Brussels were breathing poorer air than those who lived in the higher parts, contradicting the local rumour. The remainder of his journey – to Breda, Rotterdam, Delft, Amsterdam and returning through Antwerp to Brussels on 19 December – was interspersed with many more eudiometry experiments. His readings, however, were never more than a few percentage points apart whatever the local circumstances and the weather. In retrospect we can see that Ingen Housz drew, wrongly, significant conclusions from his itinerant research.[99]

It appears from these experiments that the air at sea, and close to it, is in general purer and fitter for animal life than the air on the land tho' it seems to be subject to the same inconsistencies in its degree of purity with that of the land so that we may now, with more confidence, send our patients labouring under consumptive disorders to the sea, or at least to places situated close to the sea.

On the other hand, the thesis of the times – that good and bad airs were major players in health and disease – allowed that even minor differences in quality were relevant and the Royal Society was happy to publish Ingen Housz's findings, written as a letter to Pringle, dated 22 January 1780. The paper was read to fellows at the evening meeting of Thursday 24 April. The author himself was in France, having arrived at a frozen Paris on Wednesday 29 December. He and Dominique found rooms at the Hotel Dubouloir[100] but within hours Ingen Housz was in Passy where Franklin was still in residence. It was more than two and half years since Ingen Housz had left Vienna and he was not to return for some six months, keeping up his air experiments all the while. Having set off on a long journey that had every prospect of being recreational, and, indeed, was almost frivolous for some time, it had become one of constant demand and hard work driven by an impelling desire to unravel nature's workings, to be an enlightened investigator.

14
Home – to Vienna

Two well-dressed men cross the Pont Neuf deep in conversation. They're heading for the north bank of the Seine, tailed by two servants who are shuffling along, half-sideways, carrying a heavy box between them. They pretend to ignore the frowns of the onlookers as they descend the stone steps down to embankment level and wait at the door of the hot bathhouse. It's firmly closed but they know better and knock. The two gentlemen planned this excursion some weeks ago but have been waiting for warmer weather, when there will be less fog above the hot water. However, Ingen Housz has now been called to Brussels on urgent business and they will have to make the best of the circumstances in the late spring humidity. The proprietor, impressed by their links to the Royal Academy, has promised to open up early for them on a Sunday morning when they can have a few hours of privacy.

Once undressed and in the water, they perch their box of equipment on the tiled surround of the bath and begin their trial. They've convinced each other that they will be able to collect, in glass vials held underwater, the bubbles that rise from their submerged limbs. They realise that they may not gather enough gas to test sensibly but they have a theory that, in the right conditions, they will be able to express further gas, as bubbles, from their skin, when subject to pressure or friction. They're both trying to maintain a serious attitude, an objective approach to their experiment, but when they finally realise they're not succeeding, they corpse – both start giggling at one and the same time. It's a nice day, the bath is quiet, the water warm and they enjoy each other's company. But their theory is froth, or lack of it. They giggle and shiver uncontrollably and punch each other on the arm. Having packed away the glassware in the straw nests inside the box, the giggles recede to smiles as they sink down into the warmth, trying not to swallow any of the water. Science should always be fun.

INGEN HOUSZ AND Dominique arrived in a Paris white with frost on Wednesday 29 December 1779.[1] To his relief, Ingen Housz discovered that Franklin was still at Passy. Even better, the royal physician was still a welcome guest at the Chaumont establishment and he was reinstalled there before 1780 was ten days old.[2] The American independence war was reaching its climax and much still depended on the support of the French: Franklin remained a decisive piece on the international chessboard. The French had taken him to heart, and he them: he was now, perhaps, the most famous man in France.[3]

> Franklin's own popularity was so widespread that it does not seem exaggerated to call it a mania. Mobbed wherever he went, and especially whenever he set foot outside his house in Passy, he was probably better known by sight than the King . . . (and his) undressed hanks of white hair and ostentatiously unostentatious brown coat, deliberately worn at court audiences, were expressly affected with public sensation in mind and they succeeded brilliantly.

However, Franklin's diplomatic responsibilities were beginning to be too taxing. Age and gout were taking a toll and he had recently written, to the Congress, a plea that he be replaced: he had, he pointed out, now been 'engaged in public affairs' for over 50 of his 75 years.[4] He had developed an unshakeable confidence in the American cause and was certain of victory. He was jaded, and wanted to go home, but at least the reappearance of his Dutch friend was a fillip. Philosophy rebounded up his daily agenda and there was much in the way of distracting news; joy in being reconnected, by proxy at least, with London life and friends. Ingen Housz had brought greetings from Margaret Stevenson, Franklin's old landlady and close friend in London who had asked him to take to Paris a silver milk jug that Franklin had very much longed to see again. Franklin wrote, immediately, a thank-you letter but directed it to Mary Hewson, Mrs Stevenson's daughter,[5] having, perhaps, an ulterior motive, for he invited her, in the letter, to join him at Passy where he was, he described, living 'in a pleasant village within two miles of Paris; a lofty situation with good air, with a fine large garden and neighbouring woods to walk in.' Nothing came of the whimsy seduction but at least we can see why Ingen Housz had other reasons to be happy at Passy.

Ingen Housz had brought other welcome papers and letters including one from Joseph Priestley[6] and, naturally, some copies of *Vegetables*. Franklin was delighted with the book, writing to a physician friend at Philadelphia[7] that he wished the American Philosophical Society would elect Ingen Housz a member, an honour to both parties, for his Dutch friend had 'made some great discoveries lately respecting the leaves of trees in improving air for the use of animals.' And he was enthusiastic that Ingen Housz was continuing to investigate the many nooks and crannies of his major discovery and planning a larger, French, edition of the work. Every day with Franklin was to be treasured but all through the spring of 1780 Ingen Housz continued with his enormous sequence of plant experiments, striving with his Fontana eudiometer over a water trough day after day. There appear to have been few, if any, new findings but he remained determined not to be found wanting by any hovering critics.

Ingen Housz returned to Brussels in early May[8] and, though the reason is nowhere explicit, we may assume that he finally inoculated the Duke of Arenberg's little daughter. His four-week stay in the city would certainly have given him time to find a suitable donor case and see a patient through the course of the procedure into full recovery. Two long letters, between Ingen Housz and Franklin, that crossed between Paris and Brussels during that May, appear to have been a continuation, by correspondence, of a face-to-face debate they had been having when the former was dragged away by duty and when there was a real possibility that Franklin might not be at Paris on his return. As it happened, Franklin's hopes of an early return to Philadelphia were crushed by the Congress but the question and answer sequence on electrical theory[9] was a satisfying exercise for both 'sparking' exponents: their future correspondence was to be much less philosophical and prompted, more and more, by painful personal circumstances. Ingen Housz was back in Paris by the end of the month[10] but only for a few, final, days. This time the parting from Franklin was very unlikely to be an *au revoir* and both men knew it. There was a very painful separation sometime in the first ten days of June.

V EGETABLES WAS PROVING to be a successful publication, the original London print run selling out within a year.[11] Ingen Housz had taken the opportunity to distribute a few complimentary copies to friends in the Netherlands; in Amsterdam, in Rotterdam and, of course, to Louis at Breda. At Delft, on 29 November,[12] he had delighted in giving Hoogeveen a copy and there had obviously been a conversation centred on arranging for a Dutch version of the book. It had been Hoogeveen, perhaps, who had suggested a likely translator – a near neighbour who was a young physician in Delft; someone in whom Hoogeveen had recognised gilt-edged literary talent. Ingen Housz was promptly introduced and Jacob van Breda became an instant friend. The two men were to become very regular correspondents over the next twenty years, their letters providing enormous insight into the subsequent lives of both of them. Van Breda instantly agreed to translate

Vegetables into Dutch and set about the task with impressive speed, pausing only to translate, also into Dutch, Ingen Housz's Bakerian lecture to the Royal Society – *The phenomena of the electrophorus* – by way of a sampler. Translation of the book was finished by late May 1780.[13] Van Breda was not only efficient; he was as technically superlative, as a translator, as Hoogeveen had forecast, and Ingen Housz was most impressed.[14]

> The translation you recently did, of my memoire, to the Royal Society at London, on the nature of the *Electrophorus Perpetuus*, assures me that your translation will present the conceptions of the author most faithfully and that they will be received by our fellow patriots with as much favour as the original edition was received by the British nation.

The moment Ingen Housz had finished editing a final bundle of minor additions to his original *Vegetables* text, he sent them to van Breda with a covering note to say that he was perfectly happy to have the book published unseen. It was quickly sent to the presses of van der Smout and de Groot, publishers at Delft, and the Dutch version of *Vegetables* was therefore the first translation of the work to appear in print[15] – but only just. Work on Ingen Housz's own translation into French[16] was finished but he had to leave Paris before all the pages had been printed. Didot, the printer, had to then send sample pages to Vienna in order for Ingen Housz to construct the index. And the predictable delays of having the press in Paris and the author in Vienna were aggravated by the usual frustrating wait for the French royal censor[17] to sanction actual publication. An unexpected version in German filled the void.

Ingen Housz was surprised to find, on arriving back at Vienna, that Niklas Molitor, soon to be professor of chemistry at Mainz, had somehow obtained an English copy and overseen a rendering in his own, German, language[18] and that this was already being printed at Leipzig.[19] In fact, things were free-wheeling out of control by late summer 1780. Someone, Ingen Housz was also to learn at Vienna[20], was already frantically translating his text into Italian. And there had been a letter awaiting his homecoming to tell him that professor Jean-Jacques Barbier de Tinant, at Strasbourg, to whom he had given a presentation copy only a mere month earlier, had barely read the book before making his own translation into French.[21] All of this was unsettling. As Ingen Housz saw it, his peers were queuing up to be presumptive and over-hasty translators and publishers. He had now come to see his first, hurried edition of *Vegetables*, in English, as wanting in many places and significantly flawed in others.[22] These were embarrassing imperfections that were now being broadcast in several different languages before he could launch his enlarged, and much improved, French edition.

In the meantime, Ingen Housz had returned to working on his book on smallpox inoculation; the project that had been put aside at his rural retreat in early summer 1779 when his flood of ideas on gases had breached the dam of this particular ambition. The book was eventually assembled by November 1780[23] but he was still far from happy with it. It was to moulder further. Presumably aiming at an international medical readership, he had written it in Latin but was apprehensive about the quality of his grammar; his life had become so crowded that he felt he had lost his familiarity and former agility with the strict syntax.[24] He was sure, he wrote to van Breda, that 'Mr. Hoogeveen or one of his sons would correct any errors in the Latin'. Ingen Housz had already negotiated attractive terms at a Paris publishing house – for the book to be typeset at the printer's expense, for 100 free copies to be set aside for his personal distribution among friends, and for any eventual profit to be shared fifty-fifty.[25] Hoogeveen was certainly prepared[26] to correct the manuscript: he probably saw it as a way of paying back the debt of gratitude that he owed to his star pupil for arranging the successful London launch, by subscription, of *Particles*. And van Breda was just as enthusiastic in seeking better terms from a Dutch printer.[27] But sometimes a project will stall even when circumstances seem favourable and the donkey-work is done. There are incidental references to the lethargy of various printers in subsequent correspondence but perhaps Ingen Housz himself simply lost interest in the work. This is disappointing, for the text could have given us many new insights into Ingen Housz's personal inoculation refinements and probably more intimate details of his pioneering exploits in Vienna and elsewhere. The whereabouts of the manuscript, if it has survived, is unknown and the book, had it reached the presses, would certainly have secured him a more fitting prominence in the history of man's long battle against smallpox. Perhaps, though, it is superficial to assume that Ingen Housz only lost traction. He may have been seriously distracted, having something on his mind, something to regret. The obvious possibility is the fact that he had speculated with very large sums of money whilst in London and Paris and was, perhaps, already coming under the cloud of anxiety that was now to test his emotional buoyancy for several years.

S OME TIME WHILE he was in England in the summer of 1778, Ingen Housz met an American called John Williams (as opposed to the younger Jonathon Williams, Franklin's other distant relative). Williams had been, for several years in the late 1760s, inspector general of customs at

Boston, his native town. He had then moved to London where his business activities appear to have been numerous, if somewhat furtive. His kinship with Franklin may have commended him to Ingen Housz. A friendship developed between the two men and they fell into conversation about their futures. Ingen Housz had worries about his, and one long-term strategy involved his moving to America. He wanted to secure, therefore, in the new world,[28]

> an independent fortune, on which I intended very positively to go and live when I should find my income in the old world become a precarious possession, which I apprehended may be the case when the sense of gratitude for my past services should wear out or cool.

Williams grasped the point readily enough and convinced Ingen Housz that there could be no better time to invest in America against his future needs there. The American currency had devalued by about 40% against sterling because of the War of Independence so that he could buy dollars cheaply and, when the war was settled, reflation would boost his investment. Or he could buy some Massachusetts state bonds; besides shadowing the value of the dollar, they also gave interest. Ingen Housz was convinced enough to entrust £2000 sterling (£126,000), to Williams,[29] that could be so invested on his behalf. But the Bostonian, knowing Ingen Housz to be significantly wealthy, also successfully persuaded his new friend to speculate further. Williams was entrusted with an unspecified further sum that he promised to sink into one of the many mercantile ventures, in America, that were 'certain' of outstanding profits. These were only the first of several punts that Ingen Housz was to make in America at this period in his life. In October 1778 Williams left for Nantes and thence back to Boston where he duly traded Ingen Housz's 2000 pounds into bonds worth 7000 dollars (£126,000) and the remainder of the money into an investment in the dyeing industry in Carolina.

On his return to Passy in early January 1780, Ingen Housz was in an expansive and optimistic mood; a huge number of experiments behind him, a fundamental philosophical discovery to his name and the resulting book well received. *Vegetables* was a crowning of all his successes in England. He was also flush with cash after selling the 'micropicture', given to him by Maria Theresa.[30] It had been sold, in strictest confidence, to an English aristocratic family for a considerable sum, a figure large enough, it was reported, for Ingen Housz to have bought, outright, a sizeable mansion.[31] Franklin himself was now a human lodestone for a stream of American visitors. Among these was Samuel Wharton, a prominent merchant and land speculator at Philadelphia who had been a friend of Franklin for some years.[32] Ingen Housz was introduced: it was a fateful meeting. Franklin was quite prepared to vouch for Wharton's character, which he 'confirmed in the most significant terms',[33] and Ingen Housz was seduced into two more American business ventures that did seem, admittedly, to have excellent prospects. Wharton was quite convinced, that with accelerating inflation caused by the colonists' independence struggle, capital investments in the American cause would deliver huge returns when the dollar rose again in value, as it surely would. This reinforced the recent advice of John Williams and Ingen Housz was pleased to know that Wharton was willing to buy further bonds for him on his return to Philadelphia. Ingen Housz sought, and obtained, a nod from Franklin and whatever heavy air of anxiety there might have been about sinking more money into a country that was not yet even born, was blown away by the exuberant enthusiasm of his host at Passy, le Ray de Chaumont, who 'encouraged me a great deal in this undertaking by telling me that . . . one could reasonably expect to secure in America a fortune at least 25 to 30 times worth the sum employed.'[34] A sum of 8,000 livres tournois (£21000) was therefore entrusted to Wharton.[35]

Spurred on by seeing that Ingen Housz was willing to speculate and, no doubt, riding high on his salesmanship, Wharton unwrapped a further tasty investment.[36] He had just returned from a tour of several cloth-manufacturing towns in northern France, during December 1779, where he and le Ray de Chaumont's son had been guided by the Flemish broker acting as American agent at Dunkerque, Francis Coffyn. The American independence struggle was also presenting unique business opportunities to those willing and able to invest in European goods and transport them across the Atlantic. Routine commercial trade between Britain and its American colonies had obviously stalled and demand for a wide range of consumer products in America was forcing up prices there. Wharton had conceived a syndicated scheme involving the purchase of Flanders linen that he would sell at Philadelphia on his return. The mark-up was likely to be three or four-fold; almost too good to be true. Nevertheless, Coffyn had agreed to put up a small stake as well as actually purchase the linen on behalf of the syndicate. Wharton proposed that he and Ingen Housz each take a third stake and Coffyn one sixth and before leaving Passy he recruited a young American called Edward Bancroft, working, ostensibly, as Franklin's secretary, to take the remaining sixth.[37] If Ingen Housz had known, if anyone at Passy had known, that Bancroft had a very lucrative secret life there might have been second thoughts about the whole compact. He was acting as a double agent for the English and for the

Americans and was certainly not trustworthy.[38] However, no one suspected Bancroft at the time and the deal went through. As Wharton left Passy for the coast, his parting words were a promise that he would write to each of his co-investors on leaving France, at the very moment he was back on American soil, continue to correspond as the sales proceeded and the profits come in, and that he would distribute the proceeds through the good offices of 'their common friend' Franklin.[39]

By April Coffyn had written[40] to say that fifteen bales of linen had been bought[41] for some 39,000 livres tournois (£102,000) and that he was arranging to send them around the French coast to L'Orient where his opposite number there, Jonathan Williams, Franklin's grand-nephew who was another American that Ingen Housz already knew, would arrange to buy space on ships sailing for America.[42] Everything appeared to be proceeding to plan and Ingen Housz arranged for his one third stake, of almost 13,000 livres tournois (£34,000), to be transferred from Stametz at Vienna to an Amsterdam bank[43] and thence to repay Coffyn. In addition, he arranged transfer of 500 livres tournois (£1320) to Williams at L'Orient,[44] to meet what he was told was his share of the shipping costs and the French export taxes. He then heard from Coffyn who had been told, by Wharton, now arrived at the L'Orient, that this sum was insufficient and that a further 500 livres (£1320)[45] were needed although there was some equivocation about the exact total and its purpose. Wharton promised, again, to write further with all the details of shipments and a final account of costs before he himself embarked for Philadelphia. No letter appeared. Here were disturbing little niggles, an early indication, perhaps, that this business venture might not proceed smoothly. Here was the beginning of what was to become, for Ingen Housz, an exquisitely painful silence, a morale-sapping void that was to last many years. In fact, he was never, ever, to hear directly from Wharton again and it was to be nearly a decade before he was to learn the fates of all his massive investments.

AFTER THE PAINFUL parting from Franklin, the journey back to Vienna, broken at Strasbourg, took Ingen Housz and Dominque two months and they arrived home on 11 August 1780.[46] Ingen Housz stepped from his coach outside the apartment in the Weihburggasse[47] to greet his wife of five year's standing. They had been apart for more than half of that time but without any apparent resentment. Dominique, too, was married[48] but servants had no choices in such delicate matters. The contents of the returning trunks and bags, carried into the house, were both eclectic and bizarre. Five short lengths of wire, bundled and labelled, various columns of glass tubing, packets of chemical powder, a heavy brass pistol, and various books and papers were nothing extraordinary to his old servants but a box containing two glass chambers, a large metal plate, a pad of hard resin and sundry tubes and rods, all evil-smelling, excited much speculation in the household. When assembled and demonstrated, guesswork was replaced by amazement, fear even – the ability to generate a flame at will was spectacularly new and rather awesome. Ingen Housz had travelled through Strasbourg, where he was in contact with Professor Jean Barbier de Tinant [49] from whom he had bought a so-called *Fürstemberger* lighter.[50]

For as long as humankind had used fire there had always been the problem of producing the starter flame. It was probably Alessandro Volta who had first invented the *briquet* or electric lighter. It was a curiosity with a purpose; a machine for creating an instant, temporary flame. This facility with fire is something we take for granted today. The modern device fits in one's pocket and is disposable. However, arranging that a spark enflame a controllable jet of hydrogen was a major breakthrough of applied science; it made lighting a candle so simple. It was first described, by Volta, in a letter to Jean Senebier of Geneva, about 1770 and eventually became an everyday household appliance.[51] Volta did not try to develop his own invention but even if he had it is very unlikely that this would have stopped Ingen Housz trying to improve the apparatus he now owned.

Once the novelty of the lighter had worn off, the bags had been unpacked and the travellers gained some rest, life at Vienna was resumed, though not without a brief and pleasant vacation. Ingen Housz took the opportunity, in early October, to return to Baden, some twenty four miles south of Vienna where he once again enjoyed taking the waters.[52] But even in this recuperative environment he could not resist trying, once more, to see if he could collect and assay air emanating from his skin.[53] He was still reluctant to accept that there was not a parallel transpiratory process, in animals, to the one he had investigated in plants. Quite what the other bathers thought when a self-absorbed figure emerged into the bath chamber carrying various items of glassware with which he appeared to furiously rub his skin can only be surmised. He was probably given the wide berth he valued.

At court, Ingen Housz would have learned that Maria Theresa had fallen down a palace staircase during the summer of 1779 and although she seemed to recover,[54] she had been far from well since late spring 1780: indeed, some whispered opinions were very gloomy and already predicted the worst. Her mitigating influence on her son and co-regent, Joseph, might be about to end and there were good reasons to fear a stormy political future. Nevertheless, she continued to exert herself; responding to correspondence and state papers with her exemplary efficiency despite being breathless, especially when walking, even on a flat surface, and being very oedematous. There is no reason to think that she did not give Ingen Housz an audience soon after his return; the latest views of Franklin would have been prized. Ingen Housz noticed, with much regret, the bloated body and the breathless speech and by the end of October Maria Theresa was unable to sleep flat without suffocating. She died, of congestive heart failure, on 29 November 1780.[55] Ingen Housz does not appear to have been involved in any immediate clinical sense; her terminal care falling on Anton Störck.[56]

There had been changes in the Vienna medical fraternity, in ethos and personnel but most of them were less significant than the fact that Anton de Haen had now been dead for four years and that smallpox inoculation at Vienna had finally become routine. Ingen Housz's trials and tribulations of 1768 now seemed light-years away and he was no longer shunned in society. In fact he and Agatha were now a prominent part of the Vienna establishment, as Maria Theresa had schemed, and the popular couple were welcomed at many royal celebrations[57] and glittering gatherings. Ingen Housz's opera glasses were dusted off [58] and put to regular use. Much of the focus of their particular social life would have been at the weekly supper parties held in the Jacquin household; 'very agreeable' evenings as they were recalled by one of Ingen Housz's visitors at Vienna.[59] These were increasingly musical occasions as the Jacquin children entered their teens and revealed significant musical talents. Francizca Jacquin had a natural affinity to the keyboard and, from about 1782, had regular private lessons from a certain Wolfgang Amadeus Mozart. Mozart became very friendly with Gottfried Jacquin, the younger brother, and there exist, still, delightfully intimate and roguish letters between the two young men.[60] One regular guest remembered, fondly, the Wednesday evening get-togethers all her life.[61]

> ... the household of the famous Baron von Jacquin ... was a shining beacon to the scientific world in and outside Vienna, and which, with its convivial ambience, was sought by many. There, on Wednesday evenings for as long as I can remember ... this friendly home was devoted to scholarly speeches in the father's room, and we youngsters chatted, joked, made music, played little games and conversed splendidly.... the youngsters gathered around the younger son Gottfried and his sister Franziska. She played piano expertly and was one of Mozart's best pupils, who wrote the clarinet trio for her.[62]

We cannot be certain from such evidence, it being circumstantial only, but it seems very unlikely that Ingen Housz was not, sooner or later, introduced to Mozart. If he was, his physician's eyes would have hovered on the facial smallpox scarring[63] with intense interest. There was another direct connection between Ingen Housz and Mozart. Ferdinand de Jean (Jong) was a Dutch compatriot and medical colleague of Ingen Housz, and, also, a friend of Hendrik Hoogeveen.[64] De Jean, who had visited London, came to Vienna in 1781 bearing a letter of introduction to Ingen Housz from Richard Huck.[65] A keen amateur exponent of the flute, de Jean had met Mozart in Mannheim three years before and was sufficiently affluent to commission some 'easy and short' pieces for his instrument. Mozart had obliged but rather chaotically and de Jean had refused to pay him all of the agreed fee. The supposed contract between the two men therefore ran on for some years and the so-called 'third' flute quartet, in C major (K285b), is now thought to have been completed only in 1781/2 when both men were at Vienna.[66] Did Ingen Housz act as a moderating go-between; an honest broker? We shall probably never know but Ingen Housz might even have met Mozart through another common friend – Gottfried van Swieten – son of Ingen Housz's late senior colleague at court. The younger van Swieten, who had taken on his father's role as Imperial librarian after Ingen Housz had declined the post, and who had just been appointed a Privy councillor (*Hofrat*), became an amateur impresario. He fostered Mozart and was also to patronise a youthful Ludwig van Beethoven in a slightly later era.[67]

THE YEARS 1780 to 1788 were to comprise the longest, uninterrupted time span, bar brief local vacations, that Ingen Housz spent at Vienna. The city was now his home and his wanderlust seemed to wane. They were certainly happy years for him, a conclusion sustained by his being almost totally immersed in philosophical experimentation. There were to be some unavoidable distractions but his many investigations covered a huge scope and he was soon in a position to produce two very substantial, sequential volumes that he entitled *Vermischte Schriften' (Various Papers)* in German[68] and *Nouvelles Expériences' (New Experiments)* when they were eventually published in France.[69] The next segment of his professional life was to be that of a free-lance academic having a huge research bias but with an impromptu teaching brief. He was to find the latter irritating at times but accepted that it was important. He became the one-man, unofficial, and truly amateur, department of natural philosophy of the Vienna university when the Emperor was at the height of his enthusiasm for the Enlightenment and informed, rational thought was supposed to dictate policy, especially in education. Moreover, Ingen Housz possessed a personal 'cabinet' of apparatus and instruments, and a library, worthy of any self-respecting academy. He now owned several electrical machines and a substantial telescope,[70] a microscope, a large collection of Leyden jars, an electrophore, powerful magnets, thermometers, a wide range of glassware and an enviable stock of metals and other chemicals. His library was still expanding and would, by the end of the decade, contain some 746 books.[71] A third of them were medical but another third were natural history books. It is also a reflection of the man that a further fifth of the shelf space was occupied by the classical authors, mostly in Latin. Bar the latter, the texts were in a multiplicity of contemporary languages including English, French, German and Italian as well as the expected Dutch. And without wishing to discount his domestic life with Agatha and at least some excursions into 'social' Vienna it is impossible to recount much of the next eight years of his life except for an impressive and continuous series of scientific investigations.

Ingen Housz was to be at his laboratory bench within days of his return in August 1780. In quite which order he performed the hundreds of experiments that he undertook during this phase of his life is not clear. However there is evidence that he soon had many scientific briefs open all at once. He became involved in so many overlapping projects that strict continuity can only be sacrificed for the sake of narrative coherence. It is the case, though, that Ingen Housz turned, first, to the *Fürstemberger* lighter he had brought home with him. He felt it was inefficient and dangerous; so unsatisfactory that it deserved his prompt attention.

In essence, the *Fürstemberger* apparatus consisted of two connected glass globes, one of water above one containing hydrogen. By opening a stop-cock between them water flowed down, by gravity feed, into the lower globe, displacing hydrogen along an outlet pipe and, on release into the atmosphere a spark generated by contact with a Leyden jar was thrown across the gas jet, igniting it.[72] Before long a host of Volta's scientific peers had conceived numerous improvements to his cumbersome and rather inelegant design. There had been so many modifications that by 1780, only a decade after its invention, there was a substantial book on the topic, written by Frederic Ehrmann, the professor of physics at Strasbourg and Ingen Housz was able to refer to a copy of this 'manual'.[73] The *Fürstemberger* he now owned was among the second generation of these machines and used an electrophore to make the spark. Ingen Housz was only too conscious, after

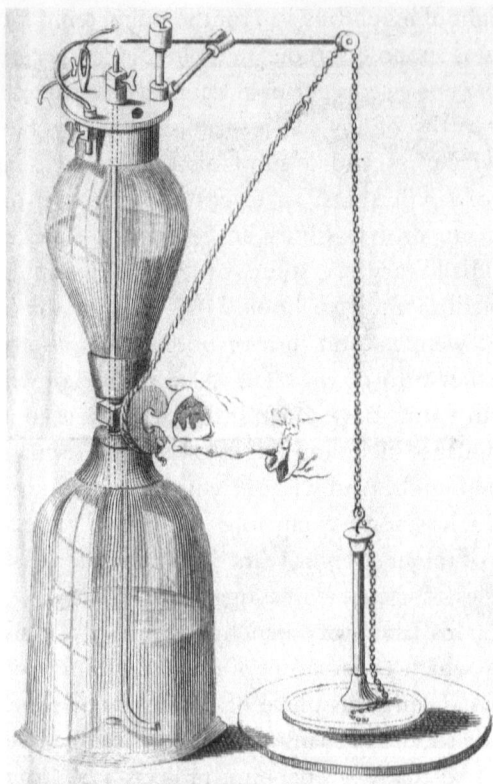

The Fürstemberger lighter as adapted by Jan Ingen Housz and Georg Pickel.

his experiments with Fontana in London, that there was a considerable risk of explosion if the hydrogen was allowed to mix with atmospheric air, which could easily happen if triggering of the spark was delayed or failed. And should ignited gases travel back down the release pipe the consequences might be catastrophic.

Ingen Housz's thoughts on the need for urgent modifications struck a cord in a new friend, an unexpected and enthusiastic assistant, someone with exactly the right talents. The 29-year-old Georg Pickel, professor of chemistry at Würzburg from 1782, had travelled to the city specifically to see Ingen Housz and learn about his 'vegetable' experiments. He was to be the first of a long series of such visitors to the expert's home and laboratory during the next eight years. Fame can have its downside and people can impose on one's time and freedom as Ingen Housz increasingly discovered. Pickel found himself very welcome, however, and not just because playing host had yet to irk Ingen Housz. The royal physician quickly recognised that Pickel was a very knowledgeable 'savant' and an ingenious and skilful technician.[74]

> This man, who is an accredited Professor in Physics and Chemistry stayed here expressly one extra year in order to work under my direction . . . He is a very nice fellow, modest, excellent chemist, outstanding electrician and now versed in the subject of air. He was of great use to me, because he was willing to conduct all kinds of experiments, as I asked for, thus freeing me of a lot of work. He has an original and inventive mind and is seldom at work with an object without proposing an improvement.

The two men worked together enthusiastically on the gas lighter[75] and their modifications produced significant improvements. A new tap was placed near the end of the gas nozzle to shut off the flame before closing the main cock. This prevented hot gases being sucked back along the outlet pipe so that it remained, always, full of (cool) hydrogen. This made the apparatus safer as did the more foolproof sparking device. But Pickel also made 'their' machine much more compliant by inventing a chain mechanism that linked, automatically, the opening of the main stopcock with the triggering of the spark, making it the prototype of the single movement, turn-of-the-thumb, cheap lighter that we use and throw away today.

Another early project, after Ingen Housz's return to Vienna, was his work on heat conduction in metals, fundamental physics always associated with his name. His apparatus clearly demonstrated differential heat passage along different metal rods and was the grandfather of a device subsequently introduced to generations of schoolchildren as 'The Ingen-Housz box.' However, it always worried Ingen Housz that the ownership of the original idea would be mis-attributed, as indeed it was. The concept that various metals would likely conduct heat at uniquely different speeds was Benjamin Franklin's. The topic had obviously inspired speculative but significant conversations when the two men were together at Passy in 1778 and again in 1780. But Franklin, still bubbling with such brilliant ideas, was very busy with politics and finding, he didn't deny, the Paris salons more diverting than experiments. With typical generosity he not only presented, to his Dutch friend, his ideas on how the theory might be tested – by watching how wax melted along hot wires – but even donated five suitable lengths of wire, one each of gold, silver, copper, steel and iron. They were all of the same calibre, having been 'drawn through the same hole'.[76]

Ingen Housz carefully stowed the wires in the baggage for the journey back to Vienna. He may have spent some of his long journey home in planning how he might fully realise Franklin's ideas, and reticence over the true ownership of the ideas did not dampen his enthusiasm for the work. By the first days of December 1780, after barely three months back in his laboratory, Ingen Housz was able to report pleasing progress, via the diplomatic bags to Paris.[77] He had even obtained two further equivalent wires, one of lead and one of tin. He had attached each of the wires to a wooden frame so that they were equidistant, each from the next, and such that they projected exactly five inches from the frame: a large comb with just seven widely spaced teeth is a helpful depiction. A purpose-designed tin box, six inches deep and wide and long enough to accommodate the seven

The Ingen Housz box for assessing heat conduction in metals.

teeth of the 'comb' was then filled with beeswax and lowered into a glazed terracotta tub full of boiling water. When all the wax had melted and was hot enough to be seen circulating in the box, the wires were dipped up to their wooden jig and then withdrawn smoothly and evenly. The wax adhering to each wire was then allowed to cool and set at room temperature, giving all seven 'teeth' a coating of equal thickness. Preparation over, the waxed 'comb' was then racked down so that each 'tooth' just touched the base of another, flat-bottomed, earthenware bath, this time containing a one-inch depth of very hot olive oil. 'Now as they had all been dipt alike at the same time in the same oil, it must follow that the wire upon which the wax had been melted the highest had been the best conductor of heat.'[78]

Although the theory was simplicity itself, Ingen Housz found the necessary observations tricky. With this prototypal apparatus the wax melted rapidly along all the wires. It was difficult to observe and record seven melting speeds all at the same time. His answer was to remove the comb and attribute 'melt' marks to lines on a piece of paper the instant he had taken the wires out of the heat. Ironically, this being an exemplar Enlightenment experiment, he appears to have used the alchemists' symbols for the metals but the shorthand was a functional one for his purpose whatever its pedigree.

Ingen Housz, as well as any member of the burgeoning scientific community, knew the paramount importance of repeating experiments over and over in order to obtain consistent results. He fought his impatience and only reported to Franklin after testing his wires at least a dozen times.[79] His friend had been correct; it was very obvious that different metals did, indeed, conduct heat at different speeds. Ingen Housz's detailed observations were not consistent, however. There was no doubt that silver conducted heat best, followed by copper, gold and then tin. Lead came last but Ingen Housz reported conflicting results between iron and steel. He had the wit to see that when the difference between two metals was so subtle a number of extenuating circumstances could come into play. Nonetheless, an extremely important property of metals had been vividly demonstrated. Ingen Housz still fretted, though, that the real credit lay with Franklin.[80]

> As I have no right of claiming any merit on this head, I will make no farther use of these observations until you, who furnished me with the materials and suggested me the method of making the inquiry, think proper.

Here was an understandable plea to Franklin that he evaluate the findings and suggest, urgently, how and where they should be published: joint authorship seemed the obvious ploy. Unfortunately, Ingen Housz did not receive any explicit comments for some eighteen months[81] and this important scientific development languished amidst some stalled correspondence. His consolation was intense if private. For the second time in his life he had been the first to witness a profound function of nature, to be the agent of a breakthrough in human knowledge.

In the meantime his own ingenuity took wing and he went beyond Franklin's early ideas of methodology. He had manufactured for himself a bespoke tin box on to the side of which were welded small, equally spaced wells that could be plugged, each, with wine corks.[82] These were then pierced to house the ends of the test wires, wax-coated as before, so that each one was sitting with an equal length within the tin cavity and a longer part projecting out into the air. The melting process was now performed by pouring hot water into the box and it was this concept that was later refined further by Jules Jamin, the French physicist of the nineteenth century. He added a heat source under the box and generally enlarged the components to increase sensitivity and this is likely to be the style of apparatus found in the cupboards of school laboratories today.

By this time Ingen Housz had made a good friend of Niklaus Molitor who, with some effrontery, had translated *Vegetables* into German. Though his academic base was Mainz, Molitor seemed to spend much of his time in Vienna where his ebullience was to have a substantial impact on Ingen Housz. Encouraged by the success of his German translation of *Vegetables*, Molitor now turned his attention to other writings of his hero. Perhaps to keep the peace or seeing no good reason not to let Molitor off the leash, Ingen Housz gave him permission to prepare a booklet, in German, that was a barely edited version of his paper on the Electrophorus, his Baker lecture read to the Royal Society on 4 June 1778, and later published in the *Transactions* of the Society.[83] Eventually, Molitor was allowed to transcribe two further papers published in England and bind them in together; one on lighting a candle with an electric spark[84] and one on inflammatory air.[85] The pamphlet was entitled *Anfangsgründe der Electricität, (Rudiments of Electricity)*[86] and it came off the presses of Wappler, the Viennese publisher in April 1781. Molitor was probably more excited by these second-hand publications than his patron who was now focussed elsewhere.

Ingen Housz had reported to Sir John Pringle, from Paris, his evidence that air quality varied at different places.[87] He extrapolated that some localities must be 'healthier' than others. However, the eudiometric readings were never entirely consistent, even in his practised hands, and the differences were marginal in any case. But the thinking was incisive enough – that breathing high-quality, 'salubrious', air was more than just invigorating, it might combat disease specifically. Oxygen as therapy can thus be traced back to the 1780s and to Ingen Housz,[88] who argued, that 'since an animal so shut up (in dephlogisticated air) lives much longer than in an equal quantity of common air, one cannot refrain from expecting the most desirable effects in many illnesses.' Impatient to try its effects, he conceived a suitable apparatus very soon after arriving home. It was assembled and in use by November 1780. Over the next weeks there were excited, though cryptic, references to a series of oxygen-breathing trials and the whole story was finally related in a dissertation written during January 1781.[89] Unfortunately, this was sent to an obscure publication.

Salomon de Monchy, physician at Rotterdam and a close friend, had been instrumental in forming the Scientific Society at Rotterdam – the *Bataafsch Genootschap der Proefondervindelijke Wijsbegeerte te Rotterdam* (Batavian Society for Experimental Philosophy in Rotterdam) – and, anxious to promote the journal of the society, appears to have promised Ingen Housz immediate publication of any suitable article. An early report on how a patient might breath pure oxygen and its potential benefits was obviously topical and important and Ingen Housz wrote to Lambertus Bicker, secretary of the society, in December 1780,[90] that space should be made for a dissertation. He was intending to send it in early January 1781 but finally decided that, having lived away from his fatherland for sixteen years, his Dutch was now too inaccurate to be committed to print. He therefore wrote it in French and sent it, on 28 March 1781, to van Breda at Delft, with a request that he translate it, for de Monchy, into good quality Dutch.[91] Van Breda again obliged with admirable efficiency and *Verhandelingen over de gedéphlogisteerde lucht, en de manier hoe men dezelve kan bekomen en tot de ademhaaling doen dienen (Dissertation on the dephlogisticated air and the way to obtain it and make it serviceable for breathing)* was promptly delivered to the society at Rotterdam. A detailed copperplate engraving showing the apparatus in use was sent directly to Bicker at Rotterdam at about the same time (see page 303).

It clearly pleased Ingen Housz that he could have something so significant published in his own country and in his native tongue. Unhappily, it meant that this key medical development was being laid before a very small readership and in a minority language. Ingen Housz began to realise, from early in the process and perhaps with some regret, that it deserved wider publicity and included a plea to van Breda[92] that he should carefully preserve the original French version so that it might, later, be published in Paris. However, it was another four years before detailed publication in the French language was organised, due to a dilatory editor there and a global indolence that seemed to reign supreme in the French printing industry. By this time the impact was lost. But if the publicity was fluffed, the experimental work was not.

Ingen Housz produced his oxygen by a conventional means – by heating 'nitre' – later called 'saltpetre' (potassium nitrate). This was readily available, having been in demand, for many years, by arms manufacturers, as one of the essential ingredients of gunpowder. The improvements Ingen Housz introduced were two-fold. Taking the trouble to obtain as pure a supply of 'nitre' as possible, he then heated it in a glass retort that he had carefully coated with at least ten layers of clay, brushed on and then dried slowly.[93] This ensured that the heat was well distributed and evenly applied;

using the more conventional technique of burying the flask in a heated sand bath had proved to be futile.[94] He reckoned he was able to produce 3000 cubic inches of 'dephlogisticated air' (oxygen) from about four ounces of 'nitre' and found it to be over 80% pure. His real inventive genius was demonstrated, however, in how he used the product of his manufacturing process. Various philosophers had already tried inhaling the gases that they produced. Scheele and Fontana had both breathed 'inflammable air'[95] and Priestley, along with two mice,[96] had tried 'dephlogisticated air' within weeks of producing it. No doubt there were other isolated 'guinea pigs' but, in general, whether or not a gas could be 'breathed' was seen only as a further way of describing its fundamental properties. Fontana, at least, had gone one stage further in trying to breathe inflammable air continuously and had nearly died in the process.[97] He had sucked the gas in from a flexible 'bladder' via the mouth while his nostrils were blocked. Ingen Housz had no hesitation in being the trial subject – submitting oneself to experimentation has long been a scientific tradition.

At first Ingen Housz tried Fontana's technique of sucking the air into the lungs through the mouth and then blowing it out again, having closed off the nostrils.[98] He found this to be extremely tiring, as did several of his friends, and there was no conceivable way of using this method for breathing continuously for any length of time. He therefore tried the opposite - delivering the oxygen through rubber tubes pushed up his nostrils while his mouth was closed. He found it unpleasant and just as ineffective: he doesn't say so but the irritation of the nasal passages from the intruding pipes was probably intolerable. By early 1781 he had settled on administration of the gas via a rubber nasal mask that was connected to a flexible chamber by a pipe with a one-way valve. His 'chambers' were prepared from cows' bladders greased and softened and the mask made to fit the face of the recipient by moulding it individually. The mouth was kept tightly shut throughout.[99] At first the tap on the supply pipe was opened for each successive inspiration and expiration so that the oxygen in the chamber became progressively contaminated with exhaled air. This obviously set a limit on the experiment for, as Ingen Housz was able to demonstrate by using his eudiometer, the quality of the air in the balloon deteriorated massively after less than a dozen breaths. He confirmed his readings by showing how some unfortunate finches were unable to tolerate it for more than a few minutes. He was perfectly aware that the second-hand air was being contaminated, in the lungs, by 'fixed air' (carbon dioxide) and, in the prevailing understanding of that time, also by 'phlogiston.' He

The first apparatus for breathing oxygen consistently.

was unable to combat the latter, he thought, but the carbon dioxide could be reduced by bubbling the exhaled air through water, especially a solution of limewater, and this was his eventual recommendation to those who wanted to try the apparatus. The 3000 cubic inches of oxygen which Ingen Housz told us he was able to accumulate would have been accommodated in a balloon of about ten inches radius and this concurs with the drawing he produced. The sketch should not be taken, incidentally, as a further portrait of Ingen Housz. It was not done by a portraitist, but by an engraver asked to produce the copperplate to accompany the dissertation to Rotterdam. The image could be the representation of an un-named laboratory assistant, of Georg Pickel, of one of the Ingen Housz domestic servants, or entirely figmental. On the other hand, the experiment being performed was entirely real and very important.

To his chagrin, Ingen Housz was not given the privilege of pioneering oxygen therapy with an actual patient disabled by significant lung disease. Maximilian Stoll, de Haen's successor as professor of medicine at Vienna, who had initially indicated that he would find, for his friend, a suitable and willing patient, finally insisted on administering the treatment himself, using Ingen Housz's technique and it appears that Ingen Housz was obliged to comply.[100] The patient was a 47-year-old Viennese baker who had developed, over seven years, laboured breathing and shortness of breath, even at rest, and chest pains at night. He had lost his speaking voice at times but he had never been subject to a cough so the exact diagnosis is unclear. He may have had chronic airways obstructive disease, as it is now called, or fibrosed lungs from long exposure to flour dust; it is difficult to say. Stoll put him on a once-daily regime, of eight to ten deep breaths of oxygen, for sixteen consecutive days in June 1781.[101] The courageous patient was soon much better, consciously less breathless, even when lying flat, and sleeping comfortably at night. It was a promising start, and though he soon relapsed after the end of the revolutionary treatment, Stoll wrote a very complimentary letter to Ingen Housz. Fontana had returned to Florence during 1780 and he, too, administered manufactured oxygen to some patients with chest disease[102] but otherwise so-called 'pneumatic medicine' then went into abeyance until it was tried on a much more significant scale at the very end of the eighteenth century. The location was Bristol in England and, amazingly, Ingen Housz was again present, though still the attentive spectator.

FELICE FONTANA HAD been staunch enough a friend to Ingen Housz to describe to him, in Paris or in London, where they had been together for many weeks, a particular discovery. He had found a means of creating a vacuum without recourse to a pump.[103] In all probability it had been an accidental finding as most scientific discoveries are. He had suddenly realised that pieces of red-hot charcoal, in cooling, absorbed several times their own volume of air. In a confined space a negative pressure was created and, in the right circumstances, a vacuum. He was best able to demonstrate the phenomenon using a glass jar of air inverted over mercury. If he plunged some red hot charcoal under some of the mercury until it was completely cool and then released a lump to float up into the inverted jar, the mercury rose up as a column as the air was absorbed, sometimes completely if the lump chosen was large enough.

On his return to Vienna in 1780, Ingen Housz was keen to apply the discovery – 'my impatience wouldn't suffer any delay.'[104] He had conceived an apparatus that would enable philosophers to utilise Fontana's discovery in a way that would be useful for performing experiments; an apparatus that would be a simple and cheap alternative to owning a pump.[105] His final design was elegant and simple to operate but it never seemed to enjoy a popularity. Even Ingen Housz admitted that he had used it 'only rarely':

he seemed driven more by the research than by the development. His 'machine' certainly worked well but required a skilled machinist to mill it. The core component was a strong copper basin that had a snugly fitting copper lid that could be clipped quickly into position –'locked on'. A pipe, fed up through the centre of the lid, was supplied with two taps that worked independently. Above the top tap was another metal disc, a further accurately turned 'lid' that actually formed the base of a very similar basin, but inverted and of glass. An iron cauldron of red-hot charcoal was lowered into the copper basin and the lid applied. Both taps were now opened and, as Fontana had discovered, all the air in the upper, glass, chamber was sucked out as the charcoal cooled. The increasing vacuum created good seals at each lid and, with the upper tap closed, the top, glass chamber, now housing a vacuum, could be detached and used when and where wanted. Afterwards the apparatus could be reassembled, the charcoal reheated, and the whole procedure repeated at will. Nothing was consumed bar the energy in firing up the charcoal and the 'machine' was therefore as efficient as it was well-mannered. Disregarding any rights to a patent, Ingen Housz published comprehensive details including some very fine diagrams. It was, however, further proof of his genius as a design engineer; his facility for manipulating natural processes so that their properties and functions were revealed.

A hackneyed description of scientists is that they are cold, calculating and arid personalities who have rational, ordered, thoughts: that they conceive ideas and submit these theories to objective testing in a structured environment, physical and emotional. This is fiddlefaddle. Highly important scientific discoveries have often been made by chance, many in the laboratory during idle moments and among the chaos on the bench. Penicillin is a prime example, the result of putting off the cleaning of dirty and discarded equipment.[106] The majority of scientists participate in 'lateral tinkering' and though much more self-disciplined than many, Ingen Housz was no exception. By the late autumn of 1780 he had some reserve jars of reasonably pure oxygen in his workroom and scattered about were wires of various metals including, presumably, some of those given to him by Franklin. Metals, he knew, would melt, and some even burn, at very high temperatures, especially in a strong blast of air. He had only to peer into the nearest forge to witness this phenomenon but what if the farrier had a supply of pure 'dephlogisticated air' (oxygen)? Ingen Housz quickly organised a mock-up. He pushed a thin coil of steel wire through a cork that would fit snugly into the neck of one of his oxygen vials. The end of the wire

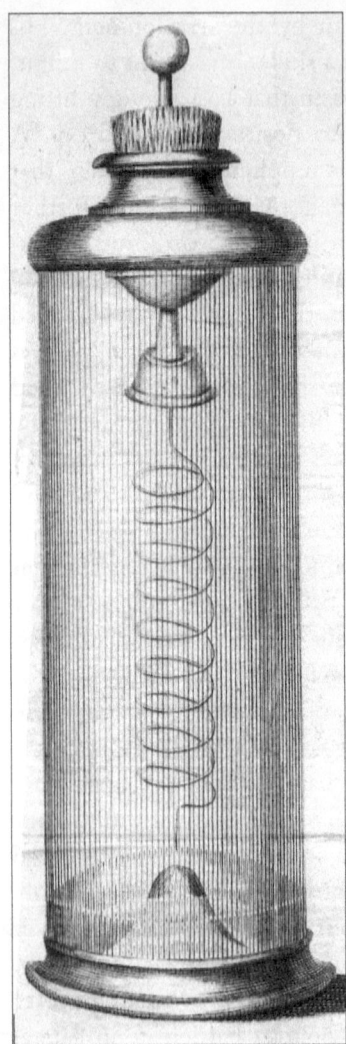

Apparatus for burning a wire in oxygen.

was taken up to red heat in a flame and then plunged into the jar, the cork sealing it in.[107] An amazing experience followed – the wire progressively burnt away to a calx but emitting, all the while, a stupendous and intense light: a 'smoak so clear that nobody can look at it without his eyes being dazzled as much as if he looks at the sun in a clear day.'[108] Ingen Housz rapidly found that there was a characteristic and unique incandescence for each metal. Iron, for instance, burned with a golden flame whilst copper gave off a blue light. Here was a wonderful natural spectacle and the burning of a length of wire suspended in a vial of oxygen carried in his pocket was to become a favourite 'party piece' for the rest of his life. Refinements included a 'cord' of different wires, seven or eight, welded end to end. These therefore burned successively giving a serial kaleidoscope of intense colours. In another demonstration he burned fine wires entwined together, quoting one instance[109] where the 'plait' consisted of iron, copper, silver and gold. The varying spectrum was spectacular enough to show to the Emperor and Ingen Housz was soon to have the opportunity. More significantly, we see, here, the beginnings of the discipline of spectroscopy, the analysis of compounds by incinerating them and detecting which wavelengths of light are emitted. But was Ingen Housz the first to realise the potential of burning metals in oxygen?[110] The phenomenon was being demonstrated at Göttingen almost simultaneously,[111] but only, it seems, because Georg Pickel, back at his alma mater, was able to tell his teacher about it. He was Georg Christoph Lichtenberg, professor of physics, and he added a memorable new 'twist' – the novelty of burning a steel watch spring; of turning time into light. The true location and 'ownership' of the original discovery that metals would burn so spectacularly in oxygen appears not to have been an issue at the time. But here began a phase, in Ingen Housz's life, of stress and irritation from having to defend his originality, the primacy of many of his other discoveries.

IN THE FACE of a mounting pile of reports on his various experiments and inventions, Ingen Housz was both motivated and motored by his new friend, Nikolas Molitor, who was still at Vienna in early 1782. The 'electrics' booklet had been well received and the young German enthusiast seemed to have also translated Ingen Housz's other assorted essays into his own tongue. Ingen Housz had clearly given his impetuous colleague access to all his papers but Molitor was not content with private study. As far as he was concerned, Ingen Housz had written a substantial volume; an important contribution to physics that was only wanting a publisher. Ingen Housz knew, of course, that publishing in the French language would guarantee a larger, universal, readership and he had long planned to send the volume he had conceived as his *Nouvelles Expériences (New Experiments)* to Paris. Achille-Guillaume Bégue de Presle, physician, scientist and prolific writer, had already promised to act as midwife there[112] and deliver the project through the presses of a respected publisher but there was still an obstacle. Ingen Housz had been anxious to have feedback from Franklin on his 'hot wires' findings and also the American's consent to a personal dedication to front the book. After twelve frustrating months, and under pressure from Molitor, Ingen Housz had to admit that he was ready to abandon his hopes of a reply from Franklin. The French manuscript pages were therefore bundled up, as they were, in the first week of January 1782 and sent to Paris.[113] For the moment, though, the articles on heat conduction and on breathing oxygen therapeutically were kept back; they could always be added at a late stage in printing. There was a satisfying sense of closure in this but also a *quid pro quo*: there was now no good reason not to let Molitor have sway. His German translations of the miscellaneous essays, bar those already published as *Anfangsgründe der Electricität (Elements of Electricity)*, were handed in to Johann Krauss, another Viennese publisher.

Krauss was very efficient – or perhaps Molitor was equally impatient with everyone – and the book *Vermischte Schriften (Various Memoires)*[114] was finished by the first week of April 1782. In octavo, it covered 415 pages and there were four copperplates. The subject matter of the fifteen dissertations included those on pointed lightning conductors, burning metals in oxygen, the hydrogen lamp developments, extracting methane from marsh mud, and the mechanics of the ether pistol that Ingen Housz had demonstrated in London. There was a resumé of his eudiometric experiments at different locations, theories why gunpowder and blasting powder were so powerful, and observations on magnetism and artificial magnets. Molitor had finally persuaded Ingen Housz to allow inclusion of some of the details on breathing oxygen as therapy and his findings on heat conduction in metals despite his still having no news from Franklin. Though disappointed by the poor quality paper Krauss had used,[115] Ingen Housz was able to enjoy the pleasure of sending two crisp copies to van Breda at Delft on 12 April[116] asking him to send on one of them to his brother Louis. At the same time he bemoaned having no news from Bégue de Presle at Paris. The silence was to prove more ominous than even the worst pessimist might have forecast.

In sending a copy of *Vermischte Schriften* to Franklin at Passy, Ingen Housz took the opportunity to plead for any news of Bégue de Presle and the fate of his French manuscript. There was none and by early May he was so disgruntled that he wanted van Breda[117] to ask among the Dutch printers if any of them were interested in publishing *Nouvelles Expériences* for he had begun to suspect that his manuscript had little chance of being set on any French press in the foreseeable future. It could be sent on from Paris. In fact his desperation gradually faded into a passive resignation but van Breda took up the challenge. He persuaded a publisher at The Hague to take on a Dutch edition of *Nouvelles Expériences* and by February 1783[118] Ingen Housz was sending copies of some of the component dissertations for van Breda to begin the translation. By that summer the young Delft doctor thought he had enough material – admirably translated according to Ingen Housz – for a volume to go to the printer.[119] To this end, Ingen Housz sent some relevant copper plates to Delft at the end of August.[120] But by the end of 1783 there was still no real progress with the original version at Paris, although Ingen Housz was now receiving small batches of printed pages for his approval. The lack of drive shown by Bégue de Presle and his collaborators was now holding up two editions of the book: Ingen Housz asked van Breda to delay the Dutch version until its French precursor was released for sale. It was asking a great deal of his friend after all the work he had done but when *Nouvelles Expériences* finally appeared, in July 1785, the Dutch edition, *Verzameling van verhandelingen (Collection of Treatises)*,[121] was soon on sale in the Netherlands. There was, at last, delayed but double success for Ingen Housz. By this time, however, he was having other problems with publications – with the publications of others.

15
Intrigues

The moment he finishes eating, Kaunitz sets about his teeth. His servants always lay a row of picks alongside his knives. Quite unabashed by his tasteless habit, he pokes and prods and even stabs mid-air with the pick when he agrees with a point made in the conversation. At least, this afternoon, there are no strangers at his dining table and no-one is embarrassed. The Emperor's planned visit to Rome, a pay-back for the Holy Father's recent descent on Vienna, is a source of much rumour – will the Emperor be able to resist further papal pressure and maintain his religious reforms? There are all sorts of opinions and forecasts but Kaunitz, rather than give anything away inadvertently, remains out of the speculations. He absentmindedly lines up his remaining, unused, toothpicks before signalling for the wine glasses to be refilled and launches into an overdue change of topic.

'So, tell us dear Ingenhausen, how are you coping; are you not disappointed that Monsieur Senebier is opposing your doctrine on airs? I dined with a merchant from Geneva last week and everyone there, and in Paris, now favours his opinions, I am told.'

It's a conversation stopper and someone drops a knife. Ingen Housz is not quite speechless but makes a couple of false starts.

'It is true . . . it is the fact . . . my understanding is, that Senebier has been doing plant experiments and I'm reading his latest book. The details suggest that his trials are on much too small a scale and he uses a eudiometer of poor repute. And he has been unable to demonstrate the opposite flow of the airs during the hours of darkness, the process that I named the influxus nocturnus.'

Ingen Housz can sense that his audience is uncomfortable and far from convinced. While they're furtively glancing one to the other, none of them will look him in the eye. He knows, instinctively, that there've been earlier conversations; in his absence. He ploughs on, pointing out Senebier's long-winded prose and disorganised arguments. To no avail; Ingen Housz can now see that jumping to the ramparts and conducting his own defence has been a tactical error. His standing as a philosopher is suddenly precarious. His words run into quicksand and there's an awkward silence until Kaunitz rescues him: 'chocolate, gentlemen?'

For Ingen Housz, the early days of 1782 were, perhaps, as unfulfilling as they were short. The manuscript of his book on smallpox[1] was still on his desk but he seemed unable to finish the project, it remaining one of those ambitions that sinks into a quicksand of overlong and indulgent planning. He had a valid excuse in playing host to a young Hungarian doctor who had come to Vienna to further his education but even after his guest had left,[2] Ingen Housz ignored his assorted pages on inoculation in favour of fresh manuscript pages on gases and plants. A completely new volume was to be based on the further fifteen hundred eudiometric experiments that he had performed in the two years since leaving England. In recent months, even in harsh winter weather, he had been the beneficiary of having the Jacquins as in-laws and been able to continue his observations within their domain – in a hot house of the royal botanical gardens.[3] He was writing parallel texts, in English and in French, and work was progressing well. It was to flounder by early February. The first major distraction was the shock of learning that Sir John Pringle had died.[4] However expected, Ingen Housz would have found it difficult not to spend many introverted hours reminiscing on the times he had spent with his friend of nearly forty years and how influential a figure he had been in his life. A fully-retired Pringle had certainly not been vigorous when Ingen Housz had last spent time with him, in Bath, in October 1779, but he had remained well enough to organise the sale of his house in Pall Mall during 1780 and remove to Edinburgh. Pringle had returned to London, however, in 1781, fearing that whatever might be wrong with his health, he was likely to die of simple exposure if he should risk another winter spent in Scotland. He took lodgings in Charles II Street near St. James' Square[5] but he had but a short period of life left. In the middle of January 1782, he suffered a major stroke while at his club and died a few days later, on Friday the 18th. He was buried in St. James' Church, Piccadilly[6] and a memorial tablet was erected in Westminster Abbey. He was sincerely grieved at the Royal Society and a group of fellows who had counted themselves to be long-standing friends and admirers, regulars at his Sunday soirées, begged Andrew Kippis, one of their own, to write a suitable biography by way of epitaph and his skilful rendition of the 'life' was used by way of introduction to Pringle's 'Six Discourses' when they were published, posthumously, in 1783.[7] Ingen Housz was very pleased with it when he read his copy, finding Pringle's character 'drawn in a masterly way'.[8]

In the autumn of 1781, news that would add to Ingen Housz's trauma began to arrive in Vienna. Joseph Priestley had, it seemed, published a new volume of *Experiments and Observations*.[9] This had gone under the press in London in spring 1781 and had been released to the bookshops by early summer. No actual copies were to be had in Austria for some weeks despite Ingen Housz's specific and urgent requests to London but the hearsay reports of the work were consistent. Ingen Housz was to be very upset when he discovered the deep duplicity of his former close friend but for the moment he knew only enough to be anxious. But there was a real fear that he was likely to be usurped, even at this stage, of his claim to having been the discoverer of the effects of sunlight on green plants, his right to what we might call the 'primacy of light.' It was a fear to be feared, as he reported to van Breda,[10] his new friend at Delft, on Christmas Day 1781 – '*Fronte politus astutam vapido servas sub pectore vulpem*' (Behind civilised outward appearance, low cunning lurks in vapid breast.)[11]

All during the frantic months at Southall Green Ingen Housz had fretted that he might lose the credit for his discoveries. The break-neck writing and printing speed of *Vegetables*, in the autumn of 1779, had brought him immense relief; the very actuality of the book had seemed to establish his precedence. Standing in the Strand with the printed pages in his hand that October had boosted his confidence but now he was not so self-assured. It would have been better, perhaps, for him to have stayed in England, had it been possible, and lobbied for face-to-face support and recognition. Besides claiming that the effects of light on plants had been his own finding, Priestley was, by all accounts,[12] peddling a new theory to explain the appearance of air bubbles on plants submerged in water. Furthermore, Tiberius Cavallo, now a Royal Society fellow and also a previously valued friend, had, likewise, written against Ingen Housz – that the air seen to collect on plants under water was the outcome of a chemical process in the water itself, the leaves merely acting as congregation points for the bubbles.[13] And even though Priestley now recognised that 'the green matter' of stagnant water (green algae) was a vegetative growth and that it, too, produced bubbles of air in sunlight, he thought this, also, to be a chemical process centred in the water; that the green matter and the light were mere 'agents' in the process.[14]

Ingen Housz knew that two such energetic and powerful opponents, both *prominenti* at the Royal Society, would convince everyone of their being correct, whatever the evidence. In spite of their well-established friendship and apparent respect for one another, Priestley had turned into a smiling assassin. This second volume of his *Experiments*[15] should have given him his first opportunity to allow Ingen Housz the credit for his main discovery regarding the driving force of sunlight on the chemical processes turning in green leaves. However, the proper ascription was a singular omission from the book and, in fact, there was an evident sin of commission. On only the second page of the preface to the book,[16] written on 24 March 1781, Priestley was effectively staking a claim for he himself having discovered the primacy of light in what we now call photosynthesis and for 'advancing' our understanding of nature a 'considerable step'.

> I have prosecuted with considerable success . . . experiments . . . in tracing the manner in which . . . growing vegetables purify noxious air viz. by means of the action of *light* upon them; having then sufficiently satisfied myself that it was *light* only and not *heat* . . . (Priestley's italics)

Readers could only conclude that it had been Priestley, and Priestley alone, who had made the discovery. The name 'Ingen Housz' eventually appeared on page 24 of the actual text, some 48 pages beyond the bald assertion in the preface, and then only in the context of some malevolent dissembling. Priestley denied the possibility that gases actually form within submerged leaves or within the 'green matter'[17] for 'it was very evident that there is no proper *production* of air . . . but only a *depuration* of the air previously contained in the water' (Priestley's italics). He then expressed his surprise that Ingen Housz was able to experiment, allegedly, on individual leaves rather than whole plants.[18] There is no matter of opinion here: it is an untruth. It is also largely irrelevant.

Trials of attribution are always difficult and Priestley himself had been under a cloud of mistrust when others had claimed that they had preceded him in isolating 'dephlogisticated air' (oxygen). Now, perhaps, he was vexed to the point of self-deception. There is certainly good evidence that he himself had not actually thought of light as being an important environmental factor in the daily activities of green vegetation. Priestley had been in Calne for the summer months of 1779, to be near his employer and patron, Lord Shelburne at Bowood House, and he had certainly been experimenting with plants, some of them under water. He, too, was observing that bubbles of 'pure air' emanated from the plants into the water, but not consistently. He had written to three friends,[19] Benjamin Franklin one of them, only a couple of weeks before having a copy of Ingen Housz's book put into his hands, by the author himself, in mid-October 1779.[20] In telling each of them of his own experiments with submerged plants, he used virtually the same phrases in all three letters. In proposing why they did not always release pure air he gave no hint of having guessed, even, the effects of light.[21]

> I have no doubt, therefore, but that all plants imbibe air in one state . . . and emit it . . . in another, and that the reason why my experiments did not always succeed with mint &c confined in small jars was, that they will not always enjoy perfect health in that confined state.

In other words, Priestley was assuming that the variability in the processes he had observed were a function of the health and the growth of the plants. Had he thought of the possibility that their exposure to light intensity was critical, he was not the sort of man to have shrunk from advertising the fact. By early 1781, some eighteen months later, he had apparently come to believe, by a selective memory process perhaps, that he had, in fact, been the first to deduce and demonstrate the effect of sunlight on plant metabolism and he tried to reinforce his claim by debunking his opponent, Ingen Housz. He was haughty enough to start a ten-page section of the book, *Of the mixture of common and nitrous air*, with a blunt denunciation of the Fontana eudiometer and its usage.[22]

> The business of this section shall be to explain a phenomenon which puzzled me at the time of my last publication in mixing nitrous and common air, and to correct a mistake of the Abbé Fontana, and of Dr. Ingenhousz, relating to the same subject.

However, he then failed to justify his condemnation of the Fontana eudiometer for he fell short of testing it properly.

Ingen Housz, in *Vegetables*,[23] had delineated, very clearly, the exact way of using the instrument in order to obtain consistent and reliable results. He and Fontana had spent many long hours together refining its use. He had even classified and listed some twenty particular points that any experimenter would need to consider, all and any of them being operating faults powerful enough to ruin the results if ignored. Although Priestley clearly recognised the need to ensure that all possible chemical interaction between the two gases had been completed – that a 'saturation point' had been reached – he recommended a blunt approach, dismissing Fontana's precise routines as 'operose and tedious.'[24] He was particularly scathing when Ingen Housz argued, on behalf of Fontana, that the Abbé's method obviated any worries about the purity or strength of the nitrous air;[25] a weaker sample simply requiring more repeated titration than a stronger one. Priestley's disdain was both arrogant and misplaced; based, as it was,[26] on a botched attempt to trial Fontana's methods, however often he claimed to have repeated them. It may be fanciful, but suppose that Priestley had been given, by Lord Shelburne's chef, the then twenty-year-old recipe for making mayonnaise and decided to try it himself. However, out of conceit or want of patience, he had failed to add the olive oil drop by drop, as is imperative, and inevitably found that his mixture was an inedible mess. Then, misusing a position of trust most irresponsibly, he had publicised his self-inflicted calamity as a defect of the recipe itself. The analogy is reasonable; Priestley failed to see that gases, just as much as liquids, can behave differently according to how they are mixed.

For Ingen Housz, all this unexpected turmoil was intensified by the fact that the issues had been taken beyond any philosophical arena and were in the public domain. A contributor to the July edition of the 1781 Critical Review,[27] a middle-brow London magazine, had used his supposed critique of Priestley's new work[28] as a vehicle to knock down and repeatedly run over Ingen Housz. Signed, cryptically, as 'C', the assessment was painfully biased. 'Knocking copy' is irresistible for any journalist and here, in Priestley, was a very well-known philosopher bluntly criticising others of his fraternity. After glowing praise of Priestley and 'the successes of his singular toil and ingenuity' there was nothing but repeated abuse for Ingen Housz who had 'flattered himself' in thinking that he had made a 'most important discovery'. 'Thankfully,' the review continued, 'for sake of mankind', the 'infallible' Priestley had been able to find time to correct Ingen Housz's 'most egregious blunder' and only the 'delicacy of friendship' had 'restrained him' from pointing out the many other 'errors' in the Dutchman's work. This vituperation put the issues into the realms of table-talk and tittle-tattle. Long-standing social reputations were as much at stake as Socratic wisdom but the consequences, for Ingen Housz, were immediate and obvious the moment that the actual books by Priestley and Cavallo arrived in March 1782.[29] As he wrote to van Breda: 'The poor boy, Ingen Housz, now lies prostrate on the ground, beaten down by these two great giants.' Ingen Housz had no need for the urgent advice that came from Richard Huck in London:[30] a second English edition of *Vegetables* was now very unlikely to sell. The new book rapidly taking shape as an ordered pile of manuscript pages on his desk was, he knew, now likely to be stillborn, in London at least, or end up remaindered and resented by any English publisher willing to take it. Karl Molitor was still at Vienna – it was another year before the Elector called him back to Mainz to take up the post of professor of medicine – and was incensed. Ingen Housz clearly had problems damping down an eruption of outrage in his disciple but managed to engineer a compromise. He finally allowed Molitor to include an evidenced protestation in the preface to *Vermischte Schriften* but only if it was free of 'expressions which could offend'.[31]

Ingen Housz himself adopted a righteous, noble even, response to the challenges; one that put personalities and personal feelings out of sight.

He could see that the 'supremacy of light' issue would, finally, turn on one person's word against another's despite him having been the first into print. However, he knew that he must otherwise move quickly to resuscitate his theories and restore his reputation. He therefore repeated, with the assistance of some of his Vienna friends as witnesses, an elegant experiment that was

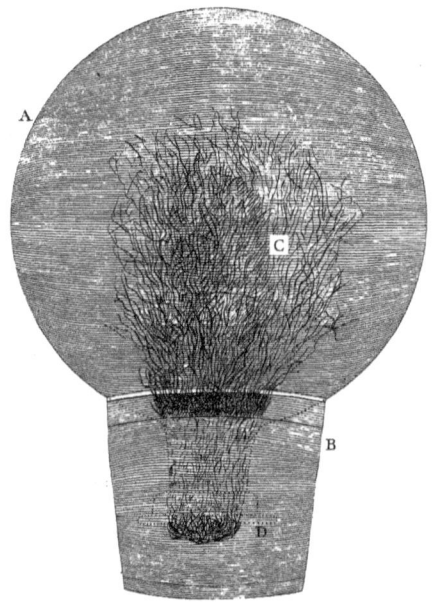

Apparatus for observing the bubbles of air that emanate from plants under water and in sunshine, together with its original key.
© *The Royal Society Library*

A. *a globular glafs vefsel containing about 160 cubic inches of water.*
B. *glafs vefsel fill'd with Mercury in which the orifice of the spherical vefsel is plunged.*
C. *a vegetable called Conferva rivularis.*
D. *a piece of wood to which the Conferva rivularis is attached to keep it in its place.*

urgently written up as a 'memo' for the Royal Society.[32] His aim was to demonstrate, unequivocally, that the bubbles of air in his experiments really were the result of chemical processes within the green parts of plants; that he was right and that Priestley and Cavallo were wrong. This was the fundamental issue, in his view, and he appears to have chosen to ignore the disparagement of the Fontana eudiometer. He set up, in a hothouse of the royal botanical gardens, a series of identical globular glass vessels, inverting them so that their necks and mouths were submerged in mercury. Each was completely filled with hot pump water that had been boiled for over two hours to expel all dissolved air. Once cooled he put a water plant called *Conferva rivularis* (Crow Silk) in two of them, lengths of woollen and silk thread in two others and nothing at all in the remaining two. They were all then put to stand in the same bright sunlight. Bubbles began to rise from the *Conferva* in the first two vessels but only from the third day of the experiment. He tested the collected gas in one of them and it was obviously 'dephlogisticated air' (oxygen). And when some of the water from each of these two jars was heated, it yielded yet more oxygen, oxygen which had, at first, dissolved in the pure water and hence the delay in bubble formation. In fact, no bubbles formed until the water had been saturated with the gas. The vessels containing inert cloth threads produced no air during three weeks of exposure to the sun and neither did the vessels having nothing in the water. Having convinced his friends that the findings were genuine and consistent, Ingen Housz then argued, against Priestley, that the air in jars one and two could only have come from within the living plants: the *Conferva* was actually producing oxygen, not purifying dissolved air. And if Cavallo's theory was correct, why was it that the suspended threads of fabric did not act as 'collecting stations' for gas bubbles that were supposedly produced by chemical processes only in the water itself?

The manuscript work was driven and intense: Ingen Housz's energy levels were as at Southall once more. There was no pacing up and down hoping for inspiration, no fighting back all those reasons for doing something else. Starting from nothing, the substantial paper, given the rather cryptic title, *Some further considerations on the influence of the vegetable kingdom on the animal creation* eventually extended over fifteen pages of the Royal Society's *Philosophical Transactions*,[33] even after harsh redaction. It was finished by the first days of April and, with absurd irony, its completion coincided with the arrival of a 'friendly' letter from Joseph Priestley.[34] This was largely an irrelevance, merely updating Ingen Housz on recent experiments with metal oxides that the Reverend doctor had burned in 'inflammable air' (hydrogen).

Ingen Housz's memo for the Royal Society needed no updating and was ready to send to London; but how? This was a document that needed a speedy but safe passage. Sir Robert Murray Keith, the British ambassador, had a diplomatic courier going to London and offered this service to his anxious friend, but the bag would only leave in mid-May.[35] It was worth waiting. A messenger would be able to carry a small parcel allowing Ingen Housz to send, with the manuscript, two presentation copies of *Vermischte Schriften*; one for the Royal Society and one for Franz Schwediauer who, as a native German speaker, would be able to understand the essays and broadcast their contents. Even if the messenger had been leaving immediately the timing was far from propitious. The Royal Society meetings schedule was always interrupted by a very long summer vacation – usually from late June until well into November – and Ingen Housz was very concerned to have his letter read to the fellows before the break. Otherwise the delay would deal a mortal blow to his publishing plans – 'a finishing stroke to my work for a whole year.'[36]

WITH THE PARCEL for London safely deposited with Sir Murray Keith, Ingen Housz was able to concentrate on a domestic matter. This was another very stressful event that spring – he had to move house. He and Agatha had lived on the Weihburggasse[37] for several years, the rent paid from the royal purse as a perk of his position at court. Now, financial strictures introduced by the Emperor included the withdrawal of all subsidised accommodation for Imperial counsellors and officials[38] and Ingen Housz was obliged to seek something more modest. His duties required that he remain in central Vienna and by late May he had found new accommodation nearby – in a north-facing upper storey apartment in the Franziskanerplatz.[39] It was to cost him nearly a thousand florins (£6,000) a year in rent, a fifth of his pension after tax.

In the meantime the London-bound parcel had arrived at its destination. Joseph Banks recognised the integrity of the arguments and tried to do his best for his friend. However, it was still early in his long reign as President and Banks was having to struggle to re-establish the society's reputation as a tower of academe rather than a gentlemen's club: showing favours to particular friends left him open to accusations of hypocrisy. Nevertheless he did his utmost, having found the parcel at his home late on the evening of Saturday 30 May.[40] At this point in late spring there remained only three formal meetings of the Society before its recess, two of which were unavailable for general papers, however important, since they were the occasions of the two annual invitation lectures.[41] Banks wrote to Ingen Housz the very next morning,[42] assuring him of his best intentions but stressing that presenting the paper so late in the season would not be easy, especially since he estimated that having it read aloud would occupy some thirty five minutes of fellows' time. Such constraints might force him to pass the manuscript on to Richard Huck for urgent private publication.[43] Finally, half of the paper was read out at the 13 June meeting of the Society and published as such in the *Philosophical Transactions*.[44] It must have had some impact but Ingen Housz might have chosen a title more indicative of his motives and its contents, and in tactical terms there was a more salient fact. He was not in London in order to defend his findings, to buttress the validity of his doctrine (which we now know to be correct) and see off

Sir Joseph Banks, President of the Royal Society. Courtesy of the Wellcome Library, London.

his detractors. His unavoidable absence allowed Priestley to drip-feed the notion, by default, that it was he who had first recognised and demonstrated the importance of the 'photo' element in what we now call photosynthesis. And, as clearly related by professor Howard Gest of Indiana University in 1988,[45] and again in 2000,[46] the consequence was that Ingen Housz was largely forgotten in this context until 1875; for virtually a century.

INGEN HOUSZ RETURNED to Vienna from a trip to Moravia[47] on Tuesday 13 August 1782. There was a bulky letter from Paris waiting for him. It was not from Bégue de Presle; there was still no news on the non-progress of *Nouvelles Expériences*. There was much consolation, though, in the packet being from Benjamin Franklin. Dated 21 June,[48] it contained a long-awaited dissertation on lightning[49] folded in with a long covering letter. Franklin reported having been ill with severe gout, his pain exacerbating the growing indolence that he shamefacedly put down to ageing. It had taken him three attempts to finish the letter, its writing extending from October 1781 to June

1782. By making repeated uncomfortable visits to his desk he had managed to respond to all the important points in the several letters that Ingen Housz had sent, dating back as far as December 1780.

Ingen Housz finally enjoyed some feedback on his hot wires experiments, Franklin having derived 'a great deal of pleasure' from the descriptions of them. Franklin was somewhat concerned that the very conductivity of the metals that were being tested might actually dictate the thickness of the coating they acquired when drawn out of the preparatory wax bath. Those that heated up most quickly might also be the ones to cool down most rapidly and acquire, therefore, a thicker coating of wax. The latter would obviously then affect the results. Here, despite all his daily pressures and distractions, was Franklin's genius in evidence. On the other hand he was totally reassuring about the challenges that now confronted Ingen Housz on his sunlight hypothesis, and the attempted debunking of his first book.[50] Opponents, he pointed out, will always

> . . . endeavour to deprive (an author of his reputation), first by disputing the truth of his experiments, then their utility and, being defeated there, they finally dispute his right to them and would give credit of them to a man that liv'd 3000 years ago, or at 3000 leagues distance. Go on, however, and never be discouraged . . . I wish you that continued success which so much industry, sagacity & exactness in making experiments, have a right to expect.

And in the third section of his letter,[51] written five months later, Franklin's commiserations turned to hard advice.

> I am sorry that any misunderstanding should arise between you and Dr. Priestley. The indiscretions of friends on both sides often occasion such misunderstandings. When they produce public altercation, the ignorant are diverted at the expense of the learned. I hope, therefore, that you will . . . go on with your excellent experiments, produce facts, improve science, and do good to mankind. Reputation will follow and the little injustices of contemporary labourers will be forgotten . . . you can always employ your time better than in polemics.

This wise counsel was not really followed by its recipient. Ingen Housz was to gnaw away at his grievances over Priestley for many years to come. Perhaps he might have been more convinced, his energy more effectively diverted, if the wisdom had been available face to face. Franklin's letter revealed that that there had even been some prospect of this. He was still begging the Congress to be released from his diplomatic duties in France in the light of his age and infirmity. If given his freedom, he had schemed[52] that he might just have enough life and energy left to 'make a tour of Italy with my grandson, pass into Germany, and spend some time happily with you, whom I have always loved, ever since I knew you, with uninterrupted affection'. Permission was refused: Congress still saw him as irreplaceable. Even in 1785, when Franklin finally returned to America, Thomas Jefferson was to make a memorable remark that will always be quoted as the measure of Franklin's remarkable, if eccentric, diplomatic triumph at Paris. Asked if he was Franklin's replacement at Paris he replied 'No-one can replace him, Sir, I am only his successor'.[53] Any private pleasures that Franklin and Ingen Housz might have derived from a blissful further meeting were subsumed to the American struggle for freedom. A long dissertation on the devastation that lightning could cause in Italy had to serve as compensation for actual travel and companionship.

IN AUGUST 1777 a severe storm had swept across the flat plains of Lombardy and the spire of one of the parish churches at Cremona had suffered a lightning strike. Summer storms in northern Italy are commonplace enough but the damage from this particular lightning bolt had been so severe that the local experts were convinced, by the degree of the devastation, that lightning could, in fact, strike twice in the same place.[54] The circumstances were certainly remarkable enough to motivate Carlo Barletti, professor of experimental physics at Pavia (Volta's teacher), to travel the forty miles to Cremona to investigate. In 1780 Barletti produced a report of his investigations in the form of a pamphlet[55] and one or more copies found their way to Paris. Although Franklin couldn't read Italian and was somewhat hampered, he had been as intrigued as everyone else by this particular lightning strike and spent considerable time constructing an explanation of Barletti's observations in the context of 'modern' electrical theory. His draft paper, in the form of a letter to 'Dr. John Ingenhauss', was very well organised. He listed 36 points in a very logical sequence, explaining rationally the consequences of the lightning strike. He invited his Dutch friend to corroborate his conclusions on this natural experiment. He was pointedly anxious because of his poor grasp of Italian. His request put Ingen Housz in some difficulty; he was unable to find a copy of Barletti's original work in Vienna and it took several weeks for one to arrive from Italy. Franklin's analysis seemed perfectly reasonable, however. The weather vane atop the steeple had taken the full force of the electrical discharge. Triangular in profile, it had been made of beaten copper coated with tin.

It had also tapered in thickness towards both the blunt and the sharp ends, the thickest portion being near the middle where it sat on a three-quarter inch spindle. This iron support had been embedded in a marble pedestal that had been shattered in the storm. The vane itself landed on the roof of the priest's house and was found to be perforated in eighteen places, the ragged edges showing that piercing had occurred from both sides. Much of the remaining copper showed marks of having been melted. Franklin explained these awesome findings using concepts that are almost present-day, on the basis of what we now call instantaneous energy release. In this phenomenon effects vary inversely with the bulk of the material. The maximum impact – heat and explosive perforation – was therefore on the thinnest parts of the vane, some parts being marginally thinner than others, even near the middle, because of its having been hammered into shape rather than moulded.

Answering his friend's request was to have a dramatic and unexpected consequence for Ingen Housz. Rather than debate theory and trade semantics, he decided to try to imitate the events at Cremona by making a miniature mock-up of the steeple and the weather vane and submitting it to a powerful electrical discharge in the laboratory. He built his model as soon as Barletti's pamphlet came to hand in October 1782.[56] He was able to reassure Franklin that his tentative reading of the paper had been perfectly correct and that re-enactment experiments were underway. By the spring of 1783 he had succeeded in imitating, perfectly, the phenomena that Barletti had reported from Cremona by firing a powerful artificial discharge through a vane-shaped sliver of metal foil. He had used a very large battery of Leyden jars to produce the 'lightning'; not his own apparatus but that belonging to a friend and near neighbour, Baron Zienmayer.[57] As the experiments continued the two friends goaded each other to spin the friction plates faster and faster, and as the atmosphere dried with the onset of summer, fate would not be tempted further and there was a serious accident. The battery suddenly discharged to its operator and Ingen Housz was knocked out.[58]

> I neither saw, heard, nor felt the explosion by which I was struck down. The flash enter'd the corner of my hat. Then it entered my forehead and passed (out) through the left hand. I lost all my senses, memory, understanding and even sound judgement. My first sensation was a peine on the forehead. The first object I saw was the post of a door. I combined the two ideas together and thaught I had hurt my head against the horizontal piece of timber supported by the posts, which was impossible as the door was wide and high. After having answered inadequately to some questions which were asked me by the people in the room, I determin'd to go home. But I was

somewhat surprised, that, though the accident happened in a house in the same street where I lodged, yet I was more than two minutes considering whether, to go home, I must go to the right or to the left hand. Having found my lodgings, and considering that my memory was become very weak, I thought it prudent to put down in writing the history of the case. I placed the paper before me, dipt the pen in the ink, but when I applied it to the paper, I found I had entirely forgotten the art of writing and reading and did not know more what to doe with the pen than a savage . . . This struck me with terror, as I feared I should remain for ever an idiot. I thaught it prudent to go to bed.

Ingen Housz had been very lucky; his injuries could have been far worse, fatal even. Artificial lightning can be as threatening as the real thing but the befuddlement of the immediate aftermath was only part of the story. Ingen Housz goes on to record, with remarkable perspicacity, further and highly significant clinical consequences, all of which he related to Franklin.

I slept tolerably well and when I awaked next morning I felt still the peine on the forehead and found a red spot on the place. My actual faculties were at that time not only returned but I felt the most lively joy in finding . . . my judgement infinitely more acute. It did seem to me I saw much clearer the difficulties of every thing, and what did formerly seem to me difficult to comprehend, was now become of an easy solution. I found, moreover, a liveliness in my whole frame which I never had observed before. This experiment, made by accident, on my self and on which I gave you at the time an account, has induced me to advise some of the London mad-Doctors, as Dr. Brook, to try a similar experiment on madmen, thinking that as I found in my self my mental faculties improved and as the world well knows, that your mental faculties, if not improved by the two strokes you received, were certainly not hurt by them, it might perhaps become a remedie to restore the mental faculties when lost but I could never persuade any one to try it.

Franklin referred back to this remarkable experience in a long letter of 29 April 1785, while he was still at Passy.[59] He, too, had been subject to unwanted and unpleasant electric shocks. He had once been helping in an experiment to see if static electricity from two large Leyden jars could reinvigorate a patient paralysed by a stroke. He had not noticed that a metal hook hanging from the ceiling was within inches of his head and his next recollection was waking up on the floor. He was quite confused for some minutes, his situation only making sense to him after the company in the room had described the events and he began to feel a sore swelling on the top of his head. He did not recall having had any feelings of euphoria, however. Nonetheless he, too, had recommended a trial of powerful electric shocks to a physician friend who specialised in epilepsy and other neurological complaints but had waited for a response in vain. The subsequent history of this line of possible therapy for depression is both patchy and obscure but sundry reports of temporary improvement in 'meloncholia' by shocking the patient into unconsciousness culminated in formal trials of 'electroconvulsive therapy' for depression in the 1930s, an area of psychiatric intervention that remains very controversial. Nonetheless, it is interesting that it is not only psychotherapy that had some of its birth pangs in Vienna.

INGEN HOUSZ SHOULD not have been surprised to find himself facing another challenge to his discovery on primacy of sunlight on plant metabolism. As he himself remarked to van Breda, quoting an American aphorism that he had probably heard from Franklin,[60] 'When two dogs fight for a bone, a third will be around, ready to make off with it'. It was from an unexpected quarter but just as unwelcome. In late February 1782, Ingen Housz had been browsing the journal of *Het Provinciaal Utrechtsche Genootschap van Kunsten en Wetenschappen* (*The Province of Utrecht Society of Arts and Sciences*), of which he had been elected a member, without his knowledge.[61] Suddenly he was alert, his eyes falling on some text that was almost a facsimile of parts of *Vegetables*. It was a long essay by a 31-year-old Amsterdam *apotheker* (pharmacist) called Willem van Barneveld. His paper was entitled *Investigation on the amount of pollution of the air and its restoration by the growth of plants*.[62] His essay, supported by experiments, discussed the 'restoration' of 'phlogisticated' (stale) air by plants, the phenomenon discovered by Priestley about 1771 and reported in 1772.[63] He had even used Priestley's 'nitrous air' test as the measure of air purity in his experiments although, in general, he had drawn conclusions by simply timing how long it took for a candle to be extinguished in his various trial airs. Unlike Priestley, however, he had recognised the influence of light on his results (or so he alleged), allowing him the same claim to fame as Ingen Housz. In one trial, for instance, a *Veronica* plant had taken six days to 'cleanse' an air sample in mixed weather conditions. He happened to repeat the trial during a very sunny spell and the air proved to have been purified in less than half the time 'about which difference he was very surprised'. Sensing that he had stumbled, purportedly, on a fundamental principle, van Barneveld had then performed a further series of experiments, using a range of common plants, in which he consistently found that stale air was revivified much more quickly if the plants were placed in bright sunlight.[64]

> Thus I had arrived at the knowledge that putrid air is restored by vegetation but not without the interference of a second working cause, namely, sunshine (and) we have reason to wonder that Mr. Priestley has not observed or communicated this.

He delivered his essay to the Society at Utrecht knowing that its trustees were particularly anxious to receive papers pertaining to human health. Since many diseases were caused, it was then believed, by having to breathe 'putrid' air, Barneveld's work was welcomed for its obvious potential. Barneveld's published dissertation was 64 pages of printed text, bound, in sequence, with other papers having submission dates from 1775 to 1779 but its gestation had obviously been an interrupted business, finally ending only in late 1780. The bulk of the script had been, according to the author, written, shortly after his experiments, in late summer 1778, and handed in at Utrecht on 23 October that year.[65] After some months it was accepted for publication but the secretary of the Society had handed the document back to van Barneveld with various suggested amendments. Van Barneveld re-worked his text in the light of these suggestions, it appeared, but also took the opportunity to extend his essay. A longer dissertation was handed back in March 1780 to which even further additions were bound in, some six months later, making the final closure date 5 September 1780.[66]

If van Barneveld's work had been the only challenge to his primacy on light in plant chemistry in those early weeks of 1782, Ingen Housz might have stayed calm and managed to suppress his disappointment. But after the onslaughts of Priestley and Cavallo this was too much. In this case he 'knew', from the outset, the exact nature of the challenge: at a single reading he drew an immediate conclusion that he was never to doubt – that van Barneveld's work was a chimera. Ingen Housz's interpretation was that the original text was original and genuine but that the late additions were plagiarism. To him, the early part of the paper was a rather woolly extrapolation of Priestley's inconsistent observations of plants 'purifying' the air. Then, tacked on, were the reports of, and conclusions from, a series of trials in which van Barneveld had made largely qualitative observations on air samples 'purified' by living plants in different light surroundings with the conclusion that the more intense the light the more the purification of the air.

It was the life history of the publication that was, to Ingen Housz, the most significant and most damning evidence that van Barneveld had simply repeated, in essence, his own experiments. Van Barneveld, as any or

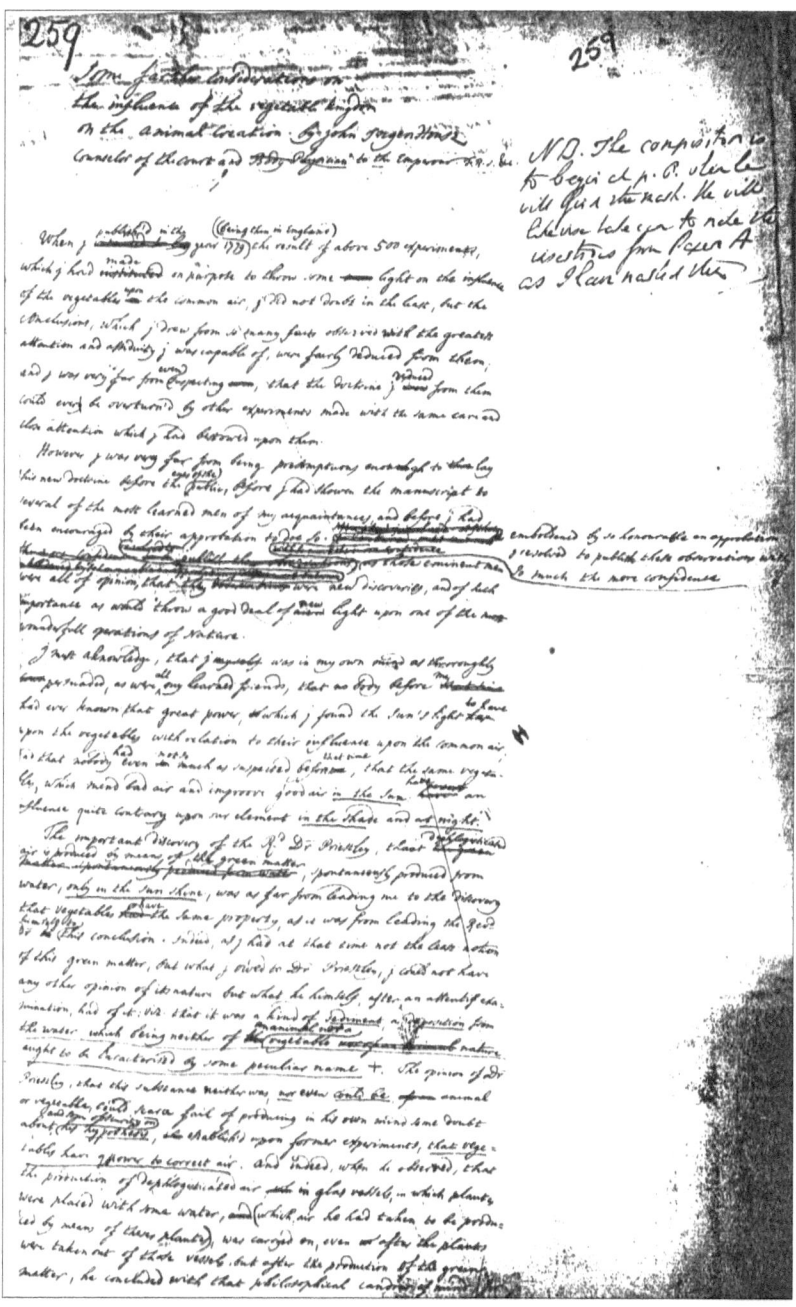

Original manuscript of Jan Ingen Housz: sent to the Royal Society in 1782.
© *The Royal Society Library.*

all of the Amsterdam philosopher fraternity, would have had access to Ingen Housz's *Vegetables* any time after November 1779 when the royal physician had been in the city for six days[67] and had presented copies to particular friends there. Ingen Housz appears to have eschewed all other possibilities including the quite likely one that he and van Barneveld had arrived at identical conclusions about plants and sunlight in total independence. The history of science is peppered with similar examples of parallel discovery, ideation and ability happening to coincide in different locations at precisely the same time.

A truculent Ingen Housz quickly conceived that, having decided to embark on a public debate with Priestley and Cavallo using the help of further experimental evidence, van Barneveld should simply be confronted as a cheat. Here there was no mileage in further trials, van Barneveld travelling along an exactly parallel track: the only question was who had started their journey first?[68] If individual dishonesty was contemptible, so was the lack of probity of the Utrecht Society. It was the Society's casual attitude to the submission of papers, or it so seemed, that really appeared to annoy Ingen Housz. In his view, van Barneveld had been a cynical manipulator but the worst crime was that the Society's officers had given him the opportunity. It was the Society that he took to task rather than their contributor. He asked van Breda, at Delft, to try to find out more, sending him a cascade of forensic questions.[69] Had the secretary of the Utrecht Society kept secure a copy of the early, 1778, part of van Barneveld's submission before handing the original back to the author? Otherwise, was it not entirely possible that van Barneveld had altered even the early parts of his essay in the hindsight of Ingen Housz's book? And had he not then had the effrontery to imitate the 'Southall' experiments? The answers to such cross-examination took some time to emerge: in the meantime yet another rival was to put in an appearance.

IN MID-FEBRUARY 1783 Ingen Housz received a small, cold parcel that was, paradoxically, much to do with sunshine. The packaging released a set of three octavo volumes that had been published the previous year – *Memoires physico-chymiques, sur l'influence de la lumière solaire pour modifier le êtres des trois règnes de la nature, et surtout ceux du règne végétal.* (Physico-chemical memoirs on the influence of sunlight in modifying the beings of the three kingdoms of nature, and above all those of the vegetable kingdom.)[70] Their author was Jean Senebier, a distinguished Protestant cleric and academic librarian. The 40-year-old Genevan was the Minister of the Saint-Evangile Church in the city, a responsibility he dovetailed with managing the republic's main library. Neither post was a sinecure and Senebier had proved his intellectual stamina in cataloguing and making sense of the enormous state archive of manuscripts. Even so he still found time for experimental science and it was this third tier of his activities that had resulted in the 1782 publication. Ingen Housz was struck, immediately, by the similarity with his own work on the influence of the light of the sun on plant activity. As he described, in a letter he was just writing to van Breda at Delft,[71] 'I have only looked into the first part and I find my own book nearly copied-out, with some new remarks'. It was the fourth challenge to his supremacy in the discovery of how plants use sunlight to 'purify' the atmosphere. This time his reaction was different; at least at first: perhaps recoiling against Priestley and Cavallo and, then, van Barneveld had drained him of emotional energy. Perhaps, though, his *sang-froid* was because Senebier's main conclusions more or less reflected his own.

In particular, Ingen Housz warmed to Senebier for he, too, argued that the bubbles rising from plants submerged under water originated within the plants themselves and didn't just assemble on their surfaces. Moreover, the Genevan pastor appeared to be staking no claim to priority. Indeed, Senebier stated, quite categorically, that he had only done his experiments during the summer of 1781, two years after Ingen Housz had been at Southall Green, and that he had even considered approaching him for technical advice before starting them. That he would have performed them the summer before, if he hadn't been ill, still left Ingen Housz well out in front and in a magnanimous mood. Senebier was unable to deny that he could have based his experiments on what he had read in *Vegetables* or on personal reports from Vienna. From September 1780, Ingen Housz had performed hundreds of further experiments, all clearly labelled and in full view; in the very public arena of the hothouses of the royal botanical gardens,[72] the province of his brother-in-law, Nikolaus Jacquin. He reported that there had been, some days, as many as a hundred 'onlookers' watching him perform his eudiometric measurements and he usually found the time to explain his actions and findings.[73] He had also given many free lectures and demonstrations.[74]

> Recently I had evening visits, three Fridays in succession, of several hundred gentlemen and ladies, mostly highly ranked, as I have done more than once before. This is to prevent them taking up my time individually. Before I started any experiments, I gave them a lecture.

Ingen Housz was even moved to compliment Senebier, indirectly, on the fabric of his books. The paper and the print were both a matter of envy[75] although Senebier had probably only called in favours owed to him by Genevan printers rather than having spent money recklessly; debts in kind earned by his influential position at the state library. But top quality oils and the best canvas don't make an artist. Ingen Housz found Senebier a difficult read: he was 'a chaotic author, very tedious, as preachers can be; saying on twenty pages what can be said in one.'[76]

Ten days after completing his first impressions and reactions[77] Ingen Housz began to examine the first volume of Senebier's *Memoires* with more vigilance. He found much to challenge; more comments than the margins could accommodate. When, a few years later, he received another book by Senebier, he took it to a printer asking that the pages all be separated and then re-stitched but only after a blank page had been interposed between each printed leaf. Knowing there would be many of them, he then had ample room to record his assiduous annotations.[78] But in early 1783, even having put other activities on hold, he remained calm and less disputatious; not that there were not endless quibbles. He was critical, scathing even, having discovered that Senebier's 'trivial' experiments used single leaves; his apparatus being on such a small scale that consistent and accurate measurements were surely impossible. Senebier should 'pack up all his equipment and give (it) to his children to play with'.[79]

Ingen Housz was not surprised, therefore, that Senebier had failed to detect, and therefore denied, the *influxus nocturnum*, the reversal of gas flow in the dark. This, too, Senebier could have corroborated if only '. . .one could convince him that he ought to take a whole plant instead of one leaf . . .'[80] However, for all his self-effacement regarding the timing of his experimental evidence, Senebier was actually claiming that his thinking, at least, had preceded that of Ingen Housz and certainly pre-dated the publication of *Vegetables* in late 1779. He alleged that he had conceived, intuitively, the supremacy of the sun in plants releasing 'dephlogisticated air' (oxygen) well before Ingen Housz. This 'fact' was corroborated, he said, by a letter to a Dutch friend, Jan-Hendrik van Swinden, written early in the summer of 1779.[81] Even this did not rouse Ingen Housz though it did provoke him enough to chase the witness. He asked van Breda to contact van Swinden, professor of physics and philosophy at Franeker in Friesland, to try to obtain a copy of the letter 'just for the fun of it'.[82] The conscientious van Breda wrote very promptly to the professor. 'You would oblige Mr. Ingen Housz and myself if you would be willing to inform me of the content of such letter . . . dated May or the summer of 1779 . . . by way of a copy or in substance.'[83] Van Swinden denied ever receiving such a document and thought it more likely that Senebier had discussed his ideas with Charles Bonnet who, after all, lived at Geneva and had been a pioneer of underwater plant experiments. But this tangential inquiry caused more trouble than it was worth, opening up yet more criticism and unwanted correspondence. Van Swinden took it upon himself to examine, in minute detail, and then cross-question, Ingen Housz's eudiometer and techniques[84] and, although we thereby learn some useful details of how Ingen Housz manipulated his apparatus, all this was, at the time, an unwelcome and troublesome eddy. It was to be late spring of 1784 before Ingen Housz really reacted to Senebier, when a not-unexpected tide of vehemence appeared.

Ingen Housz eventually learned the sobering reality of Senebier's mischief when dining with the chancellor.[85] Travellers from Geneva had told Kaunitz that Senebier's books were very popular there and that his experimental conclusions had entirely supplanted the 'erroneous' ones of Ingen Housz. Senebier's very curious statement downgrading precedence in scientific discovery; that 'new discoveries do really merit to be marked twice,'[86] suddenly fell into place. The second runner across the finishing line could thereby be awarded equal first place and, given circumstances such as local popularity, easily overshadow the real winner. And who would trouble to really compare, in detail, different versions of a novelty when they were written up in a different order, in different places and in different languages? Suddenly Ingen Housz saw all this as a 'Jesuitical' trick[87] and that his scientific reputation, even at Vienna, was likely to suffer. His rather superior disdain suddenly turned into mordant passion. Having realised that he was no longer in a private intellectual joust, Ingen Housz was certainly now unwilling to give more time to corresponding with and helping Senebier. Now he was anxious to prove that he had been ahead of Senebier and that he had been correct regarding the obverse behaviour of plants at night – that they then absorbed 'dephlogisticated air' (oxygen) and disgorged 'phlogisticated' (stale) air. Since it was now the threshold of summer Ingen Housz suggested to Marsilio Landriani, professor of physics at Milan, currently at Vienna, that he might care to witness the existence of the *influxus nocturnum* in some experiments at the royal botanical gardens.[88] Landriani, known to be a close friend of Senebier, could then take on the chore of the correspondence.

There was one other significant dispute with Senebier's conclusions – that plants release yet more oxygen in sunlight if they are submerged in water suffused with carbon dioxide. This was certainly an original finding of Senebier but seeing the experimental proof for it as flimsy, Ingen Housz refused to subscribe. He had found no such effect even though he had trialled a few plants in acidulated water. He also took the line that water in nature – be it fresh rainwater or from pond, lake or river – contained very little carbon dioxide so that Senebier's experiments were artificial and unrepresentative of nature.[89] In raising this point Ingen Housz seemed particularly blinkered: most of the plants used in his thousands of experiments had been terrestrial species submerged under water, which was equally unnatural. The eventual riposte to Senebier came in articles sent to the well-respected Parisian science publication, *Journal de Physique,* where, in two papers that were published in 1784,[90] Ingen Housz defended his primacy over this latest challenger. Responding in a journal published in the French language was obviously appropriate but it is interesting to ponder why Ingen Housz chose, after 1782, to send all his scientific submissions to Paris despite having turned down the offer of membership of the Royal Academy of Sciences there.[91] Among the eleven papers that were to appear in *Journal de Physique* between 1784 and 1789, there was not a single one in *Philosophical Transactions* of the Royal Society at London. There may have been negative feedback from his 1782 paper, his riposte to Priestley and Cavallo. Or perhaps he sensed that without his physical presence in London, his papers, read aloud by long tradition, had little impact in his absence: so much did the Royal Society thrive on intimate camaraderie at its weekly meetings and its dining clubs. But the reason for eschewing the Royal Society may have been political. The 1780s were a troubled time at the Society where Joseph Banks, in trying to restore scientific probity, had antagonised powerful forces. Perhaps Ingen Housz shied away, even as a correspondent, fearing involvement in the time-consuming issues of a power struggle. Moreover, he was already caught up in disputatious matters at Vienna, aggravation enough for someone whose priority was his laboratory bench rather than the soapbox.

MOZART'S OPERA, *Die Zauberflöte* (The Magic Flute), has always been associated with freemasonry, with its morals, its secrecy, its rituals and its symbols. Although Emanuel Schikaneder was probably not a member of a masonic lodge when he wrote the libretto, Mozart certainly was when he composed the music.[92] Freemasonry was a British innovation from the beginning of the eighteenth century and was exponentially popular, spreading rapidly across Europe and to America. In 1725 there were 52 lodges in Great Britain but within 50 years there were over 300; 87 in London alone.[93] With symbolic links back to the medieval guilds of craftsmen, new members joined as 'apprentices' and had to learn of the 'mysteries' and pass 'tests' before promotion to 'journeymen' and perhaps, eventually, 'mastership'. Traditional lodges prided themselves on brotherhood, benevolence, civilised attitudes and forward thinking which they promoted, later in the century, as an 'enlightened' outlook. The obvious paradox of being exclusively male, grounded in secrecy and devoted to pseudo-religious ritual was, and still is, disregarded.

Mozart, struggling to make his way in Vienna, had become a freemason in 1784,[94] four years after the death of Maria Theresa. Although her husband had been a member of a lodge since the 1740s,[95] she herself agreed with the views of the Vatican. A widely-ignored Papal Bull of 1738, reissued in 1751 to no better effect, had prohibited masonry. But in Vienna there was certainly an air of discouragement until the passing of the Empress, in late 1780, seemed to trigger an explosion of freemasonry in the city. *Zur wahren Eintracht* (True Concord) became the best-known and most enlightened lodge in the city. It was founded with fifteen members in 1781; by 1785 it had 197.[96] In becoming a member of *Zur Wohltätigkeit* (Beneficence), Mozart had been swimming with a tide. However, Joseph II did not follow his father's example and become a mason. Perhaps he was more in awe of Maria Theresa's long-standing opposition and less willing to antagonise her. But even after his mother's feelings were out of the equation he was, perhaps, inhibited by the negative attitude of Kaunitz who 'would have nothing to do with such nonsense'[97] On the other hand many of the lodges in Vienna rapidly postured themselves towards Joseph's enlightened views and civic reforms, stressing their humane pragmatism, charitable dispensations and keen interest in natural philosophy and new technologies. Surprisingly, Ingen Housz was never a freemason in Vienna. Indeed, despite having been a member of a lodge in London in the 1760s,[98] he preferred to stay apart and not entangle himself with people he foresaw as tedious. His attitude seems to reflect, again, his highly developed reluctance to waste his time in needless ceremony and irksome routine:[99]

> I have so many of my closest friends among them (freemasons), who tell me everything, that I never considered it worth the trouble to have myself admitted among them so as not to be obliged to have contact with people whose company was useless to me and who might inconvenience me with questions etc. They grow in number here so much that many join just to gain favours and protection; so much so that anyone not accepted into Masonry can hardly earn his bread.

The venality that ran through masonry, particularly among the ambitious Viennese, was obviously a powerful recruiting sergeant. Ingen

An intiation ceremony at an eighteenth-century Masonic lodge.

Housz was fortunate: with his substantial pension from Maria Theresa he could afford to be above the fray. And as a courtier particularly close to the Emperor he would also have been aware of a growing royal unease.

First and foremost an autocrat, Joseph II knew that his throne could never be secure if he was not at the sole apex of the triangular web of all information concerning his Empire. There should not be a 'state within the state' and secret sects and societies could not be tolerated for long. The very rapid growth of freemasonry in his territories, and especially in the cities, therefore antagonised him increasingly during the first five years of his sole reign. By 1785 his misgivings were such that he felt obliged to take action. His intervention, a series of restrictive edicts, seemed benign enough. There were to be no more than three lodges in any one city and lodge secretaries were to submit regular lists of members and their meeting dates to local magistrates.[100] The dismantling of secrecy provoked much unease: the temporal advantages of lodge membership being dwarfed by the status and value of holding civic positions in the state, all of which required, ultimately, royal patronage. Many a nobleman or high official suddenly found that they were too 'busy' to attend lodge meetings when recalcitrant membership might be seen as mutinous. The exodus was less surprising than the changes that it provoked in the nature of freemasonry in Joseph's Empire. Hardcore masons having exotic mystical beliefs gained the ascendant: prominent among them were the Rosicrucians. Rosicrucianism was an eccentric form of Christianity accentuating a total divide between the spiritual and the material: light symbolised the former as much as darkness the latter. The resurgence of alchemy, magic and cabalism[101] all contradicted the enlightened reasoning that had begun to take hold in Europe. Freemasonry was now a retrogressive focus, at least in Austria. The musicologist, Nicholas Till,[102] has even argued that *The Magic Flute* is more a Rosicrucian opera than a Masonic one and, typically, Mozart's own enlightened lodge amalgamated with two others whose members were far more inclined towards mysticism; sorcery even.

The retreat from science to superstition seemed to animate Ingen Housz further. Van Breda, who, from his immediate interest in the subject, was perhaps a member of a lodge at Delft, had asked his friend about the destiny of freemasonry in Vienna: there had been an article in a French newspaper at Leiden.[103] When he described the forthcoming suppression of freemasonry in the Empire, Ingen Housz was so certain of 'imminent thunderstorms'[104] that it might be taken as evidence that he had been personally involved in advising the Emperor. If so, his counsel must have been very negative; he was quite uninhibited in revealing his zeal – that of a reformed man:

... all the ceremonies and admission acts are nothing but ridiculous hypocrisy. They (masons) have secrets such as hieroglyphs that they do not themselves understand. They regard some as ancient Brahmin relics, others to be Christian symbols and still others the signs of Egyptian priests. They talk only of building Solomon's Temple, not out of stone, but as a mythical structure, a 'temple' of humanity, of wisdom, love, charity etc. (Their journals) ... are worth reading if only to learn about the impetuous fanaticism of people. Otherwise they are not worth to wipe one's a.

A myriad of critical and consequential events in human history have flowed from intimate conversations in secluded rooms or whispered comments in shadowy corridors. As with the debunking of Anton Mesmer and his consequent ostracism at Vienna, Ingen Housz may also have been a powerful influence, behind the scenes, in the suppression of freemasonry in late eighteenth-century Austria. We cannot know for sure, for among the black arts of politics is the skill of leaving no evidence. In his clear antagonism and distrust, however, Ingen Housz was certainly prophetic. During the next decade freemasonry would come to be seen as one of the catalysts of the French revolution and a movement to be despised.

16
Experiments and Distractions

'The towel is for you, Sir.'

Joseph II is pleased to be able to rub and dry his arms. Manipulating the eudiometer requires him to be up to his elbows in the tub of water and his arms have begun to ache with the cold. Ingen Housz takes over and lets another vial of nitrous air up into the tube. He locks it and then rotates it rhythmically as before; predicting that, this time, there will be no further visible reaction. He is right, saturation is complete and they have their final reading, the measure of purity. The Emperor nods. He's no dunce in these matters and the eudiometry details that he's read in Ingen Housz's book now make much more sense. But perhaps it's time for something less cerebral. From a box on the window-seat, Ingen Housz produces a large brass pistol. At least, it looks like a pistol, though certainly not slim and elegant. The Emperor turns it over and frowns. There's a butt with a pipe through it, a cylindrical chamber and a conical narrowing to a barrel. It has no trigger, however, and there's a nipple on the top. Ingen Housz explains how he and Abbé Fontana developed the weapon whilst together in London. It works with inflammable air rather than gunpowder, an original idea of two friends of his in Amsterdam. Though it looks cumbersome and is fired by a spark, it can be reloaded just as quickly as an ordinary pistol and is easily as powerful. It crosses his mind to tell how previous, weaker models had shattered into pieces on firing but thinks better of it. Perhaps, Ingen Housz suggests, it will be better if he fires it first. For absolute safety he won't use a ball and will wrap the weapon in a towel – they both look at the damp cloth the Emperor is still holding and he passes it over. There's a silent tension as the weapon is primed, through the feed pipe, with the ether/oxygen mixture that Ingen Housz has made earlier. Then, holding the swaddled pistol at arm's length, Ingen Housz brings a charged Leyden jar up to the side of the chamber. The spark and the very loud report seem to share the same instant. There's a ringing silence. The Emperor's eyes are wide, his pupils dilated. Both men then turn to look behind them, aware of a pounding on the stairs that's getting louder. The door bursts open and a bodyguard almost falls into the room brandishing his sword. It takes him two paces to come to a stop and two seconds to grasp the fact that his Emperor is unharmed and in no danger. He realises there's no assailant in the room and as the sword is lowered he exclaims:

'His majesty has forbidden shooting in his palace.'

The two philosophers look at each other and burst into laughter and the guard is flushed with embarrassment. The Emperor takes charge and with some difficulty, between the chuckles, thanks him for his care and invites him to return to the carriage downstairs. There may be further explosions, he explains, for he wants to fire this new type of gun himself and no-one downstairs need worry. Ingen Housz primes the pistol again and Joseph II fires it. They look at each other, wait and then the laughter breaks out again. And so it goes on for some time.

As a herald of the Enlightenment, Joseph II had an understandable interest in Ingen Housz's scientific activities. He was keen to cultivate technology and appreciated having, among all his other courtiers, someone who was not just familiar with all the latest developments in science but one of its most prolific inventors. And now that he knew the Emperor well, Ingen Housz was not abashed by Joseph II's request to visit the laboratory at his home in the Weihburgasse. In the event, a visit planned for December 1780 was postponed because of the death of Maria Theresa.[1] His mother's demise left Joseph with many matters that required his immediate close attention but he remembered his intention to visit Ingen Housz. His host was well prepared by Friday 20 April 1781, when, at seven o'clock in the evening,[2] Joseph appeared on the threshold, his carriage and his guards causing a stir in the street. The Emperor and his body physician were alone together for over three hours[3] but it is very unlikely that any personal medical matters were discussed; Joseph II was far from introspective about his health; often being ill, he couldn't afford to be. The occasion was dedicated to science. In three hours Ingen Housz would have been able to demonstrate many inventions and experiments including the use of his eudiometer and the apparatus for breathing 'dephlogisticated air' (oxygen).[4] Three hours allows time for digressions, however, and this may have been the occasion when Joseph II asked Ingen Housz about his friend, Franklin, and revealed the possibility that the American could be coming to Vienna as a delegate to a peace conference.[5] As the evening progressed, the two men became so engrossed in science that they appear to have forgotten their circumstances, both who they were and where they were. It was probably this occasion that was the origin of the story of the nonplussed bodyguard that Ingen Housz later told to his nephews and nieces at Breda. And it was perhaps on this occasion that Ingen Housz, as host, asked the Emperor to accept a small gift of a dozen self-lighting candles that he had prepared specially. They were not his original invention: the idea had come from an Italian philosopher of Turin. A hollowed-out reed containing a very small plug of phosphorus sealed within a resinous coating preventing all contact with the air was pushed into the top of a candle and checked, in a very dark place, for any immediate phosphorescence. If all was well – the phosphorus hermetically sealed in – the 'self-lighting' candles could then be safely stored until needed. Any means of admitting air to the phosphorus would light the candle automatically.[6] Ingen Housz became very adept at making these useful 'lights' and was to supply the Emperor with many more over the coming years.

Joseph II was not to be the only royal guest that Ingen Housz was to instruct and entertain in his home in Vienna. Joseph's personal diplomatic efforts, in the late 1770s, had brought the Habsburg and Russian empires closer together and in the early summer of 1780 Joseph II and Catherine the Great had met at Mohilev, midway between Warsaw and Moscow.[7] It was a meeting that was predestined to succeed. Both monarchs were euphoric and flippant, enjoying each other's company, knowing that the detailed formulation of the treaty they intended could be left to the wary and cynical diplomats in train. A mutual protection agreement was drafted. A shared fear is a good reason for near neighbours to co-operate; the joint anxiety being the prospect of Frederick the Great's next land-grabbing expedition. When Maria Theresa died, late in the year, and with her the reluctance to process the treaty, there was immediate diplomatic traffic between Vienna and St. Petersburg and the deed was signed in May 1781 – or rather, it was not signed. Both monarchs felt it wise not to publicise their compact and merely exchanged unambiguous letters of intent. Only those in the loftiest diplomatic posts knew of the agreement that Austria and Russia were now bound to come to each other's assistance should Frederick the Great turn his aggression on either of them.

Keeping the secret gave Catherine a family dilemma. Her son and heir, Grand Duke Paul, was now 26 years old and being groomed for his future reign. He would need to be informed of the pact[8] but he was a strong Prussophile and might react badly. His wife, Maria Fedorovna, was from Germany and when Paul began to lobby his mother to allow them to visit his wife's family in Würtemberg and combine it with a tour of Italy, Catherine saw an opportunity. Paul would have less chance of making trouble whilst away from St. Petersburg and in sanctioning the trip she insisted that the

EXPERIMENTS AND DISTRACTIONS

Grand Duke Paul of Russia.

couple go nowhere near Berlin but travel through Vienna. Catherine was nothing if not cunning. Her new friend, Joseph II, could then reveal to her son, innocently and harmlessly, the existence of the new treaty. Such was the scheming that eventually saw Grand Duke Paul, his wife, and an entourage carefully picked by Catherine herself, leave St. Petersburg, incognito, in September 1781. Their departure was slightly delayed for it was an anxious time for these young parents. Thomas Dimsdale, this time accompanied by his wife, Elizabeth, had returned to St. Petersburg in order to inoculate the new generation of royal children against smallpox.[9] His ministrations were again successful and as soon as it was clear that the two children were recovering safely and out of all danger, the carriages carrying their parents, the 'Count and Countess du Nord' drew away.

Joseph II was pleased to welcome his Russian guests to Vienna,[10] their arrival helping to cement the bond he had formed between the Habsburg and Russian empires. There was a political agenda but Joseph was anxious that his visitors had a memorable stay; that they were suitably entertained. There was a grand reception at the Hofburg on Thursday 29 November at which Ingen Housz was present.[11]

> ... I had the honour to spend the whole evening with their Imperial Highnesses, the Grand Duke and Grand Duchess of Russia, and other high persons of their family. The Grand Duke was very friendly, shaking hands several times as a sign of friendship (and) requested two personal copies of my dissertation on a way of breathing dephlogisticated air (oxygen). The Grand Duke wants to talk further with me and visit my house. His wife and all the others are determined to come too.

A visit to Ingen Housz's apartment laboratory was, indeed, placed on the social calendar proposed to the young Russian royal. This was not passed over and the heir to the Russian throne dismounted from a court carriage in the Weihburgasse on Friday 7 December. He stayed for an hour and a half, long enough to listen to a short speech of welcome from his host and to see one or two simple experiments. It was clear that he regretted not having more time and, on leaving, asked permission to return, bringing his wife and their entourage.[12] Ingen Housz had been honoured by the presence of his royal guest; there would have been very few Viennese private domiciles graced by his presence, but there was soon more everyday, frivolous, gratification. Reports of the visit in some German newspapers were so garbled that a breezy Ingen Housz found them very amusing. Apparently a 'vizier', a high official of the Emperor, had been making secret calls on a 'strange Irish physician.'[13] But the potential honour of a second visit by the Grand Duke seemed to evaporate. Ingen Housz probably began to think, as the days passed and no further arrangements were made, that the Russian had simply been courteous in asking to return, but not so. Around dawn on a freezing New Year's Eve, a court carriage appeared again in the Weihburgasse. It disgorged another 'illustrious traveller'; the Grand Duchess,[14] who

> honoured me by spending the early morning hours at my house without her husband. She observed everything I could show her, in a couple of hours, with extreme attention and she left very satisfied ... after conversing with me about learned subjects, mainly new discoveries in Physics.'

The Duchess showed her gratitude in a very tangible way by presenting Ingen Housz with a valuable diamond ring.[15] There were to be more visits from other royals within a few months but Ingen Housz never allowed himself to be distracted from his experiments for long. He was now spending a great deal of time looking intently at images down his microscope but even this was being interrupted by duties at the Hofburg: the Emperor was unwell and there was a dynastic struggle pending.

In the spring of 1782 Pope Pius VI visited Vienna and was there during the Easter celebrations.[16] Alarmed by Joseph II's intentions to reform a variety of religious traditions and close many spiritual institutions, Guiseppe Garampi, the papal nuncio had been sending increasingly alarmist reports to the Vatican. Many of them were opened, in transit, by the Emperor's spy ring so that Joseph was not surprised when a reaction blew up at Rome. Against the advice of many of his cardinals, and to the horror of Garampi, the Pope decided to visit Vienna and persuade the Emperor, personally, that his policies were untenable. Pius had not always led a celibate and cloistered life and the tall elegant pontiff, with his 'wide' experience of the world, was confident of his negotiating skills. He thought he could gain more by having private, man to man, talks with Joseph that would marginalize the almost apostate Kaunitz. But from the outset the Pope's tactics were in disarray. His intention to travel incognito backfired when his train of carriages was met, everywhere, by huge crowds and his journey turned into one long carnival parade. Church bells were rung all along the route, despite an Imperial ban, and the Pope's actual entry into Vienna was nothing but triumphal; anything short of a diplomatic knockout would now be an embarrassment for someone with this popular power.

At Joseph's insistence, the Pope stayed at the Hofburg, in Maria Theresa's old suite of rooms. Ingen Housz would have been in the line when all the assembled courtiers were presented to him on Friday 22 March and one might have expected that, as a Catholic, the royal physician would have documented the meeting, severally and proudly, in his letters. There is, however, a singular lack of comment bar one fleeting reference in a letter to van Breda that is mainly about airs and manuscripts. 'In a few days we expect an extraordinary and very rare phenomenon, namely the arrival of the ancient Father of Rome.'[17] Perhaps the need for confidentiality was twofold; political but also medical. One of the reasons that the Pope was to stay in Vienna for a whole month was that the Emperor was ill and for nine days the two men were unable to meet. Throughout the spring of 1782 Joseph II had been suffering from discharging eyes, from a septic conjunctivitis, such that he was unable, in early March, to read documents or write letters. Then, just as the Pope arrived, his host's eye condition deteriorated causing some very tense and highly pressured consultations among the royal doctors on how best to advise their patient. When discussions between Joseph and Pius began, the Pope found his host to be intractable and obtained only enough concessions to save face, despite some outbursts of misplaced stridency. Meanwhile, outside the Hofburg, the crowds waiting, every day, for a papal blessing, grew to be almost unmanageable. In fact the Pope's general reception so impressed him that he grew euphoric and was, finally, so outmanoeuvred by Joseph and the wily Kaunitz that he gained very little for all his efforts. Joseph had little inclination to celebrate, however. His eye infection continued and by July the torment was almost unbearable and he lacked the energy to push through many of his current reforms. Then his facial skin also became infected[18] and he developed abscesses of the scalp and lost his remaining hair: 1782 was a taxing year for the royal medical team.

OWNERS OF NEW garden ponds often despair, as spring turns to summer, when the pride and joy of their winter engineering becomes cold soup, for stagnant water, unless continually sterilised or in total darkness, will always become turgid and green. Some fresh waters do so faster than others but all seem to respond, in this way, to sunlight. The reason for these centuries-old observations had no satisfactory answer in the late eighteenth century. This want of true understanding was to fuel a revolving and exhausting dispute between Ingen Housz, Priestley and many others. The 'green matter' that they all observed was named so vaguely because everyone realised that they were ignorant of its true nature. Was it inert, a plant or even an animal? Suddenly its composition became more relevant when Priestley showed that some samples of green water could release very pure 'dephlogisticated air' (oxygen).[19] The green matter moved to the centre of the arena where the other pioneers of plant physiology were also straining to unravel the relationship between vegetation and the atmosphere. Priestley had committed himself in his 1779 book. 'This green matter can neither be of an animal or vegetable nature (appearing in containers which were closely corked), but a thing *sui generis,* extraordinary though it will seem . . .'[20] He had clearly believed, then, that the green matter was an inanimate dye and that its appearance was stimulated by the action of sunlight on water. In his next published volume on 'airs', however, in 1781, there was a *volte face.* Now he was equally adamant that the green matter was 'vegetative;'[21] after all it matched the plant paradigm of flourishing in light. But he also implied that he had, quite independently, discovered that the green matter would only release 'purified' air in sunlight: it was another claim to primacy in the sunlight/plants issue. Ingen Housz was aggrieved by these claims to priority and there was more understandable truculence when Priestley accused him, by implication, of not having taken the influence of the green matter into account when he performed his 'Southall' experiments.[22] According to Priestley this was a naive oversight that made Ingen Housz's obsessively careful and double/triple checked eudiometric readings meaningless. But despite his self-assurance he was in error. It takes several days for fresh water to go green whereas many of Ingen Housz's 1779 trials were completed within hours. Once again, on the evidence of publication dates, the only easily verifiable arbiter in such disputes, Ingen Housz easily holds the ring. In Section 22 of Part One of *Vegetables*,[23] published in October 1779, and headed 'Some remarks on the Green Matter' he refers, in several places, to its being 'vegetative' or 'vegetable matter'.

By early 1782 Ingen Housz still believed that the green matter was a microscopic plant but was agitated enough, by Priestley's predacity, to make many further observations, spending endless hours at his microscope. In true Baconian fashion he sought more evidence for the true nature of the green matter rather than trade beliefs and dispute interpretations. In the end, sadly, he managed to bamboozle himself, and his immediate circle, into an error. However, an unexpected and historically important development was to come from his persistent endeavours. At Southall Green, Ingen Housz had obtained water containing the green matter from the 'side of a stone trough placed near a spring upon the high road and always kept full for the horses.'[24] In Vienna he developed a new source, one that was private, reliable and speedy, even in winter. A small piece of raw meat or fish left in some water on the bench of a hothouse soon turned the water cloudy and offensive. Vegetable matter such as a small cube of fresh potato did the same. Down his microscope Ingen Housz observed that drops of this water were full of life and, as the water subsequently turned green, he thought that little 'insects' could be discerned; all of them livid green. Left uncovered and undisturbed, the living green matter settled on the bottom and sides of the jar, eventually drying to a green slime or 'snot.'[25] Ingen Housz took this life stage to represent some form of 'animal moss'[26] for the myriad of his little very active 'insects' were now inert and tangled among a meshwork of green filaments. That 'cross-dressers' were possible among living organisms was a cue he took from Fontana's eccentric belief (then) in the existence of *planta animalis*; that there were a few living organisms that defied the strict Linnaean division into all plants on the one hand and all animals on the other.[27]

But in reality Ingen Housz was bemused. If his version of the green matter was left in very moist, still air, so that evaporation was protracted, even in summer, it sometimes formed a kind of gelatine which he thought might mean that it was a form of *Tremella*, a variety of fungus that lacks much in the way of structural organisation.[28] He later had the notion that the green matter might be a stage in the life cycle of *Conferva rivularis* (Crow Silk) or one of the other blanket-weeds found everywhere in ponds. Both life stages, if that was what they were, certainly released 'dephlogisticated air' (oxygen) of remarkable purity in bright sunlight. However, as he confessed to van Breda in April 1783, sending him some samples in sealed glass chambers,[29] 'I have followed these changes for more than two years and hundreds of experiments, still without being capable of clearly describing the phenomena.' In fact Ingen Housz can be forgiven his frustration. Only when microscopy became more sophisticated were biologists able to develop a logical, Linnaean classification of microscopic life. In essence, the 'green matter' is a mixed suspension of plant forms that we now call green algae or *Chlorophytes*, unicellular plants with distinct green bodies within them called chloroplasts. There are hundreds of species, each favouring subtly different environments, *Conferva rivularis* being just one of the filamentous varieties. The 'blanket weed' that drives garden pond owners to distraction is a close cousin. All of them, however, flourish in sunlight and strong colonies will begin to provide their own food by photosynthesis, the other discernible product of which is, of course, oxygen. But stagnant waters may also contain other kinds of algae, flagellates, slime-moulds, and bacteria. Ingen Housz had allowed himself to be taken to the zoo wearing someone else's spectacles. It could all have been a tormenting waste of time but being the acute observer and laboratory virtuoso that he was, it wasn't.

INGEN HOUSZ FINALLY concluded that the green matter, which we now all know to be an algal overgrowth, represented populations of microscopic animals, freshwater coral perhaps.[30] After three years' study he had been deceived by consistently observing, down his microscope, active and vigorous movements that would only be expected in the animal kingdom. He should be excused for his mistake because his long grapple with this particular mystery led to a seminal advance in microscopy that proved to be critical. He had only come to his conclusions after pioneering a technical development. His description of this advance, the need for it and the advantages it conferred on the microscopist appeared as a foreword in a further volume of *Vermischte Schriften*[31] in Vienna in 1784 and, likewise, in the second volume of *Nouvelles Expériences*[32] the French version of similar texts, in 1789. Under a marvellously appropriate quote from Virgil, '*Admiranda tibi levium spectacula rerum* (The wondrous pageant before you, of a tiny world),'[33] we find, in 850 words, the clear and succinct story of his invention.

> I have often troubled my head to find a method to avoid the too rapid evaporation of the drop of water, or any other liquid, in which I wanted to observe 'insects' and I know that other observers are plagued with the same problem. . . . as long as the droplet lasts, the entire liquid and consequently everything which is contained in it, is kept in continuous motion by the evaporation. If the droplet is rather large it has a convex surface which refracts the light more or less; if it is very small it lasts hardly long enough to enable observing its contents at one's leisure. These difficulties are even greater if one uses the solar microscope, because in this case the object is placed in a brilliant cone of light that increases the temperature and accelerates the evaporation. These difficulties can be largely avoided if the droplet is placed between two flat, polished glass plates such as are used for the manufacture of mirrors. . . . all difficulties would be overcome if it were possible to find glass which is polished on both sides and sufficiently thin. However, not being able to find glass of the desired kind, I have helped myself very well . . (by placing) . . . the drop of liquid on a glass slide and then covering it with a very thin sheet of mica. But the very thin films of glass which litter the floors of all glass-blowing workshops are even better. These films flatten the droplet, make it thinner and of an even thickness. Under these films, evaporation is so slow that a droplet that would evaporate in a few minutes hardly shrinks in the course of hours.

Ingen Housz singled out these remarks and placed them before all the other articles in his new books because he was afraid, perhaps, that they might be overlooked. Sadly, his gambit failed and the slide into obscurity is unfortunate for he had invented, effectively, the microscope slide 'cover slip', the use of which rapidly became a commonplace. The significance of his creative genius, and the appropriate personal credit, was, in this case, lost until an article by Hebbel Hoff, a physiologist from Texas, appeared in the 1962 edition of *Bulletin of the History of Medicine,* close on 200 years later.[34] Other microscopists before Ingen Housz had, admittedly, recommended that specimens be forcibly compressed between plates of glass, as the filling trapped in a sandwich, but Ingen Housz appears to have been the first observer to realise the potential of having an exquisitely fine layer of fluid trapped under a very thin layer of glass. It was he who first recognised and described the advantages of even thickness of the fluid film, of the mastery over diffraction, of the control of evaporation, and of the suppression of gross fluid movements.

Another claim has been made for Ingen Housz in the important field of microscopy. It has been argued that he was the first person to observe what has become known as 'Brownian Motion'. Peter van der Pas, for instance, a Dutch engineer who migrated to California from Breda after the Second World War, made a lifelong study of his compatriot and aired this assertion at the Twelfth International History of Science Congress in August 1968, publishing a paper in 1971.[35] The justification is, however, far less convincing than it is for Ingen Housz having devised the cover slip and is almost certainly an over-enthusiastic conjecture.

Brownian Motion was named after Robert Brown, a Scottish surgeon and amateur botanist encouraged by Joseph Banks. He made his famous microscopic observations in June, July and August of 1827,[36] nearly 50 years after Ingen Housz had been peering down his own instrument at the 'green matter'. Although the powers of the respective microscopes of the two men probably did not differ fundamentally, the confidence and cumulative experience of microscopists had advanced significantly and Brown was particularly skilful. He spent the summer of 1827 looking at the pollen grains of *Clarkia pulchella*[37] suspended in water. He was not, however, focussed on the whole pollen grains but on the oblong 'granules' that he could discern within them: levels of visual perception had moved on a whole order of magnitude in half a century. Brown observed, consistently, that the intracellular granules were constantly in motion and also noted that besides moving position they always seemed to be rolling on their long axes and changing shape. Although Ingen Housz had also reported[38] the movements of small particulate matter when watching through the microscope, even of inanimate carbon particles, he was quite certain that the cause was the movement of the fluid in which the 'animacules' were suspended, the fluid flowing relentlessly towards an evaporating surface

> . . . as long as the droplet exists, its continual evaporation necessarily puts all the liquid and, as a consequence, everything floating in it, in a continuous movement, and this movement can, in many cases, make certain bodies look as if they are alive even if they are inert . . .

To Ingen Housz the significance of this gross movement was that it prevented the making of reliable observations and hence his invention of the cover slip. He also noted, however, that when taking movements to be evidence of living matter, it was important not to use compression on the cover slip.[39]

> Most animacules in a suspension, and other small insects, swim equally freely in a flattened droplet, as long as it is not covered, It is, however, the case that compression by a slip of mica or glass arrests the movement of animacules; it is therefore necessary, always, to examine the droplet without compression

before making conclusions that the corpuscles that one believes to be living, have vital movements or not.

This is clinching evidence that Ingen Housz was not observing 'Brownian motion' for compression would not stop such activity, which, in any case, was observed at an intracellular level. It is true to say, though, that Brown, working half a century later, by which time the cover slip was an habitual component of high-power microscopy, would have been taking its advantages for granted. Consequently, we can be quite certain that the gyrations and deformities he was observing were totally independent of any systemic fluid movements, exactly as he reported.[40] 'These motions were such as to satisfy me, after frequently repeated observation, that they arose neither from currents in the fluid, nor from its gradual evaporation.' Robert Brown's work is justifiably seen as ground-breaking. Not only did he go on to confirm that his phenomenon could be seen within the pollen grains of a vast number of other plants but also, remarkably, within dried pollen (some of it a century old) or in grains that he had killed by soaking them in alcohol. Moreover, as he assiduously extended his work, varying the circumstances and using other microscopes, he found that he could see similar ceaseless movement even in suspensions of very finely powdered minerals. Thus he was forced to move away from the otherwise obvious explanation for his 'movements' – that they were the 'vital squirmings' of all living matter and was honest enough to admit that he was 'unable to account' for his findings except that 'life-force' was no longer tenable. Nevertheless, his was an important breakthrough.

The true explanation of 'Brownian motion' was wanting until 1903 when Albert Einstein came up with illuminating insights. The kinetic theory of the 1870s, which supposed a constant and random motion of all molecules, had satisfactorily explained the behaviour of gases but there was a difficulty in applying it to Brown's observations. Fine particles suspended in microscopic preparations were, however small, some 10,000 times larger than a water molecule. Assuming that a visible microscopic particle was being moved by bombardment with water molecules was, in Einstein's analogy, like supposing that a cricket ball could move St. Paul's Cathedral by being thrown against it. He realised,[41] however, that

> a bump from a single water molecule would not cause a suspended pollen particle to move enough to be visible. However, at any given moment, the particle was being hit from all sides by thousands of molecules. There would be some moments when a lot more bumps happened to hit one particular side of the particle. Then, in another moment, a different side might get the heaviest barrage. The result would be random little lurches . . .

Einstein's genius went beyond innate insights. He then formulated a mathematical model for the frequencies and distances of Brownian motion, numbers that could be tested, and published them in a paper submitted in May 1905. Within months, they were verified by a German microscopist and mathematician[42], Henry Seidentopf and the actuality of atoms and molecules had been proven. So, while there is no real evidence that 'Brownian' motion should really be called 'Ingen Houszian', there is, as so often in the history of science, a traceable and highly relevant pedigree: without Ingen Housz having invented the cover-slip, the requisite high-power observations of Brown would have been impossible and atomic physics would be the poorer.

IN THE MIDDLE of November 1783 Ingen Housz received a letter from Benjamin Franklin that made him ache to be in Paris.[43] The letter itself was very short but it relayed the stupendous news that a peace treaty between England and America would be signed on the morrow. But even this was less important to Ingen Housz than the enclosure – a copy of a letter that Franklin had just sent to Sir Joseph Banks at the Royal Society in London.[44] There had been scientific developments in France that were spectacular and momentous enough to distract Franklin from his treating. And he was also using them to try to ingratiate himself in London now that peace between England and America was imminent. Franklin's letter to Ingen Housz had taken some time in finding its way to Vienna and the events it related had now featured in the newspapers. Nevertheless, Franklin was a more inspiring and discerning reporter than any jobbing journalist and had been an objective eyewitness of the scientific events. From the terrace of his house at Passy, where he had been confined with acute gout, he had been fortunate enough to observe the first flight of a hydrogen balloon. Jacques-Alexandre Charles, a young Parisian civil servant, inspired to scientific activities by meeting Franklin, had recruited Aine-Jean Robert to help make him a twelve-foot diameter globe from oiled silk.[45] The two of them had taken it to the Champ de Mars, an open green area just across the Seine from Passy, and filled it with 'inflammable air' (hydrogen) by dripping 'oil of vitriol' (sulphuric acid) on to iron filings. It took them two days: the production of hydrogen in this way, discovered by Henry Cavendish in London in 1766, had probably never been on such a large scale. By the afternoon of Wednesday 27 August an expectant crowd of some 50,000 Parisians had assembled despite heavy rain. They were jostling, fractiously, in order to see the tethered balloon and then, at five o'clock, two cannons were fired to announce the cutting of the anchor rope. The glistening wet globe was seen to rise until it seemed, to Franklin on his vantage point, to be the size of a mere orange, at which point it disappeared in the dense cloud cover. Charles and Robert must have been very satisfied with their experiment but they knew that they had not been the first to successfully launch a flying apparatus. Their competitors, however, were pioneers of a different technique, using hot air rather than hydrogen.

Two brothers, Joseph-Michel Montgolfier and Jacques-Étienne Montgolfier, paper manufacturers of Annonay, a small town some 50 miles south of Lyon, had built a kite-like box, open underneath but otherwise covered in taffeta, in November 1782 and a much bigger one a month later.[46] Both contraptions rose rapidly into the sky when suspended over a small fire of crumpled paper. By the next summer, 1783, the brothers had improved their experiment so dramatically that an invited group of local dignitaries was treated to a spectacular demonstration of flight. On Wednesday 4 June a large flaccid bag of sackcloth lined with paper and covered with a rope 'hair-net' confining some 28,000 cubic feet of air, erected itself to a tense globe when a fire was lit underneath it and then rose to 6000 feet when the restraining cables were released. News of their success soon reached Paris and Étienne left for the capital. There he recruited the help of a Jean-Baptiste Reveillon and the two of them constructed an even larger 'balloon' from varnished taffeta. This was painted sky blue with golden decorations and the signs of the zodiac: Reveillon was, after all, a wallpaper maker. This prettified apparatus was tested in a Paris park on 11 September and then allowed to fly freely from the gardens at Versailles on Friday 19 September in front of the King and Queen and a huge number of their subjects. It had been only three weeks since the Charles/Robert spectacle and balloon fever was gripping Paris. The added draw this time was to be present at the very first flight of living organisms. A sheep, a duck and a rooster were lifted aloft on a basket strapped underneath the balloon and all landed safely, from about 1500 feet, some two miles away.

The next obvious move for Montgolfier and Reveillon was to produce an even bigger balloon. It was 46 feet in diameter and 75 feet tall and by mid-October two hot-bloods – volunteers rather than pressed criminals as suggested by the King – were taking tethered flights up to 300 feet. One of them, a young doctor called Jean-François Pilatre de Rozier, together with François Laurent d'Arlandes, a nobleman officer in the French army, made the first free flight by humans on Friday 21 November.[47] They rose from near the Bois de Boulogne and again Franklin was ideally positioned. After a

The Montgolfier Balloon.

false start and a hurried repair to the canopy after hitting a tree, the two men were carried off to the east, across Paris. Franklin watched them go directly over his head, the beginning of a seventeen-minute safe flight. D'Arlandes visited Franklin later in the day to report on their experiences: they had been forced to descend only because the balloon fabric had begun to appear scorched and they therefore doused their fire. Flight was under way and a new dimension had been added to the philosophers' world; to everyone's world. Jacques-Alexandre Charles soon made his own free flight; piloting himself from outside the Palace of the Tuileries in central Paris on Monday 1 December and, only slightly more than a year later, on 7 January 1785, a hydrogen balloon took two aviators across the English Channel.

Franklin's letter to Banks, copied to Ingen Housz, was dated 30 August 1783 and therefore told only of the first flight of the Charles/Robert hydrogen balloon. Many developments had occurred in the art of ballooning while the letter had meandered its way to Vienna and the disjointed snippets of news arriving there were the source of much confusion. Ingen Housz immediately wrote to Franklin[48] asking for much more information and not just out of curiosity. There had been an immediate outburst of febrile ambition in the Austrian capital. The philosophers' circle was abuzz with a determination to fly their own balloon and was already trying to raise funds by subscription. The Emperor was in Italy so that royal patronage was not possible and they were not prepared to wait for his return. They all looked to Ingen Housz for leadership and although he appears to have taken on the mantle, his first move was a further attempt to obtain, from Franklin, more precise details of the processes involved. He wrote again on 2 January 1784: the letter[49] revealed the extent of his difficulty and bemusement. There was no grasp, even, that there were two completely different types of balloon. He needed to have, he wrote, 'a perfect knowledge of every thing belonging to the structure and management of the machine.' Ingen Housz had also received further confusing information, from van Breda,[50] of an attempt to fly a hydrogen balloon at Rotterdam. The Paris balloonists themselves had not yet published any reports, although pamphlets were supposedly being printed, and Franklin, severely curtailed by his gout, had been unable to inspect either kind of apparatus close up. In fact, as the letter he was writing proceeded, Ingen Housz seemed to sense that his American friend might not have all the answers and pleaded to know, at least, and as quickly as possible, if a Count Widmanstätten, a 'young man well versed in philosophy and mechanics' would be given access to the inside knowledge if he came to Paris.[51] Squaring up to the fact that Franklin was not a punctual correspondent, Ingen Housz then wrote directly to Charles, to Bégue de Presle and to Le Roy [52] asking all of them the same catalogue of questions. In fact Franklin did reply promptly; indeed by return of post.[53]

By the end of January Ingen Housz had his answers although, by then, Widmanstätten had already made three launches of small hot air balloons from a garden in Vienna.[54] Franklin predicted no secrecy in the affairs of the balloonists and thought it a very good idea that an 'ingenious' man visit Paris to become fully acquainted with all the techniques of 'aerostatics'. However, Widmanstätten never went to France, his place being taken by a much older man; Abbé Johann Nekrep. This important bureaucrat – head of the Vienna Oriental School (of diplomacy) – was due to visit Paris on official duties and was just as suitable to research ballooning, being a 'very good scholar and philosopher' and one of the group hoping to construct a large balloon at Vienna. Ingen Housz gave him a letter of introduction to Franklin on 10 February[55] although, at the time, it still seemed unlikely that the Viennese nobility were about to fund a serious attempt at flight. Ingen Housz seemed far from disappointed. He had always been apprehensive and now took to heart the warnings from Franklin that there was a negative side to ballooning. It had rapidly become an experiment that always attracted a huge audience: after all it could hardly be attempted in the privacy of the laboratory. That a botched attempt might damage a man's reputation was easy to predict but there was worse. Some early balloonists found that there were perils on the ground that easily matched those in the sky. As Franklin related[56]

> It is a serious thing to draw out from their affairs all the inhabitants of a great city and its environs, and a disappointment makes them angry. At Bordeaux lately a person who pretended to send up a balloon . . . not being able to make it rise, the populace were so exasperated that they pulled down his house, and had like to have killed him.

No-one foresaw it, of course, but here were clear indications of the wafer-thin tolerance and pent-up aggression of the French citizenry when condensed into a mob. If they behaved thus when a well-meaning philosopher had a failure with a balloon, how might they react when they had nothing to eat? It was a mere five years until the answer was to emerge.

One resolute maverick at Vienna, Johann Stuwer (Stubenrauch), a Bavarian fireworks expert, did assemble a substantial hot air balloon only to find Ingen Housz equally determined not to become involved.[57] When Nekrep returned from Paris, on 17 May,[58] he heard that this 56-foot balloon had failed to ascend and was therefore being enlarged. The specialist knowledge he had acquired at Paris was unwanted information: all enthusiasm for flying had sapped from his immediate friends. There were still insufficient subscribers and Ingen Housz's procrastination seemed to have been a dampener. Nevertheless, the somewhat isolated figure of Stuwer did finally launch his enlarged hot air balloon from the Prater on Tuesday 6 July 1784 in front of a crowd of over 15,000 inquisitive Viennese.[59] After this, Austrian aviation appears to have petered out and in following the interests of Ingen Housz we turn from hot air to cold air.

The winter of 1784/5 was bitterly cold in central Europe.[60] It was the second hard winter in succession and this time there was no relief until mid-April. At least the thaw came gradually as well as late. The devastating surge of water and ice floes that had swept down the Danube when everything had suddenly melted in February 1784 was not repeated in 1785: the mighty efforts to repair all the scoured and broken bridges during the previous summer had not been in vain. Yet the hiemal conditions of early 1785 were a menace and a misery to virtually everyone, even in close-quartered

Vienna; prayers were even said in all the churches, on Sunday 10 April, pleading for a thaw.[61] To Ingen Housz, however, the piercing cold presented an inviting opportunity. Only he, and perhaps the fuel merchants, favoured the conditions. Primed by the previous harsh winter, he had prepared his laboratory for an opportunistic experiment. Through the nights of 27 and 28 February[62] he and his long-suffering servants were kept from their beds for the sake of science. A thermometer in the room showed minus two degrees centigrade and although we don't learn how – perhaps by evaporating ether – Ingen Housz dropped the temperature around a tube of mercury down to minus 33. Still he was unable to achieve his goal of freezing the quicksilver. He must have been disappointed, even more so when he later learned that a team in Leipzig had managed it and that he had been only a few degrees short of success. On the other hand the exercise may be viewed as positive evidence of his servants' outstanding devotion. Their night-time stoicism was remarkable but how they must have ached for the quicksilver to solidify and for the warmth of their beds.

IN 1784 ALL the Vienna hospitals were closed and moved together on to a new site on the orders of Joseph II who had been impressed by the facilities of the Hotel Dieu, the hospital that he had seen in Paris. The location was outside the city ramparts, to the northwest, where Leopold I had built, nearly a century earlier, a 'Home for the poor and invalid'. Joseph II's *Allgemeine Krankenhaus* or 'Municipal Hospital' was formally opened on 16 August 1784[63] and we can assume that Ingen Housz would have been present. Joseph von Quarin was the first director. He was another body physician to the Emperor and an almost exact contemporary of Ingen Housz. Perhaps this was another post of administrative responsibility that Ingen Housz declined? If he had been seeking fame it would have been a missed opportunity – the hospital was on a grand scale and was to become a beacon of innovative diagnosis and treatment. He may have been involved in a minor way, however, for the *Narrenturm*, the prominent tower built to accommodate patients with mental illnesses, was surmounted by a lightning rod, remnants of which can still be seen. The extensive area given over to the original *Allgemeine Krankenhaus* is still near the medical focus of the city, the modern buildings of the new Vienna General Hospital having been opened in the same district in 1994, 210 years after the famous old buildings first admitted patients.

Ingen Housz, like the new hospital, had 'open doors', albeit on a smaller scale. As an anglophile he was always pleased to receive English travellers to Vienna and there appears to have been a succession of them from 1780 onwards. Some may actually have stayed with Ingen Housz and his wife as their personal houseguests but it was usually his laboratory bench with its associated pearls of wisdom that was their target. More likely, assuming a suitable letter of introduction, British visitors would invite themselves to stay with Sir Robert Murray Keith, their long-standing ambassador at Vienna[64]. It was one of the reasons, incidentally, that Keith found the post beyond his means, returning almost penniless to retirement in London in 1792.[65]

Jonathan Stokes, not yet 30 and a recent medical graduate of Edinburgh, visited in 1782/3.[66] He was intent on collecting any plants known to have medicinal properties, a life-long interest that evolved, slowly, towards pure botany and plant taxonomy for which he became best known when, later, he settled into practice in Worcestershire and became an active member of the Lunar Society.[67] Ingen Housz was his route to Nikolaus Jacquin and his botanical knowledge but Stokes also had much in common with the royal physician himself. Stokes's MD dissertation, submitted early in 1782, had been on 'dephlogisticated air' (oxygen) and he knew Joseph Priestley. Not that Priestley was guaranteed to be a back-slapping topic of conversation at the Ingen Housz dinner table in 1782. However, there was always levity to be found in Edinburgh nostalgia. The Jacquin family was very hospitable to Stokes and he was provided with valuable plant material, a supply that would continue for many years. A plea to thank the Jacquins arrived in a very friendly letter to Ingen Housz in October 1787.[68]

> Present . . . if you please, my respectful compliments to Mrs. Ingen Housz, to Mr. and Mrs. Jacquin and the two Master Jacquins. How happy could I be to allow myself into your agreeable circle at the Gardens but I must content myself with assuring you with how much esteem I am your sincere and obliged friend.

Some travellers to Vienna, many of whom had been given his address, were anxious to meet Ingen Housz not only for their delectation but also for a subsidiary reason. They needed him to furnish them with a letter, or letters, of introduction for the next leg of their trip. James Home, a newly-qualified Scottish doctor was a case in point, Ingen Housz writing to Giovanni Fabbroni, on Home's behalf,[69] on 27 December 1784, that '. . . he has asked me for an introduction to some of those men, celebrated for their knowledge, at Florence.' Letters of introduction were an effective lubricant in the circulation of knowledge and experience and usually exchanged freely, especially, as in this instance, when the guest passing through was pleasant company.

EXPERIMENTS AND DISTRACTIONS

Lord Wycombe, who would become the second Marquis of Lansdowne. Courtesy of the Trustees of the Bowood Collection.

In more than ten years as British ambassador, Sir Robert Murray Keith had become accustomed to the diplomatic bags from London disgorging news of compatriots he should expect to receive. Vienna was not always an objective on the 'grand tour' that young men of means were wont to make at the end of their formal education. However, a letter from Lord Lansdowne,[70] during February 1785, made a particular point of his wanting permission for his son, Lord Wycombe, to visit the city and some of the Imperial dominions.[71] Wycombe, soon to be twenty, would be leaving Christ Church College, Oxford that summer. Knowing the status of the father, Keith ensured that Kaunitz gave his blessing before writing back to Lansdowne to encourage the visit and putting himself on alert for the young man's precise travel plans. The remainder of 1785 elapsed before Wycombe, accompanied by a Major Green and a *valet de chambre*, hired at Brussels, were ensconced under Murray Keith's roof on Friday 30 December.[72] The two young men were presented to the Emperor the next day and sometime in the next few weeks Wycombe was taken to meet Ingen Housz. It appears that, for once, Ingen Housz was approached more for his medical knowledge than for his scientific prowess. Wycombe was having trouble with his ears and received valuable advice delivered with a certain drollery: but whatever the clinical circumstances, the young nobleman was favourably impressed.[73] His praise appears to have been quite genuine and not merely a diplomatic nicety of a son writing to a father about one of the parent's friends.

> It behoves me to express my obligations to Dr. Ingenhausen, from whom I have derived, I hope, some information and certainly much amusement. He approves much of the sulphur water which I am drinking, but he assures me with respect to my hearing, that the injection of warm water into my ears can only be of service where deafness is occasioned by an accumulation of wax . . .

Then, for a few short days in October 1786, Ingen Housz was again happily reconnected with England. Sir John Sinclair, Scottish knight and fellow of the Royal Society, bearing a letter of introduction from Joseph Banks himself, arrived in Vienna from Kiev and Warsaw on the 17th.[74] Widowed, in May 1784, while still a relatively young man of 30, he had set off, a year later, on a comprehensive European tour. For someone intensely interested in agriculture and determined to make it more efficient, there was much to see on a pre-industrialised continent. He was a man of boundless energy and limitless interests whose calling card showed his name superimposed on a plough.[75] Ingen Housz was obviously delighted to meet him, writing hurriedly to Banks[76] that 'I have only to regret that his very short stay in Vienna afforded me so little time to enjoy his interesting conversation.' Sinclair not only established a deep and lasting friendship with Ingen Housz that would be rekindled in the 1790s but triggered, in his host, a profound longing to return to England himself, pining to Banks that he was

> . . . much pleased to be informed, by your letter, that some of my old friends wish to see me again. The grateful remembrance of the happiness which I enjoyed during so many years in your country, by the constant civility and friendship of a great number of worthy men, renews continually my desire to return once more to Great Britain. It would be very easy to get leave to take a trip to London but I should think it mortifying to spend less than six months in a country which I love before all others.

A year later Ingen Housz's memories of England palled a little, perhaps. A young English physician arrived at Vienna bearing a letter of introduction from one of Ingen Housz's harpies— Joseph Priestley.[77] Thomas Pearson was newly qualified, the brother of a Birmingham businessman in Priestley's circle of friends. 'I beg leave to recommend to your notice a young man, a student in Medicine, travelling to improve himself. I can only say that I hope you

Alessandro Volta.

will not find him unworthy of your favour.' Pearson was, no doubt, received with all courtesy and grace. But the letter from Priestley went far beyond its introductory brief. After five years of silence, Priestley went on to rehearse his personal recollection of how the primacy of sunlight had been discovered in plant chemistry. His interpretation of events was still profoundly different to that of Ingen Housz and would have engendered many distracting thoughts. Priestley now was prepared to concede that Ingen Housz had beaten him into print but on the other hand he was still adamant that he had also made, independently and simultaneously, the same fundamental discovery. There are good grounds to suspect the interpretation as disingenuous and the air of condescension was disagreeable and misplaced.[78]

> That plants restore vitiated air, I discovered at a very early period. Afterwards I found that the air in which they are confined was sometimes better than common air, and that the green matter, which I, at first, thought to be a vegetable produced pure air by means of light. Immediately after the publication of these facts, and before I had seen your book, I had found that whole plants did the same . . . so what you did with leaves was altogether independent of what I was doing with whole plants the same summer, and the same sun operated for us both, and you certainly published before me.

Among his other guests at the apartment in the Franziskanerplatz in the mid 1780s, several wanted their regards and gratitude passed on to Agatha, particularly, when they sent in their 'thank you' letters so they may actually have been accommodated in the apartment as well as entertained in the 'laboratory'. Among these, were James Robertson, a Scottish physician who was given a letter of introduction to Franklin,[79] Samuel Vaughan, the son of one of Franklin's best friends and a political associate of Lord Lansdowne,[80] and Dominique Beck, a young physicist from Salzburg who was guided through the processes of shopping for good quality scientific equipment at Vienna.[81] Alessandro Volta from Pavia, who also stayed for some time, failed to make a good impression on Ingen Housz, bungling electrical demonstrations in which he ought to have been proficient, in the rather critical view of his host.[82] There were to be more royal visits. In the middle of August 1784, the Emperor and his brother, Grand Duke Leopold, whom Ingen Housz knew well from his times at Florence, visited together and watched their host perform various experiments and demonstrations for several hours.[83] Then, two years later, another of Maria Theresa's sons, Archduke Ferdinand, governor of Lombardy and living at Milan, paid the first of several visits, staying for more than three hours.[84] Ferdinand had been one of the two young boys that Ingen Housz had inoculated in early September 1768 and here he was, now a fit and fully-grown reigning monarch. This visitor must have stirred deep feelings in Ingen Housz but in general his hosting activities were buoyant and light-hearted. He now had a wide variety of 'entertainments' for visitors whatever their levels of scientific expertise and personal interests. Besides microscope demonstrations, any on-going plant experiments and his latest electrical apparatus, there were the self-lighting phosphorus matches, metal wires to burn in oxygen, the improved *Fürstemberger* lighter, his novel sea-compass and, of course, the ether pistol. He was becoming quite a popular 'performer' even if Agatha and his neighbours had put, in all likelihood, an embargo on firing the pistol.

FOR A MAN who had an ample pension that easily paid for all his temporal needs and who had virtually no official commitments, Ingen Housz had an extremely busy middle-age. Perhaps being so engaged in so many projects was part therapy, for his life was certainly not stress-free. The vain attempts he was to make to retrieve the substantial capital he had invested in America, let alone any of the proceeds, are recorded in a succession of increasingly desperate letters to and through Benjamin Franklin during the 1780s decade. As time rolled on all Ingen Housz really learned, in his isolation at Vienna, was the extent to which he had been duped. Having been back in Austria for some eighteen months he was expecting to hear, daily, from Williams and from Wharton. There had been nothing from either American and none of his letters to Paris asking if Franklin, Bancroft or, indeed, anyone there, had heard 'relevant' news from Boston or Philadelphia, had been answered. The silence was prolonged, painful and absolute. His paralysing isolation was only heightened by tantalizing, second-hand news from Dunkerque, from Francis Coffyn. In a letter dated 16 November 1781[85] he informed Ingen Housz that he had heard from Wharton. Most of the bales of linen had arrived in America; most but not all. Some had been confiscated by the Royal Navy when an English 'man of war' had apprehended one of the ships and taken its contents as prize goods.[86] Nonetheless, eleven of fifteen bundles had been made available to the American market and some had been sold; enough to allow Wharton to send Coffyn a bill of exchange for some 9000 Livres (£23,500). Coffyn's letter to Ingen Housz strongly implied that Ingen Housz, too, should be alert for the arrival, in parallel, of similar news and similar reward. Nothing happened and after some months, on 24 April 1782, Ingen Housz sat down, heavy in heart, to write several letters. He wrote to Wharton, to his son, Samuel Lewis Wharton, lest Wharton himself was ill or had died, and to Jonathan Williams, bundling them all up in a covering letter to Franklin at Paris. He could, he said, no longer 'forbear of being uneasy about such a long silence'.[87]

Three weeks later there was mail though it was far too soon for it to have come from America. It might have come from Paris but it was actually a further letter from Coffyn, at Dunkerque.[88] The news was mixed. Coffyn had sent a bill of exchange for 1800 Livres (£4700) to a bank in Paris that Ingen Housz might draw on. It represented a return of barely an eighth of his stake. It was a sop; a seemingly random sum chosen by Coffyn. The real news, the bad news, was that the Dunkerque agent had actually received a further 19000 Livres (£50,000) from Wharton, more than nine tenths of which he had banked for himself.[89] Coffyn was not willing to divide the spoils according to the proportionate stakes of each syndicate member because he saw the bulk of this money as his private return for having lent Wharton the very capital that the American merchant had paid into the syndicate in his own name.[90] This was shattering news to Ingen Housz. He had been totally unaware that there had been a 'deal within the deal' and that everything had not been fully transparent. By reading between the lines of Coffyn's letter it was also clear that Bancroft had been party to the inner secret. Ingen Housz's hopes of seeing a return on the venture, or of seeing his capital re-banked at the very least, had receded further. He now saw himself at the very back of the queue for any profit. Moreover, it had become clear that Wharton had forgotten, or chosen to ignore, his promise that all return traffic from America – details of sales and the profits – would be channelled via Benjamin Franklin who could then see fair play in dividing the spoils. Ingen Housz's next letter, to Wharton via Franklin[91] was, understandably, more challenging and more assertive. He was disillusioned and disappointed but still prepared, just, to believe the best of Wharton: 'I do not in the least suspect your integrity and character but . . . do not retard your answering me immediately . . . giving a proper account of all transactions.' A falsely sympathetic letter that then arrived from Edward Bancroft[92] was remarkably vague and the few details it contained strongly contradicted Coffyn's version of events. It merely confirmed Ingen Housz's suspicions that Bancroft and Coffyn had not only shared a confidence but that they were dividing the spoils among themselves. It added insult to injury just at the time that Ingen Housz learned that his pension from Maria Theresa has been devalued by a peremptory reform of Joseph II. The Emperor had decided to withdraw, across the board, the right to free lodgings from all court officials.[93] This was the unwelcome reform that had forced Ingen Housz to move from the Weihburgasse to the Franciskanerplatz but it was the principle rather than the domestic upheaval that upset Ingen Housz. He was incensed by the effective erosion of his pension, and a personal protest, written in haste, was resented by Joseph II, provoking a 'disdainful frown'.[94] The royal physician was rueing his hasty remonstration within hours. With the failure of his investments there, he now had no prospect of an independent life in America and could now ill afford to rebel at court. Whatever promises Maria Theresa had made, of a generous pension for life, he knew that it could be revoked – 'the frowns of a Monarch will ever inspire fear.'[95] His anxiety and regret, compounding his financial disappointments, turned into depression and, for some weeks, he was 'low-spirited and unfit for business'.[96]

On 8 April 1783 there was, finally, news from John Williams at Boston.[97] The 7,000 US dollars (£92,000) invested in the Massachusetts loan office in March 1779 had appreciated. The bonds were now worth about 8750 dollars (£115,000) and remained invested. This news, at least, was an encouraging boost to morale and gave Ingen Housz a little more spirit although there was no mention of the rest of the money he had given to Williams. By the end of the month he had written, yet again, to Coffyn[98] and to Wharton.[99] After six months there was one response – a letter, not from Wharton, but from his son, Samuel Lewis Wharton.[100] It told Ingen Housz that his father had replaced, in a Paris bank, the 8,000 Livres (£21,000) that had been the original stake in their first mutual venture. There were no details of how, or indeed whether, the money had been invested and Ingen Housz had every reason to suspect that Wharton had simply used it as an interest-free and unsecured loan. Nonetheless, it was back where it belonged and there was some comfort in that, at least. Wharton junior's letter[101] also referred to letters allegedly written to Ingen Housz by his father on 14 April and 29 May 1783. These had never arrived, and never did, and Ingen Housz supposed that they had been held back by the post offices at the ports of entry into France for not having the full postage paid. It was a bland assumption that allowed Ingen Housz to hold on to some remaining possibility that Wharton had some shred of morality. This was in spite of the admission, in young Wharton's letter, that his father was considerably distracted by having to defend, to Congress, his dubious behaviour in acquiring a huge tract of land from a tribe of native Americans at an insulting price.

Then, just at the turn of the year, as 1783 became 1784, Ingen Housz heard again from Francis Coffyn.[102] He had sent further small sums totalling 2400 Livres (£6250), in favour of Ingen Housz, to banks at Paris. This amount, again a figure that appeared to have been plucked out of mid-air, was a further pacifier and the letter was, otherwise, as unhelpful as all his others. He still denied that the money he had received from Philadelphia was for the 'common interests' of all their syndicate and having salted away easily enough to protect himself and Bancroft from any losses, he sidestepped any further moral responsibility by suggesting, to Ingen Housz, that all outstanding issues were between him and Wharton alone. He even suggested that Ingen Housz should consider legal recourse, perhaps through a power of attorney having jurisdiction in Pennsylvania. Ingen Housz must have felt more abused than ever. Some three and half years after joining in a financial venture with men he had considered to be trustworthy friends of, and heartily recommended by, Franklin, he had just passed the half-way mark to regaining his stake; a pathetic return that was painfully adrift from the three to four-fold mark-up that he had been led to expect and, for all he knew, had been the pleasure of his co-investors. He himself was still facing a loss of some 11,000 Livres (£29,000), the equivalent of a year of his court pension. For the rest of 1784, for the whole year in effect, there was no further news. Letters sent to Wharton[103] and to Williams[104] on 9 October elicited no response and the resumed silence continued well into 1785.

The next turning point in the sadness of Ingen Housz's American financial misadventures hinged on the arrival of a very long letter from

Franklin.[105] It surfaced at Vienna during the last week of May 1785, a month after leaving Paris. Franklin had put a lot of effort into tying up many loose ends in his relationship with his Dutch friend since he now knew that he was only days from leaving France and returning to Philadelphia. One of the threads was the latest news on Wharton. Franklin confessed that he was 'amaz'd' by his friend's 'conduct'. He had learnt from Bancroft that Wharton had already distributed or spent all of the profits from the linen sales. Franklin tried to console his friend that there were still grounds to hope for his money to return, given time. Wharton, it appeared, 'owned lands and enjoyed a profitable office'. Franklin would certainly see it as his responsibility, once back at Philadelphia, to apply personal pressure on Wharton to act honourably. Even better, why didn't Ingen Housz take some immediate leave of absence from Vienna, hasten to Paris and join his 'great friend' on the boat to America? 'You may, at least, obtain some land'. Failing that, Ingen Housz should at least send a power of attorney that Franklin could use on his behalf at Philadelphia together with any proofs of the debt. Ingen Housz knew that he couldn't go. He was still trying to regain the goodwill of the Emperor who had just refused him permission to travel, even, to Paris. He could not afford to go absent without leave and he also now knew the paltry value of his American money.[106] 'I would commit an unwarrantable imprudency to risk of losing what I still enjoy here by journeying to a very distant country.' Ingen Housz had to absorb the sadness of knowing that he was never likely to see his great friend, Franklin, again but at least the Philadelphian was going home. His presence and authority in America would, Ingen Housz hoped, at least ensure that his original stake be returned.

Ingen Housz took the trouble to formulate a full history of his American 'mercantile affairs' for Franklin to take home with him, no doubt regretting, all the time, that no proper contracts had ever been drawn up, and pleading to his friend, that he would, indeed, act for him in America. The letter, bundled up with yet further ones to Wharton and Williams, was dated 11 June 1785[107] and arrived in Paris with only days to spare before Franklin left – for Le Havre, Southampton and then Philadelphia – on 12 July. When he arrived home it was to a free and independent country, partly thanks to his own endeavours at Paris. He was a hero and immediately a very active one. His family, his personal affairs and state and national politics kept him very busy and, having been elected Governor of Pennsylvania, he was delegated to attend the United States constitutional talks. Nevertheless, he quickly took pains to investigate Ingen Housz's investments and put considerable pressure on Samuel Lewis Wharton, nominated, by his father, to deal with the matter. Wharton junior delivered some scrappy notes that accounted, he said, for Ingen Housz's remaining money. Franklin sent the papers, such as they were, to Vienna, with a very friendly covering letter, in June 1786.[108] They took almost a year to arrive, a time in which Ingen Housz had become frantic with uncertainty, writing, twice, to Franklin pleading for news.[109] In fact, there had been no further progress at Philadelphia except for disastrous developments. Samuel Wharton had suffered a stroke and his son was becoming an alcoholic. The documents scribbled by Wharton junior were a travesty; misinformation and vague pledges. Apparently there was still some unsold linen 'belonging' to Ingen Housz: this was a further vexation. Wharton had obviously stooped so low as to allocate the unsold bales to the one co-investor who was unable to challenge its impropriety – Ingen Housz. The obvious hole in the balance statement, where Ingen Housz had a right to expect to see his money, would be filled, the Whartons promised, by mortgaging 1000 acres of family land. It was just another empty promise – more to do with buying time than selling land. The 9,500 Livres (£25,000) still owed to Ingen Housz looked as far away as ever but their rightful owner didn't give up, writing to Franklin[110] yet again but in apologetic terms and including yet another letter to John Williams:

> Considering your age and high station in life I write to you on these affairs with the greatest reluctance. Your old friendship and constant kindness towards me could only engage me to it. For my part I feel, too much, to what degree all this affects me: and it makes me often incapable of applying to my philosophical labours.

Ingen Housz still assumed, at enormous distance and with only a flimsy grasp of the reality, that Franklin's continued moral pressure would bring him his reward, that

> Wharton will have been, or soon will be, induced by your persuasion and authority to fulfil his duty. If he does it not, he must be one of the greatest rogues existing . . . (and) . . . is it possible to believe that a free and virtuous people (US citizens) should chuse a man of such a black caracter as one of their magistrates?

Franklin had already intimated, however, that resorting to legal redress would be the only sensible tactic in dealing with the increasingly dysfunctional Whartons: '. . . unless you send a Power of Attorney to sue for demand, or come yourself, I am afraid you will never get any thing.'[111] Ingen Housz fought shy of making arrangements for his case to be pressed formally at Philadelphia. He was afraid of the cost implications; of throwing good money after bad, but that was not all. He was acutely aware that he lacked much in the way of real evidence. Wharton's original orchestration of the linen venture had never included a written contract and, not for the want of asking for letters, virtually nothing had ever been written down subsequently. A detailed statement of what had happened could only be his personal recollections, his word against Wharton's. Ingen Housz was also afraid, he said, that were Wharton accused, publicly, of being dishonest, he would have nothing to lose by making up fictitious accounts that would finally eclipse any possibility of repayment.

There was at least a little good news from Franklin. The American Philosophical Society, started many years before at Philadelphia by Franklin, had gratefully accepted the copies of Ingen Housz's books that he had donated and had immediately elected him a member. As Franklin reported,[112] the books were 'indeed, from the Variety of useful Knowledge contain'd in them, a most valuable Present. The Society have done themselves the Honour of electing you a Member, and I enclose their Diploma, by which the Election is certified.'

17
Shocks

Early August 1786. After several days of intense heat and lifeless air in the tight Viennese streets and alleys it's difficult to stay comfortable, even overnight. At last, on Sunday afternoon, the leaves on the trees begin to dance and it's cooler. But ominous clouds appear to the north of the Danube. A storm is brewing. Strollers in the Prater feel themselves exposed, instinctively quickening their pace and eyeing the distance to the nearest shelter: electric storms on the European land mass can be sudden, violent and vindictive. The general drift is indoors but there's an exception. On an exposed terrace above the Franziskanerplatz one Vienna resident is carrying objects out of his apartment and calling for his servants. A variety of glass jars are assembled as well as a large block of wood. He leans on his balcony, surveying the sky, sniffing the air. It is now completely overcast and everything is bristling. There's already been distant thunder. The servants hover, with false nonchalance, near the doors but Herr Ingenhauzen stands himself on the wooden block at the front of the terrace. From here he has an uninterrupted view across the square and sees that one of his near neighbours, a middle-aged woman, is sitting near a window. She is intent on her knitting. As he wonders whether he should be sociable and disturb her concentration by hailing a greeting, he sees blue sparks crackle and hiss from the extremities of her steel needles just as thunder rumbles immediately above. At first she's entertained by the spectacle but then, suddenly, realises what must be happening and throws down her knitting before bursting up out of the chair, knocking it over, and backing away into the far recesses of the room. After some seconds she looks up and, in turn, watches Ingenhauzen. He is standing erect with his sword held high above his head, as she has seen him do before during storms. Her husband is surely right, the man is mad and drawing the lightning down from the sky. He's putting everyone in danger. Herr Ingenhauzen, meanwhile, is barking orders to his servants. They run forward, in turn, hold a glass jar near to his arm and then dash back inside. The spectacle continues for some minutes but when large, isolated spots of rain begin to fall, Ingenhauzen dismounts. He knows that a wooden block, when wet, is no longer an insulator. The servants all dash out and everything is taken inside before it can be washed away; the common body language is eloquent of mass relief.

From the moment that Benjamin Franklin published his famous 1752 'key on a kite' experiment showing that lightning is the same natural force as the spark discharged from a friction electrical machine[1], there was a search for how electrical 'fluid' affected living organisms: atmospheric electricity could now be mimicked and investigated. It could certainly be harmful, fatal even; but did it have more subtle, beneficial, effects? The integral part that electricity plays in animal movement and awareness was becoming clear by the end of the eighteenth century even if the over-heated imaginations of some novelists distorted the reality, classically in the clandestine activities of a certain Dr. Frankenstein.[2] On the other hand, the influence of electricity on plant life remained controversial for several generations. Investigations began badly when some authoritative but sloppy experimenters, particularly in France, published extravagant and erroneous conclusions that were later reined in with difficulty. By the late 1770s the received doctrine was that plants needed electricity as much as they required warmth and moisture; the fundamentals of plant husbandry known for millennia. If isolated from this universal force of nature they struggled to grow and reproduce, if they survived at all. This all-prevailing concept was fully expostulated by the very respected French physicist, Pierre Bertholon, in his famous book of 1783.[3] With plant life being the ultimate source of all food, the commercial implications of the doctrine were enormous, particularly when all eighteenth century economies were based on agriculture. Everyone believed in the doctrine and wanted to believe in it. This had included Ingen Housz, but as early as 1777 he had begun to have doubts and he was among the first to enter a challenge.

Nicholas-Philippe Ledru, the part scientist, part showman/magician often to be found on the Paris streets, had published, in 1776, some experiments that he had performed with the fascinating plant, *Mimosa*[4]. Everyone will know that its leaves and stems are exquisitely sensitive to touch and wilt on slightest contact. This highly evolved protective mechanism could be evoked, according to Ledru, by the electrical discharge emanating from the point of an isolated charged conductor, such as a metal rod, held some distance from the leaf. Ingen Housz was immediately sceptical when he read of these conclusions in the 1776 volume of *Journal de Physique* sent to him from Paris[5]; suspicious enough to be motivated to repeat the experiments. In the spring of 1777 he had obtained some *Mimosa pudica* specimens – a particularly reactive variety – and repeated Ledru's trials. Far from confirming the Parisian's conclusions he found them to be built on a false premise. Perhaps Ledru was a more skilful magician than he was scientist. Ingen Housz made hundreds of observations and had to conclude that touching the leaves with charged conductors or with electrically neutral materials made no difference to the outcome. And as for the alleged reaction evoked by the very approach of the point of a conductor, mediated by the flow of 'electrical fluid' according to Ledru, Ingen Housz became convinced that this, too, was the result of the mechanical impact of charged air particles streaming out from the conductor's point. It could be imitated, very easily, by pursing the lips and blowing at the plant from a distance. Neither could Ingen Housz find any evidence that a drooping mimosa plant would 'redress' itself more promptly if electrified, as Ledru had also alleged. Contrary findings such as these gave Ingen Housz profound doubts.[6]

> ... having found these experiments badly interpreted I began to think it must be, perhaps, worth making very careful repeat examinations (of all the reported experiments) if there was to be a way of dismantling the general received opinion on the manifest influence of electrical force on plants.

These were misgivings and reservations, however, which had to be shelved.[7] 'Too busy, then, with affairs relating to my (court) duties and preparing for a voyage to England, I lost this business from sight . . .' It was in the spring of 1781, back at Vienna, that Ingen Housz returned to these issues.[8] He began a new line of investigation, examining with a very critical eye, whether electricity really did promote plant growth as Bertholon, particularly, had proposed. He used the conventional model of putting daffodil and hyacinth bulbs to grow above an electrified insulator, selecting matching bulbs elsewhere as un-electrified controls. His results were inconsistent and confusing and remained so when he repeated the trials in the early months of both 1782 and 1783.[9]

> I obtained nothing that conformed to the results that other physicists had consistently obtained . . . I saw, however, that bulbous plants were not best suited for making judgements because of the great variation, frequently observed, in the growth of their vegetation; so much so that one rarely finds three in a row that grow in a uniform manner.

Ingen Housz therefore devised experiments that used mustard seeds, repeating them, always, with cress lest there were species-specific idiosyncrasies.[10] Dozens or even hundreds of such seeds could be readily germinated in a small space and growth in his 'miniature forests' was much more rapid and uniform: multiple repetition was feasible over just a few weeks. Now it took only one spring and summer to amass consistent findings. We have many of the details of these important experiments because Ingen Housz repeated them, yet again, in the summer of 1785 in the presence of a young assistant called Schwankhard.[11] Although he remains a rather shadowy disciple, it appears that Schwankhard was a graduate of the University of Strasbourg; an essay he was to publish later that year[12] was in the form of a letter to his erstwhile teacher at the university, Professor Ehrmann. Schwankhard reported how Ingen Housz had tried, repeatedly, all the possible combinations in electrifying his seeds and seedlings. Always using controls for comparison, he had exposed the test seeds and seedlings to positive charge, to negative charge, to mild intermittent charge and to intense charge maintained over at least 20 of every 24 hours (one of the experimenters stayed up until three a.m. and the other rose at six, day after day). The results were always negative. There was no evidence, ever, that charged seeds germinated more rapidly nor that the subsequent seedlings were more prolific.

It is clear that Schwankhard was impressed by the fact that Ingen Housz not only performed scrupulously as an experimentalist, obtaining consistent and repeatable results that allowed only one interpretation, but that he could explain how others had been misled into false conclusions. In dramatic but simple trials, easily performed on his kitchen table, Ingen Housz revealed to Schwankhard the primacy of light intensity in such work. Plants, seeds even, are extremely sensitive to light and if this aspect of the environment is ignored by unwary experimenters, their trial results can become a meaningless jumble. Seedlings given just a very little more light – placed just a few inches nearer a window for instance – flourished much better than counterparts. Ironically, seeds themselves were shown to prefer darkness. Ingen Housz even demonstrated how a pot of mustard seed, set to germinate on the shady side of a piece of thin paper set parallel to a window, would emerge earlier than their exact peers on the sunny side.[13]

After his 1779 experiments at Southall Green, Ingen Housz was probably the person best attuned, in the then scientific world, to the sensitivity of plant material to light. By the summer of 1785 he was fully convinced that the received doctrine on electricity in plants was a fallacy. But he himself had once accepted it, together with the purported evidence, and nearly all reputable scientists, at least within his compass, still did. Schwankhard found his host facing a dilemma: Ingen Housz might risk unpopularity if he publicly opposed the doctrine and even ostracism and a bankrupt scientific reputation if his negative evidence was later found to be in error. On the other hand he had always subscribed to the notion that scientific evidence belonged in the public domain whatever the personal consequences, as had been urged on him by Pringle and Franklin particularly. Schwankard, with the impetuosity of youth, was much less cautious and ostensibly obtained permission, from his new mentor, to publish some of the findings. Ingen Housz gave him permission as long as he mentioned no other scientist by name and was careful to avoid insult, even by implication.[14] At least, that was the story as presented. However, Ingen Housz was playing a more devious, underhand, game. He confided the truth in a letter, written six months later, to Joseph Banks in London. 'You may have seen, in the . . . *Journal de Physique*, a paper of mine, under the name of Mr.Schwankhard, being inserted in the pamphlet of that most useful work.' Allowing his young colleague to be a stalking horse, as, seemingly, was planned, was a clever ruse. It appears, certainly, that two papers intended for the *Journal de Physique* were submitted together having obtained the agreement of de la Metherie, the editor, that he would publish Schwankhard's 'work' first[16] and sit on the next paper, author Ingen Housz,[17] for a few months. The first of these two articles appeared in the *Journal de Physique* of December 1785.[18]

The first few months of 1786 were probably a tense time for Ingen Housz and for Schwankhard. What would be the backlash to the paper in the *Journal de Physique*; would someone be able to point out a cardinal error in their work; would they look foolish if the received doctrine, that electrically charged plants were stronger, was correct after all? In fact de la Metherie wrote, from Paris,[19] that there had been 'grumblings' but no orchestrated reaction. The single written rebuff came from a Parisian physician[20] whose only real criticism was that the neophyte Schwankhard had been impertinent in challenging revered French scientists such as Pierre Bertholon and the late Jean-Antoine Nollet. Ingen Housz's follow-up article, of which he was the manifest author,[21] went unchallenged. It was a more generalised essay on the factors which, in his view, affected seed germination and plant growth, and included his dogmatic reiteration that electricity played no part in these processes. Ingen Housz was obviously much more confident and went on to make other, crucial, experiments. He had been trying, for some time, to obtain a copy of a dissertation[22] submitted, in 1782, to the Lyon Academy by an Italian professor at Turin, Francesco Gardini. Gardini's paper, appearing as a book in 1784,[23] had won a prize at the Academy which Ingen Housz took to be a signal that it was exemplary work reporting, fully, on trustworthy experimental evidence. In fact, not knowing any particulars of Gardini's findings had been a major reason for his reluctance to publish his own. It was one of the 'important reasons to use circumspection in undermining the foundations of an edifice not only venerable by its age but made equally respectable by the great men still working to confirm it and ornament it.'[24] A copy of Gardini's report finally reached Vienna some time in early 1786.[25] It served as both stimulus and template for experiments that Ingen Housz performed in the royal botanical gardens that summer.[26]

Gardini, he learned, had erected iron wires, stretched taut, across and above a monastery garden at Turin. His aim had been to explore the state of atmospheric electricity during storms and the wires were presumably insulated at their attachments, at the very least while he was making his measurements of accumulated charge. Ingen Housz noted, with acute perception, that this was not specified by the author. The wires had been in place for some years when, for three consecutive summers, the plants in the garden below them languished and gave miserable crops. The monks attributed this sterility to the presence of the wires and insisted on their removal, after which their garden began to flourish again. Their intervention proved, to them, that their suspicions about the wires had been correct and their demand justified. Unwittingly, the neurotic monks had created an experiment and Gardini used this *prima facie* evidence to further the doctrine of atmospheric electricity favouring plants. He interpreted the situation as the wires having acted as an electrical shield above the garden, conducting away all atmospheric charge from the vegetation beneath and the period of dearth in the crops was the result. That the plants rebounded, so to speak, when access to charge was restored only confirmed the doctrine that plants need electricity to prosper.

Ingen Housz had, immediately, an alternative, more human, explanation. He himself had already experienced similar antagonism whenever and wherever he had been involved in erecting lightning rods. Many people living in the vicinity, naturally conservative and anxious about these newfangled installations, would harbour fears that such metal conductors, projected into the sky, could only attract electricity during storms and were more a danger than a protection. Before too long they would begin to agitate and lobby for the removal of the rods. The monks, he felt, had behaved in just this way. They had only used what was likely to have been a coincidental lean period in their garden as an excuse to be rid of the wires. Gardini's interpretation of the events could, however, have been correct and Ingen Housz decided to repeat the experiment.[27] He had long had a copper wire, 150 feet in length, strung across part of the Vienna royal botanical gardens.[28] It was electrically isolated, at every twist and turn, by suspending it with silk loops protected from the weather. The far end was anchored to a pole fixed atop one of the tallest trees in the gardens. The near end had been brought into a building through a hole drilled in a window frame and ended in a small copper knob. This terminal was held a short distance away from a second one that was, in turn, attached to a copper wire dropped down a well as an earth. Sporadic sparks could be observed passing between the two terminals and monitored by leaving a thin piece of card between them and counting the charred holes burnt through in a set time period. Ingen Housz had been using this ingenious apparatus in the way that his friend, Giovanni Beccaria, had done at Turin – to measure atmospheric charge.[29] It 'had never entered his head' that this apparatus could have any influence over the plants that happened to be underneath it but, primed by Gardini's report, he looked more carefully. He even tried, elsewhere in the gardens, an even higher and longer copper wire sensor, this time 250 feet long. However, although it detected abundant atmospheric electricity, there was still no apparent influence on adjacent vegetation.[30] Perhaps single wires were not shield enough for the 'Gardini effect' to be discernable?

In the early summer of 1786 Ingen Housz built a large pyramidal construction using copper wires stretched from a common apex, atop a wooden pole, to four smaller stakes hammered into the ground at the corners of a square. Cross wires were then wound around horizontally such that there was now what he described, fancifully, as a 'birdcage'. The apparatus was efficiently earthed by connecting it to a substantial iron pole knocked deeply into the ground. Ingen Housz had constructed, effectively, a safe haven that could not be penetrated by atmospheric electricity. Remembering his experiences with mustard grains – that it was better to compare many small plants than a few large ones – he planted numerous specimens of a small outdoor plant called *calaminth* within his 'cage'. He then planted a matching number of *calaminth* elsewhere as controls; plants that would be fully exposed to ambient atmospheric electricity. By early June[31] they had grown enough to draw conclusions.

> The results of this experiment agreed, again, entirely with those from all my other experiments. The plants thus destitute of all electric influence grew, flowered and seeded exactly like others of the same species in many other places in the same garden.[32]

Never happy until he had triangulated any observation of nature's workings with repeated or parallel trials, Ingen Housz also tried his original, domestic, technique of seeding mustard (and cress) but this time inside a similar metal cage to the one out in the gardens. Here the wires cascaded down from a central apex, four feet high, to an iron hoop wrought for a barrel.[33] Again there were no differences, in speed of germination or rate of growth, between the confined pots of mustard and cress, shielded from all electricity, and their matching controls.

In the early part of 1787 Ingen Housz found himself facing another 'electrical' challenge. It was in a new book just published in Paris by Pierre Bertholon[34] and the contents pages would have directed Ingen Housz's immediate attention to what Bertholon pointed up as 'a very curious observation of Abbé Toaldo'.[35] Guiseppe Toaldo was professor of astronomy and meteorology at Padua and a fellow of the Royal Society. Ingen Housz therefore took the report to be one that he should again respect but took immediate steps to repeat it. Toaldo had written to Bertholon[36] describing how a local politician, a Senator Querini, had planted a row of wild jasmines in the garden of his country residence – *Altichiero* – a large villa; one of the many alongside the River Brenta as it meanders serenely through lovely countryside towards the Venetian lagoon. The jasmines were established adjacent to the villa and, naturally enough, climbed up the walls. In three seasons most had reached the first floor cornice. But two of them, growing up close to the iron chain conductor of a lightning rod on the property, in fact becoming entangled with it, had grown extremely rapidly, streaking ahead of the others. One of them was even spreading over the roof, above thirty feet high. Moreover, these two plants, Toaldo had been told, flowered more prolifically and for longer than their peers. Ingen Housz was far from impressed by the story, focussing on the hearsay involved and that a lightning conductor, by its very design and purpose, would conduct away any accumulated electricity and immediately earth it. Far from the two giant jasmines being endowed with more electricity, they probably received less. Bertholon's own, fully-confident conclusion, that here was yet more evidence that electrified plants grow more vigorously, something he himself had observed near lightning conductors, provoked Ingen Housz into immediate activity. Although Vienna was still in the painful grip of a further severe winter in February 1787, Ingen Housz descended, yet again, on the royal botanical gardens, plans spinning in his head.[37] He had decided to test the 'Quirini/Toaldo' theory but not by imitating the circumstances exactly for he was convinced that these were fundamentally flawed. Rather, he thought to try to electrify some of the deciduous trees in the gardens and compare their bud breaking, growth, flowering, and fruiting with matched controls. Encouraged by a driven Ingen Housz, one or more courageous assistants climbed to the tops of thirty young lime trees (which must have been coated with ice, making fingers dangerously numb and slippery) and tied tall poles there such that they projected well above the topmost twigs. Each pole had a copper wire twisted around it, the upper end of which projected yet further up into the sky. The lower end of each wire was wound around a strong upper branch: each tree was, itself, now a lightning conductor. Being careful that their respective branches were not able to come into contact, Ingen Housz then chose some other lime trees growing nearby and as near equivalent in size and maturity, as control comparators. The experiment suffered a false start. On his next visit to the gardens Ingen Housz discovered, to his chagrin, that his 'spiked' trees and their control counterparts had all been dug up.[38] When he learned that the Emperor himself had ordered their removal to his own personal garden, Ingen Housz knew that he could do nothing but start the experiment again. To be sure not to be further frustrated, he chose some other, more established, limes to 'spike' and also some large chestnut trees. He also moved into the top-fruit orchard and dressed, likewise, some dormant almonds, pears and plums, always earmarking suitable controls. The Viennese public, visiting the gardens, must have pondered on the sanity of the perpetrator and frowned at the significance of all the prominent numbers that adorned so many of the trees. They would certainly not have been able to see any changes occurring for many weeks: in Austria the 1787 spring was very late and particularly cold.[39] Rain and snow succeeded each other throughout March and April and it was well into May before the first leaves broke on any of the trees. Nevertheless, Ingen Housz had been at work. He had been visiting his test wires to find that the air was heavily charged during March and April, the electrical potential dropping away only in May – as the trees came into leaf. He knew, therefore, that his test trees had been exposed to significant atmospheric electricity, at least during the initiation of that year's growth. However, the experiments all gave negative results. None of the 'conducting' trees opened their leaves or flower buds any earlier than their control counterparts. Neither were their leaves or flowers bigger or better.[40]

> It seemed very clear that the conductors had contributed nothing in deploying the leaves or garnishing the trees with flowers . . . I found that among the trees chosen for observation, none were any more precocious than any of the others in the garden.

Anxious, as always, not to rely on any single set of observations, Ingen Housz had also taken measures to double up on his 1787 experiments. This time he had asked van Breda, during February, to perform similar trials at Delft and his friend was a willing collaborator.[41] By the middle of May, Ingen Housz was pleading to know his results and when they arrived was pleased to know that they corroborated his own. Accordingly, Ingen Housz was able to write that van Breda had

> . . . obtained the same results as me, that is to say, that the plants which found themselves, by chance, under a horizontal conductor offered no peculiarity

and the trees surmounted by a pointed metallic conductor didn't advance their leaves or flowers any more than other trees of the same species.[42]

Ingen Housz appears to have fully satisfied himself, by the end of 1787, that most if not all, of the so-called evidence for electricity playing a major role in plant physiology was spoof. He had been experimenting for nearly a decade and had never managed to satisfy himself of any significant effect. What he had seen, with his own eyes, and to the satisfaction of friends and colleagues around him, was how easily physicists could be led astray by paying insufficient attention to detail; by being unaware of other environmental influences in their experiments; by accepting the results of single observations and by believing the half-truths of hearsay. He himself had noticed that during the summer of 1787, in visiting the botanic gardens day after day, that one particularly tall chestnut tree that he passed regularly had grown prodigiously that summer.[43] In all modesty, he admitted[44] that if he had just happened to 'garnish this tree with a conductor' and no others he, equally, might have been duped into a false positive conclusion. His series of papers in *Journal de Physique* did, however, bring about a general rethink on the whole topic which was, thereafter, examined more carefully and systematically. Ingen Housz appears never to have received due credit for his perspicacity, courage and meticulous discipline, suffering the common fate of those who take on the thankless task of having to point out to others the errors of their ways – in the right but in the shadows.

BETWEEN 1781 AND 1787, while Ingen Housz was contentedly investigating electricity in plants in far off Vienna, there were far less arcane activities in his fatherland: there was a revolution in the Netherlands. The obscurity of this little-known revolt may relate to its failure or because it was sandwiched, historically, between the more dramatic revolutions in America and France. It was also a relatively limp affair. Nevertheless, the 'Patriot Revolt'[45] was not free of bloodletting and most certainly made its impact on Ingen Housz. However, until it reached its end game in late 1787, he made no mention of the conflicts among his countrymen in any of his letters to van Breda. As far as he was concerned, the matter in hand was science. When he did deviate into politics[46] it was only because of his contacts in the Netherlands, among them his brother, Louis, Hoogeveen, de Monchy, and van Breda himself, all of whom had become closely involved and threatened by the turmoil.[47] Other than fleeting references, however, he dedicated only one such letter to the insurrection and even then apologised.

Forgive me that my pen, for once, deviates from our normal philosophical exchange of thoughts. Without really noticing it, it happened by the fullness of my heart, considering the disasters stemming from the lust for power and revenge . . .[48]

Like many revolutions, the footings were economic. There had been a remorseless downward spiral in the financial situation of the Netherlands since the so-called 'Golden Age' of the seventeenth century. Prince Willem V, Stadholder and head of the House of Orange-Nassau was not universally loved. Neither was he respected for his stewardship of the nation and, in supporting American emancipation, he had picked up a losing hand: it had not brought the economic rewards that had been predicted. Worse still, Orangists watched with horror as republican ideals, seeded by the American negotiators led by John Adams, began to gain a foothold among the political elite.[49] The Stadholder's opponents even took the name 'Patriot' from across the Atlantic and anti-Orangist feelings turned into actions from 1781.

The federalist structure of the Netherlands – of the seven 'United Provinces' – moulded the activities of the Patriots. The rebellion, such as it was, grew in scattered towns with little unity or coordination. It was stronger in the western province, Holland, than elsewhere and also, specifically, in Utrecht. It lacked critical mass, being very half-hearted in Amsterdam, the largest pool of population. The 'Patriot Revolution' has been described as 'tame' and 'piecemeal'.[50] While there was a strong undercurrent of hypothetical revolt, the only visible results were two-fold and relatively minor. The *schutterij* of many towns – the armed watch-guard – turned themselves into 'free corps', independent of the local establishment, of the councillors or 'regents' in the city halls. This motley militia lacked consistent views and their discipline was variable.[51] Their actions were unpredictable and they could not be trusted. Nothing caught fire until 1785 when the Patriots in Utrecht locked their council in the town hall until they agreed to a list of 'democratic' demands and to hand over the reins of power. In effect, the town seceded from the state to be an independent commune. Elsewhere, rebellion was held back by hesitancy if not lethargy: a revolutionary clock was ticking but without the impetus to strike the hour.

At least the next events of any significance occurred at political headquarters – in The Hague. By September 1785 the Patriots there had a majority mandate but the Stadholder insisted on inspecting his troops, every Sunday, on the square next to the Binnenhof (parliament building).[52] This insensitivity provoked rioting and his moral authority sank to the point where he was deprived of his military command. Emasculated and humiliated, he left The Hague. He moved his family and his court to an ancestral castle near Nijmegen. The province of Holland was now, effectively, a Patriotic enclave protected by a 'black' cordon of anti-Orange strongholds. Outside the pale there were some isolated Patriotic towns such as Utrecht but the Netherlands was now divided. During the next year or so nothing was settled by a series of armed confrontations between the two sides; the only consistent experiences were looting and personal retribution. Then, as polarisation began to harden in early 1787 and civil war looked imminent, neighbouring states tried to intervene although their motives for mediating were anything but altruistic. In meeting their ambassadors Prince Willem was led to believe that there was profound discontent with the Patriot regime in Holland and that he would be supported by the bulk of the population should he return to the traditional seat of power at The Hague. He vacillated but Princess Wilhelmina, his wife, and favourite niece of Frederick the Great, was made of sterner stuff and set off for The Hague on 28 June. She and her entourage were stopped near Gouda and held in close and ill-dignified custody at an inn. The King of Prussia, the newly enthroned Frederick William II, was outraged and demanded his cousin's immediate release. A final ultimatum from Berlin on 8 September was followed, on the 13th, by a Prussian army, 26,000 strong, crossing the border.[53] It took only a month to reinstate the Stadholder to power, massed Patriot resistance melting away in the face of the military realities, even behind the 'cordon'. Many Patriots fled to France, perhaps as many as 20,000 but others went 'underground', lurking within their communities, simmering with discontent and resentment.

At a personal level, the outbreak of republicanism and its associated violence, sporadic though it may have been, had been a source of universal anxiety in the Netherlands. There was a mutual unease shared between Ingen Housz and van Breda. The Delft councillor was in the thick of events in his own community but still had time to share disquiet about Ingen Housz's family at Breda, anxiety that was proved appropriate when Ingen Housz relayed the contents of a letter from Louis dated Breda, 10 December 1787.[54]

> . . . all the residents are still living in utmost anxiety and well-founded fear of being plundered and killed by the country's own militia . . . that his family is constantly urging him to leave the anarchical land and go to Antwerp with his wife and nine children in order to escape the murderous intent of the mutinous soldiers.

And from the relative security of Vienna Ingen Housz was hoping that van Breda's tactic of declaring himself 'above the fray' was working.[55]

He was, after all, one of the Delft regents, a member of the town council. 'I hope that, by your neutral posture, you will be positioned outside the danger zone of unpleasant experiences . . .' Ingen Housz himself was far from neutral. Perhaps it is not surprising, that as a royal adviser, he would scorn republicanism. He was one of an exquisitely small minority who had ever met the Stadholder and he had even entertained, in his own home in Vienna[56], Prince Nassau-Weilburg, the Stadholder's young nephew. And he was unrestrained and hostile when it came to urging repression and reprisal.[57]

> In such ill-fated times, when the thoughtless mobs pretend to serve the Stadholder by insulting and oppressing hundreds of honest families . . . one should take the most vigorous measures in counteracting this evil. The first one to be caught . . . plundering . . . should immediately be hanged on the doorpost of the house where he committed the act of violence.

Ingen Housz viewed the Prussian invasion as the 'salvation' of his fatherland although, in this context at least, he regretted the force of arms.[58]

> An argument delivered by a metal mouth is very decisive and in this way even the bravest of heroes has been forced to hand over his belongings to a coward, the latter holding a pistol to his head. By such argument I once lost my watch and money in England.

Van Breda had certainly been at risk of losing much more than a few personal possessions. A majority of the regents of Delft had retained their loyalty to Prince Willem and as tensions rose in the autumn of 1787 they feared that their town would be attacked by Patriots. In the event, the bulk of a strong Patriot force stood off while scattered gangs entered the town to loot, even after some of them were shot and killed at the barriers. Van Breda himself was at the City Hall, never abed, for five days and nights.[59] When he finally arrived home, thankful that his household and person were unharmed, he was roused within the hour to be told of a new alarm. Someone had reported that there were Orangist troops approaching the city and a mass of soldiery, under whatever banner, spelt trouble. Van Breda saw nothing, however, through 'a good English field glass'[60] from the tower of city hall. Even so, he still found it difficult to reassure his legislative colleagues, some of whom were 'crying in their chairs'. A squad of Prussian hussars did show up, however; about 100 strong. Van Breda found himself, by default, to be the one-man reception committee. Keeping his cool, he escorted their commander around the City Hall and took him to see the great tombs in the New Church, after which the officer and his men, drilled and disciplined, rode off to cheers from the crowd.

Hendrick Hoogeveen and his family had, likewise, come through the immediate threats unscathed; safe, perhaps, but not without other trauma. Ingen Housz received a moving letter, in Latin, in February 1788. The whole story could not be told. 'Let this be sufficient: for reasons of safety I will rather confide more to your ears than to a letter.'[61] 1787 had been a terrible year for the 76-year-old Hoogeveen and his family. His wife had suffered a stroke and two grown-up sons had died.

> The wound inflicted by the death of my son, Theophilus, had not yet formed a scar when there followed the fatal end of another, Theodorus . . . you, of all people, will know how extremely heartrending that is.[62]

The second death was probably accelerated by the revolution. Less than a month before the violence had erupted, Theodorus had left home to take an administrative post in a local aristocratic household. He had therefore become a target of the 'frenzied rabble' when civility broke down. Though 'deprived of his dignity', he had escaped with his life but only because he had been identified and linked to 'old' Hoogeveen. The mob was, for some reason, well disposed to the family name and set Theodorus free. Sadly, however, he was to die within a couple of months. The same 'scum from the Hague' also exempted the Hoogeveen house and school when they sacked parts of Delft although the family, trembling within, were terrorised by the sounds of the very next house being plundered and destroyed. 'If you had been here and seen with your own eyes you would have said that a complete swarm of evil ghosts had left their subterranean residence and moved, in close formation, on Holland.'[63]

Then, within hours, Ingen Housz had in his hand a letter dated Rotterdam, 14 February 1788. Addressed 'Dear friend' it was signed 'Q.N' (*quem nosti* – 'you know who') and was a chronicle of how some of the Patriots had also suffered. The unidentified writer was Salomon de Monchy. He and his sons had declared themselves 'Patriotic' and the letter rehearsed a convincing list of reasons for their stance. They were now suffering the indignities and dangers of being on the losing side. 'Please forgive me, my dear friend, that I bother you so much about such unpleasant matters but to unload one's breast to a friend lightens the burden.'[64] De Monchy had already been stripped of his honorary professorship and his eldest son dismissed as city physician. His brother, 75 and a long-respected resident at The Hague, had been obliged to flee to Brabant. De Monchy himself was expecting to be forced into exile and was trying, with no success, to sell some properties; to convert part of his assets into cash. Meanwhile, he and his circle of friends were attempting to maintain some semblance of reassuring normality. The *Bataafsch Genootschap (Batavian Society)* was still holding regular meetings but attendances were thin and no-one had any appetite for experiments or dissertations – 'the muses are asleep'. Men seeking Enlightenment were always going to be suspected of being Patriots and many of their number had been 'persecuted' and had had to leave the city 'in order not to be murdered'.

Ingen Housz now had a complete picture in miniature: in all revolutions there is unavoidable suffering on both sides: the only assured outcome is stress, blood, and grief. His fear and loathing of political disintegration, of 'democracy' and of the 'mob', was going to be a yoke that he would carry for the rest of his life. His introduction to these certainties might have been second-hand but it was not to remain that way for long. Though far away in Vienna he found that he had actually suffered a direct personal consequence. The company in which he and Louis had invested money at Amsterdam to provide annuities in their old age had collapsed[65] and their joint capital was lost. Reinforcement of his feelings of loss came with a letter from America. Ingen Housz learned, by a letter[66] dictated by Franklin in February 1788 – the ageing American had fallen and badly sprained his arm – the unwelcome truth about some of his investments in America. Jonathan Williams had passed through Philadelphia that month. The lump sum of 7000 dollars entrusted to John Williams senior had been used to buy indigo in Carolina. Unhappily, the ship conveying the dye to Boston had been taken by the British navy and the insurance against loss at sea, taken out at Charleston, had proved worthless after the town fell to the British army in May 1780 and the town's business community had imploded. John Williams himself had lost money in the same misadventure, and the colonial bonds that he had also acquired for Ingen Housz had consistently devalued after an initial rally. Though Ingen Housz could choose to have them redeemed they were now likely to raise only about a sixth of their face value. Franklin was able to confirm all this as the true situation – he himself had loaned the dollar equivalent of £3000 sterling (£189,000) to Congress by buying bonds and these were now showing a cash value of only £500 (£31,000). He also pointed out that even British government issue bonds giving three percent interest had devalued, during the war, by nearly half.

There was news of the Wharton family. It was now common knowledge that one of Wharton's sons had shot himself while at sea and that Samuel Lewis Wharton was an incurable 'sot', constantly drunk on rum. Samuel Wharton himself, after his stroke, was now 'reduced to a state of imbecility', his speech was slurred and he had been declared bankrupt. Franklin's letter caused Ingen Housz much distress. Although Wharton was thought unable to pay creditors more than a shilling for every pound owed, it might still be worth suing – a power of attorney could be organised. Furthermore, Franklin, as Governor of Pennsylvania, had used his legal officers to advise him on American mercantile law. In the case of joint adventures, it seemed, remittances paid to one partner were understood to have been paid, effectively, to all partners. Ingen Housz might have a better chance of redress in suing Coffyn and Bancroft. The whole shebang was spinning out of control. It was impossible to see how two swindlers, one an American and the other Flemish, could be taken before a French court while the litigant was living in Austria.

If there was any pleasure to be had in reading the Franklin letter it was to be found in the very warm expressions of affection from a friend whom he now knew to be still alive and alert. However, Franklin had recently fallen into the understandable habit of writing all letters to friends as if they were certain to be his last. In surviving into his 82nd year he was flouting his own pessimism but he never lost his literary talent. Among the recurring farewells some are especially eloquent, including one to Ingen Housz:[67]

> . . . I can expect no greater felicity in this world and I begin now to feel a curiosity of knowing something of the next. To inquisitive minds, like yours and mine, the reflection that the quantity of human knowledge bears no proportion to the quantity of human ignorance, must be, in our view, rather pleasing, viz. that tho' we are to live forever, we may be continually amused and delighted with knowing something new. Adieu my dear friend, and believe that to eternity, I shall be, Yours, most affectionately.

LATE IN MARCH 1788 Ingen Housz received two letters[68] from his brother, Louis. They had been written only a week apart, strongly suggesting a sense of urgency or, at least, a situation that was changing rapidly and unpredictably. In his haste, Louis had inadvertently sealed up a small voucher or certificate within one of his envelopes, an error that provoked a rebuke for being 'slovenly' that was hardly fraternal. From this and the advice returning from Vienna we can assume that Louis's anxieties were financial. An economic recession in the Netherlands had been no surprise but when a threatened challenge to the Prussian presence in the Netherlands, by the French army, proved to be hollow, it had exposed and highlighted the near-bankruptcy of France and the total want of steel and strategy at Versailles. Investments held in French currency were also falling precipitately. Perhaps brother Louis was concerned about the family investments in France and hoped that Jan could make a visit to Breda for them to take some joint decisions. Ingen Housz wrote of an early intention to visit Paris 'in order to bring two new works to the press' and then London 'to investigate the present state of sciences.'[69] Although he did not say so, there was an obvious inference that he and his brother would have the opportunity to meet. His official plan was to leave Vienna early in May, to have two or three months at Paris and then move on to London for the winter. The Emperor's strict embargo on unnecessary foreign travel[70] had obviously been lifted or somehow surmounted and Ingen Housz's definitive travel plans were confirmed in his next letter to van Breda.[71]

Sometime in 1785 or early 1786 Ingen Housz had returned to his unfinished second editions of *Vegetables*. At that point they were neglected piles of paper, one in French and one in English; abandoned when he had found his energies diverted to having to defend his dominance as the discoverer of the primacy of sunlight in the plant kingdom. There was, he felt, still no prospect of a successful release in England and all the pages and notes were almost certainly redundant. However, publication of a new edition in France seemed a better prospect. He saw no purpose in rewriting the core of the original book but much had been added: the dedication to Sir John Pringle survived, as did the detailed advice on how best to use the Fontana eudiometer. However, he wrote a substantial new foreword in which he gave a siege history citing Priestley, Cavallo and Senebier.[72] There was also a new section on the 'green matter' which was clearly appropriate, and, with an enlarged preface, the final manuscript of only the first of two volumes was, alone, at over 70,000 words, nearly twice the size of the first, English, edition. Niklaus Molitor was no longer at Vienna but Ingen Housz now had another friend who was keen to see the royal physician's books appear in the German language; a persuasive fellow physician and friend of the Jacquin family called Johann Scherer. Quite when Scherer was given access to the new manuscript is uncertain but he seems to have moved quickly. Ingen Housz finished the first volume of this new edition on 14 January 1786[73] and the German version,[74] *Versuche mit pflanzen (Experiments with plants),* was on sale in Viennese bookshops before the end of the year. Handed back by Scherer, the French manuscript of volume one was then sent to Bégue de Presle in Paris for him to have the censors approve it for publication. Approval was issued on 15 December 1786,[75] and it was released as *Expériences sur les végétaux II (Experiments on plants, 'Vegetables II')* during 1787. In the meantime, Ingen Housz had pressed on to try to finish the remainder of the book, writing in French and releasing tranches of the follow-on manuscript, to Scherer, piecemeal. Sherer appears to have reached a point, in early 1788, where, knowing that Ingen Housz was about to depart for Paris, he thought it reasonable to have further pages printed – *Versuche mit pflanzen, Zweyter band. (Experiments on plants, second volume.)* – although these would not see the book through to the end.[76] Later, Ingen Housz sent all the remaining pages of manuscript from France as they were finished. Thus, in the German edition, a third volume of *'Vegetables II'* appeared in 1790[77] *Versuche mit pflanzen, Dritter band. (Experiments on plants, third volume).*

Ingen Housz had also compiled a second volume of *'Nouvelles Expériences' (New Experiments)*;[78] but this time he wanted to be in Paris so that the manuscript was not left lying unattended for years. Once more there was an eclectic range of topics but otherwise, he told van Breda, he had decided to tear up a substantial amount of potentially publishable material. But it was a turning point rather than a notice of retirement. 'I still see many miracles in the laws of nature that I would like to investigate if time and ingenuity allow.'[79] At a more human level, he was still wondering if he should extend his trip to include his homeland but diverting to the Netherlands was something that gave him more than a pang of anxiety. He knew, from his brother and other contacts, that there were still significant political tensions in the aftermath of the revolution. 'However pleased I would be to see my family and friends again, I doubt whether it will give me pleasure because of the continuing chaos that makes the country hazardous. Time will tell.'[80] There was to be more time than he expected. Ingen Housz was still at Vienna in mid-June. He had been badly let down by a coach builder who, having failed to meet the agreed deadline of early May, then delivered, a month late, a travelling carriage that was so defective that Ingen Housz knew that he dare not trust it to the long journey he had in mind.[81] He approached another Viennese coachbuilder even though he knew that this would mean yet further delay to his departure. If he appeared neurotic about insisting on top quality equipage, it was entirely understandable. There had been a hammer blow in the mail.

IN MAY 1788 Ingen Housz received desperately sad news about his brother. Louis, a vigorous man of 58, had been thrown from his carriage when it overturned in an accident at Princenhage, a mile or so to the west of Breda, outside an inn called *'De Bloemkool'* (The Cauliflower).[82] He had sustained a fractured femur and died an agonising death, succumbing on Sunday 27 April. He was buried at Meerle, some ten miles south of Breda, and a difficult letter was despatched to Vienna.

Louis Ingen Housz, Jan's elder brother. Courtesy of the Ingen Housz family.

A grieving Ingen Housz finally left Vienna only on the 3rd or 4th of July. By then he had acquired *litterae patentes* – a passport – from the court; a document, in Latin and dated 20 June 1788,[83] that described him as 'Imperial Chief Physician, travelling with one servant.' The servant was the devoted Dominique Tede.[84] And, at the very last minute, on Wednesday 2 July, Ingen Housz organised a letter of credit for 8000 Florins (£46,000) from his bankers, Stametz, that could be drawn on at Strasbourg, Paris, Amsterdam and London.[85] After nearly eight years in Vienna he was again travelling to western Europe. On this occasion he intended to be away no more than eighteen months but his wife, servants, in-laws and friends were probably prepared for an extension; there was a very strong precedent for his absences being longer than forecast. Farewells over, Ingen Housz raced across Germany and France: he was in Strasbourg on 13 July and in Paris by the 21st.[86]

On this trip he settled in the centre of the city despite the well-known drawbacks of the overcrowding, noise and smells,[87] finding lodgings in the Rue du Jardinet[88] on the left bank. Franklin, who had introduced him to Chaumont at Passy in 1778 and who had contrived for him an invitation to stay, twice, at Chaumont's mansion, was no longer present to facilitate such privileged luxury. Besides, Ingen Housz was still smarting from Chaumont having given him such wildly over-confident advice to invest with Samuel Wharton. Ingen Housz also knew that his business, this time, was in central Paris. He appears to have been secure and content in his lodgings[89] at the *Grand Hotel de Toulouse*, for one and half Louis D'Or per week (£84). Outside however, on the streets and in the coffee houses, he found a pernicious atmosphere.[90]

> The increasing wealth of the city . . . only heightened the gap between the rich and the poor, who were legion. The view on the street in Paris . . . was that the king was laughing in the face of increasing poverty and hardship. In the taverns, cabarets, dives and dockside eating-houses . . . the bitterness was palpable. In manners, language and style, these places were the very opposite of the Enlightened Paris of cafes and salons, where anti-government tracts and pamphlets were composed and distributed. But it was in the lower depths of society, the so-called 'dangerous classes', where truly insurrectionary hatred simmered.

Ingen Housz lost no time in making personal contact with Thomas Jefferson, Franklin's successor as American minister plenipotentiary.[91] He also dined with John Paradise, the British scholar and Royal Society fellow,[92] who had just arrived at Paris – direct from Philadelphia. Ingen Housz was therefore delighted to learn that their mutual friend, Franklin, was 'still an inhabitant of this world' and joyfully sat down to write him a letter.[93] This was taken, personally, to Philadelphia, by an American traveller the next day,[94] Ingen Housz having enclosed, within it, a power of attorney. This gave a Philadelphia lawyer the authority to sue, under Pennsylvanian law, the perfidious Samuel Wharton. Four Americans then at Paris endorsed the document including Jefferson for it was he who had, finally, persuaded Ingen Housz that it made sense to resort to legal redress, if by proxy, and in spite of his anxiety not to waste more money. 'I should not like to spend much money with little or no hope to recover what is due to me.'[95] However, Ingen Housz did not allow such transatlantic distractions to divert him from his intended exertions at Paris. Within two weeks of his arrival he already had some pages of the second volume of *'Vegetables II'*[96] 'under the presses' of Theophile Barrois, publisher and bookseller at the Quay des Augustines. And as the printed leaves appeared he asked van Breda if he might consider translating them so that a Dutch second edition might also appear.[97] It was obviously reassuring to be able to orchestrate the publishing processes personally. He had been endlessly aggravated by the delays – nearly three wasted years – when trying to steer his first volume of *Nouvelles Expériences* through the Paris presses by surrogacy. Bégue de Presle had been mainly to blame but other contacts he had in the city had been just as sluggardly and unreliable. Now he simply needed to finish the manuscript of his second volume of *Nouvelles Expériences*[98] and there was ample time. He was expecting to stay in Paris until December although he was still torn whether to go on to London directly or via his homeland. This time the hesitation centred on the affective. '. . . the emotion of seeing for a while a few friends and relatives only to depart from them for always is an obstacle that I am unable to overcome.'[99] This rare documentation of his intimate feelings was provoked, perhaps, by grief for his brother.

Platinum had intrigued Ingen Housz since his first spell in London and he knew, by an aside in a letter from Franklin in 1782,[100] that the French chemists had been making good progress in exploring the properties of this rare metal. Franklin had been present when Antoine Lavoiser had managed to melt a piece of platinum by putting it into the bowl of a hollowed-out plug of charcoal that he then brought up to red heat before directing, at it, a jet of dephlogisticated air (oxygen). Searching out the latest news on such developments had been one of Ingen Housz's subsidiary motives for travelling to Paris and it did not take long to learn of a young genius who was still a medical student. Bertrand Pelletier was the son of a leading Paris pharmacist and obviously had much in common with Ingen Housz. He was also just as enthusiastic an explorer of the natural world. He had discovered[101] that by heating platinum with powdered phosphate glass and charcoal a kind of alloy was produced which, on further protracted heating, yielded pure platinum that was malleable and therefore invaluable to the jewellery trade. This skirted the obvious risks of the method of Marc Étienne Janety, court goldsmith, who put the metal through a similar process but with arsenic salts. Ingen Housz appears to have been carrying a sizeable lump of native platinum and may have had it treated by his new friend. Pelletier may also have been the social route to meeting the other Paris chemists of that generation, of which many were destined to become household names.

In fact, Ingen Housz's first few weeks in Paris rapidly embedded him into its scientific community. The French capital was now, he found, the epicentre of progress; not London. He was able to move into a circle of philosophers, many already friends, members of the Royal Academy of Sciences at the Louvre, who were at the cutting edge, of chemistry in particular. Their unchallenged leader was Antoine Lavoisier but there were many other names that were to become famous in the history of science – Bertholet, Charles, Hassenfratz, Morveau, Le Metherie, Fourcroy, Monge and many others. The list was as long as it was glittering. Ingen Housz had moved from virtual isolation in Vienna to be submersed in talent. Dramatic and ground-breaking advances in chemistry had taken place in Paris in the previous fifteen years.

Antoine Lavoisier, a very rich tax commissioner, economist and amateur scientist, had experimented[102] with the new gas discovered, in 1766 in London by Henry Cavendish and named, by the Englishman, as 'inflammable air' (hydrogen). Lavoisier had fully expected that this gas, produced by drizzling strong acids on to metals, would fix a portion of atmospheric air when he burned it and that there would be a weight gain. The experiments gave inconsistent results but were repeated by Pierre Macquer,[103] a colleague, a year later. To be sure that the expected soot was not escaping notice, Macquer held a white porcelain dish over the flame. There was certainly no soot but he did notice droplets of water condensing on to the crockery. He misinterpreted his findings as mere atmospheric condensation but they led, in the early 1780s, to better experiments in which Lavoisier showed, repeatedly and convincingly, the formation of water when inflammable air (hydrogen) was burned in dephlogisticated air (oxygen). Then, in February 1785, again winning over a distinguished audience, he demonstrated how water, superheated by dripping it through a red-hot gun barrel, separated into the two gases we now call hydrogen and oxygen: the composition of water had been definitively established. Ingen Housz was able to learn of these exciting developments from the active participants. He found, also, that there was no longer a rationale for the existence of phlogiston – it had been shown to be entirely fictitious, its supposed actions fatuous. Far from there being a substance that was released when materials burned, it was the reverse; something was absorbed. It was now proven that the ashes remaining after combustion in carefully controlled conditions were always heavier than the original sample. And just as the understanding of chemistry was making such progress, its very terminology was being turned on its head.

Louis de Morveau, a respected chemist from Dijon, had advocated, for some years and with the growing support of Lavoisier, the construction of a completely new chemical nomenclature.[104] The Royal Academy of Sciences was frosty but he persevered and travelled to Paris, in 1786, to discuss his ideas. Lavoisier, Fourcroy and Bertholet were prepared to listen to

Antoine Lavoisier with his wife.

his arguments and found them very attractive. They adopted the principle that compounds should be named from their components and that they rebaptise all known substances with the help of the Greek dictionary. This gave them many original names and opened up a breach between their proposals and the pervasive influence of reactionary Jesuit teaching where the dominant language was Latin. Thus it was that 'dephlogisticated air' became 'oxy-gene' – a producer of acidic salts; and 'inflammable air' became 'hydro-gene' – the producer of water. Here was the 'new' chemistry, a novel nomenclature that was more: it was a profound point of departure – chemistry was now fit to become a rational study of how materials interact. The new nomenclature and intellectual framework had an irresistible consistency and was to stand the test of time, becoming the basis of modern chemistry. All this was newly minted and sparkling as Ingen Housz arrived in Paris in 1788 although, like himself, many of the scientists were deeply sceptical.

Ingen Housz, however, soon changed his views. When Hassenfratz had sent him a publication in which the new chemical names and symbols were proposed – *New method of chemical nomenclature and symbols*[105] in December 1787, Ingen Housz gave it an instantaneous *non placet*, a verdict transmitted to van Breda.[106] 'I am not pleased with any aspect of it and I hope there will be universal resistance against such folly. I made no attempt to conceal my judgement . . . in a letter of thanks (to Hassenfratz) for sending me a copy.' However, by the time Ingen Housz had been in Paris for a few months he was a convert to the 'new chemistry'. He was ducking and weaving a little and inviting readers of his second volume of *Nouvelles Expériences* to 'arrive at their own judgements' but the dismissive dogma had disappeared.[107] In his memoire on manganese, clearly written after leaving Vienna, he displayed a completely different approach to the new developments and new locution. The French chemists had made 'rapid progress' by using 'truly ingenious experiments that had led to 'fundamental improvements' in the 'basic principles' of chemistry. His conversion was not, yet, absolute perhaps but in the same memoire we find the word 'oxygene' – more than chemically symbolic in this instance.

The new principles put Ingen Housz in a quandary. As more and more pages of *Végétaux* were committed to the presses he began to realise that his text had been built on what was clearly about to become an antiquated and superseded chemical system:[108]

> At the time that I made these experiments I believed they were a contradiction to the principles of this (new) doctrine but, in seeing, more and more, with what ease the principles of this doctrine adapt themselves to all the phenomena of Physics and Chemistry, I must attribute my difficulty to the inexactitude with which I had come to understand the new principles . . .

In mid-September he was struggling to correlate his findings and observations with the new 'antiphlogistic' system, the reconciliation proving daily more urgent and unavoidable as successive printed pages of his text were served up by Barrois. At least Ingen Housz proved to be adaptable and willing to adopt new ideas. Others of his generation were more stubborn and reactionary. Priestley, in particular, and not to his credit, went to his grave clinging to the wreckage of the phlogiston theory.[109]

INGEN HOUSZ MIGHT have converted into a 'new' chemist even earlier if he had been able to maintain the debate with all his new friends but suddenly, in late October 1788, still in his lodgings in central Paris and with only Dominique to support him, he was incapacitated by serious illness.[110] A severe, spasmodic left loin pain was followed by increasing bladder instability. Being unable to wander far from his chamber pot progressed to an unremitting but spurious desire to urinate and to crescendos of terrible pain in the pubic area. Ingen Housz did not need to be an experienced physician to diagnose that a left kidney stone had found its way down the ureter into his bladder. The lengthening waves of nausea and agony only abated if he lay absolutely still in bed but this was prevented by the dominating desire to pass water, the performance of which was, itself, excruciating. Then the variable flow of 'hot' urine was replaced by blood. He knew he needed help. At any moment a combination of stone and clot might block his bladder outlet completely and he could die, within days, from renal obstruction. Dominique was despatched to ask for Count Carbury, personal physician to His Majesty Count D'Artois, the King's youngest brother. Carbury was delighted to assist and advised Ingen Housz to take, every morning, the expressed juice of a whole lemon sweetened with honey or sugar and then mixed with two chocolate-cups full of warm broth. Ingen Housz's faith in this remedy might have been a little sceptical for he knew all about the added efficacy of carbonated water from his contact with Hulme in London. However, he may have been persuaded to comply by learning, from Carbury, that van Swieten himself had once used lemon juice to cure the Doge of Venice.[111] Ingen Housz certainly gave it a good trial. A more assiduous and diligent a patient than many in the medical profession, he continued with this regime until June 1789, until well after the initial resolution of his bladder problems. In fact, there were high hopes of a full recovery when, late in November, the cigar-shaped stone, half an inch long [112]suddenly plopped into his chamber pot 'without any difficulty or strain'. He was aware that he had lost a significant amount of blood and that there was still 'gravel' in the urine but the latter was now clearer and he was far more comfortable. But, as he watched the visible shards in his pot give way to mere traces of 'red sand' and found the strength, finally, to see the last batch of pages of *Végétaux II* into print and the finished books put on sale, he developed new symptoms.

Ingen Housz now had pain in the upper abdomen, was nauseated and was repelled by food and drink. He then realised that he was jaundiced.[113] Again the diagnosis was easy – he now had 'biliary concretions'; gall-stones, in other words. One or more gall-stones can sit in the human gall bladder for decades without causing any trouble, indeed their presence can be totally unsuspected. Should they, however, form the focus of an infection in the gall bladder or fall down into the common bile duct (the drain common to liver and gall bladder) the tender and swollen tissues obstruct the outflow of bile from liver to bowel, levels of this pigmented waste product dam back into the bloodstream and the patient goes greenish-yellow. This extremely unpleasant circumstance is life-threatening if prolonged, and there was no point in any of Ingen Housz's friends trying to disguise this from him, especially when he had not properly recuperated from his urinary problems. 'I found myself attacked with a complicated disease of which my medical friends thought I could not recover.'[114] In the next four months, until around Easter 1789, Ingen Housz was confined to his Paris lodgings, indeed mostly to his bed, the jaundice fading only to reappear a further three times. He was unable to visit his publisher, witness experiments, attend meetings of the Royal Academy or, indeed, take even a short walk to breathe some fresh air. He appears to have been nursed by Dominique alone although, from the warmth with which he refers to his host family in later correspondence,[115] they also cared for him very tenderly. The practicalities of moving on to London for the winter had been completely swamped by his poor health and with the author out of action, finalisation of the second volume of *Nouvelles Expériences* had stalled.

Even for those Parisians in good health, the winter of 1788/9 was a very difficult one; the coldest since 1709.[116] There were widespread problems with all forms of transport and the vital flourmills seized up. Not that there was much grain to mill. As Ingen Housz had travelled from Strasbourg to Paris during the third week of August 1788 he would have seen the devastating effects of terrible storms that had swept through central France on 13 July. Enormous hailstones had flattened the cereal crops over vast areas and fruit and vegetables had also been damaged. The weather had then improved for six weeks but the swing to harsh sun had been too consistent and a drought effectively finished off the injured crops. It was no surprise, therefore, that each time Dominique went out to shop for essentials such as bread or milk during that autumn and winter, he found the prices higher. In fact, the cost of a four pound loaf of bread, standard in Paris, doubled between August 1788 and the following February. The price of firewood also doubled. The essentials needed to survive the extreme cold proved beyond the means of many humble Parisians, an already restive community who, more and more, suspected the rich of hoarding. The combination of hunger and anger built up into a barely suppressed aggressive resentment on the streets, where it was still snowing on the heads of those queuing for bread in April.

As Ingen Housz recovered from his illnesses and began to circulate, again, among the Paris philosophers, he overheard more and more intimations of political instability. The whole French population, of whatever stratum, had made one last act of faith in the person of Jacques Necker, appointed chief minister by the King in August 1788, to deal with the enormous national debt and somehow revive the economy. The French working classes – the 'Third Estate' (as opposed to the nobility and the clergy, the First and Second Estates respectively) expected a miracle of Necker, a Genevan Protestant, whom they saw as an outsider among the ruling classes and at least, therefore, not corrupt. Ingen Housz would also have learned, by now, of the inordinate degree of hatred that the ordinary Frenchman had towards 'his' Queen, Marie Antoinette. Many wholly believed the salacious, but false, stories that appeared in the illegally published obscene *libelles* that repeatedly circulated, of her conceit and arrogance, of her promiscuity and sexual depravity[117]. In theatres her appearance in the royal box was met with hisses and spitting and on the Paris corners the term *Austri-chienne*[118] (Austrian bitch) was everyday parlance. Attitudes like these must have given Ingen Housz a twinge of concern – he had made no secret of the fact that he was, after all, body physician to Marie Antoinette's brother. The feeling that he might be in danger in Paris was growing, daily, and it must have been a relief to take, at last, his finished manuscript of the new volume of *Nouvelles Expériences* to Barrois. This time the book was dedicated to Thomas Dimsdale and to memories of another winter, two decades earlier; to his apprenticeship in smallpox inoculation and all that had stemmed from it.[119]

> . . . time has only strengthened those sentiments that I recall of the open and unselfish way in which you made me part of your knowledge of inoculation for the smallpox and in which you presented me with the opportunities to educate myself.

He recalled how he had then been called to treat the royal family at Vienna and how, a few weeks later, Dimsdale had been asked, similarly, to go to St. Petersburg, distant and hazardous tasks that had fundamentally changed the lives of both of them. '. . . *utrumque nostrum incredibili modo consentit astrum*[120] (the stars of the two of us are unbelievably linked together). As far as the text of the book was concerned it was another very varied compendium although more than a quarter of the pages discussed only one issue – the 'green matter' – its nature, microscopic appearance and actions on gas formation in water. Ingen Housz was still unconvinced that it was a true plant material: the fact that it produced 'dephlogisticated air' (oxygen) in sunlight did not, necessarily, place it in the plant kingdom. There was a varied miscellany of other memoires, as the title again implied: issues concerning eudiometry; his work on suspending compass needles; on a new electrical machine that used taffeta; on the effects a nasty electrical shock had had on his mood and memory; reflections on the structure of gunpowder magazines; and an edited version of Franklin's long letter to him on curing smoky chimneys. Disabled by illness or not, some of the last articles were clearly newly written, their content relating to experiences he could only have witnessed in Paris. Among these were one on platinum and one on manganese that referred extensively to the 'new chemistry'. The actual date of publication is not quoted but the frontispiece clearly locates it as 1789 and it must have been before July.

On Monday 27 April 1789, the Paris streets were full of angry people[121] determined to protest that Jean Baptiste Reveillon, proprietor of a wallpaper company and balloonist, should not reduce his workers' wages to a mere seventeen Sous (0.80p) per day. The rumour was completely unfounded however, and, having surrounded Reveillon's home and factory in the Rue du Faubourg Saint-Antoine, the mob eventually relented and dispersed. The next day they were back, irrespective of the truth and far more determined. When troops arrived there was a tense stand-off that culminated, almost inevitably, in the troops having to open fire 'in self-defence'/'to restore order'. At least 300 of the tightly-packed crowd were wounded and perhaps as many killed. It was a sign of things to come and the soldiers themselves – the *Gardes Françaises* – appalled by the dilemma they had faced in having to fire on their 'own' and confronted, at close quarters, by the contorted heaps of the dead and the writhings and groans of the wounded, swore that they would never allow themselves to be put in this position again. Their reactions were to be highly significant exactly eleven weeks later.

Being nearly 60, Ingen Housz would have needed time to recover from his illnesses as they abated. His schedule was now badly awry and the second part of his manuscript of *Vegétaux II* was still expected by the printer. He eventually made his way back to the presses of Barrois with the finished manuscript sometime in spring 1789. Although not working to a particular deadline, there were time pressures. The longer he remained in Paris, the more expensive his publishing expedition became. He had already drawn most of the 8000 Viennese Florins (£46,000) covered by the letter of credit issued by Stametz.[122] He was, though, at last able to watch the printed pages of the new volume begin to stack up alongside the presses *chez* Barrois and know that publication was imminent and certain. But the situation in Paris began to look ominous in late June and it was a far from auspicious time for the successful launching of a philosophical book, indeed for any merchandise that was not essential just for daily survival. For the first time in centuries, the political hub of Paris had moved the thirteen miles from Versailles into the centre of Paris, to the Palais Royale and the Palais de Justice. This, however, invited more lobbying and protesting and on Sunday 28 June, Gardes Françaises, still chaffing from the horrific scenes of 28 April, assembled en-masse to announce, publicly, that they would, never again, fire on their own people. This mutinous declaration created immediate difficulties for the administration who had, with bread prices still higher, just decided to post the Gardes at grain and flour stores. There was now a power vacuum threatening the maintenance of law and order. Grain stores were increasingly ransacked while the Gardes stood by. Then people began to sense the presence of more and more Prussian dragoons and other foreign mercenaries, some 30,000 troops who would have no qualms in using their firepower. The universal hatred for things 'German' went up several notches: Ingen Housz could not have been unaware and must have been yet more anxious.

Then, over the weekend of 11 and 12 July, normal life in Paris imploded[123] as it became clear that there was going to be no more political progress towards a better life for the ordinary French citizen. Word spread that the King had dismissed Necker who had left for Brussels with his wife. The last hope of reform had gone with them. Commercial Paris put up the shutters. The streets emptied and the political leaders of the Third Estate, bourgeoisie and workers alike, saw that they needed to arm themselves if they were to have any hope of defending their rights, let alone improving their lot. Throughout Sunday 12 July there was a continual call to arms on the streets; troops were confronted by very angry crowds, and there was widespread looting, especially of gunsmiths' and armourers' shops. By nightfall all uniformed troops had abandoned the city-centre to its fate and to what was described as 'haphazard violence'.[124] As dawn broke on the 13th gunfire was heard and church bells were rung summoning 'citizens' to their 'duty'. Open spaces began to fill with a toxic mixture of rabble-rousers and lightly-armed but heavily vengeful Parisians. Throughout the day there was a disorganised search for arms and ammunition led by red and blue 'cockaded' ringleaders. It became clear, eventually, that the only substantial sources were Les Invalides Hospital, a home for retired soldiers that held a large stock of muskets, and the Bastille prison where the government had opted to store its gunpowder. Before 6 am on the 14th a vast crowd of

The storming of the Bastille prison, Paris, 14 July 1789.

determined 'citizens' gathered outside Les Invalides and demanded that the governor, Charles-François Sombreuil, issue to them the 40,000 muskets in his armoury. Sombreuil knew that he could not rely on his pensioners and capitulated. The crowd availed themselves, not only of the muskets, but also of twelve cannon and a mortar. Now all they needed was powder – it was in the Bastille. By mid-afternoon, the Bastille had also fallen, the totemic event of the beginning of the revolution. The seven very confused prisoners were released and paraded triumphantly through the streets. Far more sinister was that the stock of ammunition was unlocked and that the prison governor, the Marquis de Launay, was stabbed to death in the streets. His head was hacked off and hoisted on a pike. This, too, was paraded through the streets: the savagery of the French revolution had been released. How much of all this Ingen Housz witnessed for himself is nowhere recorded, but we have good insight into his personal experiences during these historic but brutal days from notes he made seven years later when in England and provoked to do so by a good friend.[125] The manuscript, obviously written at speed and repeated in parts, records his memories of a night of violently vivid dreams.

> After having, perhaps imprudently, recapitulated in my mind, during a whole day, all the horrors of the French Revolution . . . I found myself in bed with a whirl of thoughts, oppressed, and sank from reverie into rest when soon a horrid vision seized my head.

In his nightmare, events he had himself experienced in Paris became transposed on to the streets of London. He found himself, initially, bottled up within the home and shop of the tailor with whom he lodged near Golden Square and Piccadilly. It is an interesting eye-witness account of the realities of insurrection at street level even if it is only the record of a dream.

> I thought to hear a confused noise of people on the street, crying, hurling, fighting, running in every direction. By looking out of the window I saw a vast number of people, mostly working people, all with clubs in their hands, some with a dagger on the side, some with bayonets. I heard them say words such as 'no church, no king, no nobility, no parliament, no tithes, no taxes, no excise, . . . no property; all women in common' . . . I hastened to dress myself and found my landlord and his wife in the greatest consternation, weeping, crying 'we are all ruined, what will become of our poor children?' My horror increased by seeing my servant return saying, with shrieks, that there was no bread baked as all the bakers' journeymen had been out rioting the whole night. The maid came back with her face scratched, her stays torn to pieces, her bosom full of red spots and scratches: she cried out that several ruffians had handled her with the greatest rudeness and that she had only escaped because every one of them would have her and began fighting each with one another.

Ingen Housz dreamt on. He saw himself sit down to a makeshift breakfast of three eggs and gingerbread. He was forced to drink porter rather than tea but he was beginning to feel slightly stronger and less apprehensive when, still in his dream,

> I heard the journeymen tailors, almost all drunk, bursting into the house and crying 'revolution, revolution, money, money'. They said to my landlord 'we want our wages immediately as your house will be plundered, as all others.' He said he had no money in the house as nobody had paid him, as everybody hoarded up money. So they rushed up to the workshop crying out 'we will pay ourselves'. Soon after I saw them go out of the house, each with a new coat half finished, among which was one of my own . . .

Surreal juxtaposition and implausible circumstance are the seed corns of dreaming and in his nightmare Ingen Housz ventured out into the streets as an invisible witness. Nonetheless, his descriptions of what he saw have the ring of true experience and in the tightly constricted streets of eighteenth century Paris, it would have been difficult, in mid-July 1789, not to have been confronted by some very unpleasant sights, sights liable to etch their way into any sensitive psyche.

> I ventured to go out . . . I saw nothing but confusion, running, madness, distraction; every house was shut, a great deal of women without cap or hat, red-eyed, tearing their hair, running as furies over the streets, some laying down as in convulsions . . . I was struck with horror by seeing a vast rabble carrying a human head on a pike and a naked body dragged behind it, mostly by women crying out 'down with the King'. I wanted to fetch some money from [my] banking house in Charing Cross but I soon found it plundered and set fire to . . . that Newgate and all the rest of the prisons had been forced . . . that millers, coming to town with flour . . . had been met by the mob and plundered, that the most part of the flour was fallen on the road and some of it taken away in hats, handkerchiefs, pockets, etc. Here the excesses of the horrid scum shook my vital frame to such a degree that, happily, I waked from my dream which had, however, made such a deep impression on me that it was some time before I could distinguish whether the terrible scene was a reality or not.

All this retrospective distress, recorded in 1796, gives us a very clear understanding of why Ingen Housz had to leave Paris hurriedly in July 1789. In real life he was not invisible – far from it. His dress, his demeanour, his accented speech, his official status at the Austrian court, his links to the French Queen and his friendships with government officials in France such as Lavoisier; any one of these could have lost him his life in the fevered atmosphere of Paris after 11 July. Ingen Housz himself saw his indirect link to Marie Antoinette as his greatest danger.[126]

> . . . the horrid revolution, of which the bloody consequences put my own live in danger, as being body physician to the Queen's brother, made it prudent for me to fly the bloody scene.

On Saturday 11 July he had written to Brussels to request a passport allowing him to travel through the Austrian Netherlands (Belgium) to his homeland. This contingency had been too long delayed although the letter did manage to find its way out of Paris and reach its destination in what was probably one of the last unmolested diplomatic bags for some time. In Brussels a Mr. Crumpipen 'hastened' to prepare the document, joining it to his reply to Ingen Housz on the 16th.[127] By this time its intended recipient had fled from Paris. Ingen Housz left the city on Wednesday 15 July[128] and travelled by his own carriage, without the customary documentation, northeast to Brussels. Somehow, he engineered what can only be described as an 'escape' from Paris and he was very fortunate to have succeeded. From 13 July onwards many of the revolutionaries had descended on the detested customs houses that had been, since 1787, the tightly restricted access points to the city. They were the only breaches in the newly-encircling city wall built by the 'Farmers-General' (tax authorities), ironically, to a design of Ingen Housz's friend, Lavoisier. The rebels began to tear them down, such was the resentment they had generated, and it soon became impossible for traffic to pass unscathed. Ingen Housz squeezed his carriage and his party through one of them, perhaps by luck alone. But the Parisian selvedge was only the first of a series of difficulties. Outside the city lay many miles of growing uncertainty: the insurrection was not confined to Paris. Throughout the journey there was the constant anxiety that the carriage would be stopped and challenged by revolutionary elements.

Ingen Housz did not linger in Brussels where he arrived one or two days later. The atmosphere there was equally charged and he moved on very quickly towards Breda. Although he would have been thankful to have crossed the border into the Netherlands, any optimism was to be short-lived. On the very outskirts of Breda, as the tension of the journey was falling away and he began to relax in familiar surroundings, someone suddenly fired at his carriage.[129] Fortunately the ball passed through the travelling compartment high up, hitting no one. Whether it was a wanton, random shot or there had been a resentful Patriot with particular evil intent will never be known – the coachman would hardly have been likely to slow down or stop in order to investigate. But it certainly made clear to Ingen Housz that, even on his home territory, he was not out of danger. Nevertheless, he remained in Breda, with his bereaved sister-in-law and her family, for a month; presumably staying out of sight.

18
Retreat to England

He finds himself in the position he had seen so many patients adopt – lying on his right side on the bed with his knees up high under his chin. He can then retch into a bowl without having to move too much and hugging his shins gives him a brace against the pain. The spasms are ventral of the deep-seated rectal pain that is now non-stop. When the urgency to void is overpowering he just has to turn, sit and drag up his nightshirt. Nothing much happens bar a few dribbles of mucus but the sensation of a hot knife passing up the penis is electrifying. He is perspiring and gasping even before it happens, knowing it inevitable. When it's over, the rectal pain, forgotten in the crisis of the dismal drainage, comes welling back and draws him back into the fetal position. At least this time he hasn't retched at the same time as peeing. He can therefore resist taking a drink. Though he knows that taking copious fluids is the best policy – advice he's dispensed so often – it's inhuman to tempt yet more agony by filling the bladder unnecessarily.

He is wondering how long this can go on. It's now been 36 hours since the first pangs of backache turned into fixating agony and his worst fears were confirmed as he passed water that felt to be at boiling point. Since then he's been totally disabled: he has neither eaten, slept nor hardly spoken bar one-word requests to Dominique. He suddenly knows he must square up to his chamber pot again and, bracing himself against the expected spears of pain, is bemused as he feels a strange sensation from the testicles through to the tail and a one inch oblong stone spits into the pot with a spurt of bright blood. A gush of urine follows. He stares at the products of his tormented bladder but it's all a blur while he is still shaking so. By the time he's composed and his eyes focus again the blood has started to congeal and he notices that the surface of the stone itself is gritty; glistening where exposed to the air. He leans forward to put the pot on the floor. In coming up again he is light-headed and has just enough strength and self-preservation to push backwards fully onto the bed before passing out. An hour later he suddenly wakes from a dream in which he is wetting his breeches in public. It's too late; there's no time to adjust himself before passing a healthy volume of heavily blood-stained urine onto his shirt and the bed. Though embarrassing and far from normal, the experience is almost heavenly: there is no pain despite one or two obvious fragments of solid red material filtered out by the sheet. A wave of euphoria passes over him as he starts to undress. He realises how lucky he has been. He's passed his stone without having to face the torture of surgery. He takes a huge gulp of water, and another one.

T HE BRIEF TIME that Ingen Housz was to spend at Breda, in August 1789, with his widowed sister-in-law, Maria and her children, was a domestic respite and Dominique, too, making his second visit to the Eindstraat, would have been able to make himself at home. Ingen Housz's nine surviving nephews and nieces now ranged from twenty-three years down to ten and if they were demanding of their 'Uncle Jan's' attention, it would have been a pleasure, to him, to have such a homely focus. The daily pressures at Vienna, the terrors of Paris, the discomforts of his illnesses, the tensions of the recent journey; all could be forgotten as he appreciated how much Louis' children had developed since he last saw them all in 1779. He was able, once more, and this time to a larger audience, to regale them with stories from Vienna and from all his other travels. He was able to advise on their education and praise their progress. Arnold, the oldest brother, had succeeded his father as an apothecary.[1] Ingen Housz would have been able to initiate him into the 'new chemistry' that was gestating in Paris. The twenty-two-year-old Henry Ferdinand was now away at medical school in Leiden[2] – for him there were to be very specific pointers and encouragement and there was appropriate wise counsel for each of the others in the descending line of little Ingen Houszs. Most of all he was able to entertain, amuse, tease; in fact be avuncular and indulgent. There were more serious conversations with Maria. There were the family assets to consider: it looked likely that the investments they shared at Paris were about to evaporate. Then there was the demanding household budget now that Maria was bringing up the younger children alone and the family business was rudderless. Knowing of Jan's plans to travel on to England, Maria asked him to explore the best way to sell some paintings owned by the family in order to raise some cash:[3] the English art market seemed likely to pay the best prices and Jan agreed. The 'ghost at the feast' was, of course, Louis and some interviews must have been emotional, as Ingen Housz had feared. Nonetheless, there was probably more joy than sorrow – it was an intense family reunion, after a gap of nearly a decade, even if under trying circumstances.

Ingen Housz was well rested by the first days of September and pressing family business was concluded. He decided to make an excursion through Holland. Naturally, he made for Delft, greeting old friends Hendrik Hoogeveen and Jacob van Breda after what had been an intermission of nine years. Ingen Housz seemed to have regained enough strength and composure to repeatedly criss-cross the polders in order to meet these and many other acquaintances in the scientific and medical world of the province. In particular he was hoping to meet[4] Martin van Marum, director of the Teyler's Museum at Haarlem and therefore in charge of the huge new electrical machine installed there. Ingen Housz wrote to him from The Hague on Sunday 5 September.[5]

> I shall be coming to Haarlem tomorrow, Monday, by the evening boat from Leyden and I hope very much to be able to make your acquaintance . . . I dare only to ask you to grant me a few moments, at the least inconvenient moment for you, the day after tomorrow and that you would grant me the pleasure of seeing the famous Teyler machine.

The 39-year-old Martin van Marum was to become a particular friend of Ingen Housz over the next decade. Although two Christians widely separated by sect – van Marum was a Calvinist married to a Mennonite and Ingen Housz a backsliding Catholic – the two men shared a considerable list of interests, attitudes and experience. In many ways van Marum's life proceeded along lines that we can easily imagine that Ingen Housz might have followed if he had not gone off to London at the age of 33.[6] Van Marum had been born in 1750 in Delft and in attending the Latin School there he, too, had been a pupil of Hendrick Hoogeveen. Van Marum matriculated at the University of Groningen in the very north of the Netherlands, reading medicine and philosophy. It was the ancestral town of his family and he remained there after graduating, in 1773, to do postgraduate studies in static electricity. Again, following in the footprints of Ingen Housz, he also developed a disc electrical apparatus. He took up the challenge of trying to offset the frustrations of such machines: they were inefficient in damp atmospheres and the charge-generating friction pads wore down very quickly. He had some success and published his work in 1776 just as he was setting up in medical practice. This was in Haarlem, choosing the town because it was the home of a flourishing science society, the *Hollandsche Maatschappij der Wetenschappen* (the Science Society of Holland), founded in 1752. He was immediately engaged in the town's activities and its institutions. It was only a few weeks before he was elected to the Science Society and requested, by the Haarlem authorities, to give public lectures on science and be the

Martin van Marum FRS.

director of one of the Society's 'cabinets' (collections). In 1781 van Marum married Joanna Bosch. She was the only surviving child of Jan Bosch, the printer of the Science Society documents. Bosch had been a widower for many years and Joanna had dedicated her youth to looking after her father. On marriage to van Marum, a few months after her father's death, she was nearly eleven years older than her husband, and in poor health. However, she did bring a large inheritance to the marriage enabling van Marum to give up medical practice, concentrate on scientific experiments and spend long annual summer vacations travelling around Europe. Then, within three years, his standing in the Haarlem social hierarchy was totally assured.

Pieter Teyler van der Hulst, a Haarlem cloth merchant who had also owned a bank in Amsterdam, one of the richest men in the town, left to his community a number of generous endowments by the terms of his will when he died in 1778.[7] One of these was *Teylers Sichting (Teyler's Foundation)*, embedded within which was a generous bequest specifically for the founding of a *Tweede Genootschap (Second Society)* intended to study science, coins, history and drawing. By 1784 the rear of Teyler's house had been extended and re-ordered into the famous Teyler's Museum and it remains one of the most famous and oldest scientific institutions in the world. The doors opened in 1784, by which time van Marum had been elected the museum's director. His first initiative was to make it a science stronghold. He ordered the construction of a huge electrical machine, a disc generator with everything on a gigantic scale; its discs 65 feet across. Van Marum's concept was that the strong discharges from the machine would give new insights into the nature of electricity, but he was to find that not all behemoths are as powerful as one might expect. Nonetheless, this was the apparatus that Ingen Housz was wishing, particularly, to see; hopefully in action. His introduction to van Marum's persona, personality and contacts was, however, to prove to be the more significant experience. Within hours he had made a new friend for life.

From Haarlem, a contented and intellectually reinvigorated Ingen Housz made his way into the city of Amsterdam. It took little effort to reunite with Paets van Troostwyk, a man with a well-earned reputation as an amateur but dazzling scientific experimenter. He introduced Ingen Housz to Johann Deiman, his close friend; an Amsterdam physician who, like Ingen Housz, was renowned for employing, safely and successfully, the inoculation lancet. Together, van Troostwyk and Deiman had lately pioneered an elegant experiment[8] that proved to be a vital piece of the jigsaw for the 'anti-phlogiston school' busy assembling the 'new chemistry'. Their leader, Lavoisier, had been able to show, since 1785, that hydrogen and oxygen could be extracted from superheated water (steam) passed through a red-hot iron gun barrel.[9] He, and others, had also been able to demonstrate the reverse, the condensation of water when a mixture of the two gases was sparked.[10] To all appearances, no other substances were involved. However, there were vociferous and powerful sceptics who still favoured the existence of phlogiston. Joseph Priestley was certainly unconvinced, arguing that the processes in Lavoisier's experiments were more complicated than the 'progressives' realised; essentials from the metal in the apparatus, and certainly the all-pervasive phlogiston, all being involved. By every appearance, Ingen Housz, after almost a year in Paris, had retained a few doubts[11] until, that is, he spent the few days in Amsterdam in that second week of September 1789. Van Troostwyk and Deiman needed no persuasion to repeat their experiment and, in assisting them,[12] Ingen Housz was able to see for himself the clinching evidence that phlogiston was a conceptual dead-end. His Parisian friends, pushing through a revolution in chemistry that was, in its way, every bit as important as the one being played out in

French politics, had been right. Their new elemental syntax was justified and though, after years of moving in diplomatic circles, he did not become an evangelist, Ingen Housz was finally convinced. He was able to see how his two new friends had adapted a narrow glass tube, about a foot long, so that it was sealed, by heating, at one end. While the glass was molten they had introduced a platinum wire that projected some distance along the bore of the tube. Elsewhere on their bench they had used a vacuum pump to extract all the air from some distilled water. This air-free water was then used to fill up the tube. Having inverted the tube over a dish of mercury, they fed another strand of platinum up into its lower end so that the two wires were about half an inch apart, point to point. They then created an effective circuit through the mercury, a substantial Leyden jar, a powerful disc electrical machine and back to the wire sealed through the top end of the tube. When the machine was cranked rapidly bubbles appeared in the water in the tube and rose to the top. Then, as the water level dropped to expose the tip of the top wire, sparks were seen and the collected gas disappeared: the tube was suddenly full of water again. The process could be repeated at will, just as long as the experimenters had the combined energy to crank the machine. Firing sparks through gas mixtures had been tried many times by many investigators and with many gas combinations. Only hydrogen and oxygen reacted totally in this way, to form water.

Here was a compelling demonstration that water could be broken down into just two elements, hydrogen and oxygen, and then reformed by combining the gases in the right proportions. No heat had been used so that phlogiston, the supposed element of fire, was out of the equation and there was no contact with any other substance except the inert glass. Van Troostwyk and Deiman probably had some beginners' luck for when others[13] tried to repeat their experiment they found it very fickle. The calibre of the tube, the power of the machine, the capacity of the Leyden jar and the precise links in the circuit could, each and all, scuttle the effect but perhaps, for all their inspiration, they deserved their good fortune. Ingen Housz, for his part, had shown immaculate timing once again: van Troostwyk and Deiman had already submitted their seminal paper to respected journals but nothing was yet published and their visitor was among the first with the news. Ingen Housz completed his own 'circuit' and returned to Breda a day or so later,[14] again one step ahead of other *savants* and his head bursting with ideas.

Once back in his home town, Ingen Housz was confronted by less arcane matters. The family had become aware of mutterings about his presence and, after the recent events in Paris, the Dutch revolutionaries, the 'Patriots', were resurgent.[15] Ingen Housz sensed a significant danger for himself, personally; he had already been in someone's sights. But his 'royal' presence in the Eindstraat was also a potential threat for Maria and the children – they could become a scheduled target for anarchic forces if there were a further insurrection. There was clearly the need for an early and expedient decision – that he should turn a tenuous ambition into a firm travel plan and move on, to England. Dominique re-packed their essential luggage: it seemed sensible to travel by public transport. A personal travelling coach made one conspicuous; unwise in restless times, as he had already discovered. And since he was going to cross to England, the vehicle would be safer at Breda, under supervision of his family, than at a port on the North Sea. Hence the coach was left in the town for later retrieval and the two travellers moved on, in the first instance to Rotterdam,[16] where he was among known friends in early October, particularly Salomon de Monchy and Lambertus Bicker. It was Bicker, a physician and fellow inoculator, who first introduced Ingen Housz to a new prescription for treating bladder stones that was to loom large in subsequent weeks and months and, probably, save his health. This was *aqua mephitica alcalina* (alkaline soda water). Bicker showed him the source of his knowledge; a book published by William Falconer,[17] an English physician at Bath. Bicker already had a mental file of some case histories that supported the efficacy of the treatment, successes that Ingen Housz was later to report, in turn, to Falconer.[18] Ingen Housz was easily convinced that he should try the remedy for himself but it is not clear whether there was enough of it available in Rotterdam for him to start regular treatment. He certainly saw it as likely to be an improvement on Count Carbury's recipe although he reflected that this had had 'some salutary effect in checking the violence of such a complicated indisposition'.[19] He saw some hope of an actual cure of his renal problems with alkaline soda water and was euphoric about its instantaneous effects.[20]

> . . . it was to my palate the most agreeable thing I ever tasted, exciting, besides its truly delicious taste, a most enchanting sensation when it reached the stomach, which, like a true 'Nepenthe Helenae'[21] pervaded all my limbs and produced a new and desirable sensation of the most pleasurable kind.

INGEN HOUSZ AND Dominique crossed to England in October 1789, making landfall at Harwich.[22] The Essex port again evoked memories of 1768 when he had been marooned in the town for an anxious week. The remainder of the journey, from the Essex coast to London, was also familiar and would have taken another day or so. For Ingen Housz himself it became increasingly uncomfortable. Once again there was something solid wambling around in his bladder, forcing him to clench his teeth against the pain with every lurch of the coach. And just hours after arriving in London, he passed another two stones.[23] They were smaller this time, 'very hard, and composed of shining crystals' and followed by the familiar 'red sand' pebble-dashing of the chamber pot.[24] They were incentive enough for him to search, urgently, for his old friend George Baker. Now a knight and one of the King's personal physicians, Baker prescribed, as Ingen Housz had hoped, regular doses of alkaline soda water, kindly providing his patient with a pint bottle of the remedy[25] so that treatment could begin immediately. The consultation with Sir George was equally important for catching up on news.[26] There was a prompt improvement in Ingen Housz's condition and by the first days of November, he was devising his own method for making the alkaline soda water[27] using simple apparatus that could be set up in his lodgings. He needed only the slightest excuse to extemporise a laboratory bench. Otherwise, he was soon on the move around London.

Sir George Baker
FRS. Courtesy of
the Wellcome Library,
London.

There were no rooms available at any of his previous London lodgings although Dunn and Taylor, the linen merchants, were very happy to be his usual convenient *poste restante* address at 6, Fleet Street.[28] A friendly prompt sent him to enquire at 51, Jermyn Street in Westminster, 'on the corner of Duke Street', a stone's throw from Piccadilly.[29] He was in luck and rooms here were to be his London home, and laboratory, for the present. Ingen Housz did not record his accommodation costs but he was to quote the going rate for rooms in central London to van Marum a few weeks later.[30]

> One usually pays a guinea or a guinea and a half for two rooms. A servant girl will prepare tea etc. for which she expects, now and then, a small tip such as four or five shillings a month. In some of these lodging houses one can eat with the family for two shillings a day. In the soup kitchens and taverns it costs two or three shillings; five shillings per head in the best ones.

It was a comfortable stroll down to Pall Mall to pass the house of the late John Pringle, an edifice so redolent of earlier times in London, and then on east, into Charing Cross to his bank, Drummonds. Here he was pleased to find that they still held a ready balance for him of £921 (£52,000),[31] easily enough to live comfortably for many months, even in expensive London. It was certainly more than sufficient to last him until his intended return to Vienna, now likely to be summer 1790. For immediate needs he cashed £50 (£2,800), a withdrawal sum that would be repeated four times over the next year.[32] He was also given the details of the dividends the bank was receiving for him from his considerable investment in British government 3% stock – some £120 (£6,700) a year from bonds having a face value of £3949 (£221,000).[33] Despite the healthy condition of his cash balance he arranged for 400 of the bonds to be sold, realising just over £313 (£17,500) on 11 November. Perhaps he felt happier to have more ready money to cover unforeseen eventualities and with the very uncertain situation in France even British government stock looked much less secure. He was also acutely aware that, in London, no less than in Vienna, there was no escape from accelerating inflation. The purchasing power of his pension had dropped by nearly 40% since 1770 and although living modestly in London, he was still maintaining an expensive home in Vienna where Agatha remained with their domestic servants.

Having organised his health, his accommodation and his finances, Ingen Housz was now free to take up his life in London. He had still many living contacts, many old haunts and many familiar habits. It was a decade since he had last been in England and events in the world, personal, political and scientific, had all taken on a higher tempo. It was to take some weeks to reconnect with all his English acquaintances. He had a long list of contacts but was anxious not to be peremptory:[34]

> Although I have lost many of my old friends through death, I have found so many of them alive . . . that, up to now, I have not been able to see, or speak to, more than half of them. A single visit normally takes one whole day.

We can assume that Ingen Housz was impatient to see Thomas Dimsdale again. Dimsdale was still alive and living in Hertford. The Dimsdale Bank, a family venture at 50, Cornhill, in the very financial heart of the city, was run, day-to-day, by two of his sons, but their father would have been in London from time to time since he was now a Member of Parliament. However, there is no indication that he came into London that autumn[35] and this reunion may have been delayed. Much to Ingen Housz's surprise, he discovered that someone even closer was at hand. Of all the million people then inhabiting London, there was one person, another visitor, that Ingen Housz sought out immediately he learned of his presence – his young nephew by marriage, Josef Jacquin. Unexpected circumstances had brought them together in England but the fondness and familiarity that would now grow between them was to be extremely important.

Early in 1788, the 22-year-old Josef Jacquin had completed his formal education, proving his high intelligence and aptitude for an academic career. In Vienna that spring, while his 'Uncle Johann' had been contending with the aggravations of commissioning a reliable travelling coach in order to visit Paris, Josef had been planning a more adventurous trip. As a postgraduate course, he was to tour much of Europe in order to meet many of the scientists. It had been encouraged by, and sponsored by, the Emperor himself,[36] perhaps out of gratitude for all the devoted service of Nikolaus Jacquin or perhaps because he saw, in Josef, a gilded future in the university and at court that would be burnished by travel. Franz Bauer, a talented botanical artist under the wing of Josef's father, was invited to accompany the young man for it would obviously be an advantage to travel with someone who could make accurate drawings of the phenomena they might see. Bauer accepted the honour, partly, perhaps, because it gave him an opportunity to see his older brother, also an artist and working in England for John Sibthorp, professor of botany at Oxford University. Abbé Leonhard Gruber, a priest who was well-known in religious education circles in Vienna, also joined the party: middle-aged, he was already well-travelled and was seen, presumably, as a source of guidance, moral if necessary, for the two younger men. Before arriving in England their journey was to take them through lands now known as the Czech Republic, northern Germany, the Netherlands, and

Belgium. Then, after London and elsewhere in England, they would head home through France and finally Italy. They left Vienna on 4 May 1788 heading anti-clockwise around the Continent and crossed to England six months later. Franz Bauer was reunited with his brother at Oxford but most of their time was spent in London. Josef, accompanied by Abbé Gruber, was a guest at an evening meeting of the Royal Society on Thursday 27 November and he subsequently attended fifteen such meetings before the summer recess of 1789. All three Austrians met the President, Sir Joseph Banks, and were private guests at his house, admiring the 'largest herbarium in the world'. After six months in England it was probably the intention of the party to cross to France during the summer of 1789 but they demurred because of the outburst of political turmoil there. Paris, particularly, could have been dangerous for them, travelling, as they were, by way of a royal warrant. Whatever the circumstances, they remained in England for a further six months and, fatefully, Josef Jacquin was still in London when his 'uncle' arrived in mid-October.

The Royal Society had moved from Crane Court in Fleet Street in 1779. It now had apartments, including a very stately meetings room, at the purpose-built Somerset House in the Strand. It was an early port of call for an admiring Ingen Housz for he was hoping to see the secretary, Charles Blagden and ask to borrow recent editions of the *Philosophical Transactions* from the library.[37] Characteristically, Ingen Housz was anxious not to be caught unaware, in conversation, of the latest scientific developments. Blagden said that he would need to seek a licence at the next Council meeting[38] and Ingen Housz had to be patient for a few days before permission was granted.

> Leave was given to . . . Dr. Ingenhousz . . . to have such volumes of the *Philosophical Transactions* as not yet received, under the usual limitations of Council.

Fellows often had reasons to visit the Royal Society outside the evenings of formal meetings and Ingen Housz encountered Alexander Dalrymple, an acquaintance from earlier times in London, another protégé of Sir John Pringle. Dalrymple promptly invited Ingen Housz to be his guest at the next Royal Society club dinner on the afternoon of Thursday 29 October.[39] These were still being held at the Crown and Anchor Tavern, off the Strand, and there Ingen Housz was able to reintroduce himself to several old friends at once and in very congenial surroundings. However, a more singular occasion was to follow. The surprise reunion between Ingen Housz and young Jacquin was obviously a joyous event but the evening of Thursday 5 November was to be particularly memorable. During that afternoon,

Ingen Housz dined again with the Royal Society club members, this time as a guest of Sir Joseph Banks, the President.[40] Then, as darkness fell, he left the tavern with the other diners and walked the few yards west along the Strand to Somerset House to await the arrival of Josef and Franz Bauer who were to be his guests, in turn, at the evening meeting of the Royal Society in Somerset House.[41] Introducing Josef around the meetings room would have given Ingen Housz considerable pride and pleasure and when he was called on to make a formal presentation, to the Society, of one of Nikolaus Jacquin's books,[42] it was especially an honour to do so in front of the author's son. He also formally presented a copy of his own book, *Nouvelles Expériences II*. It was all so very different to that evening, a quarter of a century before, when he had shadowed the late Matthew Maty into a parallel meeting as a confused and overwhelmed guest with barely a word of English to help him understand proceedings.

For Josef Jacquin, the evening of 5 November had been a suitable grand finale for his stay in England but there was to be an unexpected and unwelcome coda before he left for France a few days later. Joseph Banks was in the throes of cataloguing and publishing all the plants at Kew – the famous *Hortus Kewensis* – and saw that Franz Bauer would be extremely useful in this enterprise for it desperately needed an 'accomplished botanical illustrator'.[43] He was so determined on the young man that he secretly offered him a salary, for life, of £300 (£17,000) per year. Bauer found the offer difficult to resist – it was more than three times the salary his brother was receiving at Oxford and Banks had been extremely flattering and persuasive. Josef had some inkling of an undercurrent but remained optimistic enough to book three seats on the coach to Dover. On their last day in London, however, Banks invited Bauer to dinner once more and the deal was struck. Josef and Abbé Gruber set off for Paris without their compatriot. Austria's loss was to be England's gain, and Francis Bauer, as he took to calling himself, was to become a worthy and famous figure in the history of Kew Gardens and its pictorial archives. He even served as art tutor to the Queen and some of the royal children. He was to remain in England until his death in 1840.[44]

Ingen Housz was also left in London, renewing his friendships and making new ones. Among the latter was a fellow of the Royal Society earning a justifiable reputation in astronomy, William Herschel. Herschel was a Hanoverian who had settled in Bath in the 1760s as a career musician. Telescopes and stargazing had taken over his life, however, and that of his sister, Caroline. By 1789 the two of them lived in a house at Slough, more working observatory than a home, and paid for by a large grant personally

Dr. William Herschel FRS. Courtesy of the Wellcome Library, London.

arranged by the King himself – who would often visit from nearby Windsor. Ingen Housz immediately craved an invitation to Slough and he spent three to four days with the Herschels, arriving there on Saturday 21 November.[45] Their house was easily found for an idiosyncratic reason: immediately behind it, and towering over it, stood huge wooden scaffoldings that supported the metal telescope tubes. Staying up late into each night with his host, thankful for clear autumn skies, Ingen Housz found the telescopes to be perfectly engineered and enormously powerful. He was overjoyed to view Uranus, the new planet discovered by Herschel in 1781, and to have unique and clear views of Saturn with her rings and moons.[46]

> The largest telescope is perfect – it is 40 feet long . . . the second is 20 feet long. Its large concave mirror reflects so much light that one dare not look at the moon directly for one runs the danger of being blinded for some time afterwards. The mirrors this man uses are made of an unusual metal, far superior to anything else, and it can be filed. I have seen all seven satellites of Saturn, even the two newly-discovered ones on the ring. I have seen several nebulae . . . at present a list of 2000 new nebulae discovered by Herschel

is being printed, each of them a system of countless worlds. Our own sun, with all the stars we see with the naked eye, form only a single such nebula. Our powers of imagination are too weak to comprehend such an almighty universe.

Back in London, an early call was to Peter Elsmly who was holding a small balance for copies of *Vegetables* sold from stock and Ingen Housz used this to offset some substantial purchases, spending almost £14 (£785). He went repeatedly to the shop of George Adams, the famous instrument maker, in Fleet Street on behalf of Martin van Marum. In his role as director of the Teyler's Museum in Haarlem, van Marum had ordered several instruments, from Adams, through the good offices of Jean Magellan, a Royal Society fellow. Van Marum was impatient for them, however, and there had been a considerable delay. Ingen Housz discovered that the 67-year-old Magellan had recently slipped on some ice, had taken a heavy fall, and was now lying incapacitated. He was so profoundly disabled that two named executors of his will were already handling his affairs.[47] Ingen Housz was able to retrieve the situation and his friend's items were soon packed off to Haarlem.

On Wednesday 4 November, only days after arriving in London, Ingen Housz was to take up the first of a long series of invitations to dine at Lansdowne House in Berkeley Square, the London house of the First

Herschel's forty foot telescope on its scaffolding.

Marquis of Lansdowne. Although the home of a politician who was officially 'retired', the Marquis had served, as Lord Shelburne, as 'Prime Minister' in 1782/3 and had been instrumental in settling the American War of Independence. His homes were still the epicentres of much social and political life. Lansdowne and Ingen Housz had known each other since the early 1770s but now became very close friends: the affiliation was two-way and entirely genuine. Perhaps the kindness and hospitality that Ingen Housz had shown to Lansdowne's son, Lord Wycombe, when the latter had visited Vienna in 1785, was pertinent. Sitting down, then, at table with Lord Lansdowne in the dining-room of Lansdowne House in early November 1789 had huge significance for Ingen Housz.[48] It was a prominent milestone in his life, being the introduction into an important circle of English Whig nobility and he was to become, very rapidly, one of the most cherished of Lord Lansdowne's immediate circle of relatives and friends. He would be a dinner guest at Lansdowne House eighteen times more in his first six months in London[49] and it was not long before there was an insistence that he join Lansdowne's family and friends at the customary autumn house party at Bowood House in Wiltshire. The other diners on the first occasion that Ingen Housz dined with the Marquis were typical guests. There was Benjamin Vaughan, one of Lansdowne's close advisers and companions who had been invaluable as a go-between in the negotiation of peace with the Americans; Jeremy Bentham, the jurist, philosopher and social reformer; and the young Samuel Romilly, who would become a famous reforming lawyer and Solicitor General. Ingen Housz was obviously a diverting and entertaining companion: comfortable in English society, he could call on many experiences and anecdotes from all corners of culture and spice them with many flavours - Gallic, Germanic, Batavian and Bohemian. He was fluent in four languages and could quip in Latin or Greek. He was always abreast of the latest news, especially those developments that we would now call technological. At the same time he would not have been overawed by the resplendence of Lansdowne House, being used to palatial surroundings, and his long service at the Hofburg had given him all the interlaced skills of consideration, discretion and obeisance that would have been sensed and valued by his host. Lansdowne had recently been widowed, for the second time: the Marchioness had died only three months before, aged 34. Bereavement can be at its most painful after such a time lapse and Ingen Housz would have had the sensitivity and experience to be particularly empathetic to his friend's grief.

With the renewed friendship of Lord Lansdowne came acquaintances with other, younger, members of his household. The late Lady Lansdowne had been born Louisa Fitzpatrick, one of four children of John Fitzpatrick, first Earl of Upper Ossory. He had died in 1758, when Louisa was only three, and their mother had remarried a year later. The stepfather was an untitled but very affluent gentleman, Richard Vernon. Together, the Vernons had three daughters, Henrietta, Caroline, and then Elizabeth. Hence, Lady Louisa had thee half-sisters and she was particularly close to Elizabeth. With young deaths being common in the eighteenth century, even among the aristocracy, family relationships could become complicated and extended groups the norm, as here. Louisa's sister, Mary Fitzpatrick, had married Stephen Fox, second Lord Holland, but their two young children, Caroline and Henry, had been orphaned. Caroline (a niece of Charles James Fox, the politician) was taken in by various relatives but then settled with Louisa who brought her up. Strangely, though almost a contemporary, Caroline was Elizabeth Vernon's step-niece but this didn't prevent them becoming close friends – for life. They both grew very fond of Louisa and remained her companions after her marriage to Lord Shelburne in 1779. When Louisa, now the Marchioness of Lansdowne after her husband's promotion, became ill, in 1789, Caroline and Elizabeth were both in their early twenties and Lady Louisa's terminal condition seemed to bring the three ladies closer together. It was a bonding so comforting to the dying Marchioness, that Lord Lansdowne, trying, in his grief, to keep his household together and demonstrate his gratitude, encouraged Elizabeth and Caroline to remain in his protection. They were invited to live, mostly, at Bowood House in Wiltshire, and at Lansdowne House in London. Elizabeth Vernon and Caroline Fox were thus soon acquainted with Ingen Housz.

On Monday 25 January 1790 they were all at dinner at Lansdowne House.[50] Ingen Housz may have impressed the Marquis considerably but his 'effect' on the 'ladies' was remarkable, and especially on Caroline Fox. She was obviously a highly intelligent and inquisitive being with a playful and wicked sense of humour. Ingen Housz became her living 'dictionary, to which I always had access' and the two of them, in discussing 'philosophy', cemented a considerable friendship[51] in some of the most pleasant surroundings in Georgian England – at Bowood House in Wiltshire.[52] These connections opened up other country houses and estates to Ingen Housz. For instance, Henrietta Vernon had married George Greville, second Earl of Warwick in 1776, and it was not long before Ingen Housz was also a guest of the Warwicks.[53]

IN A FREE-WHEELING spate of errands, visits and social events, dinners and meetings, Ingen Housz developed, then, a new life-style in London as the winter of 1789 passed imperceptibly into the spring and summer of 1790. Having no clinical or administrative responsibilities, he had the luxury of using his time entirely as he preferred and as his moods took him. Not that time lay heavily on his days; he was as energetic as ever, surprisingly so for a man still suffering echoes of recent illnesses and who was, anyway, entering his seventh decade. One central role, self-adopted and amateur in the truest sense, was that of academic 'sponge', absorbing information on a huge scale. He also kept up his expertise in medicine, visiting many of the top London physicians and surgeons. His sources were direct, personal and from every drawer of English society. He began to make regular entries in a loose-leaf commonplace book[54] that would grow, eventually, to encyclopaedic function as well as proportions. He summarised printed books that he read, recorded the views and opinions of his many intellectual contacts, and listed new scientific and commercial developments, new prescriptions and new surgical treatments. He personally traced much innovation back to its sources and personalities to overcome the fact that many useful discoveries did not gain the exposure they merited and which the advance of science deserved. In the face of these difficulties, and without bearing the epithet in any official sense, Ingen Housz became an 'emeritus', a fount of academic knowledge and experience and he made it his business always to be up-to-date. He aired his problems in a letter to van Marum in early 1790.[55]

> It is difficult to have a complete grasp of everything one hears because people don't usually speak openly about their own experiments and it is unbecoming to reveal another's work without proper consent. Hardly anyone hears more news than I do but it's never as complete as the author would give out himself.

He was regularly advising Josef Jacquin, now well ensconced in Paris, at his uncle's old lodgings.[56] There were scattered snippets of personal advice that clearly arose out of an anxiety that the young man might provoke the malignant interest of the revolutionary elements in the city but, in the main, the pointers were all educational and supplementary to the many letters of introduction Josef had taken with him. The young man was to be sure not to miss the lectures on 'airs' given by Jacques Charles,[57] the course in chemistry given by Jean D'Arcet, the demonstrations of his ingenious experiments by Bertrand Pelletier, and the skills of Marc Étienne Janety, the metallurgist who 'worked well' with platinum. Above all, time spent with Antoine Lavoisier was never wasted. In return Josef earned his advice. He found himself shopping for his uncle, for good quality manganese – there was none 'worth having' in London, and for black Martinique wax – 'so useful for joining pieces of glass tubing.'[58]

It would be a mistake to think that Ingen Housz had withdrawn from the workbench to be a passive observer of science, a mere spectator. His exploratory drive, his hands-on aptitude, had not left him. As an instance we can examine his behaviour after buying, sometime in early 1790, one of the letter-copying presses, invented by James Watt, and on sale in many shops in London. It had been Erasmus Darwin who had,[59] unwittingly, driven James Watt into designing his letter copier. Darwin had resurrected and much improved an apparatus consisting of linked rods, the pantograph – as a writer moved his pen, another loaded pen described the same movements over a parallel sheet of paper, producing a second version of the script. Watt hated being 'out-invented', especially by one of his most valued friends but he saw the pantograph as flimsy and faulty. He also realised, at once, how commercially desirable such clerical equipment would be if it could be made to be robust and reliable. Within a few months he had constructed a prototype 'copying press' based on a different principle. A wet sheet of thin absorbent paper, which we would now describe as tissue, was pressed on to a manuscript mopping up the excess ink to create a mirror image on the near transparent folio. After drying this, the copy image showed through and could be read from the other side. By comparison with Darwin's machine, it produced a blotchy script, involved using messy chemicals and was much slower. But with Matthew Boulton's encouragement Watt vigorously marketed his invention and it was soon profitable. A decade later Ingen Housz found he could buy one for 7s. 6d. (£21) and had no hesitation in taking it apart and tinkering with its components.[60] He was certain that he could improve it and, having dissected it, engaged a joiner to produce various adaptations to the timber framework and roller.[61]

Ingen Housz found that he was still having to defend his scientific reputation. Being back in London, he was to realise that Priestley's illegitimate claim to have been the person who first showed the importance of sunlight in the function of leaves still held credence. The fiery priest and amateur scientist was still held in awe in England. After some months of dismissing, as a trifling irrelevance, the repeated innuendo that Priestley held sway in this argument of precedence, Ingen Housz thought it time, yet again, to challenge his former friend. He was still aggrieved that Priestley had omitted, once more, to set the record straight in a third volume of *Experiments and Observations*, published in 1786.[62] Using the pretext of opening a debate on

Lavoisier's experiments on the composition of water, Ingen Housz wrote to Priestley in mid-1790.[63] Within the first 150 words, however, he was issuing a challenge, albeit in a subtle and diplomatic manner.

> I was, at the first reading of it (the VII section of your third volume), some what puzzled and thaught the contents of it might induce the reader to believe that you, and not I, were the first who had published that it is the light of the sun which is the cause why plants correct bad air and yield vital air tho' all those of our common friends whom I consulted on this head are of the opinion that such an assertion is by no means intended by you. If you have really published this doctrine before me I owe you the justice to acknowledge it publikely . . . and will very readily quote the volume of your works, and the page, in which you will indicate to me this doctrine is clearly and explicitely to be found. But if this doctrine was first publish'd by me, as I have till now been persuaded is the case, I will leave things as they are. The confidence I have in your liberal manner of thinking makes me hope you will not leave me long in this state of uncertainty, and that you will favour me with an answer.

He did not have to wait long for a reply but when it came it resolved nothing; Priestley was as 'Jesuitical' as ever, if not wholly disingenuous.[64]

> The discovery about which you seem to be solicitous . . . was first *published* by you, tho' *made* by me about the same time . . . (and) for your sake I wish the merit of the discovery had been greater. For me, philosophical matters are no great object. (authors' italics).

There seemed to be little mileage in taking the controversy further and Ingen Housz let it rest except for a vague plan that he might publish a second edition, in English, of *Vegetables* and take, therein, the opportunity to challenge this 'far from generous, oblique and insidious manner of behaving'.[65] In fact, the standing of Priestley was about to take a fall anyway, though for reasons entirely removed from the laboratory.

All through the spring of 1790 there was, for Ingen Housz, a pendulum swinging increasingly towards the many attractions of life in England. There are many inconsistencies and, perhaps, some mendacity in Ingen Housz's correspondence, certainly with his nephew, Josef Jacquin, after the younger man had left London for Paris. Ingen Housz's original plan – of spending late summer and autumn of 1788 in Paris and then winter in London before returning to Vienna in spring 1789 – had been totally demolished by his illnesses and by the French Revolution. However, by spring 1790, there seemed no reason for him not to travel back across Europe and return home. France, after all, was easily avoided and many regular travellers were still moving around Europe unmolested, unaffected even, by the recent political upheavals. Ingen Housz was apprehensive though. He seemed to oscillate, in correspondence, between two reasons not to travel; each was given credence and precedence in turn. There might be a counter-revolution in France and a regenerated political storm could easily cross national borders putting him in 'the way of harm' whichever route home that he might choose.[66] The other reluctance to travel stemmed from his poor health.

> It is certain that I have biliary concretions, which can end badly. Of that which I fear the most, it is of falling ill in some bad way on the road. If I was sure that that wouldn't happen I would leave the next day for Vienna.

His full recovery was, to him, a necessary contingent to long distance travel but this was going to be very difficult to define with confidence; he was still having symptoms from his gallstones; 'terrible' pains and jaundice, even, for a few days.[67] By May, however, his malaise, as a reason not to travel, had been put aside. He was excitedly planning an even more hazardous journey,[68] on the face of it, having resolved

> to make an excursion to America and visit my old friend Franklin. When I set off on the voyage I will go from here to Falmouth on 23 or 24 June to cross to Halifax and New York with the packet that departs in the first week of every month and come back again with the same transport in the coming May . . . the sea may do me some good or take me to the depth with a stone at my neck.

Unhappily, Franklin was already dead. He had led a long life of 84 years, varied, productive and fascinating, but time had eventually caught up with him during that spring. He had succumbed to a lung abscess, in bed in Philadelphia, late in the evening of Saturday 17 April.[69] The sad news, on its arrival in England, inevitably put paid to Ingen Housz's chancing his life on the ocean although he probably grieved more for his old friend than for his schedule. America was now an empty ambition and having given notice at his lodgings in Jermyn Street, Ingen Housz found himself looking for new accommodation in London. There was no immediate crisis, however: he was going to be out of London throughout the summer. By July he had many plans, none of them involving Vienna. As he told van Breda: 'I intend to make, soon, some long excursions through this island and seldom come into London for more than one or two days until October'.[70] This conflicted, however, with the information he was giving the family via Josef Jacquin. By mid-August he was hiding behind both familiar excuses not to return home immediately; the political uncertainties on the Continent on the one hand and his recurring ill health on the other.[71]

I am not without hope of finding myself in Vienna before November and pass the remaining days of my life where I have all possible comforts. I am holding, always, everything ready to leave as soon as peace is restored . . . so that I can subject myself to the risk of becoming ill on the journey. I haven't yet, however, been freed from all fear of obstructive jaundice.

Ingen Housz continued to lose friends and contemporaries; no real surprise for a man approaching his sixtieth birthday. It was giving him a somewhat fatalistic approach to his own future, putting him in mind of his great 'friend', Horace: '*Seu me tranquilla senectus, seu mors atris circumvolat alis.* Either a peaceful old age awaits me or death flies around me with black wings.'[72]

The death of Franklin had come very shortly after that of another man who had had a huge impact on Ingen Housz's life. During the summer of 1788 Emperor Joseph II had led his army into the 'field' in the war that Austria and Russia sued against the Turks. The campaign had been far from successful and for Joseph personally the harsh disappointments and primitive living conditions seemed to accelerate his tuberculosis. By the end of 1789 his health was destroyed and his reforming administration had stagnated. He struggled on with the administration of his empire even though he knew that death was near; it was the worst-kept secret in Europe. There was to be no miraculous recovery for the Emperor: he was not to be so fortunate as his senior body physician. The arterial blood that Joseph increasingly coughed up from his lungs from late in 1789 signalled the terminal phase of his decline. He died at the Hofburg, aged only 48, at 5 a.m. on 20 February 1790.[73] Ingen Housz would have received the news in London – we can assume about ten days later. Presumably he was not surprised but his reaction was, ironically, cold-blooded – what might be described, in fact, as 'Josephinian'.[74] 'The Emperor being dead, we shall soon find changes in the government. I do hope they will be for the better.'

The responsibilities of the empire fell on Leopold, Archduke of Tuscany since 1767, but it was more than a month before he arrived at Vienna, and he took up the Seals of Office with a transparent reluctance.[75] There was some respite, therefore, before a looming anxiety for Ingen Housz became acute – the future of his pension. Granted by Maria Theresa and ostensibly for life, it had been honoured in full by Joseph but now future payments were less certain, as was his very status at court. In fact, his income remained secure – for the time being. It was, though, a long-term worry that would come to haunt him.

Ingen Housz was not always a guest in London: sometimes he was host and guide. On 31 May 1790 he wrote to Haarlem. He owed van Marum an update on several pieces of apparatus ordered for Teyler's Museum.[76]

> I have been a possible 20 times on purpose to Mr. Adams to have him manufacture your eudiometer. His labours have always disappointed – I always found something to be wrong. Now one is ready at last.

He had also been scanning London accommodation for van Marum,[77] who was hoping to visit England.

> It is best first to stay in London in a hotel . . . one pays for each room with a bed two shillings and a half (half a crown); one of the best hotels for you is Bates in the Adelphi in a street called the Strand. Most of the hotels are now full because of the multitude of French who still arrive here steadily.

Ingen Housz had also had worries about the suggested timing of van Marum's trip, for he had imagined that he would be in Philadelphia by July, the month van Marum had chosen for his visit. Now, however, his transatlantic jaunt abandoned, Ingen Housz would be able to introduce his friend all around London; indeed, saw it as his responsibility. Van Marum therefore finalised travel plans and he and his wife left Haarlem in the early morning of Tuesday 6 July 1790[78] in the company of their adopted niece and a servant.[79] After a day's journey to the Dutch coast and crossing an uncomfortably rough North Sea, they were in Harwich. An hour and a half later they were sharing a private coach to London with two English travellers who had guided them through Customs. By 7 p.m. on Saturday the 10th the van Marums were resting in their reserved rooms at Bates Hotel, one of the houses in the Adam brothers' development between the Strand and the Thames – the Adelphi. Barely had they had time to unpack and relax before they were invaded by an ebullient Ingen Housz. Having known their Haarlem departure date by a letter, he had been staking out the hotel at regular intervals so as to know, immediately, of their arrival. Not a day was to be wasted – he was brimming full of immediate plans for visiting 'the scholars' of the city and introducing van Marum to them, even though the next morning was a Sunday.

Over the next three days van Marum was whisked around the metropolis in a whirlwind of introductions to, and meals with, an assortment of Royal Society fellows and scientific instrument makers, some of them as far afield as Chelsea. Van Marum must have wondered about the relentless pace and whether he could survive Ingen Housz's transparent anxiety that he meet everyone and see everything. There was certainly little indication that Ingen

Housz was recuperating from a temporary stroke that he had experienced, two weeks before, when the right side of his body had convulsed, to be followed by a 'dead coldness and feebleness'.[81] They went, twice, to the Royal Society in Somerset House in order to see the President, Sir Joseph Banks. They encountered Charles Blagden, secretary, and John Sibthorp, Oxford professor of botany, but Banks proved to be out of London. He returned during Tuesday 13 July and Ingen Housz and van Marum received an urgent invitation to dine that afternoon. But when van Marum returned to his rooms he found his wife unwell and he did not appear at the dinner table. Ingen Housz, who had lobbied for this invitation, a golden opportunity for van Marum to meet several important fellows, not to mention the President, was in high dudgeon. Before going to bed, he sent Dominique around to the van Marums' accommodation with a hastily written note.[82]

> I have made your excuse in telling of an indisposition of your wife that prevented you from joining the company that waited for you at dinner. One has to believe this indisposition to be dangerous in order for you to sacrifice this main way of your being introduced at this season among the scientists. It was asked of me whether Madame would, otherwise, have been left all alone. Not being able to confirm that, people were very surprised ... You must see my embarrassment, and consider that in writing to me of your intentions to enter into a connection with the scientists of this country, I informed them of your journey and of your wishes, and that I have prolonged my stay in the capital expressly in order to render you my services.

In spite of the friction, van Marum was offered a seat in Banks' coach the following day when a party from the Royal Society were making their annual official visit to Greenwich Observatory. Ingen Housz and John Smeaton were also to accompany Banks as passengers. Thankfully, van Marum was able to leave his wife by next morning and at last met Joseph Banks – in rather more intimate and prolonged circumstances than he might have expected, for the day out was followed by dinner. Over the next two weeks there were yet more private dinners and visits to instrument and glass shops but gradually the intensity of the networking and shopping was diluted by recreational trips shared with Mrs. van Marum and their niece. They watched a rowing boat race on the Thames, went down to the Royal Botanic Gardens at Kew, strolled around the British Museum and admired Albion Mill – a new flourmill driven by a Boulton and Watt steam engine. They all visited St. Paul's Cathedral and went down to Greenwich again. By the last few days of July Ingen Housz felt that he had done his duty to van Marum in raising his profile at the Royal Society and helping him to furnish his museum with the best quality instruments. He was also confident that his Dutch friends would now survive unfettered by his attentions. He had organised for them to go to Slough to see Herschel's wonderful telescopes and on to Birmingham to visit the factory of Matthew Boulton. Ingen Housz had an invitation in another direction, however; he was going to see his great old friend, Thomas Dimsdale, at home in Hertford.

As July turned to August, Ingen Housz took the now familiar road to Hertford and found his way to Cowbridge House, where Thomas Dimsdale was now living, with his third wife, Elizabeth. It was a much more substantial house than The Priory[83] and perhaps more fitting for someone who was now a civic dignitary. It is difficult to believe that Ingen Housz could have been in London for almost a year without seeing one of his oldest English friends. Even so, he and Dimsdale still had much to talk about, now as much as courtier to politician as doctor to doctor. Thomas had given up regular medical practice on having been elected to parliament in 1780. But after a decade as a local representative, and at the age of 78, he had just resigned his seat and his son, Nathaniel, had been successful at the hustings. Like Ingen Housz, Thomas had continued to inoculate sporadically; particular patients being the children of local nobility and of fellow MPs.

In the twenty-two years they had known each other, from the earliest days of their common bond as courageous pioneers in the field, smallpox inoculation had become an established routine across much of Europe. Dimsdale had even returned to St. Petersburg in 1781 to inoculate Catherine the Great's grandchildren, the two sons of Archduke Paul.[84] Coincidentally, he had been treating the two children just as Paul and his wife were preparing to set off on the European tour that had taken them to Vienna and to meetings with Ingen Housz. And although Dimsdale and his third wife, Elizabeth, did not travel to Russia via Vienna, they had, by chance, met Joseph II in Brussels.[85] Ingen Housz and Dimsdale were therefore in a position to share confidences about some of the most powerful personalities of the time. But both of them were still very vigorous and they would certainly not have spent long summer days sitting in Thomas's extensive library: he was equally proud of his vineyard.[86] Dimsdale was still a most generous host and 'bon viveur' and Elizabeth was an excellent housekeeper and very considerate hostess. An American visitor to their home, a fellow Quaker, was rather taken aback[87] by the sumptuosity for 'dinner was grand, with liveried servants and rich plate and china, having little ye appearance of a Quaker table.' So successful was Ingen Housz's visit that the Dimsdales were reluctant to let him go and they implored him to stay until the very last days of August.[88] And when the inevitable parting came, Ingen Housz was made to promise that he would again be their guest, at the house that Thomas and Elizabeth leased, every autumn, in Bath: he was not to be separated from his old friends for long.

INGEN HOUSZ AND Dominique left London on the Bath coach at the beginning of September.[89] A day and a half on the road saw them into Calne, the small 'wool' town a couple of hours short of Bath. A messenger was sent a mile to the south-west of the town, to Bowood Park, to bring down a chaise for them. They had made the journey eleven years before; in October 1779, and as the carriage brought them into sweeping views of the house Ingen Housz was able to appreciate the maturing beauty of the park and the mansion. The magnificent buildings, re-ordered and massively extended in the 1760s were now endowed with the patina of age, a fulfilling promise of the palatial interiors.[90] Robert Adam had pushed and pulled the interior spaces until they matched classical proportions and had adorned the south prospect with a magnificent ten-pillared portico. All around the mansion, the twenty-year-old specimen trees, vital elements in the 'English landscape' concept of Lancelot 'Capability' Brown, were now beyond adolescence, tall enough to be reflected in the fine lake that had also lost its severe outline. Bowood had emerged from the obscurity of a hunting lodge buried in Chippenham Forest.[91]

> Bowood, in the years of Lord Lansdowne's reign over it had been improved from the shabby centre of a dying agricultural estate into the princely home that it has remained ever since. The grounds, laid out by Capability Brown, had reached maturity and the large ornamental lake that the young Shelburne had constructed in 1763, soon after the property came to him, had merged into its surroundings, adding the beauty and peace that an expanse of water brings to the English countryside.

The arrival of Ingen Housz at Bowood came to be anticipated with so much fluttering female excitement that if he failed to appear when expected it was taken as an excuse to tease the Marquis. On one occasion he off-loaded his defence onto his younger son, the teenage Henry Petty, who felt obliged to write[92] to Ingen Housz that

> My father is in a great hurry dressing, and desires me to assure you that you are the innocent cause of a great deal of mischief to him as the Ladies attribute your not arriving to him, for they are very desirous of seeing you.

In autumn 1790 there was no reason to think that Ingen Housz was late and, when he did arrive, he was to discover that the occupants of the house and park were as mature and attractive as their surroundings.

Bowood House, Wiltshire, from the south. Courtesy of Devizes Museum Library.

THE LATE SUMMER gatherings at Bowood during the 1790 decade were a chance for Lord Lansdowne to show his hospitality when the house and grounds were in their best plumage. And the affability of each annual 'college' left even fonder memory traces in the guests. A descendant of two of them has given us a flavour of the circumstances.[93]

Surrounded by his guests and amongst his friends, the 'Lord of Bowood'

> . . . was the soul of kindliness. It was no gambling, foxhunting, gathering of Georgian rakes that was to be found under this hospitable roof. Besides passing visitors there were usually two or three of Landowne's permanent protégés . . . to add variety and intellectual colour to the transient politicians, relations or friends. Each day began with prayers presided over by a sleek young curate in the great hall surrounded by statues of the Greek Gods. The younger guests would then take excursions by chaise to local sights or local rides up wooded valleys and across open downland. For the more elderly, walks in the grounds . . . provided outside entertainment. Dinner was a lively meal presided over by Lord Lansdowne. After dinner, when the candles were brought out, the evenings were passed in dancing, playing charades, listening to music or simply in conversation.

These recollections effectively summon up the Bowood atmosphere but they make no mention of the fact that guests were also the beneficiaries of 'philosophical' diversions. Like Priestley before him, Ingen Housz was delighted to provide the entertainment, the sort of exciting and colourful demonstrations that are in the pedigree of those still laid on by The Royal Institution, for instance, every Christmas. At his first 'college', that autumn, he had found the room in the Diocletian wing of the mansion that Joseph Priestley had adapted as a laboratory in the 1770s was still intact and serviceable, much of the apparatus still sound.[94] For Ingen Housz, despite his personal animosity towards its previous owner, it was a haven within a haven. It helped him to take on a unique role among the houseguests, demonstrating experiments and giving simple lectures on scientific principles.[95]

> During the summer months Bowood might be considered as the emporium of talent: as the seat of learning and of science. Here the literati, and politicians of all nations, sects, and classes associated for the purpose of friendly and intellectual intercourse; to enjoy that truly delectable pleasure, 'the feast of reason and the flow of soul.

No doubt assisted by Dominique, who was very familiar with his master's gadgetry, Ingen Housz was able to amaze his audiences by such demonstrations as the spectacular burning of metals in oxygen (he always carried a phial of oxygen and a strand of iron wire in his pocket) and the disappearance of sound in a vacuum – there was an air pump in the laboratory.[96] On clear nights, he guided guests 'around' the moon with Lord Lansdowne's telescope. It is no surprise that Ingen Housz acquired the moniker of 'thaumaturge' – sorcerer.[97] It would be unfair, however, to allow an inference that all his new friends took him to be a mere trickster, a simple technician.[98]

> Since the arrival of Dr. Ingen Housz we walk every day between midday and three o'clock, making two miles during this time. It's true that he is very knowledgeable, a very fine man, who, after having lived twenty years in a Court, has lost neither simplicity nor . . . an interesting frankness; he tells me marvellous things about leaves, of the dephlogisticated air they release by day, by the light of the sun, and of the mephitic air they give off in the shade or at night. He recites to me many of the speeches of Lucretius, of Horace, of Virgil and of all the poets who have ever existed. His memory is astonishing . . .

Ingen Housz appears to have integrated easily into the ambience and rhythms of Bowood and matriculated, painlessly, to being a full member of the 'college'. He was soon reporting, to Hendrick Hoogeveen, the joys of being 'the companion of a Marquis' and 'a guest among erudite people'.[99] The house and the thousand acres of parkland held endless delights, including, now, a fashionable cascade and grotto but it was not an oasis. There was a pleasant hinterland in all directions and it was the custom that guests made regular trips to Chippenham – for the shops,[100] to the neighbouring communities and countryside including Cherhill, the village at the foot of the Marlborough Downs in order to view, close up, the white horse cut into the steep hillside. The landscape artist had been Christopher Allsup, the trusted Calne medical practitioner always called on when there was illness at Bowood,[101] a colleague whom Ingen Housz had met, at Bowood, in October 1779 and with whom friendship was rekindled.[102] Ingen Housz was also introduced to local gentry families including Lansdowne's immediate neighbours, the Money family (alias Money-Kyrles) whose property adjoined Bowood to the south, the Heneages at Compton Basset House[103] and the Meades at Blackland Park,[104] both to the east of Calne. But after nearly two months, as October drained towards November, the circle broke and guests began to disperse. Lord Lansdowne himself left, returning to London for the winter 'season', but Ingen Housz and Dominique moved on west, to Bath.

Once again, it was a return to a venue previously visited but only briefly. This time, as at Bowood, Ingen Housz would stay much longer and come to know the city more intimately. He was welcomed, by the Dimsdales, to 15, Paragon Buildings,[105] an impressive arcuate terrace that soars above the Avon valley east of the city centre. Although entering a period of some decline in popularity in the face of the trend towards seaside vacations, Bath was still an elegant watering place – literally so. The elegant eighteenth-century buildings, still in pristine yellow, the cut colour of the local oolite stone, created a refined ambience.[106] Spreading up the steep slopes above Queen Square, they culminated in the Palladian elegance of the Circus and Royal Crescent whilst across Pulteney Bridge, which suspended shops over the River Avon itself, were the generous proportions of Sydney Street and Sydney Gardens, Bath's equal to the Vauxhall pleasure gardens of London. For a small town, Bath had many other attractions for those visitors with no other business to occupy them – the Pump Room containing the head of sulphurous water, natural hot baths, the Theatre Royal, the new (Upper) Assembly Rooms, circulating libraries and numerous good quality shops. There were so many balls, concerts, dramas and card parties that there was an official master of ceremonies, though none so famous as the late Richard 'Beau' Nash. All of these well-trailed diversions awaited Ingen Housz but it came as a total surprise to learn that he was a member of the Bath Philosophical Society, indeed had been since 1779, for no-one had notified him.[107] He did not allow this flattering honour to distract him however and with his usual impetuosity and single-mindedness, he was soon pounding the streets in search of Benjamin Colborne, the Bath medical practitioner who was 'the true inventor' of the alkaline soda water, a potion somehow so appropriate for a town where people came to drink the 'waters'. Ingen Housz was also anxious to meet Doctor William Falconer, the author of the book on the remedy that he had first perused at Rotterdam over a year before.

Colborne was very hospitable and soon introduced Ingen Housz to his great friend, Falconer, physician at the Bath General Hospital. The two pioneers of alkaline soda water were able to give their new disciple a personal introduction to the remedy that included an historical background with which Ingen Housz was not unfamiliar. Kidney and bladder stones were very common in all age groups during the eighteenth century, causing distress much more frequently than today. Their prevalence probably stemmed from dietary habits and the high incidence of urinary infections. The desperate symptoms brought desperate measures. 'Cutting for the stone' was one of the earliest known major surgical procedures: patients became so distraught that they were willing to submit to a surgeon plunging a knife through their perineum in order to try to hook out the stone through the incision, all without anaesthetic. A few minutes of indescribable agony might rid people of their stone but if they did not die from immediate complications, they were often left incontinent and, in men, impotent. With such treatment barely less tolerable than the disease there had been a centuries-old search for a way of 'dissolving' stones by swallowing various concoctions.

Effective 'lithontriptics', as they were called, proved to be elusive and despite heroic trials going back to Arabic medicine a millennium before, the beginnings of true success only appeared in the 1750s. Drinking a solution of 'salt of tartar' (potassium carbonate) was the first chemical treatment that gave consistent evidence of stone erosion over and above the odd case of spontaneous recovery.[108] On the evidence that some liberated bladder stones fragmented when dropped into it, 'lye soap' then became a fashionable nostrum. This was made from mixing 'potash' (crude potassium carbonate) and 'quicklime' (calcium oxide) into a paste but it had a repellent taste and caused intense nausea and bladder pain, even if well diluted with attractive cordials. It was progress of a kind, however, and it was reassuring, for physicians, to note that the urine excreted by their treated patients could soften stones on the laboratory bench.

There had been another important development, virtually coincidental, in 1756. A physician in Saxony published his observations on the lithontriptic power of the thermal waters of Carlsbad in Bohemia. His paper was widely circulated and 'the stone' rapidly became yet another reason for 'taking the waters' at various European spas. Within the next decade it was realised that the active component of all spa waters was Joseph Black's 'fixed air' (carbon dioxide) in suspension. This insight was used by Nathaniel Hulme, the London physician, to explain the basis of his personal treatment for stones, a mixture of 'salt of tartar' (pure potassium carbonate) and weak 'oil of vitriol' (sulphuric acid). He theorised that the two chemicals so combined that 'fixed air' (carbon dioxide) was released into the solution and that patients were therefore reaping the benefit of the two known treatments, potassium carbonate and carbonated water. This credible chemical underwriting of his prescription helped convince the sceptics and the book, written on the back of it, and Ingen Housz's Latin translation, sold well when it was published in 1778.[109] Progress was helped by the fact that carbonated water had become freely available thanks to an invention of Joseph Priestley during his time as a Unitarian Minister at Leeds. Respectable trials of carbonated or 'mephitic' water were then carried out, on patients and in the laboratory, by two English physicians, William Saunders in London and Thomas Percival in Manchester. Their independent findings agreed and mephitic water, much more pleasant to take than any lye-soap preparation, rightly became the treatment of choice in the late 1770s. A period of rapid progress followed and Colborne had been one of many doctors who had treated themselves for bladder stone, in 1778 in his case. He chose to take mephitic water but added to it the then 'old-fashioned' remedy of salt of tartar. His personal experience was that the acidic water neutralised the caustic and nauseating effect of the alkaline salt and he had no difficulties taking the treatment. Restored to health, Colborne's success had inspired Falconer, a university-educated physician from a younger generation. He collected together a substantial portfolio of cases using this formulation and was soon publishing convincing reports of his successes: the two treatments seemed to be more than just additive. It was Falconer, then, who though not the inventor, was the driving force in developing what he called *aqua mephitica alkalina* (salt of tartar dissolved in carbonated water). It was a safe, efficacious and popular treatment for 'stones and gravel' that rightly became the dominant prescription for a generation.

On seeing bladder stone sufferers with Colborne, Ingen Housz had no hesitation in playing the apprentice role even though it meant that a royal physician of international renown was 'sitting at the knee' of a mere general practitioner. Colborne had been fortunate, however. The recipient of a large inheritance, he had been able to give up the daily chores of a practitioner to become a local expert on one particular disease and to a few selected patients.[110] Colborne had certainly built up an impressive litany of

Dr. William Falconer FRS. Courtesy of David Falconer.

clinical successes but Ingen Housz would have been even more pleased to find that his new friend had constructed a scientific basis for his practice, that he was an ingenious and 'enlightened' experimentalist. He had shown, for instance, that the urine of his 'gravellous' patients was nearly always acid on testing with litmus paper. However, after only a very few doses of his soda water it turned the paper blue, signifying a change to alkaline. Colborne had even noticed that those patients who presented with symptoms of a bladder stone but were found to have an alkaline urine, did not respond to his treatment. Moreover, he had repeatedly observed that urine saved from patients taking his treatment would dissolve the 'red fur' (blood-stained deposits) that lined the chamber pots of his untreated sufferers if their urine was allowed to stagnate. The cure was a 'specific' one for defined situations, a truly selective medication needing a scientific diagnosis; an early magic bullet. The deliberation of systematic principles by the two doctors was certainly rewarding for them but it was also the genesis, in part, of scientific medicine as we know it today – experimental evidence being utilised, in a common language, by all branches of the profession.

Meanwhile, William Falconer had shown Ingen Housz his own personal evidence of the efficacy of alkaline soda water. He sported a considerable collection of letters from widely dispersed physicians beginning to prescribe the therapy elsewhere in England. The correspondence was all very positive and pleasing but perhaps only the success stories merited the postage? Nevertheless, Ingen Housz was convinced enough to go on taking the medication himself; perhaps wisely. On 27 December, recording it in his commonplace book with remarkable candour and objectivity,[111] he noted that he was developing a 'cold' and then, at four o'clock in the afternoon, found himself passing very dark urine which he decided to save for observation. Sure enough, a few hours later, it had deposited some familiar 'red fur'. He also had the scientific curiosity to test it with some litmus paper that he watched turn red – his urine was acid. An hour after his evening dose of the alkaline soda water was taken, he passed some urine that was normal in appearance and tested alkaline: some clinical observations can be very close to home. Despite this transient scare, Ingen Housz was feeling very much better a week later.[112] Perhaps he would now be well enough to tackle the long journey back to Vienna? A letter from Prince Orsini-Rosenberg,[113] his great friend and Minister of State, amiable though it was, was still a prick to his conscience every time he saw it lying on his desk in London.

> I could not easily let the charms of England make you forget, entirely, this country . . . and hope that you will return to us so that we shall have the satisfaction to profit from your wisdom and where you will receive the assurances of perfect consideration . . .

19
Going Home: Not Going Home

Caroline Herschel's abrasive manner with her servants often causes squirming embarrassment for guests and today's no exception. William, quite used to his sister's tone, ploughs on with his story. His favourite telescope is still the twenty-footer that he had built in 1783 but Sir Joseph Banks had long encouraged him to make an even bigger one; twice the size, even. And when he'd shown him a design for a telescope with a forty-foot tube and mirror nearly five feet in diameter, Sir Joseph had enthusiastically shown the plan to the King. This had produced an amazing bounty of some £4000 and a succession of royal parties coming over from Windsor to watch the work progress and amuse each other by scuttling through the tube while it still lay on the ground. The final result, after many false starts – such large unwieldy mirrors are very prone to cracking – is in the garden. It dwarfs the Herschel's house and has made the dining-room very gloomy. But darkness is now their 'raison d'être' and tonight could be a good one for astronomy: with a north wind there are hopes of a clear sky. Ingen Housz is about to discover that serious star-gazing is all about leaving a warm fireside and trying not to shiver, with cold and fear, high in the night sky. He and Herschel have already inspected the huge wooden assembly that supports the giant telescope and climbed the precipitate thirty feet to the viewing platform. The timber pyramid is circular at the base and can be rotated on wheels. Traction is provided by moaning manpower. The inclination of the tube is adjusted by pulleys, also operated from the ground. But the view of the night sky is from the very top of the telescope, looking down into the mirror. For a man of his age, Herschel is fearless and agile on the long wooden ladders but they are quite a challenge for the uninitiated, even in daylight.

Saturn is Herschel's current party piece. He's already discovered that the planet has a sixth moon – first seen on Friday 28 August – and even a seventh – recognised during the night of Thursday to Friday 17/18 September. Ingen Housz is impatient to be among the very first people on earth ever to see these bodies but there's early disappointment. By seven it's properly dark but a fog has formed. It is November after all. William, however, is not to be beaten and they all wrap up warmly before going out into the garden. They watch as he disappears up the ladder to the viewing platform. He's back within a minute, exultant and bubbling like a schoolboy. Ingen

Housz should follow him up – the humidity is only a ground mist and the stars are all visible above it. The climb is even more nerve-racking in darkness and the damp ladder a little slippery. Once at the top, Herschel begins to give instructions to the ground by a tin speaking tube and Ingen Housz grabs a rail as the platform, the telescope and the whole scaffolding jerk into movement, creaking and groaning. The eeriness is enhanced by a disembodied shrieking. It takes some time before he realises that it's Caroline 'encouraging' the poor servants who are having to heave the assembly into the correct alignment. There's more ratchety, unsettling, motion but it's worth it: he's on his way to Saturn.

INGEN HOUSZ RETURNED to his new rooms, in Covent Garden, in the middle of January 1791 and settled into a London life that gave every indication of a man comfortably at home. There were many incentives to stay in England but slowly and surely, over the next months, these were superseded by disincentives to return to the Continent. London became more bulwark than bivouac as the journey home to Vienna began to look yet more perilous. He was just in time to take up an invitation to dine at Lansdowne House on Monday the 17th. There were eight at table on this occasion,[1] none of them names to be seen in the dinner book more often than that of 'Ingenhaus.'[2] In fact, he was to dine with the First Marquis of Lansdowne thirteen times before summer took everyone out of the city and on one occasion he was the only guest, enjoying a privileged *tête-à-tête* with his host.[3] Otherwise, it was the routine of visits to other friends, meetings at the Royal Society, Royal Society club dinners,[4] frequenting the scientific instrument shops and chasing down each and every latest development in science and technology, even, if it was humanly possible, to the very personalities involved. To this end, he kept up his loose-leaf scrapbook[5] and it became, effectively, a diary of his explorative activities and movements. France may have been the furnace of modern chemistry but England was undoubtedly the focus of applied science during this, the adolescent growth spurt of the Industrial Revolution.

At a Royal Society club dinner on the afternoon of Thursday 3 February Ingen Housz sat next to William Herschel, who was also a guest of the President, Sir Joseph Banks.[7] Herschel thought he might have a twenty-foot telescope for van Marum – another customer had cancelled his order at short notice.[8] Anxious not to miss the sale, Herschel wrote to van Marum a week later and their intermediary, Ingen Housz, also wrote to Haarlem.[9] His letter barely mentioned Herschel, however; it was mostly about van Marum's hopes of becoming a foreign fellow of the Royal Society. The cardinal point for van Marum to appreciate, once again, was that personal contact was such a critical part in lobbying for nomination and Ingen Housz urged that he should drop everything immediately and come over to London again. The next few weeks of spring would be ideal for their purpose since most of the fellows were now in the city and, Ingen Housz pointed out, he himself would never be in a stronger position to plead van Marum's case. On 11 February Ingen Housz had been elected, with Lord Shuldham, retired naval officer and ex-governor of Newfoundland, to the Royal Society Council.[10] Councillors served only for two years and Ingen Housz recognised that the opportunity to help his friend would soon expire. The implication of having accepted the position was glossed over. It could not have been the sincere action of a man who had any intention of returning to Vienna in the immediate future.

A visit to the home of the Dutch Ambassador[11] where Ingen Housz was still using Count Fagel, the embassy secretary, as a postal conduit, was to request delivery of another packet addressed to van Breda. He was now in a position to send to Delft a further article for his friend's forthcoming chemical journal, *Scheikundige Bibliotheek (Library of Chemistry)*.[12] In 1790 Jacob van Breda had decided to edit and publish a journal, a compendium of chemistry in the Dutch language, to be printed at Delft. He may have discussed his ideas with Ingen Housz when the royal physician passed through Holland in September/October 1789, and he certainly sought specific advice from his illustrious friend after Ingen Housz had gone to London.[13] Ingen Housz, anxious to help his friend, made multiple contributions and the first edition of *Scheikundige Bibliotheek* appeared in 1792; a substantial publication in three parts in which were presented 'the most important new discoveries and improvements' in the science of chemistry by a 'company of devotees'. There were 47 articles in all. Article 5 was in the format of a letter,[14] dated London 6 May 1790, from Ingen Housz to Jan Deckers, his physician friend at 's-Hertogenbosch. It described the formulation of the alkaline soda water, in the context of some history, and its efficacy in dissolving bladder stones and gravel. It went on to explain Ingen Housz's 'simple' method of making the medicine which he had devised shortly after his arrival in London. His interest was highly personal – having been prescribed this remedy he was to continue to take it for some years. The useful contribution was well constructed and referenced, occupying twenty-one printed pages, and invited the reader to refer to his 1778 booklet, his translation into Latin of the book on the treatment of kidney stones and other diseases by Nathaniel Hulme. As editor, van Breda was, perhaps, slightly frustrated by Ingen Housz's next contribution - another pseudo-letter to Deckers, this one dated London 25 January 1791.[15] This, too, extended to some twenty printed pages but it was, again, on alkaline soda water. At least the approach was different – there were some more detailed case histories, some of the patients being those Ingen Housz had seen and questioned with Benjamin Colborne and William Falconer themselves during his stay at Bath during the winter of 1790/1. But a third submission, a further 'letter' to Deckers written, once more, when Ingen Housz was at Bath and dated 14 December 1791, must have given van Breda some aggravation for it was yet more information on the alkaline soda water. Though van Breda must have seen it as superfluous and likely to detract from the eclecticism of his journal, he had little choice but to include it. After all, the status of the author, especially his fellowship of the Royal Society of London, easily trumped any redacting ambitions and this third supposed letter to Jan Deckers was also inserted in the first volume of the journal in 1792.[16]

Whatever van Breda's aggravations in building the text of his intended journal, they paled into total insignificance in the middle of 1791 in the face of a family disaster: his wife dying in childbirth on 14 July 1791.[17] Jacob van Breda and Anna Elsenera (van Campen) had been married for 18 years and Anna died aged only 42. Two days later van Breda would have rued rather than celebrated his 45th birthday. He was now a single parent with a newborn daughter, a son aged only three, an older daughter at sixteen and many other commitments over and above his medical practice. He must have been crushed by his loss and it is no surprise that his many distinguished contributions to the cultural life of Delft and the region were curtailed for some time. Besides being a town councillor he had become secretary of the Delft dyke conservation committee and had already given of his drive and energy in overseeing significant local drainage improvements. He was a member of the Batavian Experimental Philosophy Society at Rotterdam, of the Science Society at Utrecht, and had been honoured, in 1790, by election to the *Hollandsche Maatschapij Wetenschapen (Science Society of Holland)* at Haarlem.[18] Despite his tragic loss he appears to have rallied and orchestrated the appearance, within a year, of the first volume of his journal, *Scheikundige Bibliotheek*. Neither did he appear to have been discouraged by the growing competition. At roughly the same time several other similar independent journals dedicated to chemistry also began to roll off the Netherlands presses. There were three in Amsterdam alone and one in Utrecht.[19] However, the supreme effort of producing the first volume of *Scheikundige Bibliotheek* may have drained all his resources for although he went on to assemble one further edition, it did not materialize for some years; until 1798. Political turmoil in the Netherlands might, however, have been an equally important reason for the long intermission. When it finally appeared, the second volume was just as substantial as the first, 317 pages in total and Ingen Housz was again a multiple contributor. He had submitted three articles in 1793/4, which van Breda was glad to receive for they were more varied in content. The first item was a translation, into Dutch, of a letter[20] which Ingen Housz had received from Johann Schmeisser, a German chemist who became one of the most prominent young experimental chemists in London in the 1790s. Schmeisser had made a considerable impression on Ingen Housz and other London 'savants' and was even elected to the Royal Society in 1794,[21] Ingen Housz being one of his sponsors. The letter strongly advocated the external application of *Naphtha vitriolis martialis* – a petroleum jelly-like compound – for cramps. It was originally a remedy of Martin Klaproth, the leading German chemist of his day. In the same batch of submissions there was another lightly edited letter, to van Breda himself, dated October 1794.[22] It concerned vermicides – compounds that killed intestinal worms. Ingen Housz described *in vitro* experiments in which he had recorded the behaviour and fate of various intestinal worms in sundry fluids and how his findings had governed his medical practice. In this context, he made further reference to his 1778 Leiden translation of Nathaniel Hulme's work that had been, in part, about the treatment of worms. Finally, there was the copy of a seven-page letter from London, in November 1794, to Ingen Housz's nephew, Henry Ferdinand, now a young doctor serving with the Dutch army.[23] The letter concerned the healing of wounds and ulcerating cancers. The central point was the newly discovered phenomenon that exposed tissues and ulcers are pain-free if deprived of access to the open air, and specifically to oxygen.

MOTIVATED, ONCE MORE, by accessing the diplomatic bags that criss-crossed the Continent, Ingen Housz was also a visitor at the Duchess Street home[24] of Count von Warthausen, the Austrian ambassador in London. In April 1791 there was a package that he very much wanted delivered to Vienna, urgently and safely (and gratis).[25] Patrick Russell, a Scottish physician now working in London and fellow of the Royal Society, had just published a book on bubonic plague.²⁶ He had unique clinical experience of the disease having worked much further east, even, than his friend, Ingen Housz – in Aleppo in Syria and in India. He had witnessed, first-hand and with great personal danger, a plague epidemic in Aleppo in 1760-2. Ingen Housz bought himself a copy of the book but also persuaded the author to donate a copy to the Imperial Library at Vienna. In a covering note to the librarian,²⁷ Ingen Housz was very anxious that the evidence from the book be disseminated in Austria. As he himself had always thought the case, Russell was arguing that this pestilence was a highly contagious disease. It was endemic in the Levant and therefore there was a significant risk of outbreaks in central Europe following Turkish military incursions, some penetrating, even, to Vienna. Ingen Housz was well aware that he held, yet again, a minority view among Viennese physicians but he was convinced of the truth. The debate on the origin of the disease was no academic nicety; it was only a century since it had killed a third of the Viennese population and the Austrian medical fraternity would not accept the plague as being contagious. They had, therefore, not the slightest notion how to avoid it. This reactionary stance, Ingen Housz felt, was now flying in the face of yet more convincing evidence and therefore pig-headed and dangerous. He intimated that the book should be made available to the Emperor.²⁸

> . . . the contents will be particularly interesting for the Monarch in the present circumstances, so much so because of the danger of the spread of this terrible disease while the war with the Turks has, more than once, come almost to Vienna itself, and in consideration of the doctrine, truly worrying for humankind and that the late Professor Stoll unwisely supported – that the Plague isn't contagious . . . an abominable doctrine which, perhaps, some of his followers are actually still propagating, to their shame.

Ingen Housz would have embarked on this debate enthusiastically, even if he was to be in a minority of one, for it put his name back on the lips of all his colleagues in the corridors at the Hofburg.

LATER IN APRIL, Ingen Housz heard from Theophile Barrois at Paris.²⁹ 800 Livres Tournois (£1850) were being transferred into his account at Tourton and Ravel, profit from book sales; and Barrois was preparing to send some unsold copies across to Peter Elmsly. In spite of all the turmoil in France, there was obviously still some demand for philosophical books. Likewise, Ingen Housz was very pleased to receive a personal copy, direct from the author, of Jeremy Bentham's new book on the *Panopticon*.³⁰ It was a scheme that Bentham had been working on for some time; no doubt aired and discussed at Bowood, as elsewhere among his circle of friends. Nor is there any doubt that Ingen Housz would have expressed particular interest in the idea – it was new and had great social potential. Therefore it was in his purview. It was a revolutionary design for buildings in which there is a need to supervise the inmates, places such as prisons, hospitals and schools. In Bentham's blueprint, accommodation blocks were arranged such that they radiated out from a central hub housing the administration and viewing areas. The idea was particularly attractive for penitentiary structures for it implied low running costs – large numbers of the incarcerated could be supervised by very few staff and, therefore, cheaply. Having read it avidly Ingen Housz wrote to Bentham[31] proposing that he send it on to the *Hollandsche Maatschappij der Wetenschappen (Holland Society of Sciences)* at Haarlem. Perhaps Ingen Housz saw the possibility of reflected glory for he had just been proposed as a member,[32] an honour attributable to van Marum's nomination and lobbying.[33]

For his part, Ingen Housz was still trying to honour his friend and compatriot by pushing for his election to the Royal Society but one of van Marum's recent essays, circulating in London, had upset the all-powerful, though intensely shy and obsessive, Henry Cavendish. His was a powerful voice among the senior fellows and van Marum's possible election had come under a cloud.[34] Not that all of life was lived on the rarefied plateau of scientific societies and their intellectual infighting: walking the London streets could be physically hazardous. Two weeks earlier Ingen Housz had been attacked by four footpads but had put up sufficient resistance that he was not, finally, robbed. Great concern was expressed for his safety and well-being by his many London friends[35] but the victim himself played down the incident. His interest lay more in his latest attempts to further simplify the preparation of alkaline soda water for his personal use. He had recently refined a small scale recipe[36] using oil of vitriol (sulphuric acid) with just enough 'salt of tartar' (potassium carbonate) to make and collect exactly the right amount of fixed air within the solution. This could all be performed with modest, everyday equipment, presumably, in his case, still in the scullery or an outhouse of his lodgings.

In early July 1791 Ingen Housz left London again but not to go home. There were no signs of a return to Vienna but he did journey down into Kent.[37] He had been invited to Chevening House near Sevenoaks, now the official residence of the British Foreign Secretary but then the home of Lord Stanhope. Charles Mahon, third Earl Stanhope, was, besides being a well-connected and politically active aristocrat, an amateur scientist and a fellow of the Royal Society.[38] His first marriage had been to his cousin, Hester, daughter of Lord Chatham and sister of William Pitt. She had died young, however, and in 1781, a year later, Stanhope had married Lady Louisa Grenville, daughter of the Governor of Barbados and related to Joseph Banks. Stanhope had always been interested in natural phenomena and with the somewhat violent fluctuations in his political career he sometimes had opportunities to dedicate significant time and effort to various major scientific projects. He was one of those individuals whose very early, and politically motivated, election to the Royal Society[39] was not based on a scientific *curriculum vitae*. However he, later, earned his spurs with justification. He was particularly interested in electricity, calculating machines, and the fireproofing of buildings. Stanhope had also taken up metallurgy; in particular, methods to prevent the adulteration of minted metals used in coinage. Stanhope took his guest around his large estate, anxious to show him an eclectic series of phenomena in the orchards, fields and villages. Ingen Housz was introduced to the 'Chevening method' for preventing die-back in the pruned branches of fruit trees: a boiled mixture of chalk and tar, cooled and applied as a paste, was used to seal up sawn-off limbs.[40] At a nearby silk mill, Ingen Housz, Stanhope, and his wife all saw how the proprietor was dealing, successfully, with dry rot in the buildings by coating the exposed surfaces with a pitch distilled from coal after soaking the timbers in 'green vitriol' (iron sulphate).[41] They discussed the cause of 'dry' rot, so-called, and could agree that it was the result of damp timbers being invaded by a 'mushroom-type' vegetation (as, indeed, we now know to be the case).

Passing through London for just a day or so at the end of July[42] Ingen Housz discovered that he had accommodation difficulties. His tenancy in Covent Garden was being terminated at three months' notice. For the present there was no crisis. He was going to be out of London until the end of the year anyway but he would need to look for other rooms sometime in the depths of winter. With no hindrance, therefore, to his immediate plans, he set off for Hertford again.[43] Dimsdale and his family greeted their old friend with delight and the two physicians took their usual pleasure from discussing interesting cases. They certainly debated the efficacy of alkaline soda water – Dimsdale, too, had been attentive to the trials of the 'water' in Bath and even knew of two patients with chronic arthritis helped by the treatment.[44] Reminiscences apart, there were new observations to be made out of doors. Ingen Housz joined a party who set off to destroy a wasp's nest buried in an earth bank.[45] They set off at dusk so that the foraging wasps had all returned to the nest. One brave soul approached the entrance and set alight some gunpowder packed inside a hollow squib so that it gave off copious thick smoke. It was poked into the entrance of the nest and a freshly dug turf thrown over the whole opening. The nest was then dug out by hasty spadework while, hopefully, all the angry inmates inside were stupefied by the smoke. Ingen Housz watched as the nest was then pulled out whole and thrown into a large pail of water to drown all the inhabitants. It reminded him of his first arrival in Vienna and the physicians he had met there - a 'wasp nest of doctors'.

After spending a month at Hertford Ingen Housz returned to London to collect letters and attend to any urgent matters before going, again, to Bowood. There was very sad news among his post – one item a very unexpected shock. He learned of the death of van Breda's 42-year-old wife, Anna, who had died, shortly after childbirth, in mid-July.[46] It was some consolation that the infant, a daughter, Joanna Wijnanda, had survived. Ingen Housz's expression of grief was sincere [47] – after all Anna had been someone he had met on several occasions and he recalled some appropriate lines of Horace: *Durum sed levius fit patentia,* (It is hard, but it gets lighter by endurance).[48] However, he couldn't resist the opportunity to attack male midwives whom, he claimed, practised a 'murderous art' by which 'more people had been killed than by wars.' This was not necessarily justified and hardly likely to console his friend. Perhaps van Breda would not have been surprised that his 'scientific' correspondent then launched into the usual matter-of-fact tone of the bulk of his letters. Ingen Housz was soon back to updating van Breda on the unique properties of different inks and there was news on Joseph Priestley.[49]

The mutual grief with van Breda would have compounded that which Ingen Housz had recently shared with Nikolaus Jacquin whose wife, Katharina, had also died – on 15 January 1791, aged 55.[50] Her death announcement came as less of a jolt for she had been very ill for a considerable time.[51] If two good friends had passed on, another one, a less sincere one perhaps, had passed beyond the pale. Unlike Ingen Housz, who had been an eyewitness to the bloodlust and who had read the hatred in the eyes of the Paris mob, Joseph Priestley had favoured the French revolution and promoted it from the pulpit. He was far from alone in supporting the new French regime, for the majority of English literate society had voiced their initial approval of what was happening across the Channel. However, the cataclysmic events of July 1789, seen to hold out such promise, merely led from one kind of tyranny to another. The new 'democracy' was ever more violent, more sickening and more threatening. Public opinion underwent a *volte face* but, characteristically, Priestley did not. Here was a stick with which reactionary elements could beat him. On Monday 11 July 1791 an advertisement appeared in the *Birmingham Gazette*[52] that 'The Friends of Freedom' were to hold a dinner, on the 14th, at three in the afternoon, to celebrate the second anniversary of the storming of the Bastille prison in Paris. Some 80 guests, mostly members of various dissenting religious groups, were invited to a central Birmingham hotel. They were heckled and jostled, on their way in, by a scowling rabble. Priestley did not appear, though fully expected, and the crowd seemed at a loss and dispersed. By eight in the evening they had re-assembled to form a riotous mob that set upon various buildings in the town. The trail of destruction was far from random – Priestley's chapel was destroyed, among other non-conformist places of worship, and then the mob made directly for Priestley's own home, which was ransacked and set on fire. Fortunately, the threat of serious violence had been 'telegraphed' to several of Priestley's close friends and he was forewarned of the danger. As the mob, sworn to kill him, approached his house, he and his family were persuaded to flee, finding safe refuge, eventually, outside Birmingham, with relatives. They very quickly moved on to London. In the meantime the 'Church and King' riots continued for two further days in Birmingham while the local magistrates rejected military intervention and someone threw away the key to the fire engine hut. More chapels and private homes were gutted and anyone that the mob might choose to label as a 'd ----d Presbyterian' feared for their safety.

Priestley's response to these events was courageous but foolhardy. Within days he was planning to return to Birmingham to give a sermon based on the text 'Forgive them, Father, for they know not what they do.' His London friends dissuaded him of this madness. He was also forced to accept, passively and against his nature, that he was no longer an acceptable figure in society. Ingen Housz appears to have accepted the majority opinion for, as he told van Breda in a letter in September:[53]

> Most people here are not sorry for Priestley, as he is suspected, with others . . . of attempting to cause a general insurrection, to overthrow church and state. Those of his persuasion realised, just in time, that instead of a revolt in their own interests, they could well provoke one to the contrary. Anticipating this, some of the more prudent disappeared while others tried to prevent the festivities (of 14 July). Some hotheads persevered, however, but they took part in the celebrations trembling and sneaked off, like thieves, prematurely. This forestalled a terrible riot in London but was of no avail in Birmingham.

Lady Caroline Fox.
Courtesy of The
Hon. Mrs. Charlotte
Townshend.

In his view there was some rough justice in that *scelus in authorem redit* - 'the crime had bounced back on to the perpetrator'.[54] Poor van Breda might have agreed but would have noted the date of 14 July 1791, the day he lost his wife, as being absurdly coincidental.

At Bowood, later in September, that year's 'college' were all relieved to find that Ingen Housz was as lively and full of enthusiasm as ever despite now having passed the milestone of 60 years; that he had shrugged off any grief for lost friends and relatives and come to no harm from being mugged. His old friendships were consolidated and there were several new guests that summer. Dumont, Lord Lansdowne's secretary, wrote to a friend,[55] bleating that he had not a 'moment' of spare time. 'Day after day' he was failing to get the 'dawdling morning walkers' back to the house before lunch had gone cold, that the interval between lunch and dinner was 'lost in business' and that the evening hours after dinner were all 'occupied by the company' until at least eleven o'clock. And the anticipated arrival of Jeremy Bentham and his brother was hardly likely to make life any more serene. Ingen Housz and Lady Caroline Fox appear to have become close companions by this time, brought together by 'experimentals'[56] in what was, ostensibly, a Platonic relationship. At the same time one does wonder if intellectual affiliation and human affection gained somewhat more than their usual overlap when 'the doctor' and 'Focey', as she was known within the family,[57] were together. They were also in the habit of corresponding within hours of parting[58] and it was not long before they were exchanging gifts.[59] Lady Caroline's aunt, Anne, Lady Upper Ossory, wrote,[60] teasingly, that

> ... I am sure I should not be of the party of listeners to Dr I so often, but I daresay you have acquired a great deal of knowledge.

Jeremy Bentham was certainly indignant and his tangible jealousy was justified for he, certainly, had overt romantic aspirations towards 'Miss Fox'. Though there is no evidence that these were ever reciprocated, Bentham saw 'the doctor' as an obstacle, writing to a friend, in November 1791,[61] that

> Dr Ingenhousz ... is my lady's head philosopher. Being somewhat stricken in years, I was in hopes of being promoted to his place when Providence should please to call him away, considering that we are all mortal; but my evil star has ordered it otherwise.

And to add to his distress, Ingen Housz had also overtaken him in the affections of Lord Lansdowne who was heard to remark[62]

> that he always believed Bentham to be the most good-natured man in the world till he had made an acquaintance with Ingenhousz.

Following the precedent of 1790, Ingen Housz went on to Bath in mid-November and once again stayed with the Dimsdales in the Paragon.[63] He immediately contacted his medical friends, William Falconer and Benjamin Colborne, in order to update himself on the alkaline soda water (which he was, himself, still taking assiduously). He found that Falconer was in the midst of writing a further edition, his fourth, of his book on the remedy and was enjoined to make a contribution. Ingen Housz therefore spent several days writing a significant entry – some sixteen pages of printed text - for Falconer's book in the form of a letter to the author dated Bath 25 November 1791.[64] Otherwise, it took Ingen Housz no time at all to discover the latest novelties and personalities. Among these was a certain Doctor Graham. James Graham was a 46-year-old Scot from Edinburgh, where he had trained for medicine. After a short spell in general practice he crossed to America and became interested, in Philadelphia, in the possibilities of treating diseases with electricity. He then remained very unsettled until, back in England around 1777, he set up a new but eccentric practice in Bath, advertising cures using 'vapours' and 'applications' whether 'electrical' or 'magnetical'.[65] Fashionable among the nobility, he moved to London and set up a 'Temple of Health' in the Adelphi development in Whitehall. Here he treated patients with his extraordinarily elaborate electro-magnetic apparatus, used musical therapy, gave medical lectures and sold bizarre concoctions such as "Electrical Aether". The Temple of Health was a huge success and Graham became the talk of London, featuring in plays, poems, prints and newspaper articles.

Encouraged by his fame, Graham launched the 'Temple of Hymen', in premises in Pall Mall, where patients queued to try his large 'Celestial Bed' for £50 (£3,100) per night. This was sited under a dome of musical automata that was also lined with fresh flowers and contained a pair of live turtledoves. As naked couples cavorted, 'stimulating' fragrances and 'aetherial' gases were released and the bed frame tilted to put the couples in the 'best' position to copulate. As they began to do so their movements set off music from organ pipes, the tempo of which accelerated with their increasing ardour. All was designed, Graham alleged, to help infertile couples in the process of conception but the enterprise soon came to be a unique feature of the London sex industry rather than anything remotely therapeutic and was short-lived. He was soon in financial difficulties and, unable to keep out of debt, he returned to Edinburgh in 1784. Here he developed a new medical therapy called 'earth-bathing' which he exhibited and offered widely around Britain. He would even lecture on his treatment while buried up to his neck in earth. In late 1791 he was at Bath and Ingen Housz couldn't resist paying a shilling (£3.00) to be among the audience at a demonstration by the incorrigible quack. He describes the 'entertainment' in a letter to van Breda,[66] his tongue very firmly planted in his cheek. Having no duty to submit an official opinion, as in the judgement of Mesmer, he clearly enjoyed the display of human frailty and gullibility.

> . . . a good number of completely naked men and women, young and old, were buried up to the chin in wet soil. The doctor, himself fully naked, stepped into a big tub filled with mud in which he sat up to his chin. In this particular situation he gave a kind of lecture to the audience all around him, buried or unburied, about the *vis juveniscens* (youth-making force) of the earth-bath. This amazing diatribe was interrupted now and then by unhappy incidents. Among others a little boy, buried to the chin between two large adults, screamed that he was cold and that the clods of wet earth that fell from the shoulders of his neighbours got into his eyes and mouth and suffocated him. One of the buried ladies also became so panicky that she began trying to haul herself out of her grave but, failing this, some

'gravediggers' were needed to pull her, covered with mud, out of the earth, which made the observers laugh. This farce goes on every day. Dr. Graham even planned to bury patients in a churchyard since the earth there, having already 'sweetened' so many stinking bodies, must be favourable material; but he was prevented.

To use his own phrase, having reached the 'crumbs' of January[67] once more, Ingen Housz made the journey back from Bath to London. Although the turn of each year was falling into a routine there was, this time, an uncertainty on return to the capital: he had nowhere to live. The tenancy in Covent Garden having expired six months earlier, he had been a migrant guest at other houses, some of them very grand, ever since. He was able to find rooms above a tailor's shop in the southwest corner of Soho. Marylebone Street[68] was two roads south of Golden Square which defined the area. The Square had been a seventeenth-century development that held a reputation for respectability[69] despite the connotations of Soho in recent times: but it was already less fashionable than the more stately new estates further west such as Hanover Square or Berkeley Square. At number 35, the Williams family, his new landlords, ran the retail outlet with workshops to the rear and residential accommodation above. The father, in his late seventies, was on the point of handing the business over to his highly competent son.[70] By May 1792, Ingen Housz and Dominique were well installed in what would become a stable home. It was well located for his purpose, putting the stately homes of Westminster, Charing Cross, the Royal Society and Fleet Street all within comfortable walking distance. Among his '*poste restante*' mail at 6, Fleet Street he found the sad but not unexpected notification that his old teacher and friend, Hendrik Hoogeveen, had died during the autumn, still in post as head teacher of the Delft Latin School. He had been buried on 7 November 1791[71] and had been replaced, at the school, by his son, Jan[72]

There was also a letter[73] from Agatha who was having problems with the household servants in Vienna. Antoni was, purportedly, having eyesight difficulties and Ingen Housz, in replying,[74] was adamant that Agatha should not give him any cash 'handouts', knowing him to be 'conniving'. There was a strong implication that he intended to take a chance with his poor health and make the long journey home soon although this may have been more than a little disingenuous.[75]

> The obstruction that blew up while I was in Paris put me two finger breadths from death and although in abeyance at present could still shorten my days

for there is still something in the liver which isn't right. However, I am well enough to travel, if only the anarchy in France could be passed without personal danger.

Later that spring, he was reinforcing the impression that he intended to come home soon when writing to Josef Jacquin.[76]

> ... my health, always unsteady, is, however, good enough for me to dare to undertake the journey to Vienna. I am preparing myself and I have already refused all invitations to spend time at country houses.

Aware that he shouldn't emphasise this probable lie, he moved on rapidly to congratulate Josef, and Franziska, his sister: both were about to marry, some compensation for the loss of their brother. Gottfried had died on 24 January, a year after his mother and only seven weeks after his great friend, Mozart. There was also other less auspicious news in Ingen Housz's letter to Vienna,[77] the story of a tragedy by way of warning. There had been a fatal accident in the London shop of a Mr. Godfrey, a chemist. A flagon containing a very small amount of 'oxygenated muriate of potash' (potassium chlorate) mixed with sulphur had exploded, killing everyone in the room. Chemists were beginning to realise the incredible power of some of the pure substances they were now handling and mixing: Josef Jacquin had already written[78] to tell of a similar accident in Paris. Though no one there had been killed, the eyesight of the famous chemist, Bertrand Pelletier, had been in jeopardy for some weeks. Ingen Housz saw it as his duty to reciprocate and inform his friend, Antoine Lavoisier,[79] and thereby the Royal Academy of Sciences in Paris, of the fatal incident in London so that all the French chemists were alerted to the risk attached to these particular chemicals.

Regular dining at Lansdowne House had resumed. On Wednesday 23 May one of the other guests was Charles Talleyrand, a long-standing acquaintance of the Marquis. Uniquely, in London, Lord Lansdowne was prepared to accept the society of the famous Frenchman and not view him as a surreptitious hawker of revolution.[80] It was the first of three recorded occasions at which the famous French diplomat/politician was also a dinner guest when Ingen Housz was present. Talleyrand was in London for the second time within a few months, ostensibly on a diplomatic mission to 'avert war'. However, he had no official mandate and having narrowly escaped death as a traitor to the Jacobin cause a few weeks earlier, no-one expected him to return to Paris, least of all the fanatical republicans there. His first-hand experiences of the resurgent brutality in France could only have intensified the growing anxieties in Ingen Housz. Condemned prisoners in Paris now faced death by the 'machine',[81] the gruesome apparatus promoted by Dr. Joseph-Ignace Guillotin and thereafter always associated with his name. Motivated by the need for a means of execution that was not only instantaneous and (supposedly) painless but also egalitarian, he and a surgeon friend had commissioned the refining of an old decapitating machine. It would come to symbolise 'The Terror', the phase of the French Revolution that would be labelled as a 'holocaust for liberty'.[82] 'Guillotines' were being assembled in several public places in Paris just as Talleyrand was fleeing the city, and his departure also coincided with a phase of indiscriminate arrests and personal retributions. Here was first-hand evidence of Ingen Housz's worst nightmares and there is good evidence that he deferred, yet again, his plans to return to Austria: within days of meeting Talleyrand he was requesting more of his clothes be sent across from Breda.[83]

On 9 August Ingen Housz was in Hertford again[84] and then, later in the month, in the Midlands. On Tuesday 7 February he had dined with the Marquis at Lansdowne house. Other guests had included Lord Greville,

Charles de Talleyrand-Périgord.

Lady Warwick and the Birmingham industrialist Samuel Garbett, a close friend of the Marquis. Henrietta, Lady Warwick, was Elizabeth Vernon's older sister and already knew Ingen Housz well, having been at Bowood the previous autumn.[85] The talk must have been, for some time, of Birmingham and of industrial developments there. Ingen Housz appears to have been encouraged to visit the town and Lansdowne was to furnish him with several personal letters of introduction.[86] On arrival in Birmingham in late August he found a town that was[87]

> ... hardly a scenic stopover ... it had evolved into an urban landscape profoundly alien to most English people, a foretaste and warning of the century to come. To visitors it felt less like a town than a giant factory, or, as Edmund Burke called it, 'the grand toy-shop of Europe'. A thousand small workshops ... churned out toys and buckles, cutlery, buttons and snuffboxes, flooding the country, and increasingly the world, with smartly designed trinkets at rock-bottom prices.

Unfortunately, Samuel Garbett was ill but one of his friends was deputed to show Ingen Housz around the area. They visited a manufactory of sword blades – Ingen Housz was impressed enough to buy one with an ornamented ivory handle, together with its scabbard.[88] They then travelled to Soho, on the northern outskirts of the town, to the 'celebrated' manufactory of Matthew Boulton. Ingen Housz's reactions can only be surmised: Boulton was working on a massive export order, pressing 5 sou coins, in copper, for the revolutionary French government. Here, in effect, was an outpost of the French mint in the heart of England. Boulton, ever the shrewd businessman, had no time for niceties – however unpopular the French Revolution was becoming in England. He was punching out Jacobin coinage at the rate of at least sixty coins a minute on each of his eight steam presses.[89] The 'Soho Mint' was so productive that its activities, alone, had so vastly increased the demand for copper that it had inflated its price. Even so, Boulton was happy to take on a private commission for Ingen Housz; to fashion a piece of platinum into some singular fashion accessories. A design for some buttons was agreed on and there was sufficient metal to make fourteen which Ingen Housz described as 'very handsome'.[90] Meanwhile, in Handsworth, there was the painted glassworks of Francis Eginton. Eginton and Boulton had collaborated, a decade earlier, in trying to develop copper-plate engraving to produce colour images but the enterprise had stalled. Eginton had moved back into a more conventional career for an artist, specialising in painting glass and there are many Midlands churches that admit light through his work.

The famous physician, William Withering was away from his home – Edgbaston Hall – so that Ingen Housz was unable to worship at the altar of the discoverer of foxglove (*digitalis*) treatment for heart failure. At Tipton, to the north-west, he was able to witness the production of acids, alkalis, soaps and several different and useful compounds of lead at a chemical works although the proprietor, James Keir, was not willing to intimate his recipes. In just a few days Ingen Housz had been able to commune with many of the 'Lunar Society',[91] so-called because they met, monthly, on the Monday next to full moon in order to see their way across country, after dark, to their scattered Midlands abodes. The striking absence from the current membership was, of course, Joseph Priestley. The ruins of his Birmingham house and chapel were sombre reminders of the over-heated reactionary forces that had been unleashed against this particular supporter of the French 'democrats'.

Back in London in the first week of September, Ingen Housz found, among his new mail, a very striking letter that is worth quoting verbatim; more than an invitation, it was a plea.[92]

> My Dear Friend, There is no peace at Bowood for want of your presence. The ladys insist upon it, that you would have been here before now, if it was not for some omissions of mine and they go so far as to say that I have not half the respect for you which you deserve. Allow me to appeal to yourself, and if you have a grain of justice in your disposition, come immediately to acquit me, as you must be sensible, that no one can be more truly Yours. Lansdown.

Ingen Housz left for Wiltshire immediately[93] and Lord Lansdowne was duly 'acquitted'. The pleasant Bowood rounds of daily walks, dinners and entertainments were resumed and any possible staleness from repetition offset by carriage rides into the vicinity. The 'college' appears to have discussed, repeatedly, the current obsession in the country at large: the possibility of an English revolution prompted by the one across the channel. They were, after all, hosted by an ex-Prime Minister who still possessed his political antennae and was still in regular contact with many of the current administration. The consensus at Bowood was that, of all the various models of government, none could make every citizen entirely happy. Even a stable monarchy was imperfect but was the least bad option.[94] Government of some kind was essential in taming the otherwise 'savage beast' that was mankind, a 'wild animal in a natural rage' such as was then prowling the streets in France[95] and, as Ingen Housz had experienced first-hand, existed on the streets of London. He would have done his best, presumably, to cover up his growing disquiet. It was beginning to appear that an intimate of the British nobility might be in as much danger in England as was a courtier of a continental royalty.

The French anarchy, having lasted for four years already, still does not deter the fanatics here. The whole upper middle class here are frightened of riots and I wouldn't be surprised to witness another revolution during my life, far more horrible than the one I witnessed in France.

Small wonder that Ingen Housz was, again, thinking of crossing to Philadelphia as he left Bowood late that autumn.[96] There was some reassurance in following routine – once more he went on to Bath where the Dimsdales were in residence.[97] Benjamin Colborne was able to confirm that alkaline soda water was still enjoying a deserved success. It had even cured a woman of chronic jaundice. Ingen Housz was understandably intrigued. Might not the remedy dissolve his own gallstones? It did not help him, however, through a recurrence of his lumbago, a spell of back pain[98] that lasted for some days soon after his journey from Bowood – even a short trip on eighteenth-century roads could have painful repercussions!

THE EUROPEAN WINTER of 1792/3 was not remarkably cold or difficult but there was a universal chill all the same. It reached its peak on the morning of 21 January when Louis XVI, the deposed King of France, was taken through the foggy and shuttered Paris streets to the Place de la Révolution (now the Place de la Concorde).[99] A few minutes after 10 a.m. his coach arrived in the crowded square and he was escorted up on to the scaffold erected there. His hands tied and his hair peremptorily cut, he was strapped to a plank and pushed forward so that his neck lay in the brace positioned under the 12-inch blade of the guillotine. Within seconds he was dead, his dripping head held up to the 20,000 onlookers. By 10.30 the city gates were open again and life in the capital resumed as if it were a routine Monday. The middle and upper classes in England were appalled; more afraid of revolution than ever. Their government sensed that it was 'now dealing with a phenomenon of uncontainable barbarism and irrationality'.[100] The ongoing discussions to avert war between England and France now seemed pointless. In any case, while William Pitt and his cabinet deliberated, France declared war, on 1 February, and settled any uncertainty.[101] And through much of that winter there was a resurgence of the Patriots in the Netherlands where the French appeared about to invade. Having declared war on The Netherlands along with England, some French troops did descend on Breda and supported a Patriotic coup there that lasted several months.[102] Revolutionary fervour was on the move – north-east in this case but who knew where it might flourish, England included. There was now very little prospect that 1793 would be the year that Ingen Housz would risk crossing the Continent to return to Vienna: his carriage and bulk luggage at Breda were certainly no longer available and Austria, in a pragmatic alliance with Prussia, was now at war with France. However, it was not absolutely impossible to get to Austria from England. The tutor to the son of the British Ambassador at Vienna left London in mid-March; Ingen Housz even entrusted a letter to him.[103] If the Imperial royal physician did sense the irony, he certainly did not confess it and wrote to his relatives at Vienna, free of any apparent embarrassment, how his health was holding him back from transcontinental travel for at least another summer.[104]

> I am still always menaced by hydrops[105] and inflammation of the liver experienced as continual stomach cramps, all of which causes me anxiety. I hope that I can come to join you next year.

At the same time he was toying, yet again, with the idea of crossing to America – strangely, the seedbed of revolution now seemed the most stable place to go.[106]

As something of a rejoinder, there was a letter from Agatha[107] bearing the worrying news that the lease on their apartment in the Franziskanerplatz was being terminated and that 'they' would have to move their Vienna accommodation. Fortunately, the younger Jacquins, and their in-laws, the Schriebers, had already offered to help and a new apartment was in prospect in the Annagasse.[108] This was still in a very respectable part of Vienna – 'did not the French Ambassador live there?' It was a huge upheaval but despite the crisis, and having been away from home for almost five years, Ingen Housz still made no discernable plans to get back to Vienna. He was, however, determined to supervise the removals, if by proxy. There were very specific instructions, issued to Josef Jacquin, on how not to move the furniture.[109] Some locked chests and the armoire 'alongside my bed' were to be maintained absolutely upright in all circumstances for they contained 'bottles and other breakable things'. A certain cabinet maker was to be engaged to dismantle the bookcases and 'particular care' would be needed when moving the electrical machine for although it was supported by a strong board it was not very securely attached.[110] Perhaps it would be better for Josef to sell it, but for not less than 100 Florins (£560), and perhaps also the large telescope for, say, 20 ducats (£250). As pieces of furniture, crates and boxes were being trundled through the Vienna streets by labourers who could only frown and smirk by the fuss being made, their owner was returning to the role of literary agent some 800 miles away.

INGEN HOUSZ COULD hardly have guessed, in 1767, as he recruited subscribers for his old teacher's *Particles*, that he would come to repeat the exercise some thirty years later. However, his admiration of Hendrik Hoogeveen had not diminished with time and he rapidly leapt to the cause of another potential publication, although it was to be posthumous. When the 79-year-old Hendrik Hoogeveen had died, in November 1791,[111] his son, Jan, also a schoolmaster, had been the first choice of the Delft authorities to take over the headship of the Delft Latin School.[112] Working, then, from his father's study, Jan found himself immersed in all his father's voluminous manuscripts. Hoogeveen senior had dedicated himself to a lifetime of scholarship and had defied death by the completion of yet another literary exercise. Still on his shelves was a vast lexicon, the outcome of many years' work.[113] It was a reference work in which readers could locate Greek phrases linked to their Latin alternatives. Jan Hoogeveen saw it as comprehensive and authoritative, irrespective of filial respect. Writing, appropriately in Latin, he put a proposal to Ingen Housz in a letter of October 1792.[114] Having described the mass of his father's poetry piled up all around him he went on:

> . . . the deceased author has left another posthumous work. This is the *Analogicum*, which is in alphabetical order and contains more than 70,000 Greek words. You can see an example copied by me on the next page so that you may be able to pass careful judgement on the nature and use of the whole work. If I am not mistaken, it is a work that is still lacking in the learned world. This analogical study of the entire Greek language contains 795 pages . . . As his best friend, would you, whom the deceased author justly called the 'editor' of his *Particulae*, please try and find out how your British friends feel inclined and whether you can inspire them with the same enthusiasm and similar helpfulness to lift this work from darkness into light.

Hoogeveen junior was wise to the ways of the world. Suspecting that Ingen Housz might hesitate to be involved merely on the basis of loyalty, he buttressed his request with sound logic.[115]

> The reason why I ask you to act as my friend is twofold: the first has to do with your reputation which goes further than Africa and the Indies and which lends extraordinary weight to a recommendation for printing the work; the second reason is that words in praise of a father, however much deserved, but flowing from the pen of a son, are usually not estimated at their true value; if on the other hand praise comes from someone else, this adds considerable lustre to both work and author. And lastly, considering the

subject of the work itself, you will admit that it was written only for true lovers of languages. Consequently the work will attract very few buyers in our country.

Ingen Housz would have found the letter from Jan Hoogeveen when he returned to London in January 1793. There is no suggestion that he hovered over his decision and appears to have had not the least doubt that the work should be printed, and in England. Refreshed in his admiration for his late teacher's literary genius and seeing that the name Hoogeveen was now known in literary circles he hoped he could make some significant progress. Ingen Housz now knew many of the most influential and powerful personalities in London. He had a wide personal network of friends and was no longer just a subordinate of Pringle or Watson. The processes to publication were unchanged, however. He would need to convince a publisher and raise subscriptions to cover what would be a very large initial outlay. We do not have the privilege of knowing how his next move evolved but it was a stroke of genius. Sometime in January or February 1793, Ingen Housz sat down with a Prebendary of St. Paul's and showed him the copied manuscript pages of Hoogeveen's *Analogicum*.

The 46-year-old Reverend Samuel Parr was a colourful character and widely known.[116] He had known his namesake, Dr. Johnson, personally and was also a man of letters; he even looked like him. Parr and Hoogeveen had shared the same profession. Parr had, however, become ordained and had a Doctor of Laws degree from his alma mater, Cambridge University. He had latterly given up school teaching and become curate of a small parish near Warwick. The post was virtually a sinecure and he was often absent in London attending to Cathedral business. It gave him an income which, supplemented by his Prebend at St Paul's, allowed him the life of a gentleman. Parr was one of Ingen Housz's London 'literary' friends, able to dedicate much of his time to his beloved Greek and Latin ensconced in his vast personal library to which he could retreat from an unhappy family life. Parr could have been a severe and damning judge of Hoogeveen's work: he was renowned for being unpredictable[117] and was certainly no diplomat. This was a risk worth taking, however, and Ingen Housz's gamble paid off. Parr was an instant advocate of the *Analogicum* being published. He took the sample pages up to Cambridge within a few days and talked to his old tutor at Emmanuel, Dr. Richard Farmer, now Master of the College.[118] Farmer, who had served as Vice-Chancellor, was well connected within the University and an equally potent salesman for what he, too, saw as an important work. Within a very short time the Syndics of the University Press had been impressed enough to agree to publish the Dictionary, even meeting any costs that could not be offset from subscriptions. Jan Hoogeveen, informed of the good tidings over Easter,[119] was ecstatic. Under these very generous terms the family could afford to pay for the printing of the prospectus sheets and 250 of them were ordered; Ingen Housz organised the print run in England.[120] Moreover, Hoogeven vowed to sit down, with friends, and begin the fearsome task of making a copy of all the pages[121] as an insurance against loss of the original, not beyond possibility if it had to be entrusted to a sea voyage. This pleasing rate of progress was hardly likely to be maintained and it wasn't. However, the project never lost all momentum and was eventually completed very successfully.

The machinations with the *Analogicum* eroded much of Ingen Housz's time in the first half of 1793 and on the very point of going up to Hertford, to spend his fourth successive August with the Dimsdales, he received a letter from Bowood, where he expected to spend September and October. Lady Caroline Fox had made for him a personal gift – a neckcloth 'upon which' she had 'bestowed a great deal of pains'.[122] Lord Lansdowne had become involved in a 'heavy' discussion about its delivery – should it be sent to London or, perhaps, to Hertford; or would Ingen Housz rather 'have it wait your welcome arrival' at Bowood? The gift, and the invitation, was wrapped up, however, in anxiety. News had permeated through, even to Wiltshire, that Ingen Housz had, once again, been attacked on the London streets. The 'ladies', wrote Lansdowne.[123]

> . . . insist on knowing every particular in the greatest detail, from yourself, and how you do in all respects . . . We go to the sea in about a fortnight and shall return here by the first of September ready to receive our Dear Dr. Ingenhausen . .

The victim replied immediately but his thoughts were centred more on gratitude than his misadventures on the London streets.[124]

> I am much at a loss to find words strong enough to express my respectfull sense of gratitude to your Lordship, and to the amiable Lady who has condescended to bestow so much time to worck which in regard, as well as to the beauty of the workmanship of the hand employed in performing it, can scarze be worn by a man of any age without his providing himself with some or other antidote against the danger of self conceit or some degree of vanity ill suited to his age and plainness. I can't help feeling all ready too much pride not to consider it as a kind of profanation, if the Neckcloth should not be delivered to me by any other but the same graceful hand that made it.

Ingen Housz had been robbed in the middle of the day and within sight of a company of the King's guards; the third time in a year that he had been assaulted.¹²⁵ On one occasion he had been saved by wielding his walking stick and on another by the brave intervention of Dominique. And then, walking home to Marylebone Street, about 10 o'clock one evening, he had become aware of someone following him closely and, when he turned around, he was confronted by a 'stout ragged fellow' wielding a butcher's cleaver. Ingen Housz had been saved from serious injury only by being able to jump into a stationer's shop that was still open for selling newspapers: he had been saved by the press!

At Hertford there were, again, conversations on the alkaline soda water. The remedy had become so universally accepted and fashionable that most English physicians were now reciting personal examples of its successes in all kinds of clinical predicaments.¹²⁶ That much of this unbridled efficacy might derive largely from placebo effect did not seem to dent its popularity and there was little doubt that it genuinely helped many patients with bladder stones. It began to vex Ingen Housz that his colleagues in Austria were probably unaware of this new therapy. No doubt encouraged by Thomas Dimsdale, whose attitude had always been that medical knowledge should be freely and universally available, Ingen Housz decided to write to his friend Johann Scherer at Vienna. He would ask him to translate into German, and have published there, the three articles on the alkaline soda water that had been submitted to *Scheikundige Bibliotheek*.¹²⁷ For the present, though, it was work enough just to stay alive – even in the country there could be life-threatening problems.¹²⁸

> We have been being tried by a heat rarely experienced in this country. The thermometer was showing, around London, up to 90 degrees in the shade. Many people have died. I was in the country and in order to prevent the total collapse of the body, I bathed in the evenings in a river in which the water was very cold.

A cooler Ingen Housz was back in London for the first few days of September and then to Bowood about the 12th of the month¹²⁹ where, at last, he could receive his gift of a certain neckcloth. Proudly attired, he blended back into the sociable Bowood regime and, in private, caught up on some reading. He had brought with him copies of the recently published works of Erasmus Darwin – 'the economy and the love of plants'¹³⁰ which he judged to be 'pompous'. He took more satisfaction from a book on chemistry that had recently arrived from Vienna. It was by his protégé and near relative, Josef Jacquin and 'pleased me a great deal'.¹³¹ But his repose and pleasure in the literary world was shattered by an intrusion from the real one. Word arrived in England that Marie Antoinette, the Queen of France, had been guillotined on Wednesday 16 October. All Ingen Housz's worst fears surfaced again and the death shattered his equanimity; it was 'murder', as he saw it, of someone he had first met as a bouncy little girl in the bosom of her family almost thirty years before and whom he had known, later, as a young woman of the highest rank.[132]

> I am very downcast today and struck to the quick by much news, among which is the assassination of the Queen of France . . . I hope that all honest men will contribute of their best in order to prevent Europe being taken over by a horde of savages.

INGEN HOUSZ LEFT Bowood at the beginning of November 1793, going on to Bath, as had become his custom. This time the real motive for going further west, before returning to London, was to visit Bristol, the adopted home of a young doctor called Thomas Beddoes with whom Ingen Housz had been corresponding. Beddoes was a thirty-three-year old physician and polymath who had just opened a 'pneumatic institute' at a house in Hotwells, the Bristol hot springs resort at the foot of Avon Gorge.[133] A medical graduate of Oxford and Edinburgh, he had become one of the very first to create a paid career in science when, in 1788, he had been appointed a lecturer in chemistry at Oxford.[134] He was overlooked when the first chair of chemistry was endowed in the university, despite being the outstanding candidate, for his proclaimed political views were out of step with the academic authorities. It surprised none of his friends that he became one of the most voluble supporters of the French Revolution, at least in its early years. Otherwise, Beddoes had much in common with Ingen Housz for he was fascinated by the therapeutic possibilities of breathing selected gases such as oxygen. He would certainly have known of Ingen Housz's early experiments with oxygen in asthma at Vienna for, being multilingual and very well read, he was familiar with Ingen Housz's publications. He was also the son of a tanner/leather merchant, just as his older colleague. Both men were, again, parallel in physique although the more regularly rotund Ingen Housz probably felt tall and lithe when he first met Beddoes, the latter once having been described as 'uncommonly short and fat' by no less than Humphry Davy,[135] who was to become his apprentice.

Beddoes had moved to Bristol in early 1793 and opened his first clinic in Hotwells later that year, the intention being to combine clinical practice with medical experimentation.[136] His arrival set off concerted local protest, the established residents fearing that the spectral presence of yet

Dr. Thomas Beddoes.

more terminally consumptive patients would devalue the attractions of their 'fashionable resort'. Hotwells was becoming a place of desperation for the sick, but the location held promise of good business for Beddoes.[137]

> Hotwells teemed with local doctors and itinerant quacks who touted the hot springs a cure for gout and kidney stones, scabies and diabetes; Beddoes regarded the waters as worthless except as a magnet for a clientele who could sustain both his medical practice and his need for experimental subjects once his project took shape.

Beddoes published his first substantial medical book in 1793 – *Observations on the nature and cure of calculus, sea scurvy, consumption, catarrh and fever.*[138] It was a sizeable compendium and the 300 pages gave him the elbowroom to discuss various issues pertinent to clinical practice including what is now called photosynthesis. Beddoes had read the *Vermischte Schriften* volumes with penetration[139] and was rapidly making a name for himself as a progressive clinician developing techniques based on the 'new' chemistry and 'pneumatic' therapies. As a pioneer of oxygen administration to patients with lung diseases some fifteen years earlier, it was no surprise that Ingen Housz was intrigued enough to make a personal visit to Hotwells. The fact that Beddoes held 'democratic' views that had blocked his promotion at Oxford did not appear to be an obstacle in Ingen Housz's path – science trumped politics. Their meetings would have been clashes of avuncular prophet against impatient firebrand but any tension was probably creative, a symbiosis of experience and enthusiasm. Beddoes was full of ideas for testing, including protracted inhalation therapy with gases such as oxygen, seeing the need to try the gas for hours rather than for minutes. Ingen Housz, on the other hand, was a manual of pragmatism – on how to best produce the oxygen and the likely pitfalls when trying to administer the gas. But it was not all hot air; to Beddoes, the very presence of Ingen Housz was a social coup that would help him achieve respectability and boost his chances of raising money for his institution.[140]

As usual, the turn of the year saw Ingen Housz back on the road to London and he was there by 30 December. There were already indications that the new year – 1794 – the fifth that he had spent in England since fleeing France in 1789 was not going to be the last. It would turn into a year with an agenda. Life for Ingen Housz was about to stop being one of peripatetic amusement, of recreation, of following whims and seeking philosophical entertainment without much in the way of responsibilities. The free-wheeling enquirer was about to turn away from retirement and take on new scientific and literary projects. As he began fresh research and new writings, he was about to pass the point of no return (to Vienna).

20
Putting Down Roots

19 Floréal, An II. (Thursday 8 May 1794). Paris. As the procession leaves the Conciergerie they find that they're human skittles; fundamentally unstable and unable to steady themselves, with their hands tied behind their backs. Only by being so tightly packed in the carts can they stay upright on the journey over the uneven cobbled streets. Perhaps the sheer strain of trying to stay on your feet is a helpful distraction when you know you only have minutes to live. As the sombre train turns from the Quai de la Conference alongside the Seine into the Place de la Révolution, the twenty-eight condemned prisoners can see the scaffold and the 'machine' above the milling crowds. The minutes of remaining life are now in single figures. The mutual silence that has replaced all earlier converse is broken, once more, by the sobs and pleas of Jean-Baptiste Boullongne. He's in a dreadful state and some of the others have had their remaining dignity punctured by wetting themselves. On the whole, though, they're resigned to their common fate; they've been so, now, for several days; through their increasingly Spartan incarceration and their mock trial earlier today. Snatched, painful farewells and final letters are now only memories – from nearly a week ago. Since then they've all been in isolated limbo. Now their last journey is over and the Parisian Guard, holding back the boisterous crowd, mostly women, help or pull the men down from the carts. The mob closes round them at the base of the scaffold. It's about five o'clock and the sun low in the sky. Those of the condemned who are courageous enough to look up can see the outline of the wooden pillars, the angular blade, the release rope and, shuddering now, the large wicker basket into which their heads will roll. There is no ceremonial, no public condemnation: the bailiff simply unfolds a nondescript piece of paper and begins to read out names. Clément Delaâge is first. He mounts the five steps and practised hands lower him, face down, hands still tied behind his back, onto the wide plank. This is then pushed into position and his neck propped under the blade. He is praying.

Swishthunk.

One of the condemned notices an obscene gob of blood on the bald head of the man next to him. The blood is warm and the new owner remains oblivious to it.

'Louis Dangé-Bagneux.'

Swishthunk.

'Jaques Paulze.'
Swishthunk.
'Antoine Laurent Lavoisier.' Lavoisier's last thought as he climbs the steps are that he was foolish to have been so naive, assuming, that as a famous chemist and academician, his life was sacrosanct. Even one of his jailers had been heard to mutter the going mantra. *'The world has no need of philosophers.'*
Swishthunk.

SUMMER 1794 WOULD be the fifth one that Ingen Housz had spent in England, having fled France in the middle of 1789; and there were good indications that it was not to be the last. New scientific and literary projects might not come to displace his gentlemanly 'retirement' routines entirely but they would certainly superimpose themselves. It was to be a sacrifice worth making for, although ignored at the time, he was to publish highly important and ground-breaking insights into plant physiology.[1] His recurring annual cycle of winter and spring in London, summer in the country, autumn at Bowood and Bath and thence back to London began to break down. Deep friendships, some new and some rekindled, waiting for his health to restore, official responsibilities at the Royal Society and the beginnings of a new Hoogeveen book had all conspired, together with his growing apprehension of travelling across the European mainland, to hold him in England. Now the balance was about to tip, substantially, from dalliance to decision, from exile to domicile, as he began new research and new writings.

Ingen Housz did not seem prepared, however, to admit these realities to himself and certainly not to his family and friends in Austria. On Thursday 16 January 1794, before leaving for the weekly Royal Society meeting, he sat down to write his thanks to everyone who had helped in the house move at Vienna.[2] He also requested a parcel be made up of some of the German editions of his published books for it had been intimated that a willing carrier was about to travel to London, an international trader called Watzel. Then, following yet another tirade against freemasonry, he implied, strongly, that it was his firm intention to travel back to Vienna that summer for he only wanted the books to be entrusted to Watzel if the merchant was leaving for London 'very soon.' Otherwise they might arrive in England only 'after I have left'. This would be 'very inconvenient'. However, lest this be taken as a definite promise to return home, he added a rider: he regretted that the recent spread of continental military activity beyond the borders of France could only heighten the difficulties and dangers of European travel.[3] And yet, as if the inconsistencies of his family correspondence needed underlining, he saw nothing ironic in thanking his relatives for the bundle of recent letters that had arrived in London in the hands of a young dentist who had travelled, perfectly uneventfully, from Vienna.[4] Indeed, Ingen Housz was more focussed on the sad fate of this young traveller who, once arrived, had been confronted by a brutal truth. Although London was the largest and richest city and port in the world at that time, just dropping anchor was no guaranteed route to a fortune. 'Mr. Serre . . . appeared to believe that one only had to come to London in order to get rich, without even knowing one word of English.' Perhaps such notions of instant wealth, of golden pavements, had circulated in Vienna because of the exceptional success enjoyed by Joseph Haydn. Haydn, only sixteen months younger than Ingen Housz, had been persuaded, by violinist and impresario Johann Salomon, to visit London and write music for subscription concerts when the famous composer's patron, Prince Nikolaus Esterhazy, had died in 1790.[5] Haydn, already a celebrity, had been able to command huge fees for his services. The music he wrote in London was as inspired and popular as any of the rest of his output and he had returned to Vienna a much richer man. He was even tempted back to London for a second time in 1794/5 and was again a huge success.[6] His notable courage and stamina in traversing Europe twice in the politically uncertain times must have been an uncomfortable benchmark for Ingen Housz who spent year after year hesitating and shrinking from the journey.

January 1794 also brought the good news, from Breda, that Henry Ferdinand, Ingen Housz's 25-year-old nephew, fourth child and third son of Louis and Maria, had graduated MD from Leiden University on 26 September 1793.[7] This obviously gave Ingen Housz great pride and joy and Henry would, from now on, be plied with tit-bits of medical news that his uncle thought might advance his medical prowess and career. Another link with Ingen Housz's fatherland, the journey to the presses of Hoogeveen's analogical dictionary, was also progressing well. By February 1794 Ingen Housz was holding up to the light specimen pages from the printer and promising to send some across to Delft as soon as practicable.[8] Here was an emergent project that would need his continuing supervision. He had already recruited about a hundred potential subscribers from among his London friends and contacts and it was about this time that he began to conceive a new book of his own. He executed the plan, hatched at Hertford the previous summer,[9] to inform the physicians in Austria of the powers of alkaline soda water.[10] A bulky letter was written to his old friend and colleague Johann Scherer, in which Ingen Housz described, in some detail, the formulation, preparation and clinical use of 'the water'. Scherer realised the novelty and importance of the contents and had the letter published in the *Diaris medico Vindobensis* (*Viennese Medical Gazette*). The therapy was immediately taken up by many of the other Viennese physicians; so rapidly, in fact, and with such pleasing results, that Scherer regretted having 'sat' on the news, even for a few weeks.[11] Alkaline soda water quickly found its way into the new Viennese pharmacopaeia and a puffed-up Josef Jacquin was able to send a copy to his uncle in London the day after publication although his euphoria might have had more to do with the fact that he was a new father.[12] Scherer was also full of praise for Ingen Housz. '. . . so you see, that in our country, you still have the reputation of being a sincere physician far removed from all pretence.'[13]

Even though it was not his own invention, the impact of alkaline soda water on notoriously conservative Vienna, and on his reputation there, seemed to soften up Ingen Housz to an unexpected suggestion from Scherer;[14] that Ingen Housz translate everything he had ever written on the therapy into Latin, for publication at Vienna. It was an invitation Ingen Housz could not resist; Scherer's translation into German and publication, at Vienna, of *Vegetaux II* had, after all, been a success. Ingen Housz began to plot and scheme and events moved very quickly. By coincidence, he had been working on a new medical article, finally sent to Delft in October, for van Breda to insert into the next edition of his *Scheikundige Bibliotheek*. The topic was the treatment of intestinal worms[15] and this, too, Ingen Housz could see as valuable content in a medical book that might be published at Vienna. He also began to think of other new therapies on which he could claim some expertise. But a book, certainly one that required intensive translation work, stood a high chance of never being finished. Ingen Housz had been in this predicament before. He needed to take himself off on a 'literary' retreat where he could discipline himself and where introversion could not cause offence. Within a very short period he was writing to van Breda.[16]

> This serves only to inform you that I go for six months into the country, however not in the houses of friends, whom I have thanked for their invitations. I have taken a small house on a great heath 12 miles outside London, and so I will not have any, or nearly no, communication with philosophers. Letters however will reach me.

THE HOUSE THAT served as summer retreat for Ingen Housz in 1794 was on Twickenham Common.[17] This was then a large open area of sparsely-populated heath land to the north and west of Twickenham village. It extended to Hounslow across what is now the noisy final approach to Heathrow airport from the east. Things started very badly. It was just after Ingen Housz had left the metropolis, in the hope of undisturbed solitude and peace of mind, that devastating news arrived in London. Antoine Lavoisier had been among a large group of former patriciate civil servants and financiers (*Fermiers Général*) who had been show-trialled and guillotined in Paris on 8 May.[18] The famous French scientist had been publicly decapitated with 27 others. It took only 35 minutes to despatch all of them and the Place de la Révolution had been awash with their blood. To poor Ingen Housz it was another brutal death of a former close acquaintance: fresh evidence, if needed, that he must never tempt fate by inadvertently 'wandering' into the hands of the 'democrats'. In Jacobin France scientists were bundled together with other academicians and aristocrats and all tarred as impenitent counter-revolutionaries. Attempting the 'man of great value to mankind' defence was never going to sway a revolutionary tribunal. For Ingen Housz the peace and security of Twickenham and settling down to translating his papers into his beloved Latin should have been therapeutic but nightmarish Parisian

The guillotine in Paris.

scenes probably imposed themselves for some weeks. At least the 'healthy' environs were pleasant; similar, perhaps to much of the heath-like land around Breda and therefore reminiscent of his childhood. The weather, too, was good. 1794 was another long, hot summer.[19] Despite the shimmering heat, translating proceeded. The original 'letters to Deckers' on alkaline soda water were united into what became a fully-comprehensive dissertation on the new therapy, supplemented by detailed drawings of the apparatus used in its preparation that would be etched as copper plates. As Scherer was to put it in the preface he wrote for the book:[20]

> I hope that this treatise on the effectiveness of *aqua mephitica alkalina*, presented here in Latin at my request, . . . will be most welcome and of great use. One cannot want to know more on the history of this medicine, its nature and its preparation, its best dosage and regime, or how it dissolves chalky concretions, than one finds here.

The fact that Ingen Housz was preparing a substantial new book, in England, for publication at Vienna, was surely a smoke signal to Agatha, and to the rest of the circle of family and friends there. Surviving correspondence within the family, certainly between Ingen Housz and his abandoned wife, is very sparse but none of the extant papers contain, or refer to, an open declaration of the estrangement; perhaps there never was transparency. But the reality of the situation was inescapable. When, after an absence of six years, Ingen Housz proposed that Scherer take on the sub-editorial role and local management in the gestation of a new book to be pressed at Vienna and that the book be launched in the author's absence, the whole Jacquin circle 'knew' that the 'doctor' had no intention of leaving England. No-one who expected such high standards of himself in his own printed works could possibly delegate the later processes in a publication to a friend, however close and trusted, if they intended to be on the scene in the near future. If the new book was not going to bring Ingen Housz back to Vienna, nothing would.

Ingen Housz did not spend the whole summer hunched over a desk, however. All through the long days he was spending some time in the garden of his rented house, enjoying some fresh air whilst starting some pilot experiments. He had been party to conversations centred on the possibility that common chemicals, beginning to be readily available in bulk from reliable manufacturing processes, might have a utility in agriculture. Among the hotchpotch of ideas had been the likely advantages of sprinkling acids on to seedbeds and he could not resist neglecting his book manuscript for a few hours every day to get his hands dirty and test some of his own notions. But, unlike in a previous rural summer, one which had been totally dedicated to scientific exploration, fifteen years before and some seven miles distant, at Southall Green, the 1794 experiments were not instantly instructive; there was no *eureka* moment. The findings were confusing and inconclusive.[21] Not that Ingen Housz was despondent: testing ideas always has a momentum even if the direction of travel is unexpected. It may have been during these weeks of seclusion that he began to be able to organise what had been random ideas spinning around in his head for a very long time viz. how plants actually obtain their food and, indeed, what is their food? They were ideas that had their origins at least six years before when he was debating them with the French chemist, Jean-Henri Hassenfratz, in Paris.[22] They were to mature to huge significance over the next few months. And, never willing to waste experience, he also saw, again, how perspiration dominates inspiration in research.[23]

> Though the object of my retiring into the country was not agriculture, I made however, some experiments on that subject, as I had to my own disposition a house and garden: but as the result of those experiments was not uniform I neglected them as useless. They convinced me, however, that general conclusions from such experiments, if not often enough repeated, may produce disadvantageous errors.

Ingen Housz went on experimenting in his rented garden and translating texts into Latin until well into the autumn. The days were shortening rapidly before he asked Dominique to pack all the belongings, notes and equipment and it was Wednesday 12 November when they returned to Marylebone Street and all the distractions of London.[24] However, by the middle of January 1795, the bulk of the manuscript pages of his new book were ready for Scherer. Since they made up a packet that was too bulky to impose on the usual diplomatic channels, Ingen Housz delivered them to Count Fagel, Clerk at the Dutch Ambassador's house, back in London on a new tour of duty. He was helpful, as ever, and arranged transportation to Holland.[25] Safely on the European mainland, the manuscript would find its way, hopefully, to Vienna under a cover addressed to Prince Rosenberg, Chancellor and close friend at the Hofburg. Sending bulky objects without incurring costs was now a more acute problem for, as Ingen Housz confessed to van Marum.[26]

> ... having lost all my investments in Spain, in France and in Holland, and the best part of those I had in Germany – if not the lot – I have had to cut back as quickly as possible on my expenses.

The bundled manuscript may have been despatched free of charge but it travelled at a very hazardous time. After prevailing in a series of running skirmishes in north-east France and the Austrian Netherlands during the summer of 1794, the French army had disposed of any threat from the British and Prussian forces ranged against them and pushed up into the southern Dutch provinces by the middle of winter. The natural defences of the main rivers across the Netherlands were no hindrance as they were frozen solid and on 18 January the Stadholder fled to England on a fishing boat.[27] Aided by resurgent Patriots the French took over the country, declaring it a Republic. Nonetheless, the Ingen Housz manuscript arrived safely at Vienna, and in good time. However, its excursion was not finished. Prince Rosenberg had just left for Italy on a diplomatic mission and his office, following his instructions faithfully, sent the parcel after him, with all his other post.[28] Josef Jacquin, prompted to inquire after it at the Chancellery, was horrified to be told that it had arrived and then been sent out again. It could still be lost, even at this late stage of its journey, and he was very relieved when it suddenly arrived at his apartment two weeks later.[29] He was glad to hand it over to Scherer at the end of the first week of March. It then took five months to be published. In the meantime, Ingen Housz had been diligently writing the final chapters of the book and was able to use the goodwill of the chaplain to Prince Starhemberg, the new Viennese ambassador to London whom he had already befriended, to take them with him to Vienna early in March.[30]

Miscellanea Physico-medica – a '*Miscellany on the Treatment of Some Diseases*'[31] – was exactly that, a random digest of new therapies aimed at a medical readership. The volume was in octavo on 200 pages, the mere 36,000 words being more a reflection of the concision of the Latin language than want of content. Besides the detailed thesis on alkaline soda water, there were articles on treating fevers, on dealing with intestinal worms, on breathlessness, on skin ulcers and on oxygen being the essential gas of human respiration. The text then deteriorated somewhat into a casual catalogue of commonplace case histories. These appear to have been added by Scherer to fill in for the final batch of manuscript pages, the culminating chapters as intended by Ingen Housz, that never arrived at Vienna. The intended book was far from complete when it was committed to the presses in early August 1795. There was really nothing for the non-technical browser except, perhaps, the few pages on the wisdom of wearing woollen clothing but even this had a clinical theme: it appeared under the flag of a quote from Ovid – *principiis obsta; sero medicina paratur cum mala per longas convalvere moras* (*stop it at its start, it is late for medicine to be prepared when disease has grown strong through long delays*) – that prevention is better than cure.[32] There appear to be no

JOANNIS INGEN-HOUSZ

Sacræ Cæf. Reg. Apoft. Majeft. Confiliarii aulici & Archiatri; Regiæ Societatis Londinenfis; Academiæ Electoralis Theodoro-Palatinæ; Societatis philofophicæ americanæ Philadelphiæ; Societatis fcientiarum Haerlemenfis; philofophicæ Rotterod. Ultrajectinæ &c. &c. Socii.

MISCELLANEA
PHYSICO-MEDICA.

EDIDIT

JOANNES ANDREAS SCHERER,

M. D. Regiae Societatis Scientiarum Bohemicae aliarumque membrum.

VIENNÆ
TYPIS & IMPENSIS A. A. PATZOWSKY.
1795.

Jan Ingen Housz's one medical book (title page).

indications of the book's success or otherwise but when some travellers from Vienna brought Ingen Housz three copies to Bowood, in November 1795, he was certainly disenchanted.[33] It was his first major medical publication but it was not a book that gave him any pride and satisfaction. He feared that such a badly flawed publication would damage his reputation at Vienna rather than boost it. Nonetheless, he made sure that twenty-two presentation copies were distributed 'personally', by Josef Jacquin, among the Viennese royalty and nobility.[34] Ingen Housz also gave out one or two copies to friends in London but, to complete his embarrassment, these had been damaged in transit.[35] His disappointments upset him for some time, especially that the intended 'Contents' page had not been sewn in and that Scherer had failed to produce an index. Moreover, as readers hunted blindly in the densely printed pages they would find an unacceptable number of misprints. This time Scherer had let him down badly[36] and Ingen Housz probably felt little but *schadenfreude* when he heard, a year later, that the publisher, Patzowsky, had gone bankrupt.[37] Johann Scherer could not be blamed for other failings. Ingen Housz had to face the fact that more than half the book – a superficial revision of his three 'letters to Deckers' on alkaline soda water – was now old news, even in Vienna, and the remaining subject matter was probably too eclectic to have much impact. Having left Vienna in 1788 in order to be a hands-on midwife to a book being published many miles away, he had been seduced into the reverse situation and had re-discovered all the difficulties.

I T WAS ALMOST inevitable that Ingen Housz would become involved with a new semi-political organisation, the Board of Agriculture: he had known its driving force for nearly a decade. This board was the brainchild of Sir John Sinclair, a no-nonsense Scottish farming aristocrat and Member of Parliament. By the age of 40 he already had many achievements to his name and had been a fellow of the Royal Society for a decade. He was principally known for his comprehensive and respected economic survey of Scotland.[38] Clergy and parish officers all across Scotland had been plied with questionnaires in order to collect local economic 'statistics' (a word Sinclair introduced into the lexicon). The detailed feedback on population numbers, housing quality, household incomes, food prices, availability of work, and more, was a discomfiting description of the extent and depth of poverty in Scotland. To Sinclair it was a powerful spur to try to lift the efficiency of agricultural practices and since one of his particular interests was sheep husbandry he founded, in 1791 in Edinburgh, the British Wool Society.[39]

Sir John Sinclair FRS.

Surely, he felt, a rationalised interbreeding programme could produce strains that carried better wool and made tastier mutton. It only needed drive and organisation; and parallel improvements might be made, just as easily, in other domesticated animals. At the same time, Sinclair was well versed in financial matters and was able to advise William Pitt, the Prime Minister, on the best way out of a national liquidity crisis brought on by the outbreak of war with France in 1793. Sinclair's fiscal scheme was successful and his reward was Pitt's agreement that Sinclair set up, with a modest government grant, a 'Board of Agriculture and Internal Improvement', Sinclair himself as unpaid President. Pitt, of course, had an ulterior motive; being chief executive of an island race at war, the prospect of more home-grown food was very comforting. The Board was more a 'quango' than government department but it did eventually mature, in the nineteenth century, from an advisory body into the Ministry of Agriculture. It first met in August 1793, at Sinclair's London home[40] – a money-saving ploy typical of his frugality. Sinclair had formed around himself a group of 31 'ordinary members' who were virtually all landowning parliamentary friends, and invited nineteen of the country's high and mighty to be 'official members'. These included great officers of state, bishops and, notably, the President of the Royal Society, Sir Joseph Banks. Nonetheless, Sinclair's own ambitions for the board dominated its agenda.

To best advise government, he saw his organisation as a pool of information on new farming practices, as a nursery for research, and as a conduit for promising innovations. But his more controversial ambition, to repeat his Scottish survey in England, was punctured by the Archbishop of Canterbury and a coterie of landed aristocrats. The one salaried appointment in his gift was that of secretary to the Board. There was one obvious choice, the 52-year-old Arthur Young.[41] Young had made a life-long study of agriculture, especially in his native Suffolk, and in 1784 he had started a regular journal – the *Annals of Agriculture*. This had been an intellectual success but a business failure, falling way short of the number of subscribers it needed to make Young any income. It was also financial embarrassment that made Young hesitate before accepting the post of Board of Agriculture secretary. His main grievance was that the salary of £400 (£22,500) per year did not compensate him for having to live in expensive London in order to attend the weekly meetings of the Board.[42] But when he did accept it was almost inevitable that his *Annals* became its mouthpiece.

Away from their respective desks, Ingen Housz and Sinclair had been carrying on a lively face-to-face debate on soil fertility. The canny Scotsman recognised that Ingen Housz was an important resource for his board whatever the judgements in their sparring sessions. He wanted him to assess the growing view that the main nutriment of plants was carbon in the soil and to submit his opinions in writing.[43]

> Sir John very much insists that I give him a system of principles on vegetation of my own. I favour particular ideas on the topic that I haven't developed in my works and if I have peace of mind I will work on them.

Ingen Housz complied, sending Sir John a brief essay, 'On Manures', on 2 December 1794[44] but, as he had feared, marshalling his arguments had proved to be difficult. He put his writer's block down to the political situation; returning to the paranoid atmosphere in London had re-opened all the distress he had felt for his friends in France and the anxieties for his family in Breda. '. . . the horrors that are in progress in France and which menace our own dear Fatherland, and even all of Europe, have made such an impact on my mind that I can rarely fix my attention on scientific matters.'[45] A less painful distraction arrived in the form of a letter from Cambridge.[46]

The Hoogeveen dictionary had been 'put on the press'. Essentially this was good news but created yet more anxieties for Ingen Housz. It was now urgent that Jan Hoogeveen be pressured into writing the preface for the book and completing the brief biography on his father that he had promised to submit to the Cambridge University Press.[47] Ingen Housz seemingly offered an incentive in the form of an actual sample page from the printer.[48] This was probably a genuine promise but with the difficulties of sending it across to Delft, it could have been a well-motivated bluff. With France occupying, in effect, the United Provinces, there was now a communications hiatus between London and the Netherlands and, unless one or more letters have been lost, the next letter to van Breda or Hoogeveen is dated February 1796, a whole year later and even then only a single page that Ingen Housz persuaded a traveller to secrete on his person.[49] Ingen Housz would have been anxious for his family at Breda all through this period. No one could predict how the French would behave: the common knowledge of the atrocities they had committed after having invaded Geneva could hardly have been reassuring. Perhaps, in the end, and with some irony, he managed to cope with his worries only by distracting himself; by throwing himself into a new sphere of activity. The scratchy essay that he had given to Sir John Sinclair had been circulated among the members of the Board of Agriculture and had been well received. Sir John wanted him to work further on it, flattering the author mercilessly until he convinced him by promising to print and publish the end result.

Ingen Housz's essay for Sinclair, some 14,000 words, was finished by January 1796 but its mobilisation had been a complex process. Ingen Housz was desperate to do more than hypothesise but, to perform experiments on manures, he needed access to one or more plots of private land. He turned to his great friends, the Dimsdales, and went up to Hertford rather earlier in the summer than was his usual habit, to work in their garden at Cowbridge House.[50]. He was assisted, very ably, by Nathaniel Dimsdale, and they spent much of July and August 1795 beavering away on a plot of unused vegetable garden and tested the effects of various chemicals poured onto carefully prepared seedbeds. These might be described more as experiments with 'fertilisers' although today's fertilisers were then called 'stimulating manures.' This was in contradistinction to 'enriching' or 'feeding manures' such as rotting dung or compost. We might consider the difference as inorganic versus organic. Ingen Housz was certainly much better prepared and organised than during the summer of 1794. He and Nathaniel sowed various arable seeds such as wheat, oats, rye and barley, each in small furrows two inches deep and a yard long. The seeds were planted immediately after the furrows had been drenched in the various chemicals by means of a specially designed watering can that delivered a constant wide stream of the solution.[51] In one series they tried diluted common acids – sulphuric, hydrochloric and nitric – comparing between the acids and with control counterparts. The obvious question of which strength of any acid gave optimum effect could not be answered by these simple, qualitative experiments. However, the results were more convincing and promising than Ingen Housz had obtained at Twickenham and it appeared that sulphuric acid, at least when it combined with the chalky soil to form 'gyspum' (calcium sulphate), encouraged best growth. He was enthused enough to go back to his draft dissertation for Sir John.[52]

> Though the quantity of acid was, in these little experiments, much greater than I would have employed on a field, yet the general result was, that, very far from having hurt the vegetation of the seeds, we found the plants all very thriving, and in the most part of these spots the plants came out earlier than those which were not manured with acids. The plants thus treated were neither retarded nor weaker than the others, but rather stouter or more vigorous for the most part.

In particular he was relieved to find that none of the chemicals he and Nathaniel had used seemed to damage the plants in any discernable way; he was far from wanting to be the man who could be blamed for publicising innovative plant husbandry that might be harmful.

AFTER HERTFORD, THE former pattern of Ingen Housz's years in England re-appeared in his progressing to Bowood. After the intermission of 1794, he was warmly welcomed back and the idyllic lifestyle afforded by the Marquis's generous hospitality was soon underway –sociable meals, browsing in the library, lively conversations, carriage trips into the hinterland and daily walks on the estate to admire how Capability Brown's visions were coming to maturity. And when the Marquis was called away to London, earlier than usual, taking 'the ladies' with him but leaving one of his other friends, Baron de Baye, as company for him, Ingen Housz was seen to be so prominent a person in the household that he was asked to be 'in charge.' He dutifully reported that all was well on 1 December.[53]

> . . . unwilling to deturn your Lordship's attention from his important occupations, I will only mention that the Baron hunts after hairs and rabbits in boisterous weather and I after knowledge in the Library, both enjoying the

pleasure and happiness we owe to your Lordship's kindness.

The privileged sanctity offered by Bowood gave Ingen Housz the opportunity to return to his manuscript on manures. It took him four days to improve it to his satisfaction and it is interesting to see, from his original manuscript,[54] how the thrust of the article changed. 'On Manures' became 'On Agriculture' and then 'The Chemistry of Plants'.[55] The eventual, final title was different again but the most rapid evolution of thought was in the early stages of the writing. In the development of the article we can see how Ingen Housz had been on a significant cerebral journey. Thoughts previously orbiting differently in his mind began to dock together: thinking about manures had taken him back over two decades of a personal intellectual struggle. He had been forced to defend his discovery that plants, that all parts of plants, continually 'respire' as do animals – that is, that they constantly 'spoil' the air (absorb oxygen and release carbon dioxide). However, the green parts of plants, uniquely, and only in bright sunlight, accommodate an additional and reverse cycle, overwhelmingly and with such efficiency, that they 'purify' the air (provide our oxygen).[56] We have seen all these conclusions before; the outcome of Ingen Housz's 1779/80 experimental marathon. But now, in the 1790s, he had the luxury of thinking through these processes with the inherent clarity of the new chemistry. It appears to have made him focus on the fact that plants can actually process carbon dioxide – he himself had proven that they spoil the air at night and it was now clear that the noxious compound 'spoiling' the air at night was carbon dioxide. Now there was also evidence that carbon was the major building block of plants – 'coal is the primeval principle.'[57] He was back to manures.

The popular view, current at the time, was that Jean-Henri Hassenfratz, the French chemist, was obviously right; that plants must absorb the carbon they need, in some form of aqueous solution, from the soil in which they are growing. '. . . plants take up by their roots the coal already found in manure in a state of solution'.[58] Ingen Housz could not accept this. There were many instances of plants flourishing without their roots being even in contact with the soil – many plants in rain forests for example. Another objection was the obvious fact that large plants such as trees would soon exhaust the soil around their roots. And, in conversations with Arthur Young,[59] Ingen Housz had learned the truth about pure carbon not being good manure: charcoal, it seemed, had no influence on plant vigour when spread on the surrounding soil. Tentatively, almost apprehensively, Ingen Housz began to see the consequences of these difficulties with the doctrine.

Is it not more than probable that vegetables provide for their own subsistence by decomposing the common air and changing it with their organs into this almost general ingredient (carbon) . . . even without the assistance of their roots.[60]

The eventual logical impetus of Ingen Housz's ideas was promoted in the final versions of his article. It was very advanced for its time but a justifiable overturn of Hassenfratz's 'humus' theory. Indeed, one of Ingen Housz's propositions; that plants, and therefore all of us, obtain food from the air, was ahead of its proof. This would only come a few years later in the quantitative experiments of Nicolas-Theodore de Saussure at Geneva who was able to show that the carbon dioxide absorbed by green plant life was indeed the carbon source for their organic growth. Ingen Housz had, though, come to accept that Jean Senebier had been correct in arguing that plant chemistry was influenced by the concentration of ambient carbon dioxide and, having had the humility to accept this, and using the facility of the 'new' chemistry, he was able to complete the circle. The economy of plants was amazingly efficient –from 'carbonic acid' (carbon dioxide) in the air around them, and from water via their roots, they used the energy of sunlight to drive chemical processes that split these ingredients of their 'food and drink' into oxygen and carbon. His dissertation had turned into an account of the process now known as photosynthesis and though this brilliant exposition would be universally overlooked in favour of a last-minute afterthought pegged on to the end of his text, all the elements are there and the comprehension perfectly correct. So, too, was his grasp of another circular feedback in nature. The animal kingdom is the beneficiary of plant chemistry, as he had always argued.[61] There is an equally efficient cycle that relates all living organisms. Animals breathe the oxygen and eat the plants, or each other, while exhaling carbon dioxide and manuring the soil.

Christmas 1795 found Ingen Housz in Bath[62] although not with the Dimsdales who did not travel west that autumn.[63] He took lodgings in Pulteney Street[64] and someone called at his rooms, before the turn of the year, to make a fair copy of his long essay for it was there that 'The chemistry of plants' was finished by the addition of a final few paragraphs. This early copy of the manuscript came into possession of the Bath and West Agricultural Society and it has survived at Bath University.[65] But within days Ingen Housz had a brain-storm and extended his manuscript by a dozen short paragraphs before sending it to London. His idea was a new practice for landowners and farmers. It had long been traditional and unquestioned that agricultural land use should rotate: indeed it had been recommended by Virgil . . . *sed tamen alternis facilis labor* (but yet rotation lightens the labour).[66]

It had also become normal practice that crop rotation was interrupted, one year in every three or four, by leaving the land fallow. Therefore, cleared and tilled land was only ever productive for some three-quarters of the time, at best. While apologising for his poor English,[67] which

> could not be pardoned but in a foreigner who writes, frequently, letters in four different languages in one day and who published tracts in all these languages. Trusting your indulgence, I like to send you my writings as they flow from my pen . . .

Ingen Housz presented an idea, entirely a hunch, that he hoped could be floated publicly and tested thoroughly. He was flying a kite but his scheme might just overcome the fundamental inefficiency of leaving cultivated land fallow. It would allow perpetual land use - every field could be planted every year. His answer was a chemical one: farmers should regenerate tired land by ploughing it to bury any weeds and manure it with dung, all as usual, but then revivify it for immediate use with a thorough soaking of 'oil of vitriol' (weak sulphuric acid), the acid combining with the calcium in the soil to make 'artificial gypsum' (calcium sulphate). It was the beginnings, perhaps, of the knowledge-based use of artificial fertilisers in plant husbandry. Ingen Housz was fully aware that it was all conjecture,[68] finding an appropriate quote from Lucretius, subtly adapted, to make his point; '*per agro locae nullius ante trita solo* – by agriculture we are going over untrodden ground.'[69]

Sinclair was upbeat about Ingen Housz's dissertation, focussing immediately on the proposal for obviating the need to let land lie fallow. This was very much the kind of pragmatic, modernising content that he wanted in publications of his Board and he grasped the opportunity to have the essay printed, almost immediately, in the next official bulletin. Sinclair presented the article, under its latest title – *An essay on the food of plants and the renovation of soils* – at the next meeting of his Board. The assembled members followed, as usual, the lead of their domineering President and, in what would be a fateful decision for Ingen Housz, sanctioned publication of his essay in their next official circular. This was to be subtitled 'On the subject of manures' but Ingen Housz's article was only the third, of six, in a volume actually headed *Additional appendix to the outlines of the fifteenth chapter of the proposed general report from the Board of Agriculture*. The more philosophical parts of the tract, and the more important, were already on their way to oblivion. The title page shows the printer to have been William Bulmer, printer for the Royal Society, at Cleveland Row, St James' and the date given is 1796.[70] The cover was hardly a compelling invitation to readers and the print run was far from generous. But the Board at least ordered

No. III.

An ESSAY on the FOOD of PLANTS and the RENOVATION of SOILS.

By JOHN INGEN-HOUSZ, Body Physician to their IMPERIAL and ROYAL MAJESTIES, F.R.S. Foreign Honorary Member of the Board of Agriculture, &c. &c.

THE surest way to find out the real nourishment of organized bodies seems to be, to inquire what is the substance, without which they inevitably perish, and which alone is sufficient to continue their life. All animals require two ingredients for the continuation of their life; viz. atmospheric air and moist food, derived either from animal or vegetable substances, which food being received in the stomach, or some reservoir destined for that purpose, and being gradually digested and changed into different substances in the different organs, is applied to the whole economy of the animal body. Vegetables being deprived of progressive motion, by which means the most part of animals go in search of food, must find, in the narrow compass of space they occupy, every thing necessary for their subsistence. As they are in contact with two substances only, the earth and the atmospheric air, their nourishment must exist in either of them, or in both. The earth is necessary to the plants, as the only means to fix them stedfastly to the spot, by spreading through it their roots; but as earth contains generally moisture, salts, air, &c. nature has taken advantage from this circumstance, so that the filaments of the roots pump from the soil all that is offered to their suckers, and can be absorbed by them; but as some plants may live and thrive without being in contact with any earth, we ought to take it for granted, that the soil, or what exists in the soil, is not the only food of plants. Water is necessary to all organized beings, as without it no circulation of juices could be carried on; but from this necessity it can only be deduced, that water is a vehicle of the food, and by no means that it is the true nourishment of animals or vegetables—the less so, as it is an incontrovertible fact, that several plants can live without being in contact with water.—Thus the agave, cactus, aloe, cacalia, &c. live in the most dry rocks in the hottest climates, where it does not rain sometimes in the space of several months, and where the burning sun pierces all other plants, and even deprives the trees of all their leaves, and, what is extraordinary, the most part of such plants are full of juices. The nocturnal dew cannot give sufficient nourishment to such plants, as all other plants would also maintain themselves with it. But to be certain that those plants do not subsist by dew, we ought to consider only that some plants of that species may be kept alive in the hot-houses, either in pots, without being watered, or by hanging them up from the ceiling.

Margin notes: Best way to find out the true food of organized bodies. Why water is an ingredient necessary for all organized beings. How the most succulent plants can live in the driest rocks.

B

The food of plants (page 1). The first full outline of photosynthesis.

50 separately stitched copies of Ingen Housz's essay from Bulmer for the author to have and dispense, at his will, as presentation editions.[71] It was off the presses by the middle of February and Ingen Housz's contribution was very well received internally: on 24 February he received a letter, from Sir John Sinclair,[72] informing him that he had just been elected, unanimously, an Honorary Member of the Board of Agriculture; '... an honour to which your zeal for the improvement of agriculture so justly entitles you ...' The recipient

of the honour was clearly thrilled and proud, informing the family at Vienna that this new appellation should always be used on important documents and papers.[73] He began attending the weekly meetings of the Board in March.[74]

In the meantime, however, the readers of his 14,000 word article appeared to gloss over everything bar the proposed alternative to leaving land fallow. Even then, the few landowners who did eventually try the tactic used the gypsum instead of using organic manure and without adjusting their rotation routines. The King, 'Farmer George', was sent a copy[75] and with his well-known interest in agriculture did not just shelve it. But he, too, responded, superficially, only to the final few, pragmatic paragraphs. 'The King, having read it, requested me to demonstrate the new proposal, in his presence, in his park at Kew.' We have no further information on the details of this invitation. Like every farmer or market gardener who read the article and thought to try the scheme,[76] Ingen Housz himself found the 1796 season was too far advanced, through the agricultural year, to try it that spring. But the prospect of experimenting for the King at the prestigious farm at Kew was an too good an opportunity to miss and Ingen Housz wrote, in July, to Alexander Ramsay Robinson, the farm manager at 'Richmond Gardens Farm',[77] planning a project for early 1797.

It is a pity that the inspirational gem in 'the food of plants', the early presentation of photosynthesis, was universally overlooked. It is, at the same time, no surprise. Sadly, the fact that Ingen Housz had correctly formulated a biophysical principle was lost to posterity for 200 years – until a retired professor of microbiology published the story in 1997.[78] Perhaps Ingen Housz should have realised that such important but abstract concepts would have been better appreciated at the Royal Society but he may have assumed some attention there, anyway, with Sir Joseph Banks a member of the Board of Agriculture. Perhaps he should have reconsidered the misleading title and given a good friend chance to make some editorial suggestions. His article had been buried in a dry-as-dust document of very limited circulation, and even the limited readership was not composed of gifted lateral thinkers.

S PRING 1796 BROUGHT a letter, by hand, from Josef Jacquin at Vienna.[79] It was carried to London by a team of mining experts and managers who, staying in England for only a month, presented a very good means of reply. Anxious not to miss this opportunity to compensate, perhaps, for the bungled *Miscellanea*, Ingen Housz visited Bulmer at his premises in Cleveland Row, St. James' and pestered him for his personal examples of *The food of plants* and packed up twenty-two of them for Josef to distribute in Vienna.[80] There was a copy for Josef to share with his father and with Scherer. There was one for the Emperor, the Grand Duke (of Tuscany), Princes Rosenberg, Starhemberg and Dietrichstein and for more than a dozen other Viennese nobles. Remaining copies were distributed among Ingen Housz's many English friends and, through a contact travelling to Hamburg via Holland,[81] to his nephews at Breda, van Breda at Delft, Deckers — still in practice in 's-Hertogenbosch — and van Marum at Haarlem.[82]

It was pleasing news that Nikolaus Jacquin, Ingen Housz's brother-in-law, had retired from his university positions at Vienna to be replaced, wholly, by Josef, his eldest son, someone Ingen Housz had long cultivated for high academic rank.[83] But not all the incoming news was welcome. War with France, and the military alliance between Britain and Austria and their expeditions in Flanders from early 1794, had choked off multiple communication channels across much of Western Europe. When Ingen Housz was told, by Drummonds, that no letter of credit representing his regular court pension had arrived at the due time,[84] he consoled himself that notification was merely delayed by postal difficulties. After several more months, however, the continued non-appearance of a letter from Stametz was causing him alarm.[85] Having no remunerable position in England meant that continuing the pleasant life of a 'gentleman' was possible only if his Austrian pension payments arrived safely and on due time. The problem appears to have been resolved sometime later in 1796 but only to recur: it was a serious and on-going new anxiety for him with Europe in such a state of flux and conflict.

Ingen Housz was far from being the only person having financial anxieties during the 1790s. The war with France was affecting everyone in England and had been for some time. Among many other consequences, it had become increasingly difficult to find subscribers for Hoogeveen's analogical dictionary and almost impossible, when one did, to extract actual money from them.[86] The pleasing early progress of this project was hardly likely to be maintained, and it was not. However, the project did not lose all momentum and was eventually completed. There might have been heartache about the prospect of finding subscribers but Samuel Parr had taken up the challenge with some gusto. By late 1794 there were easily enough to meet initial production costs[87] but thereafter names were being added only in ones or twos. Parr was still working his way through his network of contacts only to hear that some of his signed-up subscribers had died and there was little hope of their families not wanting the money returned. Parr reassured Ingen

Housz that he would continue to 'do all that enthusiasm can do'.[88] He was out of action, however, for much of the rest of that year, having fallen from his horse while in Monmouthshire and 'was near losing my life'.[89]

When Parr did eventually recover and return to Cambridge, in time for Christmas 1794, rumours were circulating that the University Press was now turning away manuscripts that had missed their deadlines. Parr panicked and assumed that the Hoogeveen manuscript had not arrived, writing immediately to Ingen Housz 'that you lose not one moment in sending down the manuscript'.[90] However, had he checked, Parr would have found the papers already in Cambridge. No-one had given them to the Press, however, because he had been insistent that a 'corrector' (proof-reader) be in place before any type was set. Having agreed with an old College friend, Dr. Maltby, that he 'do the honours', Maltby had then resigned his fellowship in order to marry[91] and left the university, creating something of a 'Hoogeveen vacuum'. Dr. William Pearce, a fellow of Jesus College, Dean of Ely Cathedral, and a friend of Parr took upon himself to break the log-jam[92] and the presses began to roll during 1795. At first only sample pages were pressed for it took nearly a year to replace Dr. Maltby. Finally, a Mr. Tyson, fellow of Emmanuel College, took on the task for a fee of £50 (£2,800).[93]

Though Ingen Housz had been far less involved than he had been for *Particles*, he was now able to take pleasure from watching the pages appear and know that publication was now a certainty. Everyone had agreed that Jan Hoogeveen should be pressed to write a foreword for his father's work together with a brief biography of the author.[94] However, the fraught international situation had made it difficult to communicate with Holland and it took some time both to reach Hoogeveen at Delft and to retrieve the scripts from him. They finally arrived in 1798 [95]and were soon printed and stitched with all the other pages. The University Press noted that they had spent £202 (£11,300) on paper and printing[96] and when the first edition (*Dictionarium analogicum linguae Graecae*) was released to subscribers in 1801 it was put in the bookshops priced at 30 shillings (£84).[97] Nevertheless, it sold well and there was a further edition in 1810.

THE ON-GOING CONTINENTAL skirmishes meant that there now seemed fewer prospects than ever of Ingen Housz travelling back to Vienna. The spring of 1796 represented the seventh time that this annual question had confronted him and so negative was his response this time that he was now assuming that he would die in England. On Friday 10

June,[98] he sat down and wrote a letter to Arnold Ingen Housz, his oldest nephew at Breda. He then sealed it in a cover and walked down to Charing Cross and handed it in at Drummonds Bank with instructions that they were to send it to Breda on his demise. On opening the letter at home in Breda, come the day, Arnold would find that he was instructed to arrange a power of attorney that would be respected in England and then come across to London to collect clothes and other belongings together with a box of valuables and remaining cash deposited at Drummonds. It also told him to identify, and send on, a new Will intended for use at Vienna. Arnold would not discover a huge legacy. Ingen Housz's financial situation was now considerably worse than when he had arrived in England nearly seven years before and deteriorating all the time. His cash balance at Drummonds was down by some 40% and 'fearing a revolution in England' he had redeemed all but a few of his securities around the year 1793 and invested them in Imperial bonds issued at Rotterdam.[99] With the French republicans now in charge of the Netherlands, this nest egg was currently inaccessible if not endangered. At the very least the bonds now held no value on the open market and he was holding worthless pieces of paper for the time being, even if he could have redeemed them.

Having brought all his worldly affairs into order as best he could by the middle of June 1796, Ingen Housz repaired to the country again. He first went south from London, to the village of Betchworth in Surrey. There he was the guest of a friend from the Royal Society, James Petty. Petty owned 'Tranquil Dale', later called 'Broome Park', a very large house on the northern outskirts of the village.[100] The property had extensive tracts of land attached and the two friends planned manure experiments that Petty would perform, on some suitable plots, during the forthcoming spring.[101] Otherwise, Ingen Housz appears to have been idle and at leisure. He did write to van Marum however; a reply to a letter that had taken nearly four months to arrive from Holland. For once there is a thin thread of self-revelation. He would be able to work, he told his friend at Haarlem,[102]

> if only my spirit hadn't been dampened so much by the distressing perspective of seeing the disorganisation of social order by the new political principles . . . destructive to the public well-being. My head is full of ideas and I rhapsodise with endless notes, thoughts and memoirs but I haven't the stamina to make them into a work for the public.

Ingen Housz may have been verging on depression again during the summer months of 1796 and uncharacteristically slothful but he did keep up, at least, his usual summer routine by going up to Hertford in August and to Bowood in September.

In Hertford he was keen to know the results of that year's manure experiments. They had been performed, on his behalf, by Stephen Dolignon, a close friend and neighbour of the Dimsdales. There was further confirmation of the efficacy and safety of applying 'oil of vitriol' but still no comparative information on sulphating versus leaving land fallow: his big idea remained untested, a nagging frustration that haunted his long journey, in mid-September, to Wiltshire, to Bowood. And, as if to taunt him further, a parcel from Sir John Sinclair followed him there on 28 September.[103]

> You will herewith receive three copies of the *Hints on Vegetation*, one for yourself, the other for your two pupils, Miss Fox and Lord Lansdown . . . You will see, by p. 8, that I have inserted, in a note, your remark in behalf of philosophers proving theoretically how useful they may be in furnishing the Husbandmen with useful hints . . .

Whether Sinclair was just being jocular as the mood took him or whether he had heard that Ingen Housz was low in spirits is not clear but he did see it as his mandate to spur on the newest member of his Board.[104]

> Remember that you are not to allow their agreeable and enticing conversation (Caroline Fox and Lord Lansdown) to divert you entirely from philosophical and useful pursuits, and that you will be interrogated very closely at the next Meeting of the Board as to the new discoveries you have made and the fresh Experiments you have in agitation.

Hints on Vegetation,[105] the booklets that arrived at Bowood, were, as described by Sinclair himself, 'cursory' notes of what were then understood to be the principles of efficient plant husbandry as well as a schedule of the obvious outstanding questions for agriculturalists. Ingen Housz was quoted, though not by actual name, as a good example of someone who saw that it served no purpose to be one of those 'philosophers who drew up theories in their closets' and never got their hands dirty. In essence, Sinclair was making a plea for 'uniting philosophy and practice.' But Sinclair's personal petition to his new Board member to stay focussed on science appears to have fallen on deaf ears.

Ingen Housz was diverted from any active 'philosophy' at Bowood. However, distraction proved therapeutic; his 'busy-ness' bounced back. The recuperative powers of Bowood were as potent as usual. He was soon as sociable as ever and, it appears, very engaged with the dozen or so younger members of that year's 'college'.[106] There was new accent on youth and lots of lively activity. Henry Petty, the Marquis's son from his second marriage was now in his seventeenth year and about to go up to study at Edinburgh University before matriculating, later, at Trinity College, Cambridge. Joint host with his father, he was obviously a very dynamic and intelligent youth (and would become Chancellor of the Exchequer at the age of 25). He was fostered by Ingen Housz who took a great deal of interest in the boy's education. Another guest was Anne Garbett, grand-daughter of Samuel Garbett, the Birmingham industrialist and sometime financial adviser to Lord Lansdowne. As much friend as adviser, Garbett and his family had always been made welcome at Lansdowne House and Bowood. Anne and her brother had arrived at Bowood on 20 September[107] in the care of their father. Ingen Housz appears to have been infected by the gaiety of the company and, though it was harmless enough, he was intrigued, we might almost say entranced, by the 22-year-old Anne. The Garbetts should have left after two weeks but Anne caught a cold on a trip to see the Cherhill White Horse. Their departure was delayed by ten days. Perhaps Ingen Housz took on the role of physician, for the patient was ill enough to be confined to her room for some days. But his professional veneer appears to have cracked. Although he later, in pencil, wrote the word 'burlesque' across his copy of the letter, he wrote to her in most unusual (for him) terms three weeks after the Garbetts finally left Bowood.[108] Phrases in this bizarre letter implicate him as having had inappropriate feelings for Anne and the ostensible reason for writing the letter at all seems very flimsy at best: a contrived, and impossible, story of a minuscule bible, bought on a shopping trip to Chippenham, having followed Anne to her home and then reappearing at Bowood, all in a day.

Maybe Ingen Housz was bouncing back from a phase of exhaustion and restlessness, if not depression, and his usual rationality was subsumed to euphoric emotions. In any event it was all very short-lived and innocent but did provide the youngsters with a pretext for a prank.[109] Knowing that the 'doctor' had been away from his wife for several years and supposing his motive to be that she was a 'nagging shrew', a second Xanthippe,[110] the youngsters forged a letter. It purported to say that she had travelled to London and was planning to come down to Bowood. They surreptitiously kept a watch on their flustered prey and when they found him ordering a chaise to meet the next coach to London in Calne, they were, in 'peals of youthful laughter', forced to confront him and confess. And after the mirth there were two outstanding ironies. Firstly, Anne's eyes were far from focussed on the elderly doctor. They were fixated, instantly, and permanently, on a new arrival at Bowood whom she would not have met if her departure had not

Sir Samuel Romilly.

been delayed – Samuel Romilly. It was love at first sight for both of them.

At 39, Romilly was a very successful barrister. His peers described him as a 'powerful advocate',[111] well able to convince a jury by his 'acute and perspicuous reasoning' and with an ability to expose the 'sophistry' of his antagonist. Romilly, a regular of the 'college' throughout the 1790s, arrived at Bowood on 6 October and, as Anne recuperated from her illness and began to circulate in the company again, the two of them walked out of doors and then went for rides together. On the day that Anne left, Romilly declared his love to her and she agreed to become engaged. It was a whirlwind romance, appearing from nowhere within a week, Romilly writing that he had been 'captivated by the beauties of her person and the charms of her mind'. For her part, Anne described her husband-to-be as a 'vastly amiable man' and that she had 'never spent so happy a month in my life'. They were to marry on 3 January 1798.[112]

The second irony was that at the same time that his head was spinning with images of a 22-year-old girl, Ingen Housz was scribbling out, with great urgency and in all sincerity, a long tract that he entitled 'Advise to a young nobleman'. It was intended as an avuncular road map through life for Henry Petty.[113] It could have served as an end of term sermon at a boys' school. It praised regular meals, sobriety and physical exercise. Life-long education was strongly advocated together with the habit of writing down acquired knowledge the better to retain it accurately, for 'blundering notions are worse than total ignorance'. And it promoted self-knowledge as the best defence against being tempted into bad company, bad habits and dissipation. 'It should never be forgotten that men act in general more from passions than from reason.' But like all adolescents, Henry probably felt that he already knew everything there was to know about life: the 'School of Hard Knocks' still awaited him. The guidelines were more likely to show him, at a later date, where he had gone wrong rather than be a prospective navigational aid to happiness and morality. Nevertheless, the long essay that Henry perhaps took with him to Edinburgh, tells us much about Ingen Housz's fixated thinking at that time. Before long the diatribe started to coalesce around his obsessive fears of revolution, the same apprehensions that imposed themselves even in the 'burlesque' letter to Anne Garbett.

> . . . man in the state of nature is the only animal that hunts his own (and) continues his career of natural ferociousness even in a civilised state, by destroying . . . with impunity his fellow creatures.[114]

The fear of 'Jacobinism' had planted itself very deeply in Ingen Housz's mind.

One night, in early November, perhaps after an overindulgent or late supper, Ingen Housz retired to his bed at Bowood to have a night disturbed by a terrible nightmare. Though he woke in the middle of it, bathed in perspiration and with his heart pounding, he had a small drink of water and then gave in to sleep again and the images continued, worse, even, than before. It was the convincing and real dream referred to earlier. Revolution had broken out in England and horrific scenes he had witnessed in Paris in 1789 were translocated to the streets of London. He abreacted by writing to Nathaniel Dimsdale.[115] As an MP, Dimsdale should be reminded of the fearsome possibilities if '100,000 or more *sans-culottes* should attempt to land . . . and begin putting to death all those suspected not to be of their party.' The effect that the growing revulsion of 'Jacobin democracy' had had on Ingen Housz's personality was an emergent concern at Bowood, especially the rather tedious recitation of his fears. Caroline Fox felt obliged to warn her brother[116] and his wife, that when they visited Bowood, they should 'have compassion upon the infirmities of an old man whose head is turned with politics and revolutionary apprehensions.' It was more than the understandable repetitiveness of old age[117] and other companies might have been less tolerant; there is a very thin line between endearing eccentricity and gnawing drivel.

The eventful 1796 'college' broke up when Lord Lansdowne went to Bath to take the waters; for he, too, was beginning to suffer the tolls of age and a life full of heavy responsibilities.[118] The Dimsdales, however, appear not to have gone to their house at Bath that year: Thomas was another person who was unwell. Ingen Housz, after some resurgent clinical activity with Christopher Allsup in Calne had come to a suitable closure, travelled to Hotwells in Bristol to see Thomas Beddoes. Thoroughly recovered from his anxiety-induced turpitude, he was to find his long-standing interest in 'aerial medicine' rekindled.

INGEN HOUSZ HAD come to know Christopher Allsup very well. This Calne surgeon-apothecary was the local doctor customarily summoned by Lansdowne and his household whenever illness struck and had long been the 'Bowood Doctor'.[119] Ingen Housz and Allsup increasingly kept company, seeing patients together and trading favourite remedies.[120] Ingen Housz was particularly enamoured of Allsup's recipe for treating intestinal worms (a concoction of very finely powdered tin and a purgative) and recorded the formula for his own reference. He also kept notes of one or two unforgettable Calne patients.[121] One was a 50-year-old minister of religion/schoolteacher who lived on The Green at Calne, a Reverend Davis. He was suffering from a bladder stone that had been giving him such indescribable pain and strangury, and for so long, that he was emaciated and depressed. He had even been desperate enough to try to break up his own stone by introducing, himself, a long-bladed knife along his penis, into his excoriated bladder. Ingen Housz was delighted to be able to put an end to this decline, and heroism, by prescribing a liquid ounce of alkaline soda water per day and Mr. Davis began to improve in a very short time. Another Calne case-history, equally horrific, was that of John Highett, the 22-year-old son of a wheelwright. John's father, James, had died in 1790, to be followed to the grave by two of his daughters, both in their teens; Mary in 1794 and Sarah in 1796. All three had died of pulmonary consumption – tuberculosis – and John's widowed mother must have been grief-stricken when he, too, began to cough up blood and waste away. By the time he was introduced to Ingen Housz he was extremely weak and coughing up so much purulent phlegm that it was thought that he probably had a lung abscess. Ingen Housz was fascinated by the next part of John Highett's history and recorded it in some detail:

> He was advised by an old man who operated a lime kiln in Bowood Park

that he should inspire, every day, the smoak of the kiln during an hour. He found an almost immediate relieve by it: the spitting changed in a few days from a purulent matter into a clear transparent mucus; his cough and pain diminished and the hectic heat (fever) abated.

Ingen Housz was ever scouting for patterns of human response to disease that might reveal underlying principles. This history rekindled his interest in pneumatic medicine. Here, on the face of it, was a live example supporting the theory being peddled by Thomas Beddoes at Bristol – that tuberculosis was made worse (and might even be caused) by the breathing of well-oxygenated air. Therefore, subjecting patients to the opposite should be of therapeutic benefit and John Highett's was just the kind of case history that he was anxious to collect. Ingen Housz would oblige in reporting this instance of effective clinical folklore personally and it was not to be the last time that he would be involved in the scientific scrutiny of such country practices, a pleasing activity when they proved to have some foundation. Unhappily, John Highett's future was not so fortunate. He went on to develop a rectal abscess and despite going to London for surgical drainage at St. Thomas' Hospital, arranged by Ingen Housz and paid for by Lord Lansdowne, he died three years later, aged only 25.[122]

It was in the depths of December 1796, then, that Ingen Housz travelled from Calne to Bristol to call on Thomas Beddoes once more, the two men having kept up a regular correspondence. There had been significant progress at the clinic in Hotwells where Beddoes had been treating countless tuberculous patients.[123] It was no surprise that James Watt, the famous engineer and colleague of Matthew Boulton, had co-operated in designing and building, for Beddoes, an apparatus for delivering gases for inhalation by such patients.[124] Two of Watt's own children were consumptive, his fifteen-year-old daughter very ill.[125] Nor was it a surprise that Matthew Boulton grasped the entrepreneurial opportunity. Within weeks he was sending commercial versions of Watt's 'air machine' to his agent in London[126] for sale to the population of the stifling metropolis. And this was all long before Beddoes had been able to find convincing evidence for its efficacy in his increasingly difficult circumstances. Despite mixed results for his efforts, desperate patients were more than willing to clutch at straws and throughput at the clinic had grown to the point that the 'pneumatic institute' had outgrown the accommodation. Beddoes was actively trying to raise money for a larger, purpose-built, foundation elsewhere in Bristol. He was hoping that, given better working conditions, he could garner consistent evidence for his theory on the cause and best treatment of tuberculosis by way of inhalation therapy. Although he was in no position to subscribe to Beddoes' building fund, Ingen Housz was always a supporter worth cultivating, particularly as a close friend of the all-powerful President of the Royal Society, Banks. The advances made, at Bristol, in inhalation medicine would have convinced Ingen Housz that he had been right, nearly twenty years before, to wrestle with the early difficulties of administering oxygen to patients.

O N THEIR RETURN to London, in late January 1797, Dominique set off to retrieve any letters from their *poste restante* address, 6 Fleet Street. His master's immediate attention, however, was given to one that he had collected, himself, from the Austrian Embassy; Starhemberg having left him a message. It bore an imposing crest. It was from the new Emperor of Russia, Paul.[127] Catherine the Great had suffered a brain haemorrhage on 5 November 1796 and had died the following afternoon.[128]. Tsar Paul, as he now was, had requested copies of recent 'works' by Ingen Housz. The author wrote, immediately by paid post, to Josef Jacquin at Vienna,[129] asking him to send on copies of *Miscellanea Physico-medica* to St. Petersburg regretting, perhaps, more than ever, the book's shortfalls. He need not have worried.

The bureaucratic stranglehold of the paranoid high officialdom of the Russian court was uniquely stifling. Josef, when he received his uncle's request, in early May, instantly sought the advice of the Russian ambassador at Vienna.[130] The envoy's response was remarkable: he predicted obstacle after obstacle. Even as representative at a very powerful foreign court, Prince Andrey Razumovsky was not allowed to send letters or parcels, from third parties, to St. Petersburg. If he did choose to ignore this standing order there was another one. He was obliged to send 'philosophical' matters to the Russian Academy of Sciences where the contents would be judged as suitable or not suitable for His Majesty, the Tsar. And Jacquin should understand that virtually nothing was ever seen as suitable. Then, if he, Rasumovsky, or anyone else, tried to by-pass the system by addressing such material directly to the Tsar, it would be diverted, without question and without exception, to the academy. Winston Churchill's famous description of Russian politics – that it was 'a riddle wrapped in a mystery inside an enigma' – has a long pedigree. Josef appears to have allowed his frustration to well up into an angry outburst[131] but it was clear that there was nothing further he could do to fulfil his uncle's request.

Back in London, in February, Ingen Housz had heard that a traveller was going to Vienna through Hamburg.[132] Knowing that it would not cost him, Ingen Housz sat down to write a much longer and more considered letter to Josef Jacquin. He was rueful rather than annoyed that the only recent letter he had received from Vienna had taken ten months to arrive. Not that the news was uplifting when it did come to light: several high-ranking officials of the Habsburg court, with whom Ingen Housz had served for many years, had died. Prince Orsini-Rosenberg, a close friend, was among them. Then there was the painful news that many of the Transylvanian and Saxon mines, where he had invested on the prompting of Nikolaus Jacquin, a supposed mining expert, had been abandoned. The central European mining industry was collapsing. Ingen Housz, Jacquin and many others had been misled[133] and the list of investments that Ingen Housz had made and which had failed was now yet longer. It all added to a pall of gloom hanging over late winter at 35, Marylebone Street.[134]

> . . . here I am badly accommodated, not having a home of my own and finding myself in a life situation that, by and by, wears me out with the anxieties that there could be disaster before I die . . . I did not know that I could become so unhappy.

And in the very next letter, barely a week later, there was a sense of hopelessness.[135] Physical and emotional exhaustion had returned with a new wave of depression linked to a 'flu-like cold. '. . . I do not know what to do or where to go. I am too old and too infirm to travel anywhere . . .'[136] Reading Erasmus Darwin's latest book,[137] in which the irreligious physician had the temerity to air notions of the evolution of species, did nothing to cheer him up. Everywhere Ingen Housz looked he saw atheism and Jacobinism.[138] There was, though, some prospect that people would at last come to understand the true horrors of the French Revolution by reading another new book, one where Ingen Housz had the privilege of reading the manuscript draft, knowing the author well. The Jesuit priest, Augustin Barruel was also a refugee of the French revolution. *Memoirs serving as a history of Jacobinism* would eventually run to four volumes. The first volume alone made the point well enough as far as Ingen Housz was concerned and he felt less isolated on finding that Barruel, too, saw much of the origin of revolution in freemasonry.[139] Ingen Housz was anxious for the book to be widely distributed: 'I hope it will open the eyes of sovereigns, philosophers and statesmen, all of whom have lost the power of reasoning.' By a striking coincidence, James Petty, Ingen Housz's host at Betchworth the previous summer, was someone who could help Abbé Barruel with his subsequent volumes.[140] Ingen Housz put the two men in touch, knowing that Petty could relate his horrifying experiences of 1785 when, travelling in Europe, he had found himself entangled in a short-lived revolution in Transylvania. A description of this outbreak of human butchery, and of its equally brutal suppression, could further the impact of Barruel's publications – Ingen Housz himself had known of an aristocratic woman who had fallen into the hands of the mob and had had her breasts and hands cut off before she was finally killed. Whether or not poor Petty had spent years trying to suppress any similar flashbacks seems to have been ignored by Ingen Housz in his fixated anti-revolutionary lobbying.

The traveller passing through Hamburg was also kind enough to leave letters there for van Breda and Hoogeveen. Ingen Housz had heard from neither of them for many months.[141] His letters to them were matter of fact: there was good reason to believe that 'along with the present liberty' went the 'opening of letters'[142] and he was conscious of the need not to put his friends in danger. On a happier note, Ingen Housz had been talking to Sir Joseph Banks. In early February 1797 the royal physician had been to his house in Soho Square to admire some recent books on the 'new chemistry' and had taken the opportunity to discuss the latest situation regarding the election of foreign members to the Royal Society.[143] There was some expectation of forthcoming vacancies[144] and Ingen Housz promptly wrote to van Marum. Here was a long-awaited opportunity to honour his friend and Ingen Housz set the wheels in motion without waiting for a reply. A list of candidates was announced[145] and a ballot paper posted at Somerset House after the 1797 summer recess. Van Marum received the unanimous support necessary, several other fellows writing their names under that of the proposer, Ingen Housz,[146] and van Marum was duly elected (on the foreign list) on 19 April 1798.

A year earlier Ingen Housz had travelled to Kew where he was to spend a month setting up manure experiments. The present 'Botanic Gardens' at Kew began their existence in 1719 when Princess Caroline, the wife of the future George II, took an interest in the land around their summer residence, Richmond Lodge. She planned a landscape garden beyond the deer enclosure.[147] With the northern boundary of the territory fixed by the Thames, the gardens eventually extended a long way south to comprise what are, now, more than 300 acres of world famous cultivation. George III, in his turn, took great pleasure from being at Kew. A newer residence, in the north-east corner, now called the Dutch House, (still standing and recently refurbished in contemporaneous style) was where he, his wife, Charlotte and their burgeoning young family could live, simply and modestly, each summer, away from the intrigues and claustrophobia of Windsor and St. James' Palace. Repairing to Kew between May and October each year became a long-standing routine for the King and Queen while their children were young although, later in his life, Kew had unpleasant connotations for George himself; his doctors having incarcerated him there, strictly isolated from family and friends, when he was emotionally broken from porphyria in the late 1780s. Ingen Housz did not expect to see the King himself at Kew in February but he did encounter old friends: William Aiton junior, who had succeeded his late father in 1793 and Franz Bauer, the artist from Vienna, now ensconced in the park. Ingen Housz stayed a month, or thereabouts, presumably living in a house near the Kew farm, conveniently close to the intensively cultivated vegetable gardens. Unfortunately, despite trying to arrange things, well in advance, with Alexander Ramsay Robinson, the keeper of the Royal parks and farms at Kew and Windsor,[148] Ingen Housz did not find the conditions he wanted. He had been assuming that there would be one or more plots of land that were scheduled to lie fallow during the 1797 season. The whole object of his trial was to apply acid to one of these plots and sow it without resting it in order to show that there was a simple means of obviating the need for land to have rest periods, a highly inefficient use of cultivated land. This was not to be; none of the plots made available to him were in the right phase of rotation.

However, he took the opportunity to test, again, the effects of various concentrations of 'vitriolic acid' (sulphuric acid) on otherwise equal plots of land, land that was far less chalky than at Hertford.[149] It was all second best but he worked in a field that had not been, nor was due to be, fallow, dividing it into six equal parts. It was ploughed and manured evenly, all as usual. On one part he simply sowed some barley. On the second part he applied 'pearl-ash' – an alkali derived from garden bonfires – before sowing more of the barley. Then, on the remaining four subplots he sowed barley from the same stock having drenched each with increasing strengths of sulphuric acid.[150] At least he could hope to see different rates of growth in the acidified plots and determine if there was an optimum concentration – too much might be harmful after all. There was the un-edifying prospect of creating hostility in the farming community if his ideas were followed over-enthusiastically, grain was killed and crops lost.

However, just as Ingen Housz left Kew in late March – he had other experiments planned elsewhere – the barley had germinated well in all the trial plots and was growing vigorously. He would rely, now, on reports from

Ramsay Robinson. Unfortunately, the old adage, that if you want something done well you must do it yourself, was to apply. Robinson never really performed the regular and reliable observations necessary for the trials to be worthwhile. All that Ingen Housz learned, again, was that none of his progressively stronger applications of acid had been toxic to the growing cereals, even on a soil much less chalky than at Hertford.

S PRING 1797 SAW oscillating levels of general anxiety in England as the threat of French invasion waxed and waned. Ingen Housz was more fretful than most of the population, not least because he was better informed. His daily routines in England, now very well established, were probably therapeutic. There were Royal Society dinners and readings, committees of the Board of Agriculture, private dinners and meetings; time soon passed and all the while he was performing yet more plant physiology experiments. Some, on a small scale, were in his rooms,[151] where he was trying the effects of more chemicals on germinating seeds, while other, outdoor, trials were set up in the garden of his friend Charles Morton.[152] Dr. Morton had succeeded Matthew Maty as principal librarian at the British Museum where he, too, lived and where he had a private plot of land.[153]

The results, once again, were less informative than he had wished for. He could not hope, of course, for the real evidence he was seeking – that treating soils with chemicals could supplant any need to leave land fallow and unproductive. It was now the end of the second sowing season following making his major proposition and he still had nothing to support it; it was all very disappointing. His reputation as an agricultural innovator was still on the rack. As a consolation he had made a new friend, one that opened up a new source of graceful hospitality. A fellow physician from the Royal Society circle, Richard Brocklesby, had introduced him to John Anthony Rucker, a very successful city merchant, of German extraction who had an import business centred at Hamburg.[154] Rucker had built himself an impressive Palladian mansion, West Hill, south of the Thames at Wandsworth, then a village in pleasant countryside having views north to Fulham and Chelsea. An invitation to Ingen Housz to spend some of the summer there had been irresistible and the anticipation of long warm days in idyllic surroundings was heightened by a very particular and impelling invitation to Bowood in Lord Lansdowne's own hand.[155]

'... we do not propose opening our doors till the 1st of October to anybody but Mr. Ingenhouse to whom they are always open ... for we think you can be nowhere that you will meet with more unaffected regard.'

Ingen Housz was flattered to be the one exception from the embargo at Bowood, sensing, perhaps, that the heady exuberance of the previous year's 'college' had been too hectic a precedent for the host. He rapidly wrote his acceptance and also acknowledged Lord Lansdowne's commission that he buy, in London, an 'air machine' for use by the family at Bowood, in particular for Elizabeth Vernon who suffered from asthma. He could easily imagine the light-hearted defiance that had occurred when 'the ladies' had made it very clear to the Marquis that they expected him to exclude their 'doctor' from his prohibition. They would have been pushing on an open door, as they well knew: the relationship between Ingen Housz and Lord Lansdowne himself had become an intense friendship of deep mutual regard. It was witnessed as much by the reply as by the invitation.[156]

> I think myself much honoured by your Lordship's kind expressions towards me, in excepting me from the general restraint laid on his other friends as to the time of opening for them his doors. To such constant and unaffected demonstrations of friendship, I can possibly afford nothing in return but the sincerest sense of that gratitude, with which I have allways been penetrated. . . . However I hope to enjoy once more that very tempting happiness . . . if no sudden and unforeseen emergency forces me to quit this till now so happy a country.

Ingen Housz was back in London, from Wandsworth, for a few days at the end of July to do the necessary shopping. John Rucker had taken his advice and was also spending five pounds (£280) on an air machine; the manganese that was used to produce the oxygen, and the other chemicals, costing a mere 'trifel'. A Mr. Chippendale sold the apparatus at his shop in Salisbury Court on Fleet Street[157] – he was Matthew Boulton's agent in London.[158] In addition, Rucker wanted a good telescope, apparatus that Ingen Housz no doubt promoted, for he would then be able to court popularity, as usual, by showing hosts and other guests at West Hill around the firmament on clear nights. He hurriedly wrote to William Herschel asking for a price.[159] He was fortunate – the craftsman/astronomer had several seven-foot telescopes that were virtually ready for sale. They were certainly expensive, at 100 guineas (£5,900) for the cheapest, a price that must have made even the very wealthy Rucker blanch and hesitate. Of equal or more value to Ingen Housz personally was a sincere invitation to revisit Slough that came with the quote, an invitation he relished.

Ingen Housz was at Bowood by mid-September 1797, in time to advise Caroline Fox on the best counsel she should send to her brother, the third Lord Holland. His wife had just miscarried. In fact, the Hollands came to stay at Bowood during the latter half of October: Lady Holland was back on her feet and a restful stay at Bowood was probably seen as part of her recuperation. Henry Fox, Lord Holland, was Caroline Fox's younger brother. He had succeeded their father in 1774, inheriting Holland House in Kensington. His recent marriage, in 1797, to Elizabeth, Lady Webster, divorced by her much older husband, Sir Godfrey Webster after Holland had embarked on a passionate affair with her in Florence,[160] had been the source of much pruriency in aristocratic circles. She had borne him an illegitimate son in November 1796. The Holland marriage was to be very successful, however dubious its beginning, and Holland House, an impressive building in 500 acres of prime land to the west of London was to become a glittering social and literary centre. The new arrivals at Bowood were made very welcome – their acceptance into such 'polite' society helping to subdue the scandal.

Otherwise, the two months that Ingen Housz spent among his Wiltshire friends passed quietly. There were no over-excited letters and he spent no time moralising for the benefit of the young. Ingen Housz was now very familiar to everyone at Bowood, as was his ever-faithful servant, Dominique. Ingen Housz had no scruples in going 'below stairs'[161] and was ever on the hunt for domestic implements that he could adopt as apparatus for experiments. Even though there were few houseguests that year his demonstrations were still in demand and he had learned how to make most of the reagents he needed from basic constituents. A rare image of Ingen Housz and Dominique involved in just this sort of activity has survived[162] and shows the tall, spindly servant and his somewhat less willowy master engaged in producing gas under water in a so-called 'pneumatic trough' – probably improvised from a wash tub. The artist of the image is unknown, as is the author of the poem below it but this is well worth repeating for it gives us, delightfully, a vivid image of Ingen Housz as Bowood philosopher cum thaumaturge (conjuror).[163]

> The Sage, ambitious to instruct the Fair,
> Extracts from Cabbage Leaves his vital air,
> Shews how each Plant can boast the double use
> Of latent fire and alimental juice:
> Planets and Satellites appear at will,
> Proclaim his pow'r and verify his skill.
> His mighty thunder then he deals around,
> The very walls re-echo to the sound:
> Dominique with fear and wonder gazes,

Following the Doctor thro' his magic mazes.

Perhaps that autumn at Bowood was a peaceful interlude before an expected storm. The whole country was now in fear of imminent French invasion and Lord Lansdowne left for London much earlier than usual.[164] The prevalent air of resignation to fate permeates a letter to Josef Jacquin that Ingen Housz sent on 23 November.[165]

> ... despite my age I am in pretty good health and shall, no doubt, continue to be so until a revolution puts the whole country to fire and the sword, when I shall expect to perish alongside millions of others. Any revolution here would be far worse as the revolutionaries are ready for it whereas in France the common people were totally unaware of what was going on (until the explosion occurred).

The non-political news was also depressing for Ingen Housz. The feedback from the various manure trials was inconclusive. He already knew that his own experiments at Kew could never prove his central thesis, that sulphuric acid might replace the need to leave fields fallow, because he had not found free plots in the royal gardens at the right point in their rotation. The evidence that high concentrations of the acid were not toxic was useful enough but he had already shown this at Hertford with the Dimsdales in 1795. Otherwise, he heard only of 'mishandled' experiments – of farmers who had let their enthusiasm befuddle their thinking and simply tried sulphuric acid as a replacement for the usual organic manure such as cow dung on plots that were in the wrong stage of the rotation. The point of replacing a year lying fallow with normal preparation, acid application and then normal sowing, had been missed, almost universally.[166]

A despondent Ingen Housz returned to London on Tuesday 12 December. Caroline Fox had implored him to break his journey on the outskirts of London and visit her brother and sister-in-law at Holland House. Lady Holland, recovered from her miscarriage, was now recuperating from a new setback and Ingen Housz had been giving proxy medical advice during his last few days at Bowood.[167] The Hollands, on their way back to London in late October, had been stopped by a highwayman on Hounslow Heath. Whether they were robbed of any personal items is not known but they had at least been able to go on their way physically unharmed. Lady Holland had certainly been shaken emotionally, however; once again she had taken to her bed. Now there was an opportunity for Ingen Housz to become involved personally, for Caroline Fox pleaded with him to call and advise her brother and sister-in-law face to face; she had continuing concerns for Lady Holland. In the event, however, the ageing and tired physician did not deviate on his homeward journey. The December darkness and stormy weather turned him away from any notion of a diversion. Perhaps he felt he had nothing more to contribute to Lady Holland's recovery – he had already advised by proxy that she should not lay abed whatever the traumas of the robbery; rather, she should take plenty of exercise. This failure to call was not taken as an offence and within a few days he received a proper invitation to Holland House in the hand of Lord Holland himself.[168] This was accepted on Christmas Eve and six months later Ingen Housz was apologising for having made more visits to his new friends than might be considered good etiquette.[169] It was a signal of change, of the beginning of a time when roles would be reversed, of a phase in his life when it would be Ingen Housz himself who would need the support of friends rather than vice versa.

21
A Noisy Quietus

Way below a clock strikes three, slicing the dark dense peace of the house. It is half expected. Midway between drowsy and jumpy, he's also heard the chimes for one o'clock, and for two. The stertorous breathing is his focus. Had it been regular it might have been soporific and Dominique might now be asleep in the chair. Instead, he tenses up every time it slows down and fades, fearing, knowing, it will stop. The end is proving difficult to define. In the next quarter of an hour he's twice convinced that death has come, only to be shocked into realising that there is still life force, strengthening and quickening to a surprising vigour before waning once more. A candle begins to flicker and he crosses the room to light another one. A mere glance at the unconscious form in the bed convinces him that there has been no new gross movement, no distress, no awareness. The last few meaningless mutterings – in Dutch, he thought – had been just before midnight; the last coherent sentence or two twelve hours earlier. Instinct tells him this cannot continue for long but time has lost its momentum. Another crescendo of breathing, this time with a sinister sibilance, is followed by another one. He reaches for his pipe, his tobacco pouch, puts a plug into his pipe bowl but then hesitates with his thumb over it. The ebbing tide of sound, of sinking life, seems longer this time. Now the silence in the house is complete, the stillness absolute. Suddenly his own breathing is deep with finality. A single tear rolls down his left cheek and into the corner of his mouth. It brings a bitter taste with it, of tobacco smoke. He stands over the bed, rigid with reality, for some minutes. Four o'clock chimes. In a quarter of an hour, if nature is not still playing tricks, he will end his vigil and wake the house as instructed: Lord Lansdowne had been most particular.

INGEN HOUSZ WAS to suffer increasing torments for having a foot in two camps; for living in London and maintaining a home in Vienna. By the late 1790s he found himself charged with having to contribute to the costs of two wars. In December 1798, William Pitt, acutely aware that England was living above its war-torn means, introduced a new levy that would, forever, be linked with his name – income tax.[1] Initially this was imposed only on those who were endowed with a substantial income but Ingen Housz found himself in the net. He was forced to make a detailed and extensive appeal, in writing, to the officers of his London parish, St. James', pointing out that he was in receipt of no income in England, that he 'gained' no money by 'practising physic' as they had supposed and that he was only resident by default.[2]

> Considering that I, by living in this expensif country, am deprived of the comfort of a wife and family, comfortable house, coach and horses, which the court of Vienna has always furnished me with, I aught rather to be considered a downright madman to live here in a scanty lodging of 52 pound a year, without being properly served, in an advanced age of near three score and a half, if my constitution was such as to allow me to undertake the long and tedious and dangerous journey to Vienna: besides, my travelling equipage and luggage is still in Holland whither it is forbidden for me to goe.

It was a one-sided summary, a defence statement to a jury; all of it true (almost) and designed to evoke sympathy. It appears to have been successful. However, Ingen Housz then went on to wander from his brief. He could hardly have expected a sympathetic hearing for some of the other extenuating circumstances he recited which were simply repercussions of being an expatriate. They were, however, very pertinent to him individually and to his growing monetary predicament. The Austrian administration, even deeper than the British Treasury in the financial mire so often the consequence of war, had been even more aggressive in trying to raise revenue. It had raided its citizens' savings as well as their incomes. A third of all stocks and bonds held in the banks were to be liquidised and the cash 'loaned' to the Emperor. This, of course, caused a cascade of selling and a precipitate crash in market prices. Enormous fortunes evaporated, irrespective of what had to be gifted to the war chest. Ingen Housz, like many other investors in Austria, saw his Viennese 'nest egg' implode; his certificates in the vaults at Stametz turn to virtually worthless bundles of paper. Rather than crystallise a huge loss he chose to raise the capital he owed, technically, by selling some of his silver, some jewels 'of my wife', and other 'superfluities' at Vienna; Josef Jacquin bustling to the Mint to try to achieve the top prices.[3] Ingen Housz even asked Drummonds to send some cash, worth some 2000 Florins (£5,900),[4] to Vienna. Penury loomed; the value of his pension was being rapidly eroded by inflation even if it appeared and now the new Emperor, Francis II, followed suit in imposing an income tax - a 10% 'war tax' on all state pensions and salaries.[5] But for Ingen Housz all this was only act one of a tragic personal drama.

Early in February 1798 Ingen Housz received a letter from Vienna. It was important enough to have come through the diplomatic channels and be sent over to Marylebone Street by Starhemberg, the Ambassador. It was from Ingen Housz's brother-in-law, Professor, now emeritus, Nikolaus Jacquin. Ingen Housz unfolded news of a personal disaster in the guise of a desperate and unprecedented measure, an Imperial edict, brought on by the free-fall economy of Austria. A communal madness had descended on the Hofburg. All the earlier measures to balance the budget had proved totally ineffective and a new plan had been conceived – in some haste, it appeared. In fact, the scheme was so clumsy and deluded and the 'experts' advising the Emperor as hopeless as the sinking economy, that Francis II was forced to revoke it within weeks. However, it was in action long enough to cause an incalculable degree of individual distress with traumatising tentacles that reached as far as London. And Ingen Housz was to have no knowledge of its abrupt cancellation for many months.[6]

Because most of the provincial towns and cities in his Empire were bankrupt, and insurrection loomed, Francis II had given his authority to a statute that compelled, at a few weeks' notice, all wealthy Viennese families, and pensioners with an annual income of 600 Florins or more, to leave Vienna and go and re-settle in any provincial Austrian town of their choice. Affluence, so the theory went, would follow its proprietors. The only exceptions were the very elderly, the infirm and those who were still 'serving' the court. The latter group would, however, be expected to 'decline' their salaries voluntarily.[7] The panic and confusion this ill-conceived law created within just one Viennese family circle, the Jacquins, reveals why it could never have worked. There were too many vague definitions waiting to be interpreted to advantage. And the question of how it was to be policed was left begging. Many rich Viennese shrewdly mothballed their city apartments and moved to their retreats in the neighbouring countryside as if it were summer, a simple suburban migration to no fiscal effect. The Jacquins could have done this: their house at the royal botanical gardens was, strictly, outside the city boundary, but it was overdue for refurbishment with many broken and rotting windows.[8] Winter there would certainly not be easy. On the other hand, Josef was neither a pensioner nor yet, in his own right, rich. Could not his widowed father stay living with him in the city? What constituted a family after all? But Agatha was clearly in a state of panic, seeing no loophole available for her to exploit. She had enclosed a brief note inside her brother's letter to her husband. Her distress is palpable, even after two centuries and Ingen Housz, for his part, must have suffered the agonies of being unable to help.⁹

> What my brother has written you makes me very worried and I request that you reply by the first post, before I have to leave Vienna. What must I do and where should I go? I think going to Pressburg (Bratislava) would be best as everything can be transported there by water and not so much will get broken. Nor will it cost too much. Perhaps I should just rent a room outside the city although this is most dangerous because of thieves . . . I must leave by 15 February: do I give notice for a quarter or do I have to pay rent for a half-year?

And if this was not distressing enough, there was more bad news. Returning to the letter from Jacquin and trying to re-read it at a more controlled, rational pace, Ingen Housz found that his first impressions had been painfully correct. Lurking among the clauses of the edict was, by a 'Catch-22', the inevitable stoppage of his pension, the status of which was proving to be as critical as its value. The promise made to him, thirty years before, by a grateful Maria Theresa, that his annual monetary reward for having safely inoculated her children would be for life and carry no commitment, implied that it was a guaranteed 'annuity.' Now it was being re-classified as a 'salary' and, after a self-imposed exile from Vienna and long absence from his official position, he was being put down as a 'quiescent' courtier who no longer deserved such payment. But if he risked rushing back to his post and hoping that he would still be accredited as an active court physician and counsellor –for which there was no certainty – he would be expected to forgo his 'salary' anyway. Then, to cap it all, if he was no longer required at court he would be forced into a pointless exile away from Vienna. Small wonder he was apathetic. It took many weeks before he was calm enough to make a coherent response to the family, his mood, even then, soured by the unsolicited advice of officious strangers.¹⁰

> If Mr. Sonnenfeld believes that I should return only to have my pension income confiscated how can I come back? He surely takes me for an imbecile if he believes that I need his counsel to return . . . but if I know, that on returning, I am to be threatened with exile from the capital, my return serves me no purpose, for I prefer death to a condemnation so little merited and contrary to my conditions of engagement, without the sanction of those who called me, from the other side of the world, in order to save their august family from an affliction which had destroyed almost all their ancestors.

After a flurry of correspondence in August, an exchange that framed far more questions than answers; Ingen Housz received no post from Vienna for four months. He did not know if he still had an apartment there; if his wife was still living in the city; if his remaining personal property was secure; whether he had been struck from the court roll. He was half afraid that his Viennese connections had been totally severed;[11] that in telling his wife, rather derisorily, to throw herself on the mercy of her rich relatives, her family had decided to cut him off.

Ingen Housz appealed directly to the Chancellory at Vienna when he was confronted with another new ruling - that all courtiers must re-apply for their positions every three years. Here, too, there was a long and agonising silence. Eventually he learned, through the ever-loyal Josef, that he had lobbied unsuccessfully.[12] Presumably, he was now a non-person at court. He also learned from his nephew that a parallel appeal – that the considerable sum of money he had already donated to state coffers be accepted in lieu of the new demands – had also been denied. In fact, his diligent nephew, pulling rank as a university professor, had been to the Hofburg on several occasions to present his uncle's case in person, all to no avail. And Ingen Housz's personal pleading to an old friend, Count de Sauran, now Minister of Finance, had led only to the receipt of one of those letters that politicians are wont to send, a fawning profession of 'admiration', 'respect' and 'esteem' promising nothing.[13] In fact, for many months the only positive financial news to come Ingen Housz's way was from his manservant. Dominique revealed that he was hiding, somewhere secure, some £48 sterling (£1550) in cash and that he wished his master to deposit it at Drummonds so that it could be transferred to his small account in Vienna, also at Stametz. It was money he had saved from his spending allowance and accumulated from tips while staying, with Ingen Housz, at grand houses: his salary, £10 (£320) per quarter, was still being paid into Stametz in Vienna by Agatha.[14]

INGEN HOUSZ GAINED some respite from his money problems, in the autumn of 1798, by an academic distraction. He became embroiled in a personal correspondence that related to Edward Jenner's privately published description,[15] in September 1798, of his studies and experiments on cowpox and its apparent power to protect people against smallpox: 'vaccination' (from 'vacca', Latin for cow) as opposed to 'inoculation'. The pamphlet, universally known as *The Inquiry*,[16] was a milestone in the history of medicine and is seen by many as the very genesis of immunology, although the actual story is rather more complex than most people realise. Jenner was actually obliged to publish a second paper, in early 1799,[18] compensating for the deficiencies of the first and Ingen Housz's valuable intervention has never been recognised.

Edward Jenner, son of the local vicar, was born in 1749, at Berkeley in Gloucestershire.[17] Like Ingen Housz, he became a surgeon-apothecary but by serving an apprenticeship rather than going to a university. He did, however, enhance his medical studies by being attached to, and living with, John Hunter, the well-known London surgeon. A very warm master/pupil relationship evolved into a guiding friendship by correspondence after Jenner returned to Berkeley to set up in practice. The rigours of his profession were not sufficient intellectual stimulation for him and, with the encouragement of his famous mentor, he wrote a paper describing his discovery of the 'non-nesting' habit of the cuckoo, for which he was elected, in 1789, a fellow of the Royal Society.[19] Otherwise, he was still a relatively unknown country doctor of modest means. He had long been intrigued by the folklore that cowpox, not a rare disease in dairy counties, conferred protection against smallpox. It was the reason, it was said, that milkmaids were always 'pretty', for they

Dr. Edward Jenner FRS. Courtesy of Dr Jenner's House: Birthplace of Vaccination, UK

had never suffered the facial scarring of the dreaded disease. Jenner had been collecting case histories that supported the story and, remembering Hunter's favourite dictum[20] - 'why not try the experiment?' - performed his trial in the spring of 1796. On Saturday 14 May he inoculated the eight-year-old James Phipps, son of his gardener, with fresh cowpox serum from a patient, Mary Nelmes, the daughter of a local landowner/farmer. After the lad had fully recovered from his induced cowpox he impregnated him again, this time with fresh smallpox serum (the original type of inoculation that came to be called 'variolation', *variola* being Latin for smallpox). James remained fit and well. At no subsequent stage did he show any signs of smallpox.[21] Equally important, Jenner recorded and published his experimental findings.

It is worth noting, at this juncture, that though Jenner performed his trial perfectly well, he made no effort to repeat it and test its consistency before drafting his first paper on *Variolae Vaccinae* (literally, the smallpox of the cow) late in 1796. Nonetheless, the experiment is now one of the most famous in the history of medical science. The manuscript was sent to Sir Joseph Banks, President of the Royal Society, as a submission to the *Philosophical Transactions* of the Society.[22] It recorded Jenner's collected case histories and his single experimental finding of 1796, all supporting his view that 'the Cow-pox protects the human constitution from the infection of the Small-pox'. Banks was not medically qualified and consulted Dr Everard Home, a London physician and Royal Society fellow; paradoxically, he was the late John Hunter's brother-in-law. Home replied to the President, we now know,[23] in a letter, of 22 April 1797, advising that Jenner's work should not be published before more experimentation:

> ... that '20 or 30 children might be innoculated (sic) for the Cow pox and afterwards for the Small pox.

Banks heeded this advice and rejected Jenner's paper. Its author must have been disappointed but also very concerned that his big idea might have been stolen, that precedence of publication might be jeopardised. He knew that few secrets were secure at the Royal Society; it is even possible that Ingen Housz knew of the paper from this time. Nonetheless, Jenner did decide to embark on more experimentation despite the fact that smallpox was uncommon in his practice and cowpox even less prevalent; it might be a considerable time before he could perform, and test by later variolation, the couple of dozen experimental 'vaccinations' suggested by Home. Perhaps he could compromise – perhaps five or ten trials would suffice? And, in effect, this is what he did. After more vaccinations during the spring of 1798, though not all of them tested by variolation, Jenner extended his former manuscript. This time he eschewed the Royal Society and arranged a private printing. *An inquiry into the causes and effects of the variolae vaccinae: a disease discovered in some of the western counties of England, particularly Gloucestershire, and known by the name of the cow pox*,[24] to give it its full title, was published by Sampson Low of 7, Berwick Street, Soho. Although the dedication to his friend, the Bath physician Caleb Parry, is dated 21 June, the book only appeared on sale on 17 September 1798, for seven shillings and sixpence (£12), at London booksellers in and around Fleet Street.[25] Jenner's ploy, of reporting case histories as naturally occurring experiments that supported his hypothesis, survived the re-write. This was perfectly acceptable for the time even though much of the evidence was therefore hearsay, from friends and colleagues, many of them at some considerable distance. His true experimental findings were much improved, however: they were now from several trials of vaccination. But many propositions in the paper were still entirely speculative and there was, at best, only a hint that it was critical to diagnose cowpox accurately; it being, after all, only one of several ulcerative conditions of cows' udders.

Ingen Housz spent much of Tuesday 24 July 1798, at home in London, dealing with the latest batch of painful correspondence[26] relating to the withholding of his pension and the other difficulties created, for him, and for his wife, by the financial crisis in Austria. His whereabouts during August are uncertain but he is known to have revisited William Herschel's observatory at Slough on 11 September.[27] Perhaps he was on his way to Wiltshire where he was certainly installed, at Bowood House, by early October. One of the hottest topics of conversation in that year's 'college' was Jenner's book[28] and the other 'members' were now able to call on the renowned expertise and opinions of Ingen Housz. This may have been a bonus for them but it must have made it difficult for Ingen Housz not to become fixated on Jenner's proposition. He could certainly see the advantages of inoculating with cowpox to protect against smallpox; variolation, even after decades of refinement and cumulative experience, still carried risks, ultimately of death. With cowpox, however, human patients never died and the pocks, though large and gruesome, were always localised to the site of inoculation so there was no risk of generalised scarring, especially on the face. Equally important, perhaps, was the fact that human cowpox was far less contagious. Inoculated patients did not have to be physically isolated: how to avoid spreading the very disease you were trying to prevent was a problem solved.

However, Ingen Housz was very alarmed to find that Jenner's tract was built, in essence, only on a sparse experimental testing of the theory.

Among his peer group of 'philosophers' Ingen Housz was, perhaps, the arch exponent of repeating experiments endlessly until the evidence for a suspected effect was consistent and overwhelming. Suppose Jenner's practice caught on but proved to be, at least in some circumstances, fallacious? All the pioneering work of inoculation with smallpox serum through much of the previous century might now be undermined by an ineffective but fashionable intervention. It was possible to see, even, a consequent explosion of smallpox cases, and deaths; the resurgence of the 'speckled monster'.

At least, with Bowood situated in dairy farming country, Ingen Housz had the opportunity to learn more about cowpox and he turned, not surprisingly, to his old friend Christopher Allsup for local information. He found that Allsup had immediate knowledge of what seemed to be a very relevant case history and Ingen Housz was impelled to write to Jenner. Uniquely, this letter, the first of several that he was to write to his new medical opponent, someone he would never meet, is not in his own hand. It is in that of Lady Caroline Fox and in later correspondence with Ingen Housz she reminds him of her willingness to remain his amanuensis. [29] The letter was dated 12 October 1798.[30]

> ... I thought it my duty to inquire concerning the extraordinary doctrine contained in your publication ... the first Gentleman to whom I address'd myself was Mr. Alsop, an eminent Practitioner at Calne. This Gentleman made me acquainted with Mr. Henry Stiles,[31] a respectable Farmer at Whitley near Calne, who thirty years ago, bought a Cow at a Fair which he found to be infected with what is called the Cow Pox—this Cow soon infected the whole Dairy; and he himself, by milking the infected Cow, caught the disease which you describe, and that in a very severe way, accompanied by pain, stiffness, and swelling in the axillary glands. Being recovered from the disease, and all the sores dried, he was inoculated for the Small pox by Mr. Alsop. The disease took place: a great many Small pox came out, and he communicated the infection to his father, who died of it. This being an incontrovertible fact ... (it) ... cannot fail to make some impression on your mind and excite you to inquire farther on the subject, before you venture finally to decide in favour of a doctrine which may do great mischief, should it prove erroneous. I will make no further observations, as it is far from my wish or my intentions to enter into any controversy with a man of whom I have conceived a very high opinion—Let it suffice, to have communicated to you in a friendly way, a fact which may awaken your attention.

Ingen Housz's letter was entirely respectful and friendly but he might have hesitated if he had known that he was entering into what would be, for some time, a sterile debate that would centre on interpreting contradictory case histories, not all of them trustworthy. Nevertheless, his efforts would bring some reward for although Jenner never admitted that he might have been wrong, he did come to admit that his paper had been inadequate and he was forced to produce a follow-up article, his 'appendix'. It took many weeks, though, and a flurry of further correspondence, to arrive at this reasonably satisfactory end point.

Thirty years is a long time; memories fade and a sequence of events can become distorted but Ingen Housz confidently quoted the 'Stiles' case as an absolute contradiction to Jenner's 'doctrine'. He had been seduced, perhaps, by the fact that he had been able to corroborate the story by personal interrogation of both the variolator, Allsup, and the patient, Stiles. From the rest of the letter, however, it appears that Ingen Housz had also questioned others on the Bowood estate and in surrounding communities. He had to admit, to Jenner, of having heard that cowpox 'did sometimes', confer immunity to smallpox but that it was not always 'severe enough' to 'extinguish the susceptibility'. He was therefore confident that he had said enough to make Jenner retract his 'universal doctrine' and that any further correspondence would be superfluous. However, the situation developed rapidly and unexpectedly. Five days later, Ingen Housz set off back to London and, almost immediately after his arrival in the capital, he was again writing to Jenner.[32] The tone was somewhat more urgent and, deprived of his young 'secretary', it was written, on 23 October, in his own hand and with his eccentric spellings.

> Two days after I had the honour to communicate to you . . . on the variolae vaccinae, a favorable oportunity to inquire farther about your doctrine offered itself to me accidentally by a visit of Mr. Hastings[33] late Governor of the east indies, to the Marquis of Lansdown at his seat near Calne . . . (who) told me that . . . an authentic fact had come to his knowledge, which invalidates the infallibility of your doctrine. It was this: the son of William Beman a farmer at Adlestrop, Gloucestershire had got the cowpox on milking a diseased cow, accompanied with pain, swelling and stiffness under the arms. Two years afterwards he was inoculated for the small pox, and the disease took place in the most characteristic way. Being now returned to London, I perceive that your doctrine has made a deep impression on the mind of the public: and for that reason I think it the more my duty to inform you by the first oportunity of this second case, in hopes that the knowledge of it may awaken further your zele for the public good, and afford you the best means to correct your mistakes, if you should find to have committed one or more inadvertently: *quas aut incuria fudit, Aut humana parum cavit natura* – (I shall not take offence

at a few blots) which a careless hand has let drop, or human frailty has failed to avert.[34]

The first letter had been sent to Berkeley whereas Jenner was actually at Cheltenham, exploring the possibilities of setting up in the town as a physician.[35] He was to remain there until 30 November[36] and Ingen Housz, still unaware of this, also sent his second letter to Berkeley, subjecting the correspondence to further delay. He was making the point that there was now another 'incontrovertible' fact to 'invalidate' Jenner's general conclusion. But despite the enforced thinking time of the three days journey back to London, Ingen Housz was now blinkered. He failed to allow that the Beman case was, to him, at least third-hand. Nevertheless, it is easy to understand how his conviction was so strong because his two cases – both appearing to confound Jenner's doctrine – were obtained, as he says, from 'the two very first men to whom I addressed myself'.

Presumably, it would have been during the last days of October 1798 that Jenner eventually received, at Cheltenham, the first letter from Ingen Housz. It is not difficult to imagine his apprehension. Here was testimony, strongly contradicting the central tenet of his treatise, from a very high-ranking source, an *eminence grise*. As a renowned variolator, Ingen Housz was going to be a formidable opponent if he remained unconvinced by Jenner's thesis. Jenner, however, showed no signs of panic. His reply was supremely diplomatic without being submissive.[37]

> I shall ever consider myself as under great obligations to you, for the very liberal manner in which you have communicated a fact to me on a subject in which at present I feel myself deeply interested; a subject of so momentous a nature that I am happy to find it has attracted the attention of some of the first medical philosophers of the present age, among whom it is no compliment in me to say that I have long classed you . . . It will doubtless, in the course of time, meet with a full investigation . . . (but) . . . truth, in this and every other physiological investigation which has occupied my attention, has ever been the object which I have endeavoured to hold in view. . . . Should it appear in the present instance that I have been led into error, fond as I may appear of the offspring of my labours, I had rather strangle it at once than suffer it to exist, and do a public injury.

At this juncture Jenner brings in an important rider, one that would come to be so critical that the point must be stressed, noting particularly the applied adjectives 'perfect' and 'imperfect'.[38]

> At present I have not the most distant doubt that any person, who has once felt the influence of perfect cow-pox matter, would ever be susceptible of

that of the small-pox. But on the contrary, I perceive that after a disease has been excited by the matter of cow-pox in an imperfect state, the specific change of the constitution necessary to render the contagion of the small-pox inert is not produced,

It was now inevitable that Jenner also had to admit, if only by implication, that his publication on cowpox, his 'Inquiry', was deficient in detail. In other words, he was now seeing flaws in his book, among them a serious oversight in that he had not stressed, enough, the critical importance of making the diagnosis of cowpox accurately and, secondly, using serum from fresh lesions as inoculum. As so often in life, the devil was in the detail and he was certainly accepting Ingen Housz's point, that there was a potentially serious public health risk - 'of public harm' - should vaccination prove unreliable.[39] From a surviving fragment (undated) of a further letter from Cheltenham,[40] we learn that Jenner must have received the second letter from Ingen Housz before posting his first response. The second reply was, effectively, a supplement to his first and the two were mailed together. He went over the same ground viz. that the protection offered by cowpox against smallpox was conferred only by inoculating with the specific disease, there being, he admits, many other, and irrelevant, 'distempers' of cattle. However, he alleged that the diagnosis was not difficult and that 'people in my neighbourhood' were easily able to 'discriminate between the true and the spurious cowpox'.

The next letter in the faltering sequence was, again, from Jenner to Ingen Housz. Only ten days, at the very most, after posting his first two responses, Jenner sent what was really only a note containing one simple message,[41] some news that was to put Ingen Housz on the back foot .

> Since I did myself the honour of writing to you last, I have received some authentic information respecting one of the cases you communicated to me, which it would be wrong to withold from you. About a month ago the Rev.ᵈ Mr. Leigh of Adlestrop in this county mentioned to me the case of the son of one of his Tenants, who having had the cow pox was after wards affected with the small pox. Very soon after this Mr. Leigh call'd upon me to inform me that he had been led into an error; for on making a more minute Inquiry into the matter he found the fact to be the reverse of what was first represented to him, the boy having first had the small pox & afterwards the cow pox. I have written to Mr. Leigh to know the name of the Person in question & find it to be Beman.

Ingen Housz had been sold a pup. Of the two cases he had quoted to Jenner as evidence that he should hesitate before pushing his theory any further, one of them was worthless, a most unwelcome distraction. It was a garbled story that, once checked with real witnesses, turned out to be in reverse order, thereby undermining it as a call for revocation. Ingen Housz appears to have learned his lesson and determined not to trust any more third-hand information.

Jenner, in the meantime, must have written to a friend, Thomas Paytherus, a fellow surgeon-apothecary who had worked in Ross-on-Wye and was now practising in London. The letter must have included a plea for Paytherus to call on Ingen Housz and represent Jenner's respects and views. Ingen Housz returned to London on Thursday 13 December 1798, having been, for a couple of weeks or so, a guest of his new friend, John Rucker, on Wandsworth Common. Paytherus made an appointment to see Ingen Housz in his London rooms at 35, Marylebone Street for the morning of the 14th, writing a full report to Jenner later that day.[42] Even as a biased intermediary, he summarized his view of the situation in chilling prose. He found Ingen Housz to be a 'formidable opponent' who 'would not hear a word in defence of your opinion'. The royal physician had opened fire in an imperious mood, relating to Paytherus that he had been seeking other relevant case histories from among his physician friends at the Royal Society and that he already had several reports indicating that although cowpox did 'in many instances' render patients immune to smallpox, it 'was not with certainty ... in all cases'. But as the interview proceeded, Ingen Housz softened and Paytherus was able to inform Jenner[43] that the Dutch physician later

> ... spoke very handsomely of you' and 'desires that you will not be in haste to publish a second time on the cow-pox, but wait until you have collected a sufficient number of facts, and to secure your ground as you advance.

The meeting with Thomas Paytherus and the news that Jenner was on the point of publishing his 'appendix' provoked Ingen Housz into a further letter to Berkeley dated London, 20 December 1798.[44] It was a long and detailed essay over which he obviously took a deal of time and trouble: just a few extracts give the flavour.

> You will easily believe, Sir, that being in every respect a stranger to you, it would not be but the very favourable opinion I have conceived of your talents and character, that induced me to give myself the trouble of writing so much. . . I thaught to doe you a real service to leave you for some time to your own reflexions, till your mind, probably some what agitated by my letters, should be becalmed by farther reflexions . . . and I have still confidence enough in your good sense to expect that . . . (you will not) . . . publish the intended appendix of which you sent me a copy; and which I thaught it imprudent to lay before the public eye, such as I found it.

There was an apology for the Beman 'error' and then a reiteration of the central reservations held by the duty-bound Ingen Housz.

> As to the content of my second letter, I could not attest that fact myself, as I can the first fact: but my authority was fairly stated: and if Mr. Hastings and the Revd. Mr. Leich had been led into an error, my second letter must be considered as if it never had been written. As my only intention in communicating to you my first letter was to point out to you in a private way, what I thaught was an error in your work, to give you a fair oportunity to correct it yourself, before an other would doe it publickly and I thaught you would make this partial retraction with honour . . . and as I did not found a single person . . . who did not openly acknowledge that to make a person invulnerable from the small pox, it is required that the cow-pox should afflict with a certain degree of severity. Now, Sir, who will be a propre judge of that accurate degree of severity required? Is it the hideousness of the ulcers, the fever, the pain under the sores, or all together? I perceive clearly that it would be vain to attempt to convince you of the fallibility of the doctrine and, besides, I would find it impossible to answer fully the appendix you did me the honour to write me in the form of a long letter, without composing a whole book . . . After all this, receive, if you please, my last friendly advise; which is, that you should not be too much in hurry in publishing either my letter or your appendix, though revised, corrected, enlarged, without calmly considering your arguments and expressions . . . Wishing you a good success in your laudable endeavour to promote usefull knowledge, not locking out of sight the old proverbe *Ekhaldi bradieu* (make haste slowly). I have the honour to subscribe myself, with great esteem, Dear Sir, your humble obedient servant.

Jenner's 'Appendix' was first published as *Further observations on the variolae vaccinae, or Cow Pox*, on 5 April 1799.[45] It was a belated attempt to distinguish 'true' cowpox from what Jenner classified as four varieties of 'spurious' disease. It was a reasoned response, centred on the critical need for diagnostic accuracy, to the various critics that Jenner had found confronting him, Ingen Housz most prominent among them. Ingen Housz, in his own Damascene conversion, came to see that Jenner was articulating a real and worthwhile phenomenon, if inadequately. The perfect scientific paper has never been written and Jenner's 'Inquiry' would never be a contender, even for a consolation prize. Nevertheless, his dissertation remains extremely valuable, and rightly so, because it contained the kernel of an immensely important concept – that exposure to one infective agent, by natural contagion or by inoculation, can sometimes confer protection against another. The inherent faults of the 'Inquiry' seem to have been ignored by Jenner's early biographers. The common theme has been that the modest, well-meaning, and brilliant Jenner was harangued by a number of overweening critics, among them an arrogant and blinkered Ingen Housz. Actually, history shows that Jenner was, finally, very fortunate and that constructive criticism was not misplaced. His detractors were unjustly disregarded or discredited.[46] Nevertheless, he had the genius to perceive the significance of, and then to demonstrate, a vital phenomenon that kindled a whole new branch of medical science – immunology. But the *Inquiry* nearly buried itself because it failed to specify how cowpox was correctly diagnosed and when and how serum should be taken for inoculation purposes. The resulting confusions could easily have aborted the uptake of vaccination but it did, of course, eventually succeed in the battle against smallpox. Jenner's prediction, in his third publication,[47] that this 'scourge' would, one day, be eliminated came true: the last known case of smallpox was diagnosed in a young man in Somalia on 26 October 1977.[48] Jenner, with much justification, became rich and famous but the subsequent story for Ingen Housz was to be far less auspicious. His financial predicament remained and now there were other problems.

As the spat with Jenner died down and the worst of the 1798/9 winter passed, Ingen Housz's London life slotted back into a familiar routine; meetings at the Royal Society, dinners with friends,[49] overseeing the final stages of the printing of Hoogeveen's dictionary,[50,] and keeping abreast of all the new developments in medicine and science.[51] It was a lifestyle that was doomed. He was not well; at least he was not as robust as usual. Perhaps the first clue to his troubles was a surprising decision to decline an invitation to Holland House in January.[52] He had to admit to himself, and to Caroline Fox, that he could not face the cold nights there. He had found that he was less tolerant of the cold whilst staying at Rucker's villa on Wandsworth Common. It was an experience, an early symptom even, of looming problems.[53]

> Nothing, indeed, can be more encouraging than Holland House from ten in the morning to 12 at night in regard to the good fires, the rich library, and above all the incomparably good and respectable company, but in such a season as the present is, I might excuse, at night, to be frozen into a lump of ice in my room . . .

Ingen Housz had lost a significant amount of weight through the winter[54] and by spring he was weak, light-headed and unsteady on his feet.[55]

Holland House, Kensington.

He had even fallen out of bed on one occasion.[56] Not that he had lost one iota of his intellectual fervour. On Wednesday 6 March he had been to dinner at Lansdowne House, Berkeley Square.[57] Lord Holland was also at table. The Marquis of Lansdowne wrote, five days later, to his son, Lord Henry Petty, now an undergraduate at Trinity College, Cambridge.

> We had a fine scene with Dr. I on Wednesday. I mentioned the Duke of Portland's[58] having sent two volumes of papers about the Union (between Great Britain and Ireland – the laws were eventually passed in 1800) to the Royal Society, I suppose for the sake of writing a fine letter a propos to nothing. Holland, who was at dinner, said, very naturally, what was the Royal Society to do with the Union? At which the Dr. took fire – enter'd into a comparison of the H of L's & the Royal Society – the latter the most learned society in the whole world, capable of every thing – the other a pack of fools who had nothing but riches and titles to make up for want of understanding. It was in vain everybody cried out 'Jacobinism' – it only made him more furious.

An ever-widening gap between physical and mental powers was to be the one consistent health experience for Ingen Housz over the next weeks and months. At first he put his growing disability down to the 'severe shock' of learning that all pensioners were being 'exiled' from Vienna and that his pension had been stopped – it was now well over a year without a 'single heller' from Austria.[59] But patients can usually expect to recover from 'shocks' whereas his illness worsened as spring passed. By early June he was resigned to his ultimate fate being imminent.[60]

I feel that death is nigh, which my friends, likewise, think isn't far off. I have prepared myself for it in good time.

He was certainly now experiencing more alarming symptoms. He had a constant low-grade fever, had lost all of his characteristic 'embonpoint' – his portliness – and was becoming so wasted, pale and breathless that he was 'receiving a great deal of sympathy' from those around him. The right side of his body felt heavy and, on occasion, he could be tottery. His speech was slurred and he was chewing food clumsily. There was a mistiness in his right eye. None of this was alleviated by heavy bleedings, presumably done by Christopher Allsup, for by early June Ingen Housz was at Bowood.

A very considerate Lord Lansdowne, a true friend indeed, had suggested, just as he was about to leave London for Wiltshire at the end of the London 'season', that his old friend should also travel to the country, for the sake of his health. There he would be '. . .surrounded by people he knew and where he was sure to find friendly care and sincere concern.'[61] Soon after arriving at Bowood, Ingen Housz travelled down to Bath to consult William Falconer. Presumably he called on his good friend at his house in the Circus.[62] Falconer's opinion and prognosis were never divulged but Dr. Drew, an intimate of Lord Holland and his family, and who passed through Bowood during late July,[63] '. . . saw in this illness a general weakening which left no room for hope.'

It is reasonable to ask, at this point, whether it is possible to make a confident diagnosis of this seemingly terminal illness. The usual answer in these historical situations is no; any rash attempt to do so being littered with medical pitfalls when everything in the discipline has changed so much in two centuries, even the very terminology of symptoms. Nonetheless, the progressive loss of physical powers in Ingen Housz although he retained his formidable intellectual faculties until the very end might be a diagnostic clue. So, also, is the fact that at least two doctors were able to predict, confidently, that death was inevitable. Diabetes mellitus, for which there was then no treatment, is one obvious possibility but a longer period of confusion might have been expected in the lead up to death and there was, in the late eighteenth century, a primitive understanding of the disease which had already been given the name we still use. A specific mention of this diagnosis might have been expected somewhere in the personal archive but there is none. Moreover, Ingen Housz was very well aware,[64] that if he had been diabetic, he would have been thirsty from the very outset of his illness and would have been passing copious volumes of urine, a clinical feature to which there is no reference in his descriptions. Perhaps he had a cancer of some sort – disturbed vision and right-sided weakness would suggest tumour, or tumours, in the brain which could, of course, have been secondary to a cancer elsewhere. He could also have had blood cancer in the form we now call lymphoma. This can progress slowly but, without specific treatment, death is inevitable. In lymphoma, any doctor, even Ingen Housz for himself, would have been able to read only one prediction into the rock hard swollen lymph glands that one can easily feel. There was no hope of long survival even if the disease had no name. The type of lymphoma we now call non-Hodgkin's disease, the commonest variant, is more frequent in men than in women and there is a considerable peak of incidence in late middle-age. Disease of this kind would certainly explain the early and long-standing symptoms of mild, recurring fever, pallor and tenacious weight loss. However, the extreme dryness of the mouth that Ingen Housz describes so vividly near the end of his life, from a failure of salivary flow and, very often, an oral fungal infestation, is a concomitant of the dying process that is extremely common and not a diagnostic clue. It is certainly a most distressing situation: the patient develops more than just a sticky sore mouth – speech and swallowing become very difficult without frequent drinks and these very circumstances came to dominate his last few weeks of life.[65]

> ... I continue to be very thirsty and my throat to be dry so that I must have allmost continuously some blackcurrant jelly on my tongue to form the saliva and to prevent my drinking too much.

In early July Ingen Housz began to arrange to travel back to London, much preferring not to become a burden,[66] but the Marquis was insistent that he remain at Bowood rather than perish in lonely isolation, waited on only by the 'faithful and honest'[67] Dominique. In the end, Ingen Housz saw that the humane kindness and wisdom of Lansdowne made sense despite the obvious *sine qua non* – that he would be far removed from all aspects of his former London life. But the offer of continuing hospitality and support, no matter what difficulties might arise, was irresistible, as were the Arcadian charms of Bowood. After all, his accommodation in London was very functional – lodgings in which there was room to 'barely turn around'[68] – and in the heat of summer most of his other close friends would be out of reach at their own country retreats. However, Ingen Housz insisted on making one return journey to London before he became too frail to tolerate it: there were affairs to put in order and final arrangements to make. He broke his journey at Holland House, perhaps in both directions but certainly on Wednesday 3 and Thursday 4 July where his name can be seen in the dinner book.[69] There is no record of whom he might have called on in London in order to

The First Marquis of Lansdowne. Courtesy of the Trustees of the Bowood Collection.

say goodbye: not having a last meeting with the Dimsdales would have been very upsetting. He may have tried to disguise his predicament from many of his good friends, anxious to protect them from any distress by way of a reminder of their own mortality. In letters written in August he certainly presents a disingenuous outlook and there is no reason to think that he was actually in denial. He certainly tied up all loose ends in his London life; settling all debts, locking his cupboards, tidying his papers and his rooms. It would have been a raw parting from John Plaggenborg,[70] his landlord, and his family: here there could have been no refuting his behaviour as that of a doomed man making a last tidying of his worldly affairs. He took away only the few things he might need; some twenty guineas or so, his watch, a few more clothes and some personal documents. Everything else of value was deposited at Drummonds.

As July merged into August Ingen Housz was back at Bowood to embark on the last full month of his life. He remained very weak but alert; sensitive to events in the house and in the outside world. Far from being introspective and neurotic he was more worried about 'learned friends' at Milan than himself.[71] There had been a significant battle between the French and a joint force of Austrians and Russians near the city [72] with troops from both sides on the rampage. Nor did he take to his bed. On 28 July he spent some time with the cooks in the Bowood kitchen, helping to make raspberry jam.[73] On Sunday 4 August he assembled, with the rest of the household, under the portico at the front doors of the mansion to wish 'God speed' to the 'ladies' – Caroline Fox and Elizabeth Vernon were going to the Midlands to stay with Lord and Lady Warwick, Miss Vernon's relatives.[74] It must have been a poignant parting, the ladies sensing that their old friend had a very short time to live: it was now inescapable that Ingen Housz was dying, that his future was about care, not cure.

Ingen Housz appears to have written few letters that summer but they were all perfectly coherent; one to van Marum for instance. He did not write to van Breda, even after receiving a letter from his friend at Delft in late July.[75] In fact, as van Breda reminded him by way of mild reprimand, he had last written in March 1797. Ingen Housz wrote to his wife but the actual letter seems to have disappeared. We can gauge, from Agatha's reply, that it must have been posted sometime in late June or early July, perhaps when he was in London. It seems more than likely that he tried to shield her from the truth by giving an open prognosis. When she replied, on 13 August,[76] her letter was dominated by financial woes and recited a list of problems with the servants. It is very likely that Ingen Housz had died before Agatha's letter arrived at Bowood and perhaps it is as well, for she was in no better shape than her husband.

> I have received your letter with great sorrow seeing that you are sick again. I have been sick all through the winter. I have had a fever every day for the last six months, coughing and unable to sleep. I have not taken any medicine but I have got better although there is nothing left of me but skin and bones. My worries for you caused me to lose my health . . . it is a year ago that I received your previous letter and I wish that you would write to me. To know that you are well and that I share in your friendship is my only consolation.

By 31 August, however, it was obvious to all that Ingen Housz was very near the end of his life. Lady Holland was quite taken aback by his change of appearance when she and her family arrived at Bowood for a three-day stay on their way to Devon.[77]

> Poor old Ingenhousz is dying rapidly. He is shrunk to a skeleton. Ld. L, with great humanity and feeling, affords him an asylum (and) the beholding of a

dying man is a painful spectacle. He talks with intrepidity of death, but he'las, where can the courage come from? The subject is a painful one.

DEATH CAME SHORTLY after three in the morning of Saturday 7 September. If any death can be easy, at the very end this was one. Ingen Housz was 68. There are three contemporary records of the last days and hours of Jan Ingen Housz. Each of them is moving and instructive, each slightly different to the other two and all three tell us something new. It is therefore well worth giving all three accounts and in the order in which they were written – from the day of the death, from the next day and from the day after that, Monday 9 September, the day of the funeral.

The first account is by the Reverend Étienne Dumont, Lord Lansdowne's secretary. Written on Saturday 7 September, only a few hours after Ingen Housz had died, it was part of a letter sent, by Dumont, to his close friend and regular member of the Bowood summer 'colleges' who knew 'the doctor' well, Samuel Romilly. Romilly, now married and immersed in his legal practice,[78] had remained in London that summer. Ingen Housz had asked, from his bed and sensing that the end was very near, to see Dumont. Dumont was an ordained pastor of the Swiss Reformed Church and far from being a Catholic. He was, however a 'man of God' and there are plenty of pointers that although Ingen Housz had, by this time, lapsed from the ceremonial practices of his family faith, he still held a sincere religious belief. He therefore saw Dumont, not an especially close friend, as an appropriate person to comfort and counsel him in his final sensate hours.[79]

> When I arrived at Bowood, on August the 15th, Dr. Ingen Housz seemed very weak, but he had the same taste for society; he performed, with the same vivacity, his experiments for all the new company; he had the same fire when he spoke of the Jacobins and the Masons but he gradually weakened; for eight days the decline was marked. He didn't come down to dinner anymore; he didn't want company in his room; he suffered at no point, he never showed fear, he was calm. Yesterday he begged me to spend time with him in his room. I told him of the interest everyone had in his situation . . . the conversation tired him; at all times he moistened his tongue with vinegar. He offered me his hand as I left him but the scene of this serenity, of this unaffected courage at the approach of last moments and this sign of friendship, towards me to whom he had never previously been generous, left me . . . particularly affected. He begged me to testify to all his gratitude to Lord Lansdowne, in comparing his situation with that which might have been, in his meagre

rooms, and said that he took to his heart all the kindness that he had received. Lord Lansdowne went to see him but four or five hours had made a lot of difference; the remarks were disjointed, his eyes wandered, the words no longer made sense. He fell into a kind of drowsiness, and tonight he took his last breath, not like a man who was dying but like a man who ceases to live – I saw him, his features didn't show the least alteration – truly death is one of the biggest mysteries of life.

The second letter in which the circumstances of the sadness at Bowood were related was written by Caroline Fox who, with Elizabeth Vernon, had returned to Bowood from Warwick Castle sometime in the afternoon of 7 September. It was an agonizing arrival but she was composed enough to inform Lady Holland the next day.[80]

Miss Vernon & I returned yesterday & found, as you feared, our poor old friend the Doctor not almost, but altogether, extinguished. Indeed I was much shocked. His faithfull good Dominique had the attentions to send us word by the postillions and horses which met us at Malmesbury & thereby prevented our learning it abruptly. I am so much indebted to him for many hours of amusement as well as of solid instruction that I shall always remember him with gratitude & affection & recollect with satisfaction the many instances I have had of his confidence and esteem – poor old man, peace be with him. He dined below on Wednesday but on Thursday, I understand, he was confined to his room: he spoke with great firmness, clearness & composure of the arrangement of his affairs, expressing with effusion his sense of Ld. Lansdown's goodness & the comforts with which he had surrounded him – the next day he fell into a sort of stupor & expired without a groan at half past three yesterday morning.

The third account was that of the Marquis of Lansdowne himself. On Monday 9 September he retired to his study with Dumont in attendance and dictated, in French, consolation letters to Breda and to Vienna, informing relatives of their loss. Although the letters to Agatha Ingen Housz and to Josef Jacquin are unavailable or have been lost, the long letter to Ingen Housz's nephews in The Netherlands has survived at Breda.[81] The prose is beautifully crafted and would have provided as much warmth as condolence: the actual description of the dying process is written in very compassionate terms.

No-one believed his death was so imminent. He had consulted several doctors ... (and they) ... saw in his condition a general debility that had no cure. His strength gradually petered out but his intellectual faculties did not wane at all; he enjoyed the Society's gatherings almost until the last moment. He showed the same liveliness and the same pleasure in repeating some of his scientific experiments in front of a large party. He was always calm and serene, showing great strength, without any display of vanity, and composure born of true courage and wisdom. After several hours sleep he drew his last breathe without ever waking.

Later in the letter we learn that Ingen Housz and Lord Lansdowne had had a 'very confidential conversation' in which they had planned, in a wholly frank and courageous way, the administration of the former's death. Having promised to write to inform relatives, Lansdowne had also agreed to organise Ingen Housz's funeral according to his wishes – 'to be buried quietly, with propriety and without show' – and began to make arrangements early on the day of death. Ingen Housz had wanted burial within 'the church of the parish in which I will die' according to his 1796 will.[82] Fortunately, Lansdowne was in a position to organise this without too much difficulty for he owned the unobstructed rights to a sizeable burial vault under St. Mary's Church, Calne, the nearest church to Bowood.[83] It was a structure associated with the ownership of the largest house in the town, Castle House, which Lansdowne had acquired in 1797.[84] In 1799 there was no tenant in the house.

The interior of St. Mary's Church, Calne as in 1799.
Courtesy of Ray Downham MBE.

Vault deposition inevitably meant that a lead-lined coffin would have to be constructed[85] and presumably orders were sent out to a plumber and a carpenter to join forces. Work must have proceeded over the weekend for the funeral service was on Monday, 9 September. Lansdowne acted on his own initiative in planning to 'erect a simple monument on his grave', as he had promised to relatives.[86] This, it seems, was never commissioned, for within days of the death of his friend, the Marquis himself was ill and remained very much under par until his own demise in 1805. The family in Breda did, however, have the benefit of Lansdowne's sincere eulogy, full of warm feelings and genuine respect.[87]

> My longstanding friendship with your worthy and respected parent, Dr. Ingen Housz, makes me feel deeply saddened at the painful news I have to give to you . . . if something had hastened his end, this could only have been his extreme anxiety about the political situation. He spoke about all his family in Holland in a most affectionate way and he spoke to me at length about Mrs. Ingen Housz, his wife, with feelings of tenderness and pity, and with the greatest respect; he trusted her completely and was deeply touched by her devotion to him and by all the ways in which she had proved this trust and devotion. He also spoke to me about Mr. Jacquin with regard and affection . . . His name will live on through his writings and his sense of honour and integrity, which he showed in all aspects of his life, will assure him the sorrow and esteem of everyone who knew him.

Lord Lansdowne also wrote to the Dimsdales and asked Dumont to arrange press notices of the death. These appeared in the *Gentleman's Magazine* of October,[88] the *Bath Chronicle*,[89] and perhaps elsewhere.

INGEN HOUSZ'S FUNERAL, then, was held on Monday 9 September[90] when a byre brought the encased body down from Bowood to Calne and the heavy coffin was hefted into the church. The vicar, the Rev. Thomas Greenwood[91] read the prayers to the small congregation of mourners that was led, not by Lord Lansdowne who was indisposed, but by Lord Henry Petty, his son.[92] The only others we know to have been present were Étienne Dumont, the guild stewards of Calne – William Savory and Samuel Viveash,[93] Dominique, and the 'medical people who had attended him' during his last illness – Christopher Allsup and perhaps his partner, Allin Wayte.[94] There were, of course, no close relatives, no one from his homeland and no-one from Vienna. It was, indeed, a simple service and, sombre ceremonials over, the undertakers moved in and began their struggle to lower the five hundredweight casket into its subterranean resting place. The grief at Bowood was cerebral and prolonged. Ingen Housz may have died many miles away from any actual family but the sincere and genuine friends who had been with him that summer were still being consoled, for many weeks, by letters from their own relatives and friends. However, Anne Garbett, now Mrs. Romilly, wrote by return of post.[95]

> Most sincerely do I lament with you my dear Miss Fox on the death of our valued old friend. I was indeed shocked to hear of his departure for I had flattered myself from the account we had of the benefit he received from the air of Bowood that he would have been spared to us a few years longer ... I am certain it happening at Bowood would give pain to Lord L. yet I cannot help rejoicing that he closed his eyes in a place he was so partial to and amongst friends for whom he had so great a regard.

So, also, did Lady Holland who had gone on to Saltram in Devon just before the fateful weekend.[96]

> Indeed I am very sorry for the poor old Doctor. He is really a loss to us all, but to you especially. It must be a shock to Ld. L. I wonder he did not go to Bath, or anywhere to change the scene for a few days.

Dominique was inconsolable.[97]

> Ld. Lansdowne has desired Dominique will consider this (Bowood) as his home while he remains in England and return to it if Vienna should not be comfortable to him. Poor man – he says 'je n'ai autre chose a faire qu'a fumer, promener et pleurer' (I have nothing to do except to smoke, to go for walks and to cry) and he does so, poor fellow, the whole day – no wonder, for it is thirty years since he has lived with his poor master.

Dominique had not been forgotten, however, by his late master. He had arranged for his salary to continue for four months and that Drummonds should defray him the expenses of travelling back to Vienna where he would find long-term provision.[98] In a codicil to his will at Vienna, written 12 July 1798, Ingen Housz had left him a life annuity of 150 Viennese florins, to be increased to 250 on Agatha's death.[99] If poor Dominique did have a difficulty, it was in taking a decision about his future. He had a wife in Vienna but he was in a quandary about going 'home', especially when Lord Lansdowne implored him to stay at Bowood. Dominique's popularity in the household was certainly no insincere attempt to offset his grief, the true respect for him being very transparent in a letter of introduction that Lord Lansdowne was pleased to give him.[100]

> ... I do not believe an honester or more disinterested man exists, of which I am so well persuaded that I have offer'd to him to pass the remainder of his

days here . . . there is not a single person in my family, or who frequents my house, who does not join in wishing him well . . .

The Marquis of Lansdowne also dictated, on the day of the funeral, a letter to Drummonds, the London bankers of his late friend. They, in turn, wrote to Breda informing Ingen Housz's nephews that, as bankers to the deceased, they held a strongbox of possessions, a worthwhile balance in their uncle's account, some £3720 (£120,000),[101] and the latest will. The letter also reminded them of the instructions, of their late uncle, that they should come to London to collect his effects and arrange transfer of the estate. On 24 September Drummonds reimbursed Lord Lansdowne for the cost of the Ingen Housz funeral, £25 2s. 8d. (£809),[102] a considerable sum for a funeral of that era, confirming the construction of a lead-lined coffin and, therefore, vault deposition.

As letters of condolence to his close friends began to arrive at Bowood, Ingen Housz's near relatives at Breda were also receiving post. They appear to have supported each other in the face of the unexpected bad news, and Maria, Louis' widow, wrote a very sincere and moving letter to Agatha at Vienna.[103] The two oldest nephews,[104] Arnold, a qualified pharmacist aged 33, who had taken over his late father's business jointly with his mother,[105] and his next youngest brother, Peter, aged 32, prepared to obey their late uncle's wishes and travel to England. It was not a simple journey as England and the French-run Batavian Republic were at war. They applied for passports to travel to Denmark without revealing their true destination.[106] Even this preliminary caused them anxiety for they were half afraid that, in tangling with officialdom, they might be scooped up and conscripted for military service. Fortunately, their passports were issued without question and having crossed the border into Hanoverian territory, they obtained further passports there to sail to England and boarded a ship at the port of Embden.[107] They arrived in London in mid-November. The Dimsdales,[108] father and son, assisted them personally whereas Lord Lansdowne was at Bowood and on the point of going down to Bath to 'take the waters' for his health. Nevertheless, he wrote immediately:[109]

> I am very sorry to find myself at such a distance upon your arrival . . . and if you would think of returning to Holland before I return to London, I hope you will doe me the favour to come and spend some time with us at Bath where . . . I shall have great pleasure in manifesting the respect and esteem I have for your uncle's memory by shewing it to his representatives.

The Marquis stressed that his solicitor was on stand-by to help them in any legal entanglements.[110] In fact, the legalities in England appear to have resolved themselves without too much hassle. Arnold was able to buy, in his mother's name, larger premises in Breda for their pharmacy, in March 1800,[111] suggesting, perhaps, that some of his uncle's money from England had come through.

Issues were not to be so simply resolved at Vienna even though a certified copy of the death was obtained from Thomas Greenwood, the Calne vicar, and sent to Austria in December 1799. Agatha Ingen Housz appears to have been obliged to leave the apartment in the Annagasse: Vienna records show her living at 23, Wipplingerstrasse in 1799, probably a more modest abode.[112] She continued unwell and died just over a year after her husband – on Thursday 23 October 1800.[113] She was 65. She died at an institution called the Burgerspitalinshaus on the Kantnerstrasse and was buried on 8 November,[114] probably in the nearest cemetery but there is no record.

Josef Jacquin was left to administer the Ingen Housz estate in Vienna. It was much diminished, proving to be modest for a man of Ingen Housz's status – some 29,000 Austrian Florins,[115] the equivalent, now, of about £86,000. Much of this was raised from sale of his effects for, unlike in England, where he had been living very modestly, there was a deal of furniture, fittings, silverware, ornaments, jewels, scientific instruments and a personal library of well over 700 books valued, alone, at 1160 florins (£3,400). The legal wrangling lasted longer than his scientific reputation; it took until 1827 to finalise.[116] But nearly thirty years after the actual death our journey is now definitively over. There were no children and even those personal chattels that had only sentimental value, the accretions of a life, platinum buttons included, were now widely dispersed or disposed of. Thankfully, much of the documentation was preserved and survives. But life was truly extinct. In one sense, though, there is more to the story and one, final, chapter will consider how Ingen Housz himself was lost to history.

22
The Black Hole of History
How Jan Ingen Housz was Forgotten

The spectacles of Jan Ingen Housz. Courtesy of the Breda Museum.

DISCUSSING WHY JAN Ingen Housz was forgotten begs the question of why he should be remembered. Otherwise, the whole issue is irrelevant outside the intimate boundaries of family history. We advocate that he deserves a more prominent place in history for one main reason. We then add a list of subsidiary grounds for recognition (in no particular order of significance). The imperative for his fame is that it was Jan Ingen Housz who discovered the role of light in plant biochemistry; that it is sunlight that drives the process we now call photosynthesis. In addition:

   a.   he was an exemplary practitioner of the scientific method; of rigorous repetition of experiments that produced quantitative data, and therefore a figurehead of the Enlightenment;
   b.   he was a talented design engineer, devising several ground-breaking pieces of apparatus including the cover slip for high-power microscopy;
   c.   he was a significant pioneer in smallpox inoculation, a stepping stone to the elimination of this scourge of mankind and, in this context, he put Edward Jenner and his 'vaccination' improvement back on track;

d.   he was an excellent communicator, as shown by the extent of the names index in this book, and he was the author of a literary style that bears the hallmarks of modern scientific parlance;
   e.   he was a vigorous and brave antagonist of charlatans;
   f.   he was a true polymath, as shown by the breadth of his passions and interests.

So, why should he have been more fêted when alive and why was he forgotten after death? For clarity of purpose, in the face of the obscurity Jan Ingen Housz suffered for nearly two centuries, we should think, perhaps, only about the part he played in the history of photosynthesis. That this was forgotten was not wholly accidental: there were several strong threads of conspiracy.

As a human being Jan Ingen Housz was certainly multi-faceted; he would not qualify as a polymath if he weren't. But he was also unsettled. In the main, he worked in four countries; was fluent in their languages and mostly comfortable in their cultures. While we can admire his adaptability and linguistic talents they served him ill in the long term. He left traces everywhere but no beacons. He has never inspired conventional historians because he eschewed politics and the odd fossiker who might have stumbled on some of his documents would have shied away on sniffing the major medical and scientific components. Such are the consequences of the universal divide in education. Then why have the historians of science or medicine not paid more interest? They discover another intellectual electric fence – we are back to the multiple languages and the fact that the archive is widely dispersed throughout Europe. So we may say that our man fell down between the disciplines.

Though purely hypothetical and indulgent, it is instructive to consider how a Nobel committee might have awarded credit for the early unravelling of photosynthesis and nominated for the prize. Joseph Priestley, certainly, would have been a nominee for it was he who had shown the ability of plants to sometimes revivify the air and it was also he who had, effectively, invented the eudiometer, a means of measuring the purity of that air. The committee would then have faced a disputatious sequel, one in which the true claim to fame belonged to Ingen Housz but they may have accepted Priestley's assertion that it was he himself who had first shown light to be pertinent. Hopefully, they would have interpreted the published evidence correctly but there are recent, real, precedents for this not always being the case. The 1959 prize for chemistry, for the discovery of the structure of DNA, is a good example; the committee's potential embarrassment at overlooking the signal part played by Rosalind Franklin was averted only by her untimely death. Without her unique and amazing X-ray diffraction images of the DNA molecule, the notion that it was a double helix would have been based on nothing but fantasy. In science, the backroom genius who makes the breakthrough is always in danger of being subsumed to the *prima donna* who capitalises on the development of the idea.

Ingen Housz was not very clubbable. He obviously revelled in being a fellow of the Royal Society and bathed in the warmth of its dining club but this fraternisation did not extend to other association. He took against freemasonry, against ceremonial religion, against active membership of many other scientific societies despite the honours they bestowed on him and remained determined, always, to be a 'spear-carrier'. If true scientific distinction built on his discoveries pushed him into some prominence this was the only route to fame that he wanted but it was to let him down. It meant that he had no integrated fan club. Odd contemporaries, themselves mavericks for various reasons, such as Pringle, Franklin or even Maria Theresa, certainly promoted him and wanted to do more but their impetus did not survive their declines. They were all older. Of a younger generation of admirers, we might have expected to find children or grandchildren but, of course, there were none. His nephews and nieces at Breda certainly revered his name and he remains exalted in the family today as 'Oom Jan' (Uncle John) but they never understood, perhaps, his true claim to fame. The one biographical picture arising from within the family memories and archive places him as a royal physician, the personal doctor to Emperor Joseph II, but misses the point as regards his excellence as a scientist.

Finally, there is a series of events and eventualities that is difficult to synthesise as fluent prose and we have to resort, again, to a list of random order.

§ *Experiments upon vegetables* – the one book for which Jan Ingen Housz should be famous was written using what is now a dead syntax, condemned, almost, to total obscurity by the overthrow of the phlogiston school.

§ Jan Ingen Housz rejected nearly all official posts and titles he was offered and, though his motives were entirely honourable, he therefore lost his place on the 'hard-drive' that is the permanent record of high officialdom.

§ Jan Ingen Housz's central archive only came into unity in the 1940s, a century and a half after his death, when his diligent and generous near-descendants donated much of what was in their possession to be joined with the artefacts held by the Breda Museum. This was, however, in a time of total war and subsequent austerity and the archive has been little disturbed.

§ No-one attempted a substantial and comprehensive biography until 1905 when Julius Wiesner, with colleagues, produced a 'life' of Jan Ingen Housz on the occasion of the 1905 International Botanical Conference at Vienna. However, Wiesner's considerable book was inevitably selective and wanting, for it had to rely on material actually published by Jan Ingen Housz, on such archive of primary sources that was, then, in Vienna, and on fading family recollections from the Netherlands.

§ Jan Ingen Housz's rightful prominence as a missionary for smallpox inoculation was lost, as it was for all that generation of courageous inoculators, after the establishment of Jenner's vaccination technique. Prototypes often suffer the fate of being left to gather dust and lie, discarded, as yesterday's junk.

§ The life span of the eudiometer was a short and sorry one. Jan Ingen Housz was probably the most skilful user of the best contemporary eudiometer – that of Fontana. But when the instrument lost favour, sinking under a reputation for being temperamental and inaccurate, the name of its best exponent also faded.

§ Jan Ingen Housz was undoubtedly a monarchist and his grief as royalty began to lose their authority and, in some instances, their heads, was genuine. But he was not, by any means, part of the *ancien régime* as some have so classified him, thereby shutting the door on him as a figure of the Enlightenment.

§ The process of fame is fickle and no-one ever used the name Jan Ingen Housz as semaphore in naming a law of nature. 'Ingen Housz's Law' – that green leaves produce oxygen but only in bright sunlight, schoolboy mnemonic GLOS perhaps, could well have been coined and taken to heart by generations of children learning science, but wasn't. If it had, all biochemists would know his name and his discovery. It didn't and they don't.

§ Jan Ingen Housz may have been an iconoclast but he was not a rabble-rouser and certainly not a revolutionary. There was never any prospect of his name becoming known for infamy.

§ Jan Ingen Housz was served badly by history. During the 1780s he focussed virtually all of his publishing activities in Paris. He therefore wrote in French, the then most widely used language in the literate world. Unfortunately, there then came an abrupt and massive crevasse in French life and culture and his work was lost in the intellectual flagellation of the Revolution. And when, after the turmoil in France had died down and a powerful reference work was compiled and published there – the *Biographie Universelle* – Ingen Housz was given second place to Priestley, to someone who had been an overt republican.

§ Jan Ingen Housz spent much time and effort expressing an updated summary of his understanding of what is now known as 'photosynthesis' in the middle of the 1790s, in the years immediately before he died. *An Essay on the Food of Plants* was a comprehensive and, put against modern understanding, worthy appraisal of his work and discoveries, well placed in context. However, it only appeared in an obscure government publication having very limited circulation to a scientifically illiterate readership.

§ It is a feature of the history of botany that after important discoveries in plant biochemistry in the late eighteenth century, this facet of the discipline lost precedence to an obsession with taxonomy, the classification of species. Jan Ingen Housz's name and role was resurrected, as was plant physiology, by Julius Sachs, the famous German botanist of the second half of the nineteenth century. However, Ingen Housz's name never seemed to lodge itself in the collective consciousness of succeeding generations of botanists. A similar transitory fate befell his name when it was used as an appellation in the identification of plant genera, where applying the name of famous scientists was a strong tradition. *Ingenhouszia* was used by Carl Meissner in 1837 to name a fairly obscure genus of mallow and was also applied to one or two other uncommon plants. However, the assiduous reader of plant labels is far more likely to stumble across the terms *Banksii* or *Darwinii* among the popular plants for sale at a nursery or garden centre. Other scientists have, however, given Ingen Housz some recent recognition. 'Project Ingenhousz' was a joint scientific venture, based in the Arizona desert, that brought together a whole range of experts having some aspect of life science and sunlight in their research domains.

U NHAPPILY, WE CONCLUDE by relating that the few memorials to Jan Ingen Housz have been jejune and, in some cases, lost. The citizens of Breda conceived a statue and a half-sized mock-up was sculpted. The actual statue was never carved, however, and the model lurks in a cupboard in the Breda Museum. Also in Breda, there is a square named after the town's hero: the *Dr. Jan Ingen-Houszplein* is in a nondescript area of the city and has been neglected and weed-strewn. Admittedly, it was once the site of a flourishing fruit and vegetable market but this has gone and the symbolism is somehow poignant. Likewise, there is a street named after Jan Ingen Housz in Vienna – *Ingenhouszstrasse*. This, too, is off the beaten track in an undistinguished area of anonymous tall buildings and offices. Rather more prominent is the memorial to the three Dutch doctors engaged by Maria Theresa – Gerard van Swieten, Nikolaus Jacquin and Jan Ingen Housz. The three bronze busts, unveiled in 1905, are together in an arcade of the University of Vienna that is lined with much other statuary. At least it puts Jan Ingen Housz among equally deserving company. For the next accolade we return to the Netherlands. To boost morale during the German occupation of their country, between 1940 and 1945, the Dutch government commissioned a series of new postage stamps depicting famous Dutchmen. In 1941 one of these was Jan Ingen Housz, 'physicist'. After the war there was a desire, among some of his countrymen, to erect a permanent memorial to Ingen Housz in St. Mary's Church, Calne. For over 150 years there had been no tablet or other marker at his resting place. Eventually, a stone plaque was

Dutch postage stamp of 1941.
Courtesy of Hans Ingen Housz.

unveiled during an evening service in the church on Sunday 18 November 1956. Sadly, however, its site is on a wall of a side chapel hidden from the nave: the majority of worshippers and visitors remain unaware of its existence. More obvious are two more recent memorials in Calne. One is a Blue Plaque erected by the Calne Civic Society. It is on the wall of Church House, the administrative centre of the parish next to the church, and was unveiled during celebrations in the town on 7 September 1999, two hundred years after his death. The other is a bronze replica of his 1779 Royal Society medallion inserted into the pavement at the entrance to the town's library. It is burnished by the feet of the readers visiting the building and is an apt link to learning. We can only hope that our book, likely to find its way into the same library, might also burnish his reputation, a reputation that certainly needs re-appraisal.

Sample page of Jan Ingen Housz's original 1779 'photosynthesis' experimental notes. On very thin paper and difficult to reproduce but his rational organisation, careful recording, and calculations may be seen. Courtesy of Stadsarchief, Breda.

Notes

The following frequently-used bibliographical sources are abbreviated as shown:

BGA: Breda, Gemeentearchief (Municipal Archive). Section IV, 16; 'Ingen Housz File'. Documents in section 'A' came into the archive from Breda Museum; those in section 'B' were handed in by the Ingen Housz family in 1944.

FB: 'Familieboek'. A collection of personal papers and written family recollections held within the Ingen Housz family. Copied prior to destruction of the original by fire in 1944.

GF: Godefroi M. Dr Jan Ingen-Housz. Geheimraad en Lijfarts van Z.M.Kaiser Jozef II van Oostenrijk. Records of the Provincial Association of Arts and Sciences of North-Brabant. 's-Hertogenbosch, July 1875. The life of Ingen Housz as told by a Netherlands physician who was a friend of a near descendant of Ingen Housz. Personal papers and family recollections (familieboek) were used to compile this document, its value now much enhanced because the original archive was destroyed by fire at Arnhem in September 1944. Translated into English and newly annotated in 2004 – Ingen Housz, J-M, Beale N, Beale E. The life of Dr Jan Ingen Housz (1730-1799) Private Counsellor and Personal Physician to HM Emperor Joseph II of Austria. *Medical Biography 2005*; 13: 15 – 21.

HKB: The Hague, Koninklijke Biblioteek (The Royal Library); File 133, B24.

HRNH: Haarlem, Ryksarchief Noord-Holland, File 529.

HVTS: Haarlem, Archieven van Teyler's Stichting (Archive of the Teyler Foundation).

IHLD: Breda, Gemeentearchief (as above). A, 6. Jan Ingen Housz's letters diary 1774 - 1791.

ÖNB: Österreichische Nationalbibliothek, Vienna. Abendländischen Handschriften (Austrian National Library. Handwritten documents section). Series Nova, 4061.

POBF: The Papers of Benjamin Franklin. http://franklinpapers.org/franklin/

WJ: Wiesner J. Jan Ingen-Housz: Sein Leben und sein Wirken als Naturforscher und Artz. Wien, Carl Konegen 1905.

Notes to Chapter 1

1. Later called Zaltbommel.
2. FB.
3. Ibid.
4. 1701-1714, in which several European nations combined to stop France uniting with Spain in a single Bourbon kingdom that would have been overwhelmingly powerful.
5. FB.
6. Churchill, W. S. *Marlborough: his Life and Times*. London, Harrap 1938. Vol. IV, pp. 248 – 251.
7. Buried at Bommel as was his father, also an advocate. FB.
8. Rae P. *The History of the Late Rebellion, rais'd against King George by the Friends of the Popish Pretender*. London, Millar 1746. p. 172.
9. Plumb J. H. *Sir Robert Walpole. The King's Minister*. London, The Cresset Press 1960. p. 359.
10. Strictly 'First Lord of the Treasury'.

The title 'Prime Minister' did not come into official use until the late nineteenth century.
11. Rae P. *The History of the Late Rebellion, rais'd against King George by the Friends of the Popish Pretender*. London, Millar 1746. p. 327. Passim.
12. Baynes J. The Jacobite rising of 1715. London, Cassell 1970. pp. 117-125.
13. Ibid, pp. 139-153.
14. FB.
15. FB.
16. We have had to rely on circumstantial evidence: when the only surviving transcript of *Familiboek* (FB) was typed out in the 1930s, the typist, perhaps finding the name difficult to read, omitted the location of the battle.
17. Galbraith J. The battle of Glenshiel. *Transactions of the Gaelic society of Inverness 1927-1928*. Vol. 34, 280-313. passim
18. '...the wounds, the scars of which we see on his lofty forehead ...' From commemorative poem, in Latin, on the occasion of the graduation of Jan Ingen Housz as MD from the University of Louvain, 24 July 1753. Reproduced in Reed, H.S. Jan Ingen Housz, physiologist. *Chronica Botanica 1949*, 11; opposite p. 301.
19. FB.
20. Ibid. Widow Ingen Housz moved, with her other children, to Tilburg in 1722.
21. He died on 12 July 1764 – *Nederland's Patriciaat. 59e Jaargang 1973*. The Hague, Centraal bureau voor Genealogie.
22. Rehm G. *De Bredase apothekers van de 15e tot het begin van de 19e eeuw. (The apothecaries of Breda from the 15th century to the beginning of the 19th century.)* Breda, Geschied- en Oudheidkundige Kring van Stad en Land van Breda 'De Oranjeboom' 1961. pp.84-87.
23. He became a registered citizen of Breda on 23 November 1730. Ibid.
24. Ibid.
25. Ibid.
26. Ibid.
27. And, in fact, buy it outright on 29 January 1729, giving it in trust to his daughter and any children that might outlive her. Rehm G. *De Bredase apothekers van de 15e tot het begin van de 19e eeuw. (The apothecaries of Breda from the 15th century to the beginning of the 19th century.)* Breda, Geschied- en Oudheidkundige Kring van Stad en Land van Breda 'De Oranjeboom' 1961. p. 84.
28. Ibid. p. 86.
29. Stadsarchief Breda: digitale stamboom. Register of Roman Catholic Church, Brugstraat, Breda 1701-1757, p. 178.
30. Stadsarchief Breda: digitale stamboom. Burial register of the Great Church, Breda 1727-1729, p. 69r.
31. Stadsarchief Breda: digitale stamboom. Register of Roman Catholic Church, Brugstraat, Breda 1701-1757, p. 182.
32. IHLD. Note inside front cover.
33. Ibid.
34. *Nederland's Patriciaat. 59e Jaargang 1973*. The Hague, Centraal bureau voor Genealogie, p. 111.
35. IHLD. Note inside front cover.
36. Ibid.
37. *Nederland's Patriciaat. 59e Jaargang 1973*. The Hague, Centraal bureau voor Genealogie, p. 111.
38. Stadsarchief Breda: digitale stamboom. Burial register of the Great Church, Breda 1731-1733, p. 60r.
39. Thomas, D. *Collected Poems 1934 – 1952*. London, Dent 1952. p. 24.
40. Rehm G. *De Bredase apothekers van de 15e tot het begin van de 19e eeuw. (The apothecaries of Breda from the 15th century to the beginning of the 19th century.)* Breda, Geschied- en Oudheidkundige Kring van Stad en Land van Breda 'De Oranjeboom' 1961. p. 86.
41. The so-called 'Latin School' evolved into what is now called the 'Gymnasium' in continental Europe and 'Grammar School' in the UK; the prime purpose, of all, being the preparation of their pupils for studies at a University (where all tuition was once in Latin).
42. GF. p. 3.
43. Van Boven M. (ed.). *Ach lieve tijd. (Ah, good old times)*. Zwolle, Waanders 1985.

Part II, The people of Breda and their education. p. 255.
44 Luyendijk-Elshout A.M. (editor) *Wandelen met Boerhaave in en om Leiden (Walking with Boerhaave in Leiden)*. Caecilia Foundation, Leiden 1994 p. 28.
45 Ibid. p. 29 and passim.
46 Van Boven M. (ed.). *Ach lieve tijd. (Ah, good old times)*. Zwolle, Waanders 1985. Part II, The people of Breda and their education. p. 254.
47 Luyendijk-Elshout A.M. (editor) *Wandelen met Boerhaave in en om Leiden (Walking with Boerhaave in Leiden)*. Caecilia Foundation, Leiden 1994 p. 28. passim.
48 Hoogeveen, H. *Doctrina Particularum Linguae Graecaeto*. (Dictionary of the particles of the Greek language.) Lugduni Batavorum, 1769. Praefectio.
49 Van Boven M. (ed.). *Ach lieve tijd. (Ah, good old times)*. Zwolle, Waanders 1985. Part II, The people of Breda and their education. p. 254.
50 Hoogeven H. *Dictionarium Analogicum Linguae Graecae*. Cambridge University Press, Cambridge 1801. p. 14.
51 FB. – voted to citizenship of Breda in 1730.
52 Schama S. *Patriots and Liberators. Revolution in the Netherlands 1780 – 1813*. Collins, London 1977. p. 26.
53 Ibid, p. 28.
54 Ibid.
55 In fact Arnold would have needed to serve a four-year apprenticeship (according to one author) so it may have been that the family apothecary business started after Louis had undergone such a training and began his professional life in premises owned by his father.
56 Burnby J. *A study of the English Apothecary from 1660 to 1760*. London, Wellcome 1983. pp. 22 –27. Thompson C. *The mystery and art of the apothecary*. Detroit, Singing Tree Press 1971 pp. 93-99. Trease G. *Pharmacy in History*. London, Baillière, Tindall and Cox 1964. pp. 122-9.
57 The Hague, Centraal bureau voor Genealogie. *Nederland's Patriciaat. 59e Jaargang 1973*. p. 111.
58 Anna Catrina, 1688–1773; Elisabet, 1696–1777; Hedwig, 1700–1772: FB.
59 Whitworth R. *William Augustus, Duke of Cumberland*. London, Leo Cooper 1992. p. 29.
60 Wheatcroft A. *The Habsburgs*. London, Penguin 1996. p. 218.
61 Gillespie C (Ed. in chief). *Dictionary of Scientific Biography*. New York, Scribner's 1970-1980. Vol. 7, p. 11.
62 GF. p. 7.
63 Singer D. Sir John Pringle and his circle – Part I. Life. *Annals of Science 1949*. Vol. 6, p. 128 passim.
64 An alternative explanation is given by the famous biographer of Samuel Johnson – James Boswell, who was son of Lord Auchinleck, a contemporary and friend of Pringle. He alleges that John Pringle gave up medical studies in Edinburgh in order to travel to Leiden to learn commerce but, while there, attended a lecture of Boerhaave and was inspired to take up his clinical studies once more.

Notes to Chapter 2

1   Ingen Housz, J. *Experiments upon vegetables, discovering their great power of purifying the common air in the sunshine*. London, Elmsly 1779. Dedication to Sir John Pringle. p. v.
2   GF. p. 3.
3   Ibid.
4   Reed H. Jan Ingen Housz, plant physiologist: with a history of the discovery of photosynthesis. *Chronica Botanica 1949*, Vol. 11, opp. p. 301, passim.
5   Catholic University of Leuven. Encyclopædia Britannica. 2009. http://www.britannica.com/EBchecked/topic/349388/Catholic-University-of-Leuven (last accessed 24/03/11.)
6   Personal communication.
7   Catholic University of Leuven. Encyclopædia Britannica. 2009. http://www.britannica.com/EBchecked/topic/349388/Catholic-University-of-

Leuven (last accessed 24/03/11.)
8 Reed H. *Jan Ingen Housz, plant physiologist: with a history of the discovery of photosynthesis. Chronica Botanica 1949*, Vol. 11, opp. p. 301, passim.
9 That is, published on one side of a large piece of paper as a notice or flyer.
10 The earliest of these, and the likely source of all the others, is the family reminiscence put together by Dr. M.J.Godefroi, a friend of a near descendant of Ingen Housz (see 2. above). Although a very valuable record it has the historiography wrong in several other respects and gives no evidence of Ingen Housz being at Paris or Edinburgh at this early time.
11 GF. p. 4.
12 Ibid.
13 's-Hertogenbosch Stadsarchief. *Huizen en Gebouwen van 's-Hertogenbosch door van Sasse van Ysselt*. Vol.3, p.581.
14 Maurice D. *The Marlborough Doctors*. Stroud, Alan Sutton 1994.
15 GF. p. 5.
16 Rehm G. *De Bredase apothekers van de 15e tot het begin van de 19e eeuw. (The apothecaries of Breda from the 15th century to the beginning of the 19th century.)* Breda, Geschied- en Oudheidkundige Kring van Stad en Land van Breda. 'De Oranjeboom' 1961.
17 HVTS. 2045. Letter from Jan Ingen Housz, Vienna, to Jacob van Breda, Delft, 3 August 1786.
18 Ingen Housz J. Improvements in electricity. *Philosophical Transactions of the Royal Society of London 1779*, Vol. 69, pp. 661-673.
19 GF. p. 5.
20 Marsh A. *A history of the borough and town of Calne*. Calne, Heath 1903. p. 194. This recollection is mis-timed, by Marsh, to when Ingen Housz was living, much later in life, at Bowood House in Wiltshire, a property well isolated from any neighbours.
21 The Hague, Centraal bureau voor Genealogie. *Nederlands Patriciaat 1973*, Vol. 59. p. 111.
22 Wiesner J. *Jan Ingen-Housz: sein leben und sein wirken als naturforscher und artz*. Vienna, Carl Konegen 1905. p.19.
23 The Hague, Centraal bureau voor Genealogie. *Nederlands Patriciaat 1973*, Vol. 59. p. 111.
24 GF. p. 5.
25 BGA. B, 14a. Legal deed between Louis and Jan Ingen Housz, drawn 24 September 1764, that gave Louis the power to handle Jan's affairs in his absence abroad.

Notes to Chapter 3

1 Royal Society of London. Journal Book, Copy 25 (1763 – 66). p. 324.
2 Ibid.
3 Ibid.
4 Nichols J. *Biographical and literary anecdotes*. London, Bowyer 1782.
5 Caygill M. *The story of the British Museum*. London. British Museum Press 2002. p. 14.
6 Royal Society: Sackler Archive Resource, NA 4514.
7 Ibid.
8 Wilson D. *The British Museum. A History*. London, British Museum Press 2002. p. 28. Maty succeeded Gowin Knight as Principal Librarian in 1772.
9 Royal Society of London. Journal Book, Copy 25 (1763 – 66). p. 324.
10 BGA. B, 14a. .
11 GF. p. 7.
12 HKB. 1.
13 Ibid.
14 Hoogeveen, H. *Doctrina Particularum Linguae Graecaeto*. Lugduni Batavorum 1769. Praefectio.
15 Sonntag O (Ed). *John Pringle's correspondence with Albrecht von Haller*. Basel. Schwabe-Verlag 1999. p. 109. Singer D. Sir John Pringle and his circle. Part 1. Life. *Annals of Science 1949*; Vol. 6: p. 143.
16 Hopkins D. *The greatest killer. Smallpox in history*. Chicago, University of Chicago Press 2002. p. 59.
17 Tomalin C. *Samuel Pepys. The unequalled self*. London, Viking 2002. p. 234.
18 Ackroyd P. *London: the biography*.

London, Chatto and Windus 2000. p. 103.
19 Hughes R. *The fatal shore. A history of the transportation of convicts to Australia, 1787–1868.* London, Collins Harvill 1987.
20 Ibid. p. 20.
21 Ibid. p. 21.
22 The London 'season' was, by tradition, coincidental with the sitting of parliament – late autumn until June but, because of hunting, many people returned to the city only in January each year. London was certainly uncomfortable and malodorous during summer heat.
23 Schama S. *The embarrassment of riches. An interpretation of Dutch culture in the golden age.* London, Collins 1987. p. 597.
24 Porter R. *The greatest benefit to mankind. A medical history of humanity from antiquity to the present.* London, Harper Collins 1997. p. 303. Letter from William Heberden to Thomas Percival, 1794.
25 Prebble J. *Culloden.* London, Secker and Warburg 1961. p. 174.
26 Kippis A. *The Life of the author – being a preface to six discourses delivered by Sir John Pringle.* London, The Royal Society 1783. p. xiv.
27 Royal Society of London. Journal Book, Copy, XVIII (1742-45).
28 Royal Society of London. Archives; EC/1745/13.
29 Pringle J. *Observations on the Nature and Cure of Hospital and Jayl Fevers.* London, Millar & Wilson 1750.
30 Sir Godfrey Copley FRS died in 1709 leaving money to the Royal Society to support scientific research. In 1736 the Society decided to use the fund to purchase a gold medal, annually, to be awarded to outstanding contributors to 'knowledge'.
31 Pringle J. *Observations on the diseases of the army, in camp and garrison.* London, Millar, Wilson & Payne 1752.
32 Bacon F. *Novum Organum 1620.*
33 Singer D. Sir John Pringle and his circle. Part 1. Life. *Annals of Science 1949*; Vol. 6: p. 138.

34 J. S. G. Blair, 'Pringle, Sir John, first baronet (1707–1782)', *Oxford Dictionary of National Biography*, Oxford University Press, 2004; online edn, Oct 2007. http://www.oxforddnb.com/view/article/22805 (Last accessed 24 March 2011).
35 Ibid.
36 Hibbert C. *George III. A personal history.* London, Penguin Books 1998. p. 36.
37 Ibid. p. 44.
38 Ibid. p. 45.
39 Tomalin C. *Samuel Pepys. The unequalled self.* London, Viking 2002. p. 10.
40 Singer D. Sir John Pringle and his circle. Part 1. Life. *Annals of Science 1949*; Vol. 6: pp. 159-160.
41 Pottle F (Ed). *Extracts from Boswell's London Journal 1762 - 1763.* Heinemann, London, 1950. p. 277. We also get a slightly different perspective on Pringle from Boswell – that the doctor could be "sour" (p. 48) but since Pringle had been asked by Boswell's father, a contemporary and friend still in Scotland, to keep an eye on the young man's morals and behaviour in London, perhaps we should brush this off as bias tinged with guilt.
42 GF. p. 8
43 University of Edinburgh Journal 1936-7, Vol. 8, pp. 57-71.
44 Ibid. p. 59.
45 Ibid. p. 58.
46 GF. p. 8.
47 Ibid.
48 Wilson D. *The British Museum. A History.* London, British Museum Press 2002. p. 30.
49 ÖNB. Fol. 19.
50 http://www.georgianindex.net/London/Fleet_st.html (Last accessed 24/03/11).
51 Dis-assembled 1878, it was recently re-erected in Paternoster Square, adjacent to St Paul's Cathedral.
52 ÖNB. Fol. 7.
53 ÖNB. Fol. 5.
54 Ibid.

Notes to Chapter 4

1. BGA. A, 4. Collection of medical remedies and their formulations.
2. Powdered mercury.
3. 57% proof alcohol.
4. A resin of very disagreeable odour and taste.
5. Porter D, Porter R. *Patient's progress. Doctors and doctoring in Eighteenth-century England*. Cambridge, Polity Press 1989. p. 163.
6. Royal College of Physicians of Edinburgh. Pringle Annotations, 8.443.
7. Whytt R. *Observations on the nature, causes, and cure of those disorders which have been commonly called nervous, hypochondriac, or hysteric*. Edinburgh, Balfour 1765.
8. HKB. 1.
9. French R. *Robert Whytt, the soul, and medicine*. London, Wellcome Institute of the History of Medicine 1969. p. 8.
10. HKB. 1.
11. Farrand M. *(A restoration of) The autobiography of Benjamin Franklin*. Berkeley CA, University of California Press 1949. pp. 53–58.
12. Brands H. *The first American*. Doubleday, New York 2000. p. 272.
13. Ibid. p. 360.
14. Cooke A. *America*. London, British Broadcasting Corporation 1973. pp. 117-118.
15. HKB. 1.
16. ÖNB. Fol. 18.
17. The home of widowed Margaret Stephenson who was more a close friend of Franklin than a landlady. He had lodged here during his earlier sojourn in London and became very close to the family. The house still exists and is the only building remaining in the world in which Franklin ever lived. It has been restored as a museum and education centre by the Friends of Franklin House.
18. Österreichische Nationalbibliothek, Abendländischen Handschriften, Series Nova, 4062.
19. ÖNB. Fols. 12,13.
20. ÖNB. Fol. 14.
21. ÖNB. Fol. 19.
22. Ingen Housz J. Improvements in electricity. *Philosophical Transactions of the Royal Society of London 1779*, Vol. 69, pp. 661-673.
23. Allan Chapman, 'Ramsden, Jesse (1735–1800)', *Oxford Dictionary of National Biography*, Oxford University Press, 2004; online edn, Jan 2008 http://www.oxforddnb.com/view/article/23105 (Last accessed 24 March 2011).
24. Hackmann W. *Electricity from glass. The history of the frictional electrical machine 1600-1850*. Alphen aan den Rijn, Sijthoff & Noordhoff 1978. pp. 143,144.
25. Hoogeveen, H. *Doctrina Particularum Linguae Graecaeto*. Lugduni, Batavorum, 1769. Praefectio (Preface).
26. Seager J. *Hoogeveen's particles: abridged and translated into English*. London 1829.
27. ÖNB. Fol 4.
28. Ibid.
29. Ibid.
30. ÖNB. Fol. 6.
31. ÖNB. Fol. 8. The caves were probably Treak Cavern, famous for the Blue John (Bleu-Jaune) stone that was already being mined there: and the larger Peak Cavern, also known as The Devil's Arse; the entrance was reputed to be Hades' back door. Brighton T. *The discovery of the Peak District*. Chichester, Phillimore 2004 pp. 37, 73.
32. ÖNB. Fol. 8.
33. ÖNB. Fol. 5.
34. ÖNB. Fol. 8.
35. ÖNB. Fol. 10.
36. Ibid.
37. ÖNB. Fols. 12,13.
38. ÖNB. Fol. 12.
39. ÖNB. Fol. 14.
40. Ibid.
41. ÖNB. Fol. 19.
42. ÖNB. Fol. 18.
43. GF. p. 6.
44. ÖNB. Fol. 19.
45. POBF. Letter from Jan Ingen Housz, London, to Benjamin Franklin, Passy, 5 October 1778
46. ÖNB. Fol. 34.

47 One of the pairs of volumes is now at the British Library at shelfmark 67E2, part of 'The King's Collection'. The twin leather-bound copies each bear the emblem of George III in gold leaf and Hoogeveen has signed them personally. They have their place of honour among all the other tomes, donated by George IV, at the genesis of the British Library. They are among all the books that clothe the shelves of the giant bookcase towering up through the middle of the new building at Euston.
48 ÖNB. Fol. 34.
49 ÖNB. Fol. 20.
50 POBF. Letter from the Earl of Morton, London, to Benjamin Franklin, London, 21 May 1767.
51 ÖNB. Fol. 20.
52 Hoogeveen, H. *Doctrina Particularum Linguae Graecae*. Lugduni, Batavorum, 1769.
53 ÖNB. Fol. 35.
54 Breda Gemeentearchief. Breda Commissieboeken (The committee books of Breda's official appointments 1637-1811.).
55 ÖNB. Fol. 35.
56 Hoogeveen, H. *Doctrina Particularum Linguae Graecaeto Lugduni Batavorum*, 1769. Praefectio (Preface).
57 ÖNB. Fol. 35.
58 Ibid.
59 University of Amsterdam. Hs. Died 15Bg1. Letter from Gerard Meerman, Rotterdam, to Jan Ingen Housz, London, 24 February 1767.
60 Meerman G. *De l'invention de l'imprimerie*. The Hague, Paris and London, 1765.
61 ÖNB. Fol. 17.
62 ÖNB. Fol. 21.

Notes to Chapter 5

1 Macaulay T. Firth C. Ed. *History of England*. London, Macmillan 1914. p. 2468.
2 Hopkins D. *The greatest killer. Smallpox in history*. Chicago, University of Chicago Press 1983.
3 Ibid. pp. 1-3.
4 A genetically determined illness in which there is an inability to metabolise porphyrins, break-down products from red blood cells, resulting in severe abdominal pains and hallucinations. George III is the most famous sufferer.
5 Grundy I. *Mary Wortley Montagu. Comet of the enlightenment*. Oxford, Oxford University Press 1999. p. 113.
6 Plumb J. *Sir Robert Walpole. The King's minister*. London, The Cresset Press 1960. p. 80.
7 Grundy I. *Mary Wortley Montagu. Comet of the enlightenment*. Oxford, Oxford University Press 1999. p.113. The couple had eloped to be married in 1712.
8 Ibid. p. 114.
9 Ibid. pp. xvii-xxiii.
10 Now called Edirne.
11 Now called Istanbul. No longer the capital city.
12 Who was, ironically, to succumb to smallpox in 1726.
13 Halsband R. (Ed). *The complete letters of Lady Mary Wortley Montagu. Volume 1*. Oxford, Clarendon Press 1965. pp. 338-9.
14 Grundy I. *Mary Wortley Montagu. Comet of the enlightenment*. Oxford, Oxford University Press 1999. p. 65.
15 Ibid. p. 101.
16 Woodward had also been a fellow of the Royal Society but expelled by the council after a tiff with Hans Sloane.
17 This had not been the first correspondence to the Royal Society conveying such news but Timoni's letter was best known to Woodward: he had read it to a Royal Society meeting.
18 Grundy I. *Mary Wortley Montagu. Comet of the enlightenment*. Oxford, Oxford University Press 1999. p. 162.
19 Maitland C. *Mr. Maitland's account of inoculating the small-pox*. London 1722.
20 Glynn I, Glynn J. *The life and death of smallpox*. London, Profile Books 2004. p. 52.
21 Halsband R. (Ed). *The complete letters of Lady Mary Wortley Montagu. Volume 1*.

Oxford, Clarendon Press 1965. p. 339.
22. Grundy I. *Mary Wortley Montagu. Comet of the enlightenment*. Oxford, Oxford University Press 1999. p. 210. She subsequently recovered well and was, when mature, to marry, and bear 11 children by, Lord Bute, young George III's confidante and first Prime Minister.
23. Ibid. p. 213.
24. Ibid. p. 215.
25. Fisher R. *Edward Jenner 1749–1823*. London, Andre Deutch 1991. p. 14.
26. Zwanenberg D. The Suttons and the business of inoculation. *Medical History* 1978; Vol. 22: p. 76.
27. Glynn I, Glynn J. *The life and death of smallpox*. London, Profile Books 2004. p. 48.
28. Simon Schaffer, 'Watson, Sir William (1715–1787)', *Oxford Dictionary of National Biography*, Oxford University Press, 2004 http://www.oxforddnb.com/view/article/28874 (Last accessed 25 March 2011).
29. Pulteney R. *Historical and biographical sketches of the progress of Botany in England, from its origin to the introduction of the Linnaean system. London 1790. Volume 2*. p. 295.
30. The two eldest sons of George II, the former being the father of George III. We have already referred to the later relationship between Cumberland and John Pringle.
31. Pulteney R. *Historical and biographical sketches of the progress of Botany in England, from its origin to the introduction of the Linnaean system. London 1790. Volume 2*. p. 319.
32. Geikie A. *Annals of the Royal Society club*. London, Macmillan 1917.
33. Watson W. *An account of a series of experiments instituted with a view of ascertaining the most successful method of inoculating the small-pox*. London, 1768.
34. Wedd K. *The Foundling Museum*. London, The Foundling Museum 2004. p. 7.
35. Ibid. p. 8.
36. Ibid. p. 10.
37. Ibid. p. 12.
38. Ibid. p. 21.
39. Initially, the children had been received into more modest accommodation in Hatton Garden.
40. Wedd K. *The Foundling Museum*. London, The Foundling Museum 2004. p. 14.
41. Ibid. p. 13.
42. Foundling Hospital Archives – LMA – A/A4/A/3/5/6 – Subcommittee minute book: 19 July 1766.
43. Foundling Hospital Archives – LMA – A/A4/A/3/5/6. Subcommittee minute book, 18 September 1765.
44. Ingen Housz J. *Lettre de Monsieur Ingen Housz, Docteur en Médicine, à Monsieur Chais, Pasteur de L'Église Wallonne de la Haye . . .* Amsterdam, Van Harrevelt 1768.
45. Foundling Hospital Archives – LMA – A/FH/A18/8/. Inoculation books.
46. McLure R. *Coram's Children*. New Haven CT, Yale University Press 1981. p. 193.
47. Some foster mothers refused to part with their little charges, preferring to adopt them.
48. Foundling Hospital Archives – LMA – A/FH/K02/009. General minutes: 22 January 1766.
49. Ibid.
50. McLure R. *Coram's Children*. New Haven CT, Yale University Press 1981. p. 213.
51. Foundling Hospital Archives – LMA – A/FH/A3/5/6. Subcommittee minutes book: 7 March 1767. Four guineas is now worth £270 and £1 11s 6p is now worth £100.
52. Inoculating seems not to have happened during the winters – there are no recorded cases for January to April inclusive in most years. It may be, though, that children were not brought back to the Foundling Hospital during the months of worst winter weather and there were therefore none to inoculate.
53. Watson W. *An account of a series of experiments instituted with a view of ascertaining the most successful method of inoculating the small-pox*. London, 1768.
54. It was only in 1767 that William

Heberden published his outstanding bedside descriptions that led to the consistent differentiation of smallpox and chickenpox.
55 Foundling Hospital Archives – LMA – A/FH/A18/8/. Inoculation books.
56 GF. p. 8.
57 Ibid. p. 9.
58 Maloney W. *George & John Armstrong of Castleton*. Edinburgh and London, Livingstone 1954. p. 49.
59 Armstrong G. *An essay on the diseases most fatal to infants*. London, Cadell 1767.
60 Maloney W. *George & John Armstrong of Castleton*. Edinburgh and London, Livingstone 1954. p. 56.
61 GF. p. 6.
62 Hopkins D. *The greatest killer. Smallpox in history*. Chicago. University of Chicago Press 1983. p. 55.
63 Zwanenberg D. The Suttons and the business of inoculation. *Medical History 1978*; Vol. 22: p. 73.
64 Ibid.
65 Daniels C. De kinderpok-inenting in Nederland (The child smallpox vaccination in the Netherlands). *Nederlandsch Tijdschrift voor Geneeskunde 1875*; Vol. 11, 17-223. pp. 88,89.
66 Ibid.
67 Dimsdale T. *The present method of inoculating for the small-pox*. London, Owen 1767.
68 ÖNB. Fol. 35.
69 Webster N. *The great north road*. Bath, Adams and Dart 1974. p. 61.

Notes to Chapter 6

1 Dimsdale T. *The present method of inoculating for the small-pox*. London, Owen 1767.
2 Now Port Hill House, Farquhar Street & Bengeo Street, Hertford. Heath C. *The book of Hertford*. Chesham, Barracuda 1975. p. 25. The Dimsdale family themselves lived at The Priory House – personal communication.
3 Register of baptisms, marriages and burials of Little Berkhamsted, Hertfordshire. Hertfordshire Record Office, D/P 20/1/3: an appended contemporaneous note.
4 Ingen Housz J. *Lettre de Monsieur Ingen Housz, Docteur en Médicine, à Monsieur Chais, Pasteur de L'Église Wallonne de la Haye . . . Amsterdam*, Van Harrevelt 1768.
5 T. E. Kebbel, 'Dimsdale, Thomas (1712–1800)', rev. Andrea Rusnock, *Oxford Dictionary of National Biography*, Oxford University Press, 2004 http://www.oxforddnb.com/view/article/7675 (Last accessed 25 March 2011).
6 Ibid.
7 GF. p. 9.
8 Hopkins D. *The greatest killer. Smallpox in history*. Chicago and London, University of Chicago Press 1983. p. 249.
9 Francis Franklin was Benjamin Franklin's only legitimate son. Brands H. W. *The First American*. New York, Doubleday 2000. p. 155.
10 Ibid.
11 The 'Chinese' technique may have worked because the virus particles lost some potency when dried, but no-one is sure.
12 Ingen Housz J. *Lettre de Monsieur Ingen Housz, Docteur en Médicine, à Monsieur Chais, Pasteur de L'Église Wallonne de la Haye . . . Amsterdam*, Van Harrevelt 1768.
13 This population total includes the residents of the overlapping community, Bayford; a hamlet on the opposite side of the valley.
14 Ingen Housz J. *Lettre de Monsieur Ingen Housz, Docteur en Médicine, à Monsieur Chais, Pasteur de L'Église Wallonne de la Haye . . . Amsterdam*, Van Harrevelt 1768.
15 Edinburgh. Royal College of Physicians. Pringle J. Medical Annotations Vol. 8. p. 418.
16 Ibid.
17 Register of baptisms, marriages and burials. Little Berkhamsted, Hertfordshire UK. Hertfordshire Record Office, D/P 20/1/3: an appended contemporaneous note.
18 Ingen Housz J. *Lettre de Monsieur Ingen*

Housz, *Docteur en Médicine, à Monsieur Chais, Pasteur de L'Église Wallonne de la Haye* . . . Amsterdam, Van Harrevelt 1768..
19 Crankshaw E. *Maria Theresa*. London, Constable 1983. pp. 265-267.
20 Ibid.
21 Ibid. p. 260.
22 Who was later sent, at the age of 14, to marry the Dauphin, at Paris. She became Queen Marie Antoinette of France, in 1774, when Louis XV died – of smallpox.
23 Wheatcroft A. *The Habsburgs. Embodying empire*. London, Viking 1995. p. 226.
24 Crankshaw E. *Maria Theresa*. London, Constable 1983 p. 263.
25 Ibid. p. 271.
26 Ibid.
27 Now Bratislava.
28 Crankshaw E. *Maria Theresa*. London, Constable 1983. p. 273.
29 Ibid.
30 Ibid.
31 Russell J. *Nelson and the Hamiltons*. Harmondsworth, Middlesex, Penguin 1972. p. 13.
32 Maria Theresa is reported to have felt directly responsible for this death. She had taken Marie Josepha with her to the Capuchin vault, the Habsburg mausoleum under central Vienna, to pay respects to the remains of the other Josepha, late second wife of Joseph (death 'nine') unaware that the sarcophagus had not yet been sealed. However, there are good reasons, applying modern knowledge of incubation periods etc., that the Empress need not have reproached herself.
33 Slang name for smallpox that appeared in the eighteenth century.
34 Crankshaw E. *Maria Theresa*. London, Constable 1983. p. 275.
35 Beales D. *Joseph II. 1. In the shadow of Maria Theresa 1741-1780*. Cambridge, Cambridge University Press 1987. p. 138.
36 Ibid. p. 158.
37 Gentleman's Magazine 1768, Vol. 38, February. p. 75 passim.
38 Ibid.
39 Sonntag O. *John Pringle's correspondence with Albrecht von Haller*. Basel, Schwabe 1999. p. 109.
40 Edinburgh. Royal College of Physicians. Pringle J. *Medical Annotations Vol. 8*, p. 443.
41 Chaise C. *Brief van den Heere Chais, Predikant van de Walsche Gemeente in 'sHage aan den Heere Sutherland*. 's-Gravenhage, Du Mee (January) 1768.
42 Daniels C. De Kinderpok-inenting in Nederland (The inoculation of children against smallpox in the Netherlands.) *Nederlandsch Tijdschrift voor Geneeskunde 1875*; Vol. 11: 17-223. pp. 88,89.
43 Ibid. p. 70.
44 Chais C. Essai apologétique sur la méthode de communiquer la petite vérole par inoculation. *Verhandeling der Hollandsche Maatschappij der Wetenschappen, Deel 1, 1754*.
45 British Library Tracts, T.301,(2★). Letter from Matthew Maty, Doctor of Medicine, Principal Librarian of the British Museum and Secretary of the Royal Society, London, to Monsieur Chais, Pastor of the Walloon Church at The Hague, Netherlands, 27th March 1768.
46 Chaise C. *Brief van den Heere Chais, Predikant van de Walsche Gemeente in 'sHage aan den Heere Sutherland*. 's-Gravenhage, Du Mee (January) 1768.
47 Sutherland A. *Antwoord van den Heer Sutherland, Medicine Doctor, op den brief van myn Heer Chais*. 's-Gravenhage, Du Mee (February) 1768.
48 Ingen Housz J. *Lettre de Monsieur Ingen Housz, Docteur en Médicine, à Monsieur Chais, Pasteur de L'Église Wallonne de la Haye* . . . Amsterdam, Van Harrevelt 1768.
49 Edinburgh. Royal College of Physicians. Pringle J. Medical Annotations Vol. 8. p. 157..
50 Vienna. Österreichisches Haus-Hof-Staatsarchiv (England Section) F.157. Letter from Prince Kaunitz, Vienna, to Count Seilern, London, 4 March 1768.
51 GF. p. 12. More likely to have been Friday 25 March.

52 Vienna. Österreichisches Haus-Hof-Staatsarchiv (England Section) F.156. Letter from Count Seilern, London, to Prince Kaunitz, Vienna, 29 March 1768. passim.

53 British Library Tracts, T.301.(2*). Letter from Matthew Maty, Doctor of Medicine, Principal Librarian of the British Museum and Secretary of the Royal Society, London, to Monsieur Chais, Pastor of the Walloon Church at The Hague, Netherlands, 27th March 1768.

54 W. W. Webb, 'Saunders, Richard Huck- (1720–1785)', rev. Jeffrey S. Reznick, *Oxford Dictionary of National Biography*, Oxford University Press, 2004 http://www.oxforddnb.com/view/article/24703 (Last accessed 26 March 2011). London. Wellcome Trust Library. Charles Gordon papers, p. 79. Charles Gordon was preparing a biography on Sir John Pringle but died before it was completed. The extracts are from the Medical Annotations of Pringle held by the Royal College of Physicians, Edinburgh.

55 Sonntag O. *The correspondence between Albrecht von Haller and Charles Bonnet*. Huber, Bern 1983. p. 754. Letter from von Haller, Berne, to Charles Bonnet, Geneva, 14 May 1768. Daniels C. De Kinderpok-inenting in Nederland (The inoculation of children against smallpox in the Netherlands.) *Nederlandsch Tijdschrift voor Geneeskunde 1875*; Vol. 11: 17-223. p. 45.

Notes to Chapter 7

1 GF. p. 12.
2 GF. p. 13.
3 Ibid.
4 The 'inner court', literally translated; being the central yard of the government buildings at The Hague.
5 GF. p. 13.
6 Ibid.
7 Vienna. Haus- Hof- Staatsarchiv, F.45. Letter from Thaddäus Reischach, The Hague, to Prince Charles Alexander, Brussels, 15 April 1768.

8 GF. p. 13.
9 Ibid.
10 Ibid.
11 They married on 30 April 1764 at Tournhout. The Hague, Centraal bureau voor Genealogie. *Nederland's Patriciaat. 59e Jaargang 1973*. p. 111.
12 The Hague, Centraal bureau voor Genealogie. *Nederland's Patriciaat. 59e Jaargang 1973*. p. 112.
13 GF. p. 12 passim.
14 The language always employed in court and diplomatic circles.
15 BGA. A 7. Ingen Housz finances note book 1767-1789.
16 The value of 1 ducat (a gold coin) was worth 4.5 Austrian florins and there were 10 florins to the English pound. Therefore 49 ducats was worth about £21 – this would represent, now, about £1350.
17 Phillips D. *The Great Road to Bath*. Newbury, Countryside Books 1983. p. 87.
18 BGA. A 7. Ingen Housz finances note book 1767-1789. Note inside front cover. Ratisbon is now Regensburg.
19 Leppmann W. *Winckelmann*. London, Victor Gollancz 1971. p. 258 passim.
20 Ibid. p. 24 passim.
21 Ibid. p. 220.
22 Ibid. p. 221.
23 Ibid. p. 235.
24 Ibid. p. 255.
25 Beales D. *Joseph II. 1. In the shadow of Maria Theresa 1741-1780*. Cambridge, Cambridge University Press 1987. p. 259.
26 Howard S. Albani, Winckelmann and Cavaceppi. The transition from amateur to professional antiquarianism. *Journal of the History of Collections 1992*; Vol. 4, pp. 27-38.
27 Leppmann, W. *Winckelmann*. London, Victor Gollancz 1971. p. 258.
28 Stoll H. *Tod in Triest*. Berlin, Union Verlag 1968. p. 593.
29 In this pre-mechanical era boats were only able to travel downstream on the Danube, a river that has a very substantial fall. Boats were therefore made knowing that they would be

broken up downstream: many travelled only as far as Vienna where they were scrapped and used as firewood. Pezzl, J. *Skizze von Wein – Sketch of Vienna, six instalments 1786–1790*. As edited and reproduced in Robbins Landon, H. *Mozart and Vienna*. London, Thames and Hudson 1991, p. 143.
30 BGA. A 7. Ingen Housz finances note book 1767-1789. Note inside front cover.
31 Ibid.
32 ÖNB. Fols. 23,24.
33 BGA. A 7. Ingen Housz finances note book 1767-1789. Note inside front cover.
34 HKB. 2. Letter to Jan Ingen Housz, Vienna, from Richard Huck, London, 1 June 1768.
35 Pezzl, J. *Skizze von Wein – Sketch of Vienna, six instalments 1786–1790*. As edited and reproduced in Robbins Landon, H. *Mozart and Vienna*. London, Thames and Hudson 1991, p. 64.
36 Ibid. p. 65
37 BGA. A 7. Ingen Housz finances note book 1767-1789. Fol. 1.
38 HKB. 2. Letter to Jan Ingen Housz, Vienna, from Richard Huck, London, 1 June 1768.
39 Ibid.
40 BGA. A 7. Ingen Housz finances note book 1767-1789. Note inside front cover.
41 GF. p. 16.
42 HKB. 2. Letter to Jan Ingen Housz, Vienna, from Richard Huck, London, 1 June 1768.
43 Ibid.
44 Ibid.
45 GF. p. 16.
46 HKB. 2. Letter to Jan Ingen Housz, Vienna, from Richard Huck, London, 1 June 1768.
47 Van der Korst J. *Een dokter van formaat. (A doctor of stature.)* Amsterdam, Bert Bakker 2003. p. 19.
48 Ibid. p. 63 passim.
49 Ibid. p. 55.
50 Ibid. p. 252.
51 Lesky E. *Archiv dür Österreicheische Geschichte, Vol. 122. (Die Pocken)* Vienna, Rudolf M. Rohrer 1959. p. 142.
52 HKB. 2. Letter to Jan Ingen Housz, Vienna, from Richard Huck, London, 1 June 1768.
53 Ibid.
54 BGA. A7. Ingen Housz finances notebook 1767-1789.
55 GF. p. 16
56 Edinburgh. Royal College of Physicians. Medical Annotations of Sir John Pringle, Vol. 8. p. 593. Letter from Gerard van Swieten, Vienna, to Sir John Pringle, London, 7 September 1768.
57 Ibid.
58 Ibid.
59 Ibid.
60 GF. pp. 15-17. Letter from Jan Ingen Housz, Vienna, to Johann Deckers, 's-Hertogenbosch, May or June 1768.
61 Ibid.
62 Ibid.

Notes to Chapter 8

1 Kugler G. *Schönbrunn Palace. The State Apartments. Schloss Schönbrunn Kultur- und Betriebsges.* Vienna, Christian Brandstätter 1995.
2 BGA. A 7. Ingen Housz finances note book 1767-1789. Note inside front cover.
3 Crankshaw E. *Maria Theresa*. London, Constable 1983. p. 273.
4 Glynn I & J. *The life and death of Smallpox*. London, Profile Books 2004. p. 83.
5 Ibid. p. 82.
6 GF. p. 15.
7 Ibid.
8 Glynn I & J. *The life and death of Smallpox*. London, Profile Books 2004. p. 83.
9 WJ. p. 35. Wiesner refers to 65 in one citation, and 200 in another – p. 47.
10 Beales D. *Joseph II. 1. In the shadow of Maria Theresa 1741-1780*. Cambridge, Cambridge University Press 1987. p. 246.

11 GF. p. 17.
12 Crankshaw E. *Maria Theresa*. London, Constable 1983. p. 260.
13 GF. p. 17.
14 Daniels C. De Kinderpok-inenting in Nederland (The inoculation of children against smallpox in the Netherlands.) *Nederlandsch Tijdschrift voor Geneeskunde 1875*; Vol. 11: 17-223. p. 47. *Mozart: Briefe und Aufzeichnugen 1762-1775*. International Mozart Foundation, Salzburg p. 386. Letter from Leopold Mozart, Vienna, to Lorenz Hagenauer, Salzburg, 6 Aug 1768.
15 Now a dense urban suburban district.
16 WJ. p. 35.
17 Daniels C. De Kinderpok-inenting in Nederland (The inoculation of children against smallpox in the Netherlands.) *Nederlandsch Tijdschrift voor Geneeskunde 1875*; Vol. 11: 17-223. p. 47.
18 WJ. pp. 34/5.
19 Leppmann, W. *Winckelmann*. London, Victor Gollancz 1971. p. 6.
20 ÖNB. Fol. 22. Letter from Jan Ingen Housz, Vienna, to Hendrik Hoogeveen, Delft, 20 May 1768 (but must have been 20 June).
21 Ibid.
22 Leppmann, W. *Winckelmann*. London, Victor Gollancz 1971. p. 3 passim.
23 Ibid. p. 10.
24 Daniels C. De Kinderpok-inenting in Nederland (The inoculation of children against smallpox in the Netherlands.) *Nederlandsch Tijdschrift voor Geneeskunde 1875*; Vol. 11: 17-223. p. 47.
25 WJ. p. 35.
26 Edinburgh. Royal College of Physicians. Medical Annotations of Sir John Pringle, Vol. 8. p. 593. Letter from Gerard van Swieten, Vienna, to Sir John Pringle, London, 7 September 1768.
27 WJ. p. 35.
28 Ibid.
29 Ibid.
30 HKB. Fol. 3. Letter to Jan Ingen Housz, Vienna, from Thomas Dimsdale, London, 10 July 1768.
31 Ibid.
32 HKB. Fol. 5. Letter to Jan Ingen Housz, Vienna, from Salomon de Monchy, Rotterdam, 6 October 1768.
33 HKB. Fol. 4. Letter to Jan Ingen Housz, Vienna, from Thomas Dimsdale, St. Petersburg, 8 September 1768.
34 Alexander J. *Catherine The Great: life and legend*. New York, Oxford University Press 1989 p. 145.
35 Ibid. p. 31.
36 Ibid.
37 Ibid. p. 144.
38 Ibid.
39 Ibid. p. 145.
40 Literally 'head officer of the court' – personal governor and adviser in this context.
41 Alexander J. *Catherine The Great: life and legend*. New York, Oxford University Press 1989. p. 100.
42 Ibid. p. 146.
43 Ibid. p. 147.
44 Ibid. p. 148.
45 Ibid.
46 ÖNB. Fols. 23,24.
47 WJ. p. 36.
48 Mercuric chloride – a mild laxative in small dosage.
49 GF. p. 16.
50 WJ. p. 31 passim.
51 Ibid.
52 Ibid.
53 Ibid. p. 36.
54 Ibid. p. 39 passim.
55 http://de.wikipedia.org/wiki/Ober_Sankt_Veit
56 Personal communication: letter from Thomas Dimsdale, St. Petersburg, to Henry Nicols, London, 27 October 1768.
57 Edinburgh. Royal College of Physicians. Medical Annotations of Sir John Pringle, Vol. 8, p. 593. Letter from Gerard van Swieten, Vienna, to Sir John Pringle, London, 7 September 1768.
58 Personal communication: letter from Thomas Dimsdale, St. Petersburg, to Henry Nicols, London, 27 October 1768. passim.
59 Personal communication: letter from Thomas Dimsdale, St. Petersburg, to

Henry Nicols, London, 27 October 1768. Dimsdale quotes £550 per annum here and our text figures assume an exchange rate (which fluctuated irregularly) of one Austrian florin being equivalent to two shillings i.e ten Aus. Fl. to the pound sterling.

60 Daniels C. De Kinderpok-inenting in Nederland (The inoculation of children against smallpox in the Netherlands.) *Nederlandsch Tijdschrift voor Geneeskunde* 1875; Vol. 11: 17-223. p. 47.

61 His Vienna Bankers. BGA. A 7. Ingen Housz finances note book 1767-1789. Fol. 1.

62 Pezzl, J. *Skizze von Wein – Sketch of Vienna, six instalments 1786–1790*. As edited and reproduced in Robbins Landon, H. *Mozart and Vienna*. London, Thames and Hudson 1991, p. 116.

63 WJ. p. 37.

64 Hoogeveen, H. *Doctrina Particularum Linguae Graecae*. Lugduni, Batavorum, 1769. Praefectio (Preface).

65 Vienna, Hofkammerarchiv, 1536. Royal Decree on Inoculation, 29 July 1769.

66 Alexander J. *Catherine The Great: life and legend*. New York. Oxford University Press 1989. p. 147.

67 Personal communication: letter from Thomas Dimsdale, St. Petersburg, to Henry Nicols, London, 27 October 1768.

68 HKB. Fol.7. Letter from Thomas Dimsdale, St. Petersburg, to Jan Ingen Housz, Vienna, 16 November 1768.

69 HKB. Fol.8. Letter from Thomas Dimsdale, St. Petersburg, to Jan Ingen Housz, Vienna, 10 December 1768.

70 Ibid.

71 BGA. A 7. Ingen Housz finances note book 1767-1789. Fol. 2.

72 Fenner F, Henderson D, Arita I, Ježek Z, Ladnyi I. *Smallpox and its eradication*, Geneva, World Health Organisation 1988, p. 1062.

Notes to Chapter 9

1 The so-called Theresian Academy – founded by the Empress in 1751 and based at the Schloss Wiener-Neudstadt some 30 miles south of Vienna. It is still the foremost centre for the training of military officers, the Austrian equivalent, perhaps, of Sandhurst.

2 Daniels C. De Kinderpok-inenting in Nederland (The inoculation of children against smallpox in the Netherlands.) *Nederlandsch Tijdschrift voor Geneeskunde* 1875; Vol. 11: 17-223. p. 48.

3 Edinburgh. Royal College of Physicians. Medical Annotations of Sir John Pringle, Vol. 8; p. 157. Letter from Jan Ingen Housz, Vienna, to Sir John Pringle, London, 14 December 1768. The annotations actually record 800 but this is almost certainly in error or refers to a total that included inoculations performed after Ingen Housz returned from Italy.

4 Ibid.

5 BGA. A 7. Ingen Housz finances note book 1767-1789. Fol. 2.

6 Ibid.

7 Daniels C. De Kinderpok-inenting in Nederland (The inoculation of children against smallpox in the Netherlands.) *Nederlandsch Tijdschrift voor Geneeskunde* 1875; Vol. 11: 17-223. p. 48.

8 Ibid.

9 BGA. A 7. Ingen Housz finances note book 1767-1789. Fol. 2.

10 GF. p. 20.

11 Beales D. *Joseph II: In the shadow of Maria Theresa, 1741-1780*. Cambridge, Cambridge University Press 1987. pp. 255-258.

12 Acton H. *The Bourbons of Naples*. London, Methuen 1956. p. 129.

13 IHLD. Abstract of a letter to Francis Millman, London, from Jan Ingen Housz, Vienna, 25 August 1774. He died of a strangulated umbilical hernia – an abdominal rupture nearly always linked to obesity.

14 Van der Korst J. *Een dokter van formaat. (A doctor of status)*. Amsterdam, Bert Bakker 2003. p. 146.

15 Ibid. p. 149.

16 BGA. A 7. Ingen Housz finances note book 1767-1789. Fol. 2 verso.

17 Ibid.
18 Fine material, usually of silk, bearing dainty floral patterns, originally designed for 'Madame' Pompadour, mistress of Louis XV.
19 da Mosto F. *Venice*. London, BBC Books 2004. p. 11.
20 Black J. *The grand tour in the eighteenth century*. Stroud, The History Press 2009. pp. 43-44.
21 Beales D. *Joseph II: In the shadow of Maria Theresa, 1741-1780*. Cambridge, Cambridge University Press 1987. p. 256.
22 F. M. O'Donoghue, 'Lane, William (1746–1819)', rev. Jill Springall, *Oxford Dictionary of National Biography*, Oxford University Press, 2004 http://www.oxforddnb.com/view/article/16001 (Last accessed 27 March 2011).
23 Ibid.
24 GF. p. 20.
25 Personal communication.
26 Von Arneth A. (Ed.) *Maria Theresia und Joseph II. Ihre correspondenz sammt Briefen Joseph's an seinen Bruder Leopold. Ester Band 1761 – 1772.* Wien, von Carl Gerold's Sohn 1867. Letter from Emperor Joseph, Florence, to Maria Theresa, Vienna, 12 April 1769.
27 Crankshaw E. *Maria Theresa*. London, Longmans 1969. p. 140.
28 Little Berkhamsted, Hertfordshire. Register of baptisms, marriages and deaths, Hertfordshire County Record Office, D/P20/1/2, Memorandum. Daniels C. De Kinderpok-inenting in Nederland (The inoculation of children against smallpox in the Netherlands.) *Nederlandsch Tijdschrift voor Geneeskunde* 1875; Vol. 11: 17-223. p. 48.
29 Acton H. *The Bourbons of Naples*. London, Methuen 1956. p. 128.
30 Ibid. p. 115.
31 Ibid. p. 137.
32 Ibid. p. 141.
33 Ibid. p. 131.
34 Ibid. p. 77 passim.
35 Cooley A, Cooley M. *Pompeii: a source book*. London, Routledge 2004. p. 200. He was fully recovered and out of quarantine by 7 April.
36 The same William Hamilton (1730-1803), later knighted, who is, unjustly and almost exclusively, now remembered as the ageing husband of Emma, the second Lady Hamilton, cuckolded by Nelson. These notorious developments were more than thirty years into the future, Nelson starting his affair with Emma after arriving at Naples fresh from his victory over the French fleet at Aboukir Bay, the Battle of the Nile, in 1798.
37 Constantine D. *Fields of fire. A life of Sir William Hamilton*. London, Weidenfeld and Nicolson 2001. p. 21 passim.
38 Ibid. pp. 19,20.
39 Ibid. p. 22.
40 Ibid.
41 Ibid. p. 20.
42 Ibid. p. 34.
43 Ibid. p. 35.
44 Royal Society, Certificates of Election. GB17; EC1766/14.
45 Constantine D. *Fields of fire. A life of Sir William Hamilton*. London, Weidenfeld and Nicolson 2001. p. 21.
46 Acton H. *The Bourbons of Naples*. London, Methuen 1956. p. 49.
47 From Corinth and Athens, although some were probably made slightly later by Greek colonists of Southern Italy. There was no evidence that they had been made by the Etruscan race of Northern Italy.
48 Wilson D. *The British Museum*. London, The British Museum Press 2002. p. 47.
49 Constantine D. *Fields of fire. A life of Sir William Hamilton*. London, Weidenfeld and Nicolson 2001. p. 37.
50 Ibid.
51 Ibid. p. 4.
52 Ingen Housz did not inoculate the disabled prince, Philip – he was to die of smallpox some eight years later, the cause of much panic at Caserta and the hasty inoculation of Ferdinand and Caroline's children (not by Ingen Housz).
53 Cooley A, Cooley M. *Pompeii: a source book*. London, Routledge 2004. p. 200.
54 Von Arneth A. (Ed.) *Maria Theresia und Joseph II. Ihre correspondenz sammt*

Briefen Joseph's an seinen Bruder Leopold. Ester Band 1761 – 1772. Wien, von Carl Gerold's Sohn 1867. Letter from Emperor Joseph, Florence, to Maria Theresa, Vienna, 12 April 1769.
55 Ibid.
56 WJ. p. 38.
57 Von Arneth A. (Ed.) *Maria Theresia und Joseph II. Ihre correspondenz sammt Briefen Joseph's an seinen Bruder Leopold. Ester Band 1761 – 1772*. Wien, von Carl Gerold's Sohn 1867. Letter from Emperor Joseph, Florence, to Maria Theresa, Vienna, 22 May 1769.
58 Ibid. Letter from Emperor Joseph, Florence, to Maria Theresa, Vienna, 24 May 1769.
59 Ibid. Letter from Emperor Joseph, Florence, to Maria Theresa, Vienna, 22 May 1769.
60 Ibid. Letter from Emperor Joseph, Florence, to Maria Theresa, Vienna, 16 May 1769.
61 Ibid. Letter from Emperor Joseph, Florence, to Maria Theresa, Vienna, 24 May 1769.
62 Ibid.
63 Ibid. Letter from Emperor Joseph, Florence, to Maria Theresa, Vienna, 22 May 1769.
64 GF. p. 23.
65 Ibid. p. 24.
66 BGA. A 7. Ingen Housz finances note book 1767-1789. Fol. 2 verso.
67 GF. p. 20. Letter from Jan Ingen Housz, Florence, to Johann Deckers, 's-Hertogenbosch, 10 May 1769 (date wrong – perhaps 10 June).
68 During M, Didi-Huberman G, Poggesi M. *Encyclopaedia anatomica*. Koln, Taschen 2004. p. 8 passim.
69 Ibid.
70 GF. p. 20. Letter from Jan Ingen Housz, Florence, to Johann Deckers, 's-Hertogenbosch, 10 May 1769 (date wrong – perhaps 10 June).
71 GF. p.23. Letter from Jan Ingen Housz, Florence, to Johann Deckers, 's-Hertogenbosch, July 1769.
72 HKB. Fol. 34. Letter from Ingen Housz, Hertford, to Joseph Jacquin, Paris, 21 August 1790. The sequin was a small Italian coin.
73 Royal Society, Certificates of Election. EC/1769/6.
74 Ibid.
75 BGA. A 7. Ingen Housz finances note book 1767-1789. Fol. 2 verso.
76 Vienna, Hofkammerarchiv, 1536. Royal Decree on Inoculation, 29 July 1769.
77 IHLD. Abstract of a letter from Jan Ingen Housz, Vienna, to Count Bolza, Dresden, 4 November 1776.
78 Gentleman's Magazine September 1769. p. 457. en.wikipedia.org/wiki/Brescia (Last accessed 24 March 2011).
79 http://cscs.umich.edu/~crshalizi/White/air/rod.html (Last accessed 24 March 2011).
80 BGA. A 7. Ingen Housz finances note book 1767-1789. Fol. 3.
81 Ibid.
82 Ibid.
83 Ibid.
84 WJ. p. 39.
85 Von Arneth A. (Ed.) *Maria Theresia und Joseph II. Ihre correspondenz sammt Briefen Joseph's an seinen Bruder Leopold. Ester Band 1761 – 1772*. Wien, von Carl Gerold's Sohn 1867. Letter from Emperor Joseph, Florence, to Maria Theresa, Vienna, 24 May 1769.
86 HKB. Fol. 7. Letter to Ingen Housz, Vienna, from Thomas Dimsdale, St. Petersburg, 16 November 1768.
87 BGA. A 7. Ingen Housz finances note book 1767-1789. Fol. 4.
88 WJ. p. 39.
89 Braunbehrens V. *Mozart in Vienna 1781-1791*. London, Deutsch 1990. p. 302.

Notes to Chapter 10

1 BGA. A 7. Ingen Housz finances note book 1767-1789.
2 Ibid.
3 Ibid.
4 Ibid.
5 *Hortus Botanicus Vindobonensis*, written by Nikolaus Jacquin and colour-illustrated by Francis von Scheidel was published in three parts from 1770 to

1776. It is a description of plants at the Vienna University's Botanical Garden, of which Nikolaus Jacquin was the second, and most significant, curator. In folio; only 162 copies were published.
6   Ibid.
7   Ibid.
8   Girard G. *Correspondence entre Marie-Thérèse et Marie-Aintonette.* Paris 1933.
9   ÖNB. Fol. 37. Letter from Jan Ingen Housz, Vienna, to Hendrik Hoogeveen, Delft, September 1770.
10 BGA. A 7. Ingen Housz finances note book 1767-1789. Note inside front cover.
11 Fraser A. *Marie Antoinette: the journey.* London, Weidenfeld & Nicolson 2001. p. 10.
12 Ibid. p. 34.
13 Ibid. p. 43.
14 Ibid. p. 48 passim.
15 ibid. p. 60.
16 Ibid. p. 98.
17 Arneth A, Geffey M. *Correspondence secrete entre Marie-Thérèse et le Cte. de Mercy-Argenteau.* Paris, Mesnil 1874. Vol 1, pp. 119/120. Letter from Compte de Mercy-Argenteau, Versailles, to Maria Theresa, Vienna, 23 January 1771.
18 Girard G. *Correspondence entre Marie-Thérèse et Marie-Aintonette.* Paris 1933. p. 38.
19 Dunlop I. *Marie Antoinette: a portrait.* Sinclair-Stevenson, London 1993. p. 3.
20 Ibid. p. 4.
21 Ibid.
22 Arneth A, Geffey M. *Correspondence secrete entre Marie-Thérèse et le Cte. de Mercy-Argenteau.* Paris, Mesnil 1874. Vol. 1, pp 119/120. Letter from Compte de Mercy-Argenteau, Versailles, to Maria Theresa, Vienna, 23 January 1771.
23 Ibid.
24 Arneth A. *Maria Theresia und Marie Antionette. Ihr Briefwechsel wahrend der Jahre 1770-1780.* Paris, Wien 1865. p. 17. Letter from Maria Theresa, Vienna, to Marie Antoinette, Versailles, 10 February 1771.
25 Arneth A, Geffey M. *Correspondence secrete entre Marie-Thérèse et le Cte. de Mercy-Argenteau.* Paris, Mesnil 1874.

Vol. 1, pp 119/120. Letter from Compte de Mercy-Argenteau, Versailles, to Maria Theresa, Vienna, 23 January 1771.
26 BGA. A 7. Ingen Housz finances note book 1767-1789. Fol. 4 verso.
27 Sontag O. *Correspondence between Albrecht von Haller and Charles Bonnet.* Bern, Huber 1983. p. 914. Letter from Albrecht von Haller, Berne, to Charles Bonnet, Geneva, 25 October 1770.
28 Sontag O. *John Pringle's correspondence with Albrecht von Haller.* Basel, Schwabe 1999. pp. 155,156. Letter from Sir John Pringle, London, to Albrecht von Haller, Berne, 19 March 1771.
29 BGA. A 7. Ingen Housz finances note book 1767-1789. Fol. 4 verso.
30 *Nederland's Patriciaat. 59e Jaargang 1973.* The Hague, Centraal bureau voor Genealogie, p. 112.
31 BGA. A 7. Ingen Housz finances note book 1767-1789. Fol. 4 verso.
32 Ibid.
33 Ibid.
34 Ibid.
35 Royal Society. Club Meetings Book 5, (1743-1772).
36 Ibid.
37 Royal Society. Fellows' details.
38 Royal Society. Club Meetings Book 5, (1743-1772).
39 Royal Society. Charter Book for 1771.
40 Royal Society. Club Meetings Book 5, 1743-1772.
41 Ibid.
42 London. Wellcome Trust Library. Charles Gordon papers, p. 116.
43 Johnson, S. *The invention of air.* London, Penguin 2009. p. 17.
44 Brands H. *The First American.* Doubleday, New York 2000. p. 200.
45 A report of the committee appointed by the Royal Society, to consider of a method for securing the powder magazines at Purfleet. Royal Society. *Philosophical Transactions, Vol. 63 (1773-1774).* pp. 42 – 48.
46 Brands H. *The First American.* Doubleday, New York 2000. p. 445.
47 BGA. A 7. Ingen Housz finances note book 1767-1789. Fol. 4 verso.

48 ÖNB. Fol. 27. Letter from Jan Ingen Housz, London, to Hendrik Hoogeveen, Delft, May 1771.
49 BGA. A 7. Ingen Housz finances note book 1767-1789. Fol. 5.
50 With a copy of a Franklin article, explaining how it worked, and some barley seeds, for Jacquin, to be sent on to Vienna.
51 Vienna. Haus-Hof-Staatsarchiv. Handwritten letters, Vol.3, Fols. 158,159. Letter from Jan Ingen Housz, London, to the Court at Brussels, 28 April 1771.
52 ÖNB. Fols. 25,26. Letter from Jan Ingen Housz, London, to Hendrik Hoogeveen, Delft, May 1771.
53 ÖNB. Fol. 37. Letter from Jan Ingen Housz, London, to Hendrik Hoogeveen, Delft, September 1770.
54 ÖNB. Fols. 25,26. Letter from Jan Ingen Housz, London, to Hendrik Hoogeveen, Delft, May 1771.
55 POBF. Letter from Benjamin Franklin, London, to John Canton, London, 12 May 1771.
56 Bloomington, IN, USA. Indiana University, Lilly Library. The Jonathan Williams Manuscripts. Journal of a tour made with Benjamin Franklin and others in 1771, passim.
57 Brighton T. *The discovery of the Peak District*. Chichester, Phillimore 2004. p. 40.
58 Uglow J. *The Lunar Men*. London, Faber and Faber 2002. p. 71 passim.
59 Royal Society. Certificates of Election: EC/1766/12.
60 Priestley, J. *The History and Present State of Electricity, with original experiments*. London 1767.
61 Priestley J. *Experiments and observations on different kinds of air*. London, Johnson 1774.
62 Williams in error in recording return to London on 27 May.
63 BGA. A 7. Ingen Housz finances note book 1767-1789. Fol. 5.
64 ULB Bonn. Autographensammlung, Jacquin, Botaniker. Letter from Nikolaus Jacquin, Vienna, to Jan Ingen Housz, London, 28 August 1771

65 Edinburgh. Royal College of Physicians. Medical Annotations of Sir John Pringle, Vol. 9, p. 77.
66 Ibid. p. 208.
67 Ibid. p. 278.
68 Ibid. p. 274.
69 Ibid. p. 186.
70 Ibid. p. 73.
71 Ibid. pp. 73 & 110.
72 Edinburgh. Royal College of Physicians. Medical Annotations of Sir John Pringle, Vol. 7, p. 7.
73 Ibid.
74 Ibid.
75 Royal Society. Club Meetings Book 5, (1743-1772).
76 ÖNB. Fol 27. Letter from Jan Ingen Housz, London, to Hendrik Hoogeveen, Delft, May 1771.
77 ULB Bonn. Autographensammlung, Jacquin, Botaniker. Letter from Nikolaus Jacquin, Vienna, to Jan Ingen Housz, London, 28 August 1771.
78 POBF. Letter from Sir John Pringle, London, to Benjamin Franklin, London, 12 December 1771.
79 Schiebinger L. *The mind has no sex? Women in the origins of modern science*. Cambridge, Mass., Harvard University Press 1989. p. 28.

Notes to Chapter 11

1 BGA. A 7. Ingen Housz finances note book 1767-1789. Fol. 5 verso.
2 Arneth A, Geffey M. *Correspondence secrete entre Marie-Thérèse et le Cte. de Mercy-Argenteau. Paris,* Mesnil 1874. Vol I, p. 279. Letter from Compte de Mercy-Argenteau, Versailles, to Maria Theresa, Vienna, 29 February 1772.
3 Ibid.
4 Beales D. *Joseph II: In the shadow of Maria Theresa, 1741-1780*. Cambridge, Cambridge University Press 1987. p. 377.
5 Arneth A, Geffey M. *Correspondence secrete entre Marie-Thérèse et le Cte. de Mercy-Argenteau*. Paris, Mesnil 1874. Vol I, p. 279. Letter from Compte de Mercy-Argenteau, Versailles, to Maria

Theresa, Vienna, 29 February 1772.
6   Marie Antoinette's failure to conceive was only resolved when Joseph II visited and told the King the 'facts of life'. In fact the Emperor proved to be just the matter-of-fact sexual adviser that the naive young man, now the King, proved to need. It was a feature of his high isolation that no-one had ever told him the nature of normal sexual intercourse and he had no concept that he needed to ejaculate during the activity in order to get his wife pregnant. Joseph II reported this surprising discovery to his brother by a letter of 9 June 1777 (Beales D. *Joseph II: in the shadow of Maria Theresa 1741-1780*. Cambridge, Cambridge University Press 1997. p. 374.) and, shortly after walking in the gardens of Versailles with his brother-in-law, Louis XVI fathered a child.
7   Glasgow University, Hunter Collection. Letter from Jan Ingen Housz, Vienna, to William Hunter, London, 18 April 1772.
8   BGA. A 7. Ingen Housz finances note book 1767-1789. Fol. 5 verso.
9   Glasgow University, Hunter Collection. Letter from Jan Ingen Housz, Vienna, to William Hunter, London, 18 April 1772.
10 HKB 9. Letter from Jan Ingen Housz, Vienna, to Hendrik Hoogeveen, Delft, 25 May 1772.
11 Van der Korst J. *Een dokter van formaat. (A doctor of stature)*. Amsterdam, Bert Bakker 2003. p. 276.
12 BGA. A 7. Ingen Housz finances note book 1767-1789. Fol. 6.
13 Ibid. Fol. 6 verso.
14 Ibid. Fol. 6.
15 IHLD. Abstract of a letter from Jan Ingen Housz, Vienna, to his bankers at Paris, 12 May 1774.
16 BGA A, 7. Ingen Housz finances note book 1767-1789. Fol. 4.
17 BGA B. 18. Will of Jan Ingen Housz, London 10 June 1796.
18 BGA A, 7. Ingen Housz finances note book 1767-1789. Fol. 6.
19 There seems no way of knowing why the oldest child, Maria Theresia, and the youngest, Ferdinand III, were not inoculated at this time.
20 BGA A, 7. Ingen Housz finances note book 1767-1789. Fol. 6.
21 Glasgow University, Hunter Collection. Letter from Jan Ingen Housz, Vienna, to William Hunter, London, 18 April 1772.
22 BGA A, 7. Ingen Housz finances note book 1767-1789. Fol. 5 verso.
23 Glasgow University, Hunter Collection. Letter from Jan Ingen Housz, Vienna, to William Hunter, London, 18 April 1772 passim.
24 Opium powder dissolved in alcohol. A very popular eighteenth-century analgesic that could easily disorientate patients.
25 BGA A, 7. Ingen Housz finances note book 1767-1789. Fol. 6.
26 HKB 45. Letter from Jan Ingen Housz, London, to Josef Jacquin, Vienna, 30 December 1793.
27 Ibid.
28 HKB 9. Letter from Jan Ingen Housz, Vienna, to Hendrik Hoogeveen, Delft, 25 May 1772.
29 Ibid.
30 Ibid.
31 Van der Korst J. *Een dokter van formaat. (A doctor of standing.)* Amsterdam. Bert Bakker 2003. p. 276 passim.
32 BGA. A 7. Ingen Housz finances note book 1767-1789. Fol. 6.
33 HKB 10. Letter from Sir John Pringle, London, to Jan Ingen Housz, Florence, 6 November 1772 passim.
34 BGA A, 7, Ingen Housz finances note book 1767-1789. Fol. 6.
35 Laugier was to die of a strangulated umbilical hernia in early 1774. BGA. A6. Ingen Housz letter diary. Abstract of a letter from Jan Ingen Housz, Vienna, to Francis Milman, London, 25 August 1774.
36 University of Amsterdam, Hs. Died 12E. Letter from Jan Ingen Housz, Como, to Nikolaus Jacquin, Vienna, 21 August 1772.
37 Flora Austriacae, sive Plantarum Selectarum. Nicolai Josephi Jacquin.

Vienna, Kaliwoda 1773.
38 University of Amsterdam, Hs. Died 12E. Letter from Jan Ingen Housz, Como, to Nikolaus Jacquin, Vienna, 21 August 1772.
39 Ibid.
40 BGA A, 7, Ingen Housz finances note book 1767-1789. Fol. 6.
41 Vienna. Österreichische Nationalbibliothek. 6/97-2. Impfprotokol; alter Bestand. Journal de l'inoculation de Son Altesse Royal l'Archduc Francois.
42 Ibid, passim.
43 Ibid.
44 HKB 10. Letter from Sir John Pringle, London, to Jan Ingen Housz, Florence, 6 November 1772 passim.
45 Vienna. Österreichische Nationalbibliothek. 6/97-2. Impfprotokol; alter Bestand. Journal de l'inoculation de Son Altesse Royal l'Archduc Francois.
46 HKB 9. Letter from Jan Ingen Housz, Vienna, to Hendrik Hoogeveen, Delft, 25 May 1772.
47 HKB 10. Letter from Sir John Pringle, London, to Jan Ingen Housz, Florence, 6 November 1772.
48 Ibid.
49 Sonntag O. *John Pringle's correspondence with Albrecht von Haller*. Basel, Schwabe & Co 1999. p. 259 (note 3).
50 HKB 10. Letter from Sir John Pringle, London, to Jan Ingen Housz, Florence, 6 November 1772.
51 Alexander Drummond, died 1782. Probably this individual – if so he stayed in Italy, becoming personal physician to William Hamilton and his wife at Naples and died after being thrown from a horse on his way to visit a sick patient in August 1782 (Constantine D. *Fields of fire. A life of Sir William Hamilton*. London, Weidenfeld and Nicolson 2001. pp. 106,113.)
52 Pringle J. *Six Discourses delivered before the Royal Society on the annual assignments of the Copley Medal*. London, Strahan & Cadell 1783. p. 62.
53 Walsh J. Of the Electrical property of the Torpedo. In a letter from John Walsh FRS to Benjamin Franklin LLd, FRS. *Philosophical Transactions of the Royal Society 1773-4*, Vol. 63, pp. 461-480 passim.
54 Ibid.
55 Royal Society. Fellows' details.
56 Ingen Housz J. Extract of a letter to Sir John Pringle, Bart, FRS containing some experiments on the Torpedo, made at Leghorn, January 1, 1773. *Philosophical Transactions of the Royal Society 1775*. Vol. 65, pp. 1-4 passim.
57 Ibid.
58 Berlin, Staatsbibliothek G21779: Ingenhouss. Letter from Jan Ingen Housz, Vienna, to Sir William Hamilton, Naples, 7 April 1773.
59 Ibid.
60 Ibid.
61 ÖNB 28. Letter from Jan Ingen Housz, Vienna, to Hendrik Hoogeveen, Delft, August 1773.
62 Berlin, Staatsbibliothek G21779: Ingenhouss. Letter from Jan Ingen Housz, Vienna, to Sir William Hamilton, Naples, 7 April 1773.
63 Ibid.
64 Ingen Housz J. Extract of a letter to Sir John Pringle, Bart. P.R.S. containing some experiments on the Torpedo, made at Leghorn, January 1, 1773 *Philosophical Transactions of the Royal Society 1775*. Vol. 65, pp. 1-4.
65 In November 1772. Singer D. Sir John Pringle and his circle. Part 1. Life. *Annals of Science 1949*: Vol. 6; p. 142.
66 Pringle J. (Kippis A. Ed.) *Six Discourses delivered before the Royal Society on the annual assignments of the Copley Medal*. London, Strahan & Cadell 1783. p. 62.
67 BGA A, 7, Ingen Housz finances note book 1767-1789. Fol. 6 verso.
68 ÖNB 28. Letter from Jan Ingen Housz, Vienna, to Hendrik Hoogeveen, Delft, August 1773.
69 BGA A, 7, Ingen Housz finances note book 1767-1789. Fol. 6 verso.
70 ÖNB 28. Letter from Jan Ingen Housz, Vienna, to Hendrik Hoogeveen, Delft, August 1773.
71 Ibid.
72 Ibid.

73 Hartmann P. *Elfenbeinkunst (Micropictures)*. Wien, Hartmann 1999. Item no. 1. The jewel was recently (1998) valued at two million dollars US.
74 HVTS 2043. Letter from Jan Ingen Housz, Vienna, to Jacob van Breda, Delft, 20 February 1788.
75 HKB 59. Letter from John Needham, Brussels, to Jan Ingen Housz, via Vienna, 25 September 1773.
76 University of Amsterdam, Hs.Y47. Letter from Jan Ingen Housz, Florence, to Johann Deckers, 's-Hertogenbosch, 9 July 1769.
77 HVTS 2043. Letter from Jan Ingen Housz, Vienna, to Jacob van Breda, Delft, 20 February 1788.
78 Ibid.
79 BGA A, 7, Ingen Housz finances note book 1767-1789. Fol. 6 verso.
80 Ibid.
81 IHLD. Abstract of a letter from Jan Ingen Housz, Vienna, to Louis Ingen Housz, Breda, 7 May 1774.
82 BGA A, 6. Letter from Jan Ingen Housz, Vienna, to Johann Deckers, 's-Hertogenbosch, 7 May 1774.
83 IHLD. Abstract of a letter from Jan Ingen Housz , Vienna, to Sir John Pringle, London 12 May 1774.
84 POBF. Letter from Benjamin Franklin, London, to Jan Ingen Housz, Vienna, 30 September 1773.
85 IHLD. Abstract of a letter from Jan Ingen Housz, Vienna, to Sir John Pringle, London, 13 January 1775.
86 POBF. Letter from Benjamin Franklin, London, to Jan Ingen Housz, Vienna, 30 September 1773.
87 A report of the committee appointed by the Royal Society, to consider of a method for securing the powder magazines at Purfleet. Royal Society. *Philosophical Transactions, 1773*. Vol. 63, pp. 42 –48.
88 Priestley J. Observations on different kinds of air. *Philosophical Transactions of the Royal Society 1772*. Vol. 62, 147-264.
89 POBF. Letter to Benjamin Rush, July 14 1773.
90 Japikse C. *Fart proudly: writings of Benjamin Franklin you never read in school.* Berkeley, California, Frog Books 2003. p. 14.
91 Golinski J. *Science as public culture: chemistry and enlightenment in Britain, 1760-1820.* Cambridge. Cambridge University Press 1992. p. 120.
92 Fontana F. *Descrizione e usi di alcuni stromenti per misurare la salubrita del aria. (Description and uses of some instruments for measuring the salubrity of the air.)* Firenze. Gaetano Cambiagi Stampatore granducale 1775.
93 Ingen Housz J. Easy methods of measuring the diminution of bulk, taking place upon the mixture of common air and nitrous air; together with experiments on platina. *Royal Society Philosophical Transactions 1776*. Vol. 66, pp. 257-267.
94 Ibid pp. 259-260.
95 Ibid.
96 Ibid pp. 259-260.
97 McDonald D, Hunt L. *A history of platinum and its allied metals.* London, Johnson Matthey 1982. p. 33.
98 Ibid. p. 37.
99 Royal Society. Fellows' details.
100 Lewis W. *Commercium philosphocotechnicum: the philosophical commerce of arts.* London, Baldwin 1763.
101 Ingen Housz J. Easy methods of measuring the diminution of bulk, taking place upon the mixture of common air and nitrous air; together with experiments on platina. *Philosophical Transactions of the Royal Society 1776*. Vol. 66, p. 262.
102 Ibid.
103 Ibid. p. 264.
104 Brands H. *The First American*. New York, Doubleday 2000. p. 341.
105 Ibid. p. 452 passim.
106 POBF. Letter from Benjamin Franklin, London, to Jan Ingen Housz, Vienna, 30 September 1773.
107 An octagonal building, just north of Whitehall on ground now occupied by the Cabinet Office, that was, originally, in the sixteenth century, an arena for cockfighting. It was then adapted as a theatre and became, in the eighteenth century, the meeting room of the Privy

Council, the royal advisers.
108 Brands H. *The First American*. New York, Doubleday 2000. p. 452 passim.
109 POBF. Letter from Benjamin Franklin, London, to Jan Ingen Housz, Vienna, 18 March 1774.
110 Brands H. *The First American*. New York, Doubleday 2000. p. 493.
111 IHLD. Abstract of a letter from Jan Ingen Housz, Vienna, to Richard Huck, London, 6 March 1774.
112 IHLD. Abstract of a letter from Jan Ingen Housz, Vienna, to Mr. Collard, London, 26 April 1775.
113 IHLD. Abstract of a letter from Jan Ingen Housz, Vienna to Edler von Geerham, Nagybania, 30 June 1775.
114 IHLD. Abstract of a letter from Jan Ingen Housz, Vienna, to Louis Ingen Housz, Breda, 12 May 1774.
115 IHLD. Abstract of a letter from Jan Ingen Housz, Vienna, to Louis Ingen Housz, Breda, 18 July 1774.
116 IHLD. Abstract of a letter from Jan Ingen Housz, Vienna, to Louis Ingen Housz, Breda, 3 November 1774.
117 IHLD. Abstract of a letter from Jan Ingen Housz, Vienna, to Sir John Pringle, London, 12 May 1774.
118 IHLD. Abstract of a letter from Jan Ingen Housz, Vienna, to Louis Ingen Housz, Breda, 18 August 1775.
119 IHLD. Abstract of a letter from Jan Ingen Housz, Vienna, to Louis Ingen Housz, Breda, 24 May 1775.

Notes to Chapter 12

1 Crankshaw E. *Maria Theresa*. London, Constable 1983. p. 203.
2 Buranelli V. *The wizard from Vienna*. London, The Scientific Book Club 1977. p. 31.
3 Ibid. p. 39 passim.
4 Ibid. p. 76.
5 Michell J. *A treatise of artificial magnets; in which is shewn an easy and expeditious method of making them*. Cambridge, Bentham 1750.
6 Buranelli V. *The wizard from Vienna*. London, The Scientific Book Club 1977. p. 63.
7 Knight G. An account of some magnetical experiments shewed before the Royal Society. *Philosophical Transactions of the Royal Scoiety 1744*. Vol. 43. pp. 161-166.
8 Ingen Housz J. Easy methods of measuring the diminution of bulk, taking place upon the mixture of common air and nitrous air; together with experiments on platina. *Philosophical Transactions of the Royal Society 1776*. Vol. 66, p. 262.
9 Buranelli V. *The wizard from Vienna*. London, The Scientific Book Club 1977. p. 64.
10 IHLD. Abstract of a letter from Jan Ingen Housz, Vienna, to Sir John Pringle, London, 13 January 1775.
11 Mesmer A. *Mémoire sur la découverte du magentism animal*. (*Memoire on the discovery of animal magnetism*.) Geneve et Paris, Didot le jeune 1779.
12 Ibid. p. 22
13 IHLD. Abstract of a letter from Jan Ingen Housz, Vienna, to Sir John Pringle, London, 13 January 1775.
14 IHLD. Abstract of a letter from Jan Ingen Housz, Vienna, to Dr. Richard Huck, London, 20 January 1775.
15 Buranelli V. *The wizard from Vienna*. London, The Scientific Book Club 1977. p. 8.
16 Draper T. *A struggle for power: the American Revolution*. London, Abacus 1997. p. 503.
17 POBF. Letter from Jan Ingen Housz, Vienna, to Benjamin Franklin, Philadelphia, 15 November 1776.
18 Petz-Grabenbauer M. *Der Botaniker Nikolaus Joseph Freiherr von Jacquin*. Diplomarbeit (Dissertation). Wien, Archiv der Universitat. 885/94. p. 112.
19 Maria Theresa's personal involvement in arranging the marriage of Ingen Housz is something of an urban myth in Vienna but accepted as true by experts on the Jacquin family such as Dr. Petz-Grabenbauer of Vienna.
20 ÖNB. Fol. 28. Letter from Jan Ingen Housz, Vienna, to Hendrik Hoogeveen, Delft, November 1775.

21 Vienna. University Archive. Med 43.1, 204.
22 BGA. B, 20. Ingen Housz marriage contract of 24 November 1775.
23 Ibid.
24 Vienna, University Archive. Med 43.1, 204.
25 ÖNB Fol. 28. Letter from Jan Ingen Housz, Vienna, to Hendrik Hoogeveen, Delft, November 1775.
26 IHLD. Abstract of a letter to Louis Ingen Housz, 25 November 1775.
27 Koninklijke Biblioteek, The Hague. 133 B24, 11. Letter from Richard Huck, London, to Jan Ingen Housz, Vienna, 17 December 1775.
28 POBF. Letter from Jan Ingen Housz, Vienna, to Benjamin Franklin, Philadelphia, 15 November 1776.
29 Brands H. *The First American*. New York, Doubleday 2000. p. 528.
30 POBF. Letter from Benjamin Franklin, Passy, to Jan Ingen Housz, Vienna, 12 February (completed 6 March) 1777.
31 Vickery A. *Behind closed doors. At home in Georgian England*. New Haven. Yale University Press 2009. p. 104.
32 IHLD. Abstract of a letter from Jan Ingen Housz, Vienna, to Louis Ingen Housz, Breda, September 1776.
33 The Hague. Centraal bureau voor Genealogie. *Nederland's Patriciaat. 59e Jaargang 1973.* p. 113.
34 IHLD. Abstract of a letter from Jan Ingen Housz, Vienna, to Louis Ingen Housz, Breda, 12 November 1776.
35 IHLD. Abstract of a letter from Jan Ingen Housz, Vienna, to Louis Ingen Housz, Breda, 4 April 1775.
36 IHLD. Abstract of a letter from Jan Ingen Housz, Vienna, to the Supreme Burgrave of Bohemia, Prague, 19 April 1776.
37 IHLD. Abstract of a letter from Jan Ingen Housz, Vienna, to Mr. Collard, London, 1 October 1776.
38 POBF. Letter from Jan Ingen Housz, Vienna, to Benjamin Franklin, Philadelphia, 15 November 1776.
39 Ingen Housz J. Electrical experiments, to explain how far the phenomena of the electrophorus may be accounted for by Mr. Franklin's theory of positive and negative electricity. *Philosophical Transactions of the Royal Society 1778, Vol. 68;* 1027-1048.
40 Ibid. p. 1029.
41 Ibid.
42 Ingen Housz J. Electrical experiments, to explain how far the phenomena of the electrophorus may be accounted for by Mr. Franklin's theory of positive and negative electricity. *Philosophical Transactions of the Royal Society 1778*. Vol 68, pp. 1028 passim.
43 IHLD. Abstract of a letter from Jan Ingen Housz, Vienna, to Mr. de Mayern, Dischingen, 26 December 1776.
44 Ibid.
45 IHLD. Abstract of a letter from Jan Ingen Housz, Vienna, to Count Hochenward, Linz, 26 December 1776.
46 POBF. Letter from Jan Ingen Housz, Vienna, to Benjamin Franklin, Paris, 4 January 1777.
47 Ibid.
48 Ibid.
49 Sprigge T. (Ed.). *Collected works and correspondence of Jeremy Bentham*. London, Athlone Press 1968. Vol. 2. p. 184.
50 POBF. Letter from Jan Ingen Housz, Vienna, to Benjamin Franklin, Paris, 29 January 1777.
51 IHLD. Abstract of a letter from Jan Ingen Housz, London, to Mr. Cuypers, Delft, 20 February 1778.
52 Ingen Housz J. A ready way of lighting a candle, by a very moderate electrical spark. *Philosophical Transactions of the Royal Society 1778*. Vol.68, pp. 1022-1026.
53 Ibid.
54 Beales D. *Joseph II. 1. In the shadow of Maria Theresa 1741-1780*. Cambridge. Cambridge University Press 1987. p. 367.
55 POBF. Letter from Jan Ingen Housz, Vienna, to Benjamin Franklin, Paris, 2 April 1777.
56 British Library. MS 35511; Fol. 294. Letter from Jan Ingen Housz, Ratisbonne, to Sir Robert Murray

Keith, Vienna, May 1777.
57 IHLD. Abstract of a letter from Jan Ingen Housz, Dischingen, to Messrs. Stametz, Vienna, 28 June 1777.
58 It would also appear that secreted, somewhere, in the luggage or on his person was the micro-picture that Maria Theresa had presented to him on his return from Italy in summer 1773.
59 British Library. MS 35511; Fol. 294. Letter from Jan Ingen Housz, Ratisbonne, to Sir Robert Murray Keith, Vienna, May 1777.
60 POBF. Letter of 29 December 1777 from Jan Ingen Housz, London, to Baron von Pichler, Vienna, paraphrasing a letter, of 21 December 1777, from Benjamin Franklin, Paris.
61 British Library. MS 35511; Fol 294. Letter from Jan Ingen Housz, Ratisbonne, to Sir Robert Murray Keith, Vienna, May 1777.
62 Ingen Housz J. Journal du voyage 1777. The Hague. Koninklijke Bibliotheek, 133 H34. passim.
63 Ibid. p. 13.
64 Ibid. p. 19 passim.
65 Ibid. p. 22.
66 Ibid. Unmarked page between 23 & 24.
67 British Library. MS 35512; Fol. 106. Letter from Jan Ingen Housz, Brussels, to Sir Robert Murray Keith, Vienna, 6 August 1777.
68 Schiff S. *Dr. Franklin goes to France.* London, Bloomsbury 2005. p. 2.
69 Ibid. p. 49.
70 Ibid. p. 51. Rutledge, J-J. *The Englishman's fortnight in Paris: or the art of ruining himself there in a few days.* London, Durham and Kearsly 1777.
71 Schiff S. *Dr. Franklin goes to France.* London, Bloomsbury 2005. p. 15.
72 POBF. Letter from Jan Ingen Housz, Paris, to Benjamin Franklin, Passy, 30 August 1777.
73 Ingen Housz J. *Experiments upon vegetables; discovering their great power of purifying the air in the sunshine and of injuring it in the shade and at night.* London, Elmsley 1779. p. 128.
74 BGA. B, 10. Letter from John Williams, Nantes, to Jan Ingen Housz, Paris, 14 October 1777.
75 Ibid.
76 IHLD. Letter from Jan Ingen Housz, London, to Dromillat and Co., Madrid, 14 November 1778.
77 Ingen Housz J. Journal du voyage 1777. The Hague. Koninklijke Bibliotheek, 133 H34. Unmarked page between pp. 23 & 24 passim.
78 Beales D. *Joseph II. 1. In the shadow of Maria Theresa 1741-1780.* Cambridge. Cambridge University Press 1987. p. 368.
79 Ingen Housz J. Journal du voyage 1777. The Hague. Koninklijke Bibliotheek, 133 H34. p. 28 passim.
80 POBF. Request for safe passage for Dr. Ingen Housz written by Benjamin Franklin, Passy, 17 October 1777.
81 Österreichische Nationalbibliothek, Abendländischen Handschriften, 6/96 − 3. Benjamin Franklin's description of his ailments written out. 17 October 1777.
82 POBF. Letter from Jan Ingen Housz, London, to Benjamin Franklin, Passy, 6 March 1778.
83 Ingen Housz J. Journal du voyage 1777. The Hague. Koninklijke Bibliotheek, 133 H34. p. 32.
84 Ibid. p. 33.
85 Ibid.
86 POBF. Letter from Jan Ingen Housz, Delft, to Benjamin Franklin, Passy, 14 December 1777.
87 The Hague. Centraal bureau voor Genealogie. *Nederland's Patriciaat. 59e Jaargang 1973.* p. 113.
88 Ingen Housz J. Journal du voyage 1777. The Hague. Koninklijke Bibliotheek, 133 H34. p. 35.
89 Ibid.
90 *Verhandelingen van het Bataafsch Genootschap der proefondervindelyke wysbegeerte de Rotterdam. Vierde deel. (Dissertations of the Batavian Society of experimental philosophy, third volume.)* Reinier Arrenberg 1779. p. xiv.
91 Ibid. p. xvii.
92 Ingen Housz J. Journal du voyage 1777. The Hague. Koninklijke Bibliotheek, 133 H34. p. 40.

93 Ibid. p. 41. passim.
94 Ibid. p. 44.
95 POBF. Letter from Jan Ingen Housz, Delft, to Benjamin Franklin, Passy, 14 December 1777.
96 Ibid.
97 Ingen Housz J. Journal du voyage 1777. The Hague. Koninklijke Bibliotheek, 133 H34. p. 47.

Notes to Chapter 13

1 BGA. B, 1. Jan Ingen Housz. Memo book 1778 – 1791. p. 1.
2 Sprigge T. (Ed.). *The correspondence of Jeremy Bentham. Vol. 2: 1777-1780.* London, Athlone Press 1968. p. 183.
3 BGA. B, 1. Jan Ingen Housz. Memo book 1778 – 1791. p. 1.
4 Ibid.
5 Ingen Housz J. On the degree of salubrity of the common air at sea, compared with that of the sea-shore, and that of places far removed from the sea. *Philosophical Transactions of the Royal Society 1780.* Vol. 70, p. 361.
6 Sprigge T. (Ed.). *The correspondence of Jeremy Bentham. Vol. 2: 1777-1780.* London, Athlone Press 1968. p. 223. Sprigge T. (Ed.). *The correspondence of Jeremy Bentham. Vol. 2: 1777-1780.* London, Athlone Press 1968. p. 246.
7 Ibid.
8 Glasgow University. Hunter papers H335, Letter no. 478. Letter from Jan Ingen Housz, Vienna, to William Hunter, London, 17 April 1782.
9 Schofield R. *The enlightened Joseph Priestley: a study of his life and work from 1773-1804.* University Park PA, Pennsylvania State University Press 2004. p. 5.
10 POBF. Letter from Jan Ingen Housz, London, to Benjamin Franklin, Passy, 6 March 1778.
11 Schofield R. *The enlightened Joseph Priestley. A study of his life and work from 1773-1804.* University Park PA, Pennsylvania State University Press 2004. p. 10.
12 Beale N. *Joseph Priestley in Calne.* East Knoyle, Salisbury, Hobnob Press 2008. p. 56.
13 POBF. Letter from Jan Ingen Housz, London, to Benjamin Franklin, Passy, 6 March 1778.
14 Sir John Pringle still lived in Pall Mall.
15 Royal Society – Club Meetings Book 5, (1772-1779).
16 POBF. Letter from Jan Ingen Housz, London, to Benjamin Franklin, Passy, 6 March 1778.
17 Ingen Housz J. Electrical experiments, to explain how far the phenomenon of the electrophorus may be accounted for by Dr. Franklin's theory of positive and negative electricity. *Philosophical Transactions of the Royal Society 1778,* Vol. 68, pp. 1027-1048.
18 Hulme N. *A safe and easy remedy proposed for the relief of stone and the gravel, scurvy and gout.* London, Robinson and Elmsly 1778.
19 Schofield R. *The enlightenment of Joseph Priestley. A study of his life and work from 1733-1773.* University Park PA, Pennsylvania State University Press 1997. p. 256.
20 Hulme N. (Ingen Housz J, translator). *Nova, tuta, facilisque methodus curandi calculum, scorbutum, podagram.* Luzac and van Damme, Leyden 1778.
21 BGA. A, 6. Ingen Housz letter diary. Abstract of a letter from Jan Ingen Housz, London, to Lussac and van Damme, Leyden 1 March 1778.
22 Ingen Housz J. Electrical experiments, to explain how far the phenomenon of the electrophorus may be accounted for by Dr. Franklin's theory of positive and negative electricity. *Philosophical Transactions of the Royal Society 1778.* Vol. 68, pp. 1027-1048.
23 Ibid. p. 1046.
24 Pringle J. (Kippis A, Ed.) *Six discourses delivered by Sir John Pringle Bart. when President of the Royal Society.* London, Strahan and Cadell 1783. p. 237.
25 POBF. Letter from Jan Ingen Housz, London, to Benjamin Franklin, Passy, 5 October 1778.
26 Pringle J. (Kippis A, Ed.) *Six discourses delivered by Sir John Pringle Bart. when*

President of the Royal Society. Strahan and Cadell, London 1783. p. lvii.
27. POBF. Letter from Jan Ingen Housz, London, to Benjamin Franklin, Passy, 15 June 1778.
28. POBF. Letter from Jan Ingen Housz, London, to Benjamin Franklin, Passy, 5 October 1778.
29. BGA. B, 1. Jan Ingen Housz. Memo book 1778 – 1791. p. 2 passim.
30. BGA. A 6. Ingen Housz letter diary. Abstract of a letter from Jan Ingen Housz, London, to Louis Ingen Housz, Breda, 20 February 1778.
31. Personal communication.
32. BGA. B, 1. Jan Ingen Housz. Memo book 1778 – 1791. p. 4.
33. Ibid. p. 2.
34. Ibid. p. 3.
35. POBF. Letter from Jan Ingen Housz, London, to Benjamin Franklin, Passy, 5 October 1778.
36. Ibid.
37. Ibid.
38. Ingen Housz J. *Experiments upon vegetables; discovering their great power of purifying the air in the sunshine and of injuring it in the shade and at night.* London, Elmsly 1779. p. 278. BGA. A 6. Ingen Housz letter diary. Abstract of a letter from Jan Ingen Housz, London, to Nikolaus Jacquin, Vienna, 25 September 1778.
39. Priestley J. *Experiments and observations relating to various branches of natural philosophy; with a continuation of the observations on air. The second volume.* Birmingham, Pearson and Rollason, for J. Johnson, London 1781. p. 180.
40. Ingen Housz J. *Experiments upon vegetables; discovering their great power of purifying the air in the sunshine and of injuring it in the shade and at night.* London, Elmsly 1779. p. 150.
41. BGA. A 6. Ingen Housz letter diary. Abstract of a letter from Jan Ingen Housz, London, to Mr. Velasco, Brussels, 23 October 1778.
42. Fontana F. Experiments and observations on the inflammable air breathed by various animals. *Philosophical Transactions of the Royal Society 1779.* Vol. 69, pp. 337-361.
43. Ingen Housz J. Account of a new kind of inflammable air or gass, which can be made in a moment without apparatus, and is as fit for explosion as other inflammable gasses in use for that purpose; together with a new theory of gun-powder. *Philosophical Transactions of the Royal Society 1779.* Vol. 69, pp. 376-418.
44. Priestley J. *Experiments and observations relating to various branches of natural philosophy.* London, Johnson 1779. p. 474.
45. Ibid.
46. Ingen Housz J. Account of a new kind of inflammable air or gass, which can be made in a moment without apparatus, and is as fit for explosion as other inflammable gasses in use for that purpose; together with a new theory of gun-powder. *Philosophical Transactions of the Royal Society 1779.* Vol. 69, p. 407.
47. Ingen Housz J. Improvements in electricity. *Philosophical Transactions of the Royal Society 1779,* Vol. 69, pp. 661-673.
48. Ingen Housz J. Improvements in electricity. *Philosophical Transactions of the Royal Society of London 1779.* Vol. 69, p. 662.
49. Ibid. p. 667.
50. IHLD. Letter from Jan Ingen Housz, Paris, to Peter Elmsly, London, 1 April 1780.
51. Ingen Housz J. On some new methods of suspending magnetical needles. *Philosophical Transactions of the Royal Society 1779.* Vol. 69, pp. 537-546.
52. Gubbins D. *Encyclopedia of Geomagnetism and Paleomagnetism.* Dordrecht, Springer Press 2007. p. 67.
53. Ibid.
54. BGA. A 6. Ingen Housz letter diary. Abstract of a letter from Jan Ingen Housz, London, to Count Orsini-Rosenberg, Vienna, 9 May 1779.
55. BGA. A 6. Ingen Housz letter diary. Abstract of a letter from Jan Ingen Housz, London, to Agatha Ingen Housz, Vienna, 14 May 1779.
56. BGA. A 6. Ingen Housz letter diary.

Abstract of a letter from Jan Ingen Housz, London, to Mr. Velasco, Brussels, 29 June 1779.
57 Ingen Housz J. Easy methods of measuring the diminution of bulk, taking place upon the mixture of common air and nitrous air; together with experiments on platina. *Philosophical Transactions of the Royal Society 1776.* Vol. 66, pp. 259-260.
58 Ingen Housz J. *Experiments upon vegetables; discovering their great power of purifying the air in the sunshine and of injuring it in the shade and at night.* London, Elmsly 1779. p. 161 passim.
59 Ibid. p. 181.
60 Priestley J. Observations on different kinds of air. *Philosophical Transactions of the Royal Society 1772.* Vol. 62, pp. 147-264.
61 Ibid. p. 166-167.
62 Ingen Housz J. *Experiments upon vegetables; discovering their great power of purifying the air in the sunshine and of injuring it in the shade and at night.* London, Elmsly 1779. Preface, p. xv.
63 Priestley J. Observations on different kinds of air. *Philosophical Transactions of the Royal Society 1772.* Vol. 62, pp. 147-264.
64 Scheele C. *Chemische abhandlung von der luft und dem feuer (Chemical treatise on fire and air).* Upsala and Leipzig, Bergman 1777.
65 Phlogiston (Greek phlox = flame) was a 1660s concept that found favour for more than a century. Supposedly invisible but universal, phlogiston was released by burning, or on respiration, and 'spoilt' the surrounding air. Thus, when oxygen was first isolated it was named 'dephlogisticated air' for it appeared to have the ability to enhance burning 'by having a greater ability to absorb phlogiston'.
66 Schofield R. *The enlightenment of Joseph Priestley. A study of his life and work from 1733 to 1773.* University Park, PA. Pennsylvania State University Press 1997. p. 163.
67 Pringle J. *Six discourses delivered by Sir John Pringle Bart. when President of the Royal Society.* London, Strahan and Cadell 1783. p. 29.
68 Ibid. p. 34.
69 Bonnet C. *Recherches sur l'usage des Feuilles dans les Plantes, et sur quelques autres Sujets relatif a l'Histoire de la Vegetation.* Göttingen et Leiden, 1754.
70 Hales S. *Vegetable staticks; or an account of some statical experiments on the sap in vegetables – also a specimen of an attempt to analyse the air.* London, Innys 1727.
71 BGA. B, 1. Jan Ingen Housz. Memo book 1778 – 1791.
72 Ibid. p. 8.
73 POBF. Letter from Jan Ingen Housz, Brussels, to Benjamin Franklin, Passy, 18 November 1779.
74 The parish tithe rents book for 1779, that might have proved helpful, is missing from the series held at the London Metropolitan Archives.
75 BGA. A 11. Experimental notes 1779. p. 38.
76 Ibid.
77 Ingen Housz J. *Experiments upon vegetables; discovering their great power of purifying the air in the sunshine and of injuring it in the shade and at night.* London, Elmsly 1779. Preface, p. xxxii.
78 BGA. A 11. Experimental notes 1779.
79 Ingen Housz J. *Experiments upon vegetables; discovering their great power of purifying the air in the sunshine and of injuring it in the shade and at night.* London, Elmsley 1779. Preface, p. 42.
80 Breda, Gemeentearchiev IV, 16A, 11. Experimental notes 1779. p. 7.
81 Bath City Library. Bath Chronicle 6 May 1779, p. 3; 30 September 1779, p. 3
82 BGA. A 8. Ingen Housz J. Manuscript notes entitled 'On Dr. Priestley.'
83 BGA. A 11. Experimental notes 1779. p. 24. William Aiton was the Head Gardener at Kew.
84 Ingen Housz J. *Experiments upon vegetables; discovering their great power of purifying the air in the sunshine and of injuring it in the shade and at night.* London, Elmsley 1779. Preface, p. 32.
85 Ibid. Preface, p. 15.
86 Spoehr H. The development of conceptions of photosynthesis since

Ingen-Housz. *The Scientific Monthly 1919*. Vol. 9, pp. 32 – 46.
87 BGA. A 6. Ingen Housz letter diary. Abstract of a letter from Jan Ingen Housz, Ratisbonne, to Messrs. Tourton and Bauer, bankers, Paris, 17 June 1777.
88 It is true to say, however, that Ingen Housz's understanding of the flow of carbon dioxide in this process only came in later years of his life.
89 Ingen Housz J. *Experiments upon vegetables; discovering their great power of purifying the air in the sunshine and of injuring it in the shade and at night.* London, Elmsley 1779. Preface, p. 34.
90 BGA. A 13. Letter from Peter Elmsly, London, to Jan Ingen Housz, Vienna, 26 October 1781.   HVTS. 2243. Letter from Jan Ingen Housz, Vienna, to Jacob van Breda, Delft, 30 March 1782.
91 Magiels G. *From sunlight to insight. Jan Ingen Housz, the discovery of photosynthesis and science in the light of ecology.* Brussels, Brussels University Press 2010. p. 156.
92 Breda, Gemeentearchiev. Van Hal file. Letter from Peter Elmsly, London to Jan Ingen Housz, Paris, 7 March 1780.
93 Ibid.
94 HKB. 13. Letter from Lord Mahon, Chevening House, Kent, to Jan Ingen Housz, London, 21 October 1779.
95 Schofield R. (Ed.) *A scientific autobiography of Joseph Priestley: selected scientific correspondence.* Cambridge, Mass., MIT Press 1966. pp. 180. Letter from Joseph Priestley, Calne, to Giovanni Fabbroni, London, 17 October 1779.
96 Hudson K. *The Bath and West. A bicentenary hisory.* Bradford-on-Avon, Moonraker Press 1976. p. 24.
97 BGA. A 6. Ingen Housz letter diary. Abstract of a letter from Jan Ingen Housz, London, to Nikolaus Jacquin, Vienna, 15 December 1778.
98 Ingen Housz J. On the degree of salubrity of the common air at sea, compared with that of the sea-shore, and that of places far removed from the sea. *Philosophical Transactions of the Royal Society 1780.* Vol. 70, pp. 354-377

passim.
99 Ibid. p. 375.
100 POBF. Letter from Jean-Baptiste Le Roy, Paris, to Benjamin Franklin, Passy, 11 (probably) January 1780.

Notes to Chapter 14

1   Ingen Housz J. On the degree of salubrity of the common air at sea, compared with that of the sea-shore, and that of places far removed from the sea. *Philosophical Transactions of the Royal Society 1780.* Vol. 70, p. 375.   POBF. Letter from Jean-Baptiste Le Roy, Paris, to Benjamin Franklin, Passy, 11 Jan 1780.
2   Ibid.
3   Schama S. *Citizens.* London, Viking 1989. p. 43.
4   Brands H. *The first American.* New York, Doubleday 2000. p. 591.
5   POBF. Letter from Benjamin Franklin, Passy, to Mary Hewson, London, 10 January 1780.
6   POBF. Letter from Benjamin Franklin, Passy, to Joseph Priestley, London, 8 February 1780.
7   POBF. Letter from Benjamin Franklin, Passy, to Thomas Bond, Philadelphia, 16 March 1780.
8   POBF. Letter from Jan Ingen Housz, Brussels, to Benjamin Franklin, Passy, 3 May 1780.
9   Ibid, and: POBF. Letter from Benjamin Franklin, Passy, to Jan Ingen Housz, Brussels, after 3 May 1780.
10 BGA. A 6. Ingen Housz letter diary. Abstract of a letter from Jan Ingen Housz, Paris, to Francis Coffyn, Dunkerque, 31 May 1780.
11 Breda, Gemeentearchiev IV, Van Hal file. Letter from Peter Elmsly, London, to Jan Ingen Housz, Paris, 7 March 1780.
12 Ingen Housz J. On the degree of salubrity of the common air at sea, compared with that of the sea-shore, and that of places far removed from the sea. *Philosophical Transactions of the Royal Society 1780.* Vol. 70, p. 370.

13 HVTS 2243. Letter from Jan Ingen Housz, Paris, to Jacob van Breda, Delft, 5 June 1780.
14 Ingen Housz J. (van Breda J, translator). *Proeven op plantgewassen: ontdekkende derzelver zeer aanmerkelyk vermogen om de lucht des Dampkrings te zuiveren, geduurende den Dag, en in de Zonneschyn; en om gemeene lucht des nachts, en wanneer zy in de schaduw zyn, te bederven. (Experiments upon vegetables etc.).* Delft, Van der Zmout & De Groot 1780. Preface.
15 Ibid.
16 Ingen Housz J. *Expériences sur les végétaux, spécialement sur la propriété qu'ils possèdent a un haut degré, soit d'améliorer l'air quand ils son au soleil, soit de le corrempre la nuit, ou lorsqu'ils sont à l'ombre; auxquelles on a joint une méthode nouvelle de juger du degré de salubrité de l'atmosphère. (Experiments upon vegetables etc.)* Paris, Didot le jeune 1780.
17 Ingen Housz J. *Nouvelles experiences et observations sur divers objets de physique. (New experiments and observations on a variety of subjects in physics.)* Paris, Barrois le jeune 1785. p. 426.
18 Ingen Housz J. (Translated by Scherer A.) *Versuche mit pflanzen: wodurch entdeckt worden, dass sie kraft besitzen, die atmosphärische luft beim sonnenschein zu reinigen, und im schatten und des nachts uner zu verderben, und nebst einer methode, die reinigkeit der atmosphäre genau abzmumessen. (Experiments with vegetables etc.)* Leipzig, Weygand 1780.
19 HVTS 2243. Letter from Jan Ingen Housz, London, to Jacob van Breda, Delft, 21 August 1780.
20 Ibid.
21 Ibid.
22 HVTS 2243. Letter from Jan Ingen Housz, London, to Jacob van Breda, Delft, 1 November 1780.
23 Ibid.
24 Ibid.
25 Ibid.
26 HVTS 2243. Letter from Jacob van Breda, Delft, to Jan Ingen Housz, Vienna, 29 January 1781.
27 Ibid.
28 POBF. Letter from Jan Ingen Housz, Vienna, to Benjamin Franklin, Passy, 11 June 1785.
29 POBF. Letter from Jan Ingen Housz, Vienna, to Benjamin Franklin, Passy, 8 April 1783.
30 Gemmologisches Labor Austria, 'Connoisseur Collection'. Firmenbuch-Nr. Wien, November 2000. p. 4.
31 Ibid.
32 Brands H. *The first American.* New York, Doubleday 2000. p. 367.
33 POBF. Letter from Jan Ingen Housz, Vienna, to Samuel Wharton, Philadelphia (via Benjamin Franklin, Passy), 12 June 1782.
34 POBF. Letter from Jan Ingen Housz, Vienna, to Benjamin Franklin, Passy, 11 June 1785.
35 BGA. A 6. Ingen Housz letter diary. Abstract of a letter (including 'a note of my American affairs') from Jan Ingen Housz, Vienna, to Benjamin Franklin, Passy, 11 June 1785.
36 Ibid.
37 Ibid.
38 Brands H. *The first American.* New York, Doubleday 2000. p. 608.
39 POBF. Letter from Jan Ingen Housz, Vienna, to Benjamin Franklin, Passy, 11 June 1785.
40 BGA. A 6. Ingen Housz letter diary. Letter from Jan Ingen Housz, Passy, to Louis Ingen Housz, Breda, 11 April 1780.
41 POBF. Letter from Jan Ingen Housz, Vienna, to Benjamin Franklin, Passy, 29 August 1781.
42 POBF. Letter from Jan Ingen Housz, Vienna, to Benjamin Franklin, Passy, 11 June 1785.
43 BGA. A 6. Ingen Housz letter diary. Letter from Jan Ingen Housz, Passy, to Hermann Volkmar, Amsterdam, 11 April 1780.
44 BGA. A 6. Ingen Housz letter diary. Letter from Jan Ingen Housz, Passy, to Mr. Nesbitt, L'Orient, 5 March 1780.
45 BGA. A 6. Ingen Housz letter diary: Letter from Jan Ingen Housz, Passy, to Francis Coffyn, Dunkerque, 13 May 1780.

46 HVTS 2243. Letter from Jan Ingen Housz, Vienna, to Jacob van Breda, Delft, 21 August 1780.
47 BGA. A 6. Ingen Housz letter diary. Abstract of a letter from Jan Ingen Housz, Vienna, to Messrs. Tourton and Bauer, Paris, 9 January 1782.
48 BGA. B 4. Letter from Jan Ingen Housz, London, to Agatha Ingen Housz, Vienna, 24 July 1798.
49 HVTS 2243. Letter from Jan Ingen Housz, Vienna, to Jacob van Breda, Delft, 21 August 1780.
50 Fürstemberger was an amateur physicist at Basel and had been inspired by meeting Volta in late September 1777.
51 Until supplanted by the 'Dobereiner's Device' in 1823 in which hydrogen was ignited by passing it over powdered platinum when it fired without the need for a spark.
52 Ingen Housz J. *Expériences sur les végétaux, spécialement sur la propriété qu'ils possèdent a un haut degré, soit d'améliorer l'air quand ils son au soleil, soit de le corrempre la nuit, ou lorsqu'ils sont à l'ombre. (Experiments upon vegetables etc. Second edition).* Paris, Barrois le jeune 1787. p. 149.
53 Ibid.
54 Beales D. *Joseph II. In the shadow of Maria Theresa. 1741-1780.* Cambridge, Cambridge University Press 1987. p. 480 passim.
55 Ibid. p. 482.
56 Crankshaw E. *Maria Theresa.* London, Constable 1983. p. 337.
57 Robbins Landon H. *Mozart and Vienna.* London, Thames and Hudson 1991. p. 179.
58 BGA. B 16. Ingen Housz Memo book 1778-1791. p. 4.
59 HKB. Fol. 24. Letter from Jonathan Stokes, Kidderminster, to Jan Ingen Housz, Vienna, 31 October 1787.
60 Spaethling R. (Ed.) *Mozart's letters, Mozart's life.* London, Faber and Faber 2000. pp. 384 & 393.
61 Pichler C. *Denkwürdigkeiten aus meinem leben (Memorable events of my life). Volume 1. 1769-1798.* Wien, Stabt 1844. pp. 179-180.

62 Trio, K 498, for clarinet, viola & piano, August 1786. Nicknamed 'Kegelstatt' (bowling alley) because Mozart supposedly composed the entire work in a single day, whilst bowling with friends.
63 Zegers, R. Mozart and smallpox. *Clinical and Experimental Ophthalmology* 2007, Vol. 35: pp. 372–373.
64 HVTS 2243. Letter from Jan Ingen Housz, London, to Jacob van Breda, Delft, 2 June 1781.
65 BGA. A 6. Ingen Housz letter diary. Abstract of a letter from Jan Ingen Housz, Vienna, to Richard Huck, London, 10 April 1781.
66 http://www.musicweb-international.com/classrev/2007/june07/Mozart_flute_e072.htm (Last accessed 30 March 2011).
67 Schmidt-Görg J, Schmidt H. *Ludwig van Beethoven.* London, Pall Mall Press 1970. p. 150.
68 Ingen-Housz J. *Vermischte schriften, physisch-medizinisch inhalts. Uibersesst und herausgegeben von Niklas Karl Molitor. (Various papers on physics and medicine, translated and edited by Niklas Karl Molitor.)* Wien, Krauss 1782.
69 Ingen Housz J. *Nouvelles expériences et observations sur divers objets de physique. (New experiments and observations on various subjects in physics).* Paris, Barrois le jeune 1785.
70 BGA. B 13. Letter from Josef Jacquin, Vienna, to Jan Ingen Housz, London, 21 December 1793.
71 Vienna. Stadt- & Landes-archiv: 2; 2318. 1799.
72 Brenni P. *Volta's electric lighter and its improvements.* Proceedings of a conference held in Pognano sul Lario, Italy June 2003. http://sci-ed.org/Conference-Pognana/Brenni.pdf (Last accessed 23 March 2011)
73 HVTS 2243. Letter from Jan Ingen Housz, Vienna, to Jacob van Breda, Delft, 1 November 1780.
74 HVTS 2243. Letter from Jan Ingen Housz, Vienna, to Jacob van Breda, Delft, 4 October 1781.
75 HVTS 2243. Letter from Jan Ingen

Housz, Vienna, to Jacob van Breda, Delft, 1 November 1780.
76 POBF. Letter from Jan Ingen Housz, Vienna, to Benjamin Franklin, Paris, 5 December 1780.
77 Ibid.
78 Ibid.
79 Ibid.
80 Ibid.
81 POBF. Letter from Benjamin Franklin, Passy, to Jan Ingen Housz, Vienna, 21 June 1782.
82 POBF. Letter from Jan Ingen Housz, Vienna, to Benjamin Franklin, Paris, 5 December 1780.
83 Ingen Housz J. Electrical experiments to explain how the phenomenon of the electrophorus may be accounted for by Dr. Franklin's theory of positive and negative electricity. *Philosophical Transactions of the Royal Society 1778.* Vol. 68, pp. 1027-1048.
84 Ingen Housz J. A ready way of lighting a candle by a very moderate electrical spark. *Philosophical Transactions of the Royal Society 1778.* Vol. 68, pp. 1022-1026.
85 Ingen Housz J. Account of a new kind of inflammable air or gass, which can be made in a moment without apparatus and is as fit for explosion as any other inflammable gases in use for that purpose. *Philosophical Transactions of the Royal Society 1779.* Vol. 69, pp. 376-418.
86 Ingen-Housz J. *Anfangsgründe der Electricität, hauptsächlich in Beziehung auf den Elektrophor; nebst einer leichten Art, vermittelst eines elektrischen Funkens das Licht anzuzünden, und einem Briefe in Betref einer neuen entzündbaren Knallluft, mit Anmerkungen. Aus dem Englischen übersetzt von Niklas Karl Molitor.. (The rudiments of electricity, principally in relation to the electrophore; with a means of making a flame using an electric spark and a letter concerning a new inflammable explosive air, with notes. Translated from the English by Niklas Karl Molitor).* Wien, Wappler 1781
87 Ingen Housz J. On the degree of salubrity of the common air at sea, compared with that of the sea-shore, and that of places far removed from the sea. *Philosophical Transactions of the Royal Society 1780.* Vol. 70, pp. 354-377.
88 Ingen Housz J. *Nouvelles expériences et observations sur divers objets de physique. (New experiments and observations on various subjects in physics).* Paris, Barrois le jeune 1785. p. 192.
89 Ingen Housz J. (Van Breda J, translator). Verhandelingen over de gedéphlogisteerde lucht, en de manier hoe men dezelve kan bekomen en tot de ademhaaling doen dienen. (Dissertation on the dephlogisticated air and the way to obtain it and make it useful for breathing.). *Verhandelingen van het Bataafsch Genootschap der proefondervindelyke wysbegeerte te Rotterdam 1781, Vol. VI, pp. 107-159.*
90 IHLD. Abstract of a letter from Jan Ingen Housz, Vienna, to Lambertus Bicker, Rotterdam, December 1780.
91 HVTS 2243. Letter from Jan Ingen Housz, Vienna, to Jacob van Breda, Delft, 28 March 1781.
92 Ibid.
93 HVTS 2243. Letter from Jan Ingen Housz, Vienna, to Jacob van Breda, Delft, 30 January 1782.
94 HVTS 2243. Letter from Jan Ingen Housz, Vienna, to Jacob van Breda, Delft, 9 May 1781.
95 Fontana F. Experiments and observations on the inflammable air breathed by various animals. *Philosophical Transactions of the Royal Society 1779.* Vol. 69, pp. 337-361.
96 Priestley J. *Experiments and observations on different kinds of air.* Vol. II. London, Johnson 1776. (Second edition) p.102.
97 Fontana F. Experiments and observations on the inflammable air breathed by various animals. *Philosophical Transactions of the Royal Society 1779.* Vol. 69, p. 347.
98 Ingen Housz J. (Van Breda J, translator). Verhandelingen over de gedéphlogisteerde lucht, en de manier hoe men dezelve kan bekomen en tot de ademhaaling doen dienen. (Dissertation on the dephlogisticated air and the way to obtain it and make

it useful for breathing.). *Verhandelingen van het Bataafsch Genootschap der proefondervindelyke wysbegeerte te Rotterdam 1781, Vol. VI,. passim.*
99 Although a later illustration shows the mask over the mouth. (Ingen-Housz J. *Vermischte schriften, physisch-medizinisch inhalts. Uibersesst und herausgegeben von Niklas Karl Molitor.(Various papers on physics and medicine,translated and edited by Niklas Karl Molitor.)* Wien, Krauss 1782. Plate 1.
100 Ingen Housz J. (Van Breda J, translator). Verhandelingen over de gedéphlogisteerde lucht, en de manier hoe men dezelve kan bekomen en tot de ademhaaling doen dienen. (Dissertation on the dephlogisticated air and the way to obtain it and make it useful for breathing.). *Verhandelingen van het Bataafsch Genootschap der proefondervindelyke wysbegeerte te Rotterdam 1781, Vol. VI. p. 159.*
101 Ibid. p. 260.
102 Ingen Housz J. *Nouvelles expériences et observations sur divers objets de physique. (New experiments and observations on various subjects in physics).* Paris, Barrois le jeune 1785. p. 201.
103 Ibid. p. 125.
104 Ingen Housz J. *Nouvelles expériences et observations sur divers objets de physique. (New experiments and observations on various subjects in physics).* Paris, Barrois le jeune 1785. p. 127.
105 Ibid.. p. 126.
106 Maurois A. *The life of Sir Alexander Fleming.* Harmondsworth, Middlesex, Penguin 1959. p. 137.
107 Ingen Housz J. *Nouvelles expériences et observations sur divers objets de physique. (New experiments and observations on various subjects in physics).* Paris, Barrois le jeune 1785. p. 404.
108 POBF. Letter from Jan Ingen Housz, Vienna, to Benjamin Franklin, Passy, 8 December 1781.
109 HVTS 2243. Letter from Jan Ingen Housz, Vienna, to Jacob van Breda, Delft, 11 December 1781.
110 Cohen E. Wie heeft de verbranding van een horlogeveer in suurstofgas het eerst uitgevoerd? (Who was the first to publish on the burning of a watch spring in oxygen?) *Chemisch Weekblad 1907,* Vol. 48, pp. 787-798.
111 HVTS 2243. Letter from Jan Ingen Housz, Vienna, to Jacob van Breda, Delft, 17 August 1782.
112 BGA. A 6. Ingen Housz letter diary. Abstract of a letter from Jan Ingen Housz, Vienna, to Begue le Presle, Paris, 11 January 1782.
113 Ibid.
114 Ingen-Housz J. *Vermischte schriften, physisch-medizinisch inhalts. Uibersesst und herausgegeben von Niklas Karl Molitor. (Various papers on physics and medicine, translated and edited by Niklas Karl Molitor.)* Wien, Krauss 1782.
115 HVTS 2243. Letter from Jan Ingen Housz, Vienna, to Jacob van Breda, Delft, 4 May 1782.
116 HVTS 2243. Letter from Jan Ingen Housz, Vienna, to Jacob van Breda, Delft, 12 April 1782.
117 HVTS 2243. Letter from Jan Ingen Housz, Vienna, to Jacob van Breda, Delft, 4 May 1782.
118 HVTS 2244. Letter from Jan Ingen Housz, Vienna, to Jacob van Breda, Delft, 5 February 1783.
119 HVTS 2244. Letter from Jan Ingen Housz, Vienna, to Jacob van Breda, Delft, 24 September 1783.
120 HVTS 2244. Letter from Jan Ingen Housz, Vienna, to Jacob van Breda, Delft, 31 August 1783.
121 Ingen Housz J. (Van Breda J, translator). *Verzameling van verhandelingen, over verschillende natuurkundige onderwerpen. (Collection of treatises on various topics in physics).* 's-Gravenhage (The Hague), van Cleef 1785.

Notes to Chapter 15

1 BGA. A 6. Ingen Housz letter diary. Abstract of a letter from Jan Ingen Housz, Vienna, to Mr. Camper, The Hague, 16 January 1782.
2 Ibid.
3 Ingen Housz J. Some further

considerations on the influence of the vegetable kingdom on the animal creation. *Philosophical Transactions of the Royal Society 1782.* Vol. 72, p. 426.
4 J. S. G. Blair, 'Pringle, Sir John, first baronet (1707–1782)', *Oxford Dictionary of National Biography*, Oxford University Press, 2004; online edn, Oct 2007. http://www.oxforddnb.com/view/article/22805 (Last accessed 31 March 2011).
5 POBF. Letter from Jan Ingen Housz, Vienna, to Benjamin Franklin, Passy, 8 December 1781.
6 Which suffered very severe damage during the Blitz, in May 1940, so that the site of the grave cannot now be traced.
7 Kippis A. *The life of Sir John Pringle. In: Six discourses delivered by Sir John Pringle.* London: Strahan and Cadell 1783.
8 POBF. Letter from Jan Ingen Housz, Vienna, to Benjamin Franklin, Passy, 15 August 1783.
9 Priestley J. *Experiments and observations relating to various branches of natural philosophy, with a continuation of the observations on air. The second volume.* Birmingham, Pearson and Rollason for Johnson, London 1781.
10 HVTS 2243. Letter from Jan Ingen Housz, Vienna, to Jacob van Breda, Delft, 25 December 1781.
11 Persius, Satire V.
12 HVTS 2243. Letter from Jan Ingen Housz, Vienna, to Jacob van Breda, Delft, 25 December 1781.
13 Ibid.
14 Priestley J. *Experiments and observations relating to various branches of natural philosophy; with a continuation of the observations on air. The second volume.* Birmingham, Pearson and Rollason, for J. Johnson, London 1781. p. 29.
15 Priestley J. *Experiments and observations relating to various branches of natural philosophy; with a continuation of the observations on air. The second volume.* Birmingham, Pearson and Rollason, for J. Johnson, London 1781.
16 Ibid. Preface p. iv..
17 Ibid. p. 29.
18 Ibid.
19 Schofield R. (Ed.) *A scientific autobiography of Joseph Priestley: selected scientific correspondence.* Cambridge MA: MIT Press 1966. pp. 175,176. Letter from Joseph Priestley, Calne, to Giovanni Fabbroni, London, 5 September 1779. Ibid. pp. 177,178. Letter from Joseph Priestley, Calne to Radcliffe Scholefield, Birmingham, 14 September 1779. Ibid. pp. 175-179. Letter from Joseph Priestley, Calne, to Benjamin Franklin, Paris, 27 September 1779.
20 BGA. A 8. Various handwritten memoires of Dr. Ingen Housz, 'On Priestley'.
21 Schofield R. (Ed.) *A scientific autobiography of Joseph Priestley: selected scientific correspondence.* Cambridge MA: MIT Press 1966. pp. 175,176. Letter from Joseph Priestley, Calne, to Giovanni Fabbroni, London, 5 September 1779.
22 Priestley J. *Experiments and observations relating to various branches of natural philosophy; with a continuation of the observations on air. The second volume.* Birmingham, Pearson and Rollason, for J. Johnson, London 1781. p. 180.
23 Ingen Housz J. *Experiments upon vegetables; discovering their great power of purifying the air in the sunshine and of injuring it in the shade and at night.* London, Elmsley 1779. p. 155.
24 Priestley J. *Experiments and observations relating to various branches of natural philosophy; with a continuation of the observations on air. The second volume.* Birmingham, Pearson and Rollason, for J. Johnson, London 1781. p. 191.
25 Ibid. p. 184.
26 Ibid. p. 186.
27 *Critical Review 1781*, Vol 52, July 1781. pp. 134 & 180.
28 Priestley J. *Experiments and observations relating to various branches of natural philosophy, with a continuation of the observations on air.* The second volume. Birmingham, Pearson and Rollason for Johnson, London 1781.
29 HVTS 2243. Letter from Jan Ingen

Housz, Vienna, to Jacob van Breda, Delft, 30 March 1782.
30 Ibid.
31 POBF. Letter from Jan Ingen Housz, Vienna, to Benjamin Franklin, Passy, 24 April 1782.
32 Ingen Housz J. Some further considerations on the influence of the vegetable kingdom on the animal creation. *Philosophical Transactions of the Royal Society 1782.* Vol. 72, p. 426-439. passim.
33 Ibid.
34 HTVS 2243. Letter from Jan Ingen Housz, Vienna, to Jacob van Breda, Delft, 30 March 1782.
35 British Library. Add MS 8095, Fol. 83. Letter from Jan Ingen Housz, Vienna, to Sir Joseph Banks, London, 11 May 1782.
36 Ibid.
37 BGA. A 6. Ingen Housz letter diary. Abstract of a letter from Jan Ingen Housz, Vienna, to Messrs. Tourton and Bauer, Paris, 9 January 1782.
38 POBF. Letter from Jan Ingen Housz, Vienna, to Benjamin Franklin, Passy, 27 November 1782.
39 HKB. Fol. 16. Letter from Franz Schwediauer, Paris, to Jan Ingen Housz, Vienna, 1 October 1782.
40 Australia, State University of New South Wales. Mitchell Library, Series 74, CY 3683, Fols. 309-311. Letter from Sir Joseph Banks, London, to Jan Ingen Housz, Vienna, 21 May 1782.
41 Ironically, the Bakerian Lecture that year was to be given by Cavallo.
42 Australia, State University of New South Wales. Mitchell Library, Series 74, cy 3683, Fols. 309-311. Letter from Sir Joseph Banks, London, to Jan Ingen Housz, Vienna, 21 May 1782.
43 British Library. Add MS 8095, ff. 84,85. Letter from Jan Ingen Housz, Vienna, to Sir Joseph Banks, London, 6 May 1782.
44 Ingen Housz J. Some further considerations on the influence of the vegetable kingdom on the animal creation. *Philosophical Transactions of the Royal Society 1782.* Vol. 72, p. 426-439.
45 Gest H. Sun-beams, cucumbers and purple bacteria. Historical milestones in early studies of photosynthesis. *Photosynthesis Research 1988*, Vol. 19, pp 287-308.
46 Gest H. Bicentenary homage to Dr. Jan Ingen Housz, MD (1730-1799), pioneer of photosynthesis research. *Photosynthesis Research 2000*, Vol. 63, pp 183-190.
47 HVTS 2243. Letter from Jan Ingen Housz, Vienna, to Jacob van Breda, Delft, 14 August 1782.
48 POBF. Letter from Benjamin Franklin, Passy, to Jan Ingen Housz, Vienna, 2 October 1781 and completed and sent on 21 June 1782.
49 Ibid.
50 Ibid.
51 Ibid.
52 Ibid.
53 Schiff S. *Dr. Franklin goes to Paris.* London, Bloomsbury 2005. p. 381.
54 Arago F. *Meteorological essays.* London. Longmans 1855. p. 84.
55 Barletti C. *Analisi di un nuovo fenomeno del fulmine ed osservazione sopra gli usi medici della elletricità. (Analysis of a new explosive phenomenon and observations on the use of electricity in medicine.)* Pavia, Bianchi 1780.
56 POBF. Letter from Jan Ingen Housz, Vienna, to Benjamin Franklin, Passy, 2 October 1782.
57 HVTS 2244. Letter from Jan Ingen Housz, Vienna, to Jacob van Breda, Delft, 17 October 1783.
58 POBF. Letter from Jan Ingen Housz, Vienna, to Benjamin Franklin, Passy, 15 August 1783.
59 POBF. Letter from Benjamin Franklin, Passy to Jan Ingen Housz, Vienna, 29 April 1785.
60 HVTS 2244. Letter from Jan Ingen Housz, Vienna, to Jacob van Breda, Delft, 15 December, 1783.
61 HVTS 2243. Letter from Jan Ingen Housz, Vienna, to Jacob van Breda, Delft, 6 July 1782.
62 Van Barneveld W. Proeve van onderzoek omtrent de hoeveelheid van bederf, t'welk in onzen dampkring ontstaat,

nevens deszelfs verbetering door den groei der plantgewassen. (Investigation on the amount of pollution of the air and its restoration by the growth of plants.) *Verhandelingen Provinciaal Utrechts Genootschap van Kunsten en Wetenschappen 1781, pp. 408-472.*

63 Priestley J. Observations on different kinds of air. *Philosophical Transactions of the Royal Society 1772.* Vol. 62, pp. 147-264.

64 Translated from the Dutch and quoted in: Wassink E. On the discovery of the light factor in photosynthesis. *Mededelingen van de Landbouwhogeschool te Wageningen/Nederland 1958*; Vol. 58, pp. 1-10.

65 Wassink E. On the discovery of the light factor in photosynthesis. *Mededelingen van de Landbouwhogeschool te Wageningen/Nederland 1958;* Vol. 58, p. 5.

66 Wassink E. On the discovery of the light factor in photosynthesis. *Mededelingen van de Landbouwhogeschool te Wageningen/Nederland 1958;* Vol. 58, p. 6.

67 Ingen Housz J. On the degree of salubrity of the common air at sea, compared with that of the sea-shore, and that of places far removed from the sea. *Philosophical Transactions of the Royal Society 1780.* Vol. 70, pp. 372.

68 HVTS 2243. Letter from Jan Ingen Housz, Vienna, to Jacob van Breda, Delft, 4 May 1782.

69 Ibid.

70 Senebier J. *Memoires physico-chymiques, sur l'influence de la lumière solaire pour modifier le êtres des trois règnes de la nature, et surtout ceux du règne végétal.* Geneva, Barthelemi Chirol 1782.

71 HVTS 2244. Letter from Jan Ingen Housz, Vienna, to Jacob van Breda, Delft, 26 February 1783.

72 HVTS 2244. Letter from Jan Ingen Housz, Vienna, to Jacob van Breda, Delft, 8 October 1783.

73 Ibid.

74 HVTS 2244. Letter from Jan Ingen Housz, Vienna, to Jacob van Breda, Delft, 5 February 1783.

75 HVTS 2244. Letter from Jan Ingen Housz, Vienna, to Jacob van Breda, Delft, 26 February 1783.

76 Ibid.

77 HVTS 2244. Letter from Jan Ingen Housz, Vienna, to Jacob van Breda, Delft, 5 March 1783.

78 Senebier J. *Expériences sur l'action de la lumière solaire dans la végétation.* Geneva and Paris. Briand 1788. Ingen Housz's bound copy of the book, printed and blank pages alternating, still exists, in private ownership, in The Netherlands.

79 HVTS 2244. Letter from Jan Ingen Housz, Vienna, to Jacob van Breda, Delft, 26 February 1783.

80 Ibid.

81 HVTS 2244. Letter from Jan Ingen Housz, Vienna, to Jacob van Breda, Delft, 26 February 1783.

82 Ibid.

83 HVTS 2244. Letter from Jan Ingen Housz, Vienna, to Jacob van Breda, Delft, 22 April 1783 – undated addendum to this letter in handwriting of Jacob van Breda.

84 HVTS 2244. Letter from Jan Ingen Housz, Vienna, to Jacob van Breda, Delft, 8 October 1783.

85 BGA. A 13. Memoire: 'On dispute with Senebier' written at Bath by Jan Ingen Housz, 18 November 1790, and given, in person, to M. Huiller of Geneva.

86 HVTS 2244. Letter from Jan Ingen Housz, Vienna, to Jacob van Breda, Delft, 5 March 1783.

87 HVTS 2244. Letter from Jan Ingen Housz, Vienna, to Jacob van Breda, Delft, 8 October 1783.

88 HVTS 2244. Letter from Jan Ingen Housz, Vienna, to Jacob van Breda, Delft, 14 August 1783.

89 HVTS 2244. Letter from Jan Ingen Housz, Vienna, to Jacob van Breda, Delft, 22 October 1783.

90 Ingen Housz J. *Observations sur la vertu de l'eau imprégnée d'air fixe, de différens acides, & de plusieurs autres substances, pour en obtenir, par le moyen des plantes & de la lumière du soleil, de l'air déphlogistiqué.* (Observations on the power of water impregnated with fixed air, of

different acids, and many other substances in obtaining, by means of plants and sunlight, dephlogisticated air.) Journal de Physique, Vol XXIV, May 1784. pp.337-348. Ingen Housz J. *Reflexions dur l'economie de vegetaux. (Reflections on the economy of plants.)* Journal de Physique, Vol XXIV, May 1784. pp. 443-455.
91 HVTS 2244. Letter from Jan Ingen Housz, Vienna, to Jacob van Breda, Delft, 2 April 1785.
92 Braunbehrens V. *Mozart in Vienna 1781-1791*. London, André Deutch 1990. p. 232.
93 Porter R. *Enlightenment: Britain and the creation of the modern world*. London, Penguin Books 2001. p. 37.
94 Braunbehrens V. *Mozart in Vienna 1781-1791*. London, André Deutch 1990. p. 232.
95 Beales D. *Joseph II. In the shadow of Maria Theresa. 1741-1780*. Cambridge, Cambridge University Press 1987. p. 34.
96 Braunbehrens V. *Mozart in Vienna 1781-1791*. London, André Deutch 1990. p. 233. Ibid. p. 238.
97 Beales D. *Joseph II. In the shadow of Maria Theresa. 1741-1780*. Cambridge, CUP 1987. p. 477.
98 HVTS 2244. Letter from Jan Ingen Housz, Vienna, to Jacob van Breda, Delft, 21 January 1785.
99 HVTS 2244. Letter from Jan Ingen Housz, Vienna, to Jacob van Breda, Delft, 24 December 1785.
100 Braunbehrens V. *Mozart in Vienna 1781-1791*. André Deutch, London 1990. p. 245.
101 Mystical interpretation of the scriptures.
102 Till N. *Mozart and the enlightenment: truth, virtue and beauty in Mozart's operas.* London, Faber 1992. p. 298.
103 HVTS 2244. Letter from Jan Ingen Housz, Vienna, to Jacob van Breda, Delft, 21 January 1785.
104 HVTS 2244. Letter from Jan Ingen Housz, Vienna, to Jacob van Breda, Delft, 5 January 1785.

Notes to Chapter 16

1 POBF. Letter from Jan Ingen Housz, Vienna, to Benjamin Franklin, Passy, 5 December 1780. Maria Theresa died on 29 November at 21.00.
2 IHLD. Abstract of a letter from Jan Ingen Housz, Vienna, to Hendrik Hoogeveen, Delft, 21 April 1781.
3 Ibid.
4 Ibid.
5 POBF. Letter from Jan Ingen Housz, Vienna, to Benjamin Franklin, Passy, 7 April 1781.
6 POBF. Letter from Jan Ingen Housz, Vienna, to Benjamin Franklin, Passy, 28 January 1783. HVTS 2244. Letter from Jan Ingen Housz, London, to Jacob van Breda, Delft, 16 June 1783.
7 Alexander J. *Catherine the Great: life and legend*. Oxford, Oxford University Press 1989. p. 242 passim.
8 Ibid. p. 245.
9 Cross A. *An English Lady at the court of Catherine the Great: the journal of Baroness Elizabeth Dimsdale, 1781.* Cambridge, Crest Publications 1989.
10 Beales D. *Joseph II. Against the world, 1780-1790*. Cambridge, Cambridge University Press 2009. p. 132.
11 HVTS 2243. Letter from Jan Ingen Housz, London, to Jacob van Breda, Delft, 1 December 1781.
12 HVTS 2243. Letter from Jan Ingen Housz, London, to Jacob van Breda, Delft, 11 December 1781.
13 HVTS 2243. Letter from Jan Ingen Housz, London, to Jacob van Breda, Delft, 2 January 1782.
14 Ibid.
15 IHLD. Abstract of a letter from Jan Ingen Housz, Vienna, to Franz Schwediauer, London, 20 March 1782.
16 Beales D. *Joseph II: against the world, 1780-1790*. Cambridge, Cambridge University Press 2009. p. 214 passim.
17 HVTS 2243. Letter from Jan Ingen Housz, London, to Jacob van Breda, Delft, 6 March 1782.
18 HKB 18. Letter from Franz Schwediauer, Paris, to Jan Ingen Housz, Vienna, 25 November 1782.
19 Priestley J. *Experiments and observations relating to various branches of natural*

philosophy, with a continuation of the observations on air, the second volume. Birmingham, Pearson and Rollason for Johnson, London 1781. p. 16.
20 Priestley J. *Experiments and observations relating to various branches of natural philosophy, with a continuation of the observations on air.* London, Johnson 1779. p. 342.
21 Priestley J. *Experiments and observations relating to various branches of natural philosophy, with a continuation of the observations on air, the second volume.* Birmingham, Pearson and Rollason for Johnson, London 1781. p. 23.
22 Ibid. p. 24.
23 Ingen Housz J. *Experiments upon vegetables; discovering their great power of purifying the air in the sunshine and of injuring it in the shade and at night.* London, Elmsly 1779. pp. 89-92.
24 Ibid. p. 91.
25 HVTS 2244. Letter from Jan Ingen Housz, London, to Jacob van Breda, Delft, 22 April 1783.
26 Ibid.
27 Ibid.
28 HVTS 2244. Letter from Jan Ingen Housz, London, to Jacob van Breda, Delft, 7 November 1783.
29 HVTS 2244. Letter from Jan Ingen Housz, London, to Jacob van Breda, Delft, 22 April 1783.
30 HVTS 2244. Letter from Jan Ingen Housz, London, to Jacob van Breda, Delft, 7 November 1783.
31 Ingen-Housz J. *Vermischte schriften, physisch-medizinisch inhalts. Uiberfesst und herausgegeben von Niklaus Carl Molitor.* Wien, Wappler 1784. Vol. 2.
32 Ingen Housz J. *Nouvelles expériences et observations sur divers objets de physique.* Second Volume. Paris, Barrois le jeune 1789.
33 Virgil Georgicon Book IV, line 3.
34 Hoff H. Jan Ingen-Housz and the cover slip. *Bulletin of the History of Medicine 1962*, Vol. 36, pp. 365-8.
35 Van der Pas P. The discovery of the Brownian motion. *Scientarium Historia 1971*, Vol. 13, pp. 27-35.
36 Brown R. A brief account of microscopical observations made on the particles contained in the pollen of plants. *Edinburgh Philosophical Magazine 1828*, Vol. 4, pp. 161-173.
37 American hardy annual named after a William Clark who, on the first overland expedition across America to the Pacific coast, in 1804/6, discovered the plant in Idaho.
38 Ingen Housz J. *Nouvelles expériences et observations sur divers objets de physique. Second Volume.* Paris, Barrois le jeune 1789. p. 2.
39 Ibid p. 5.
40 Brown R. A brief account of microscopical observations made on the particles contained in the pollen of plants. *Edinburgh Philosophical Magazine 1828*, Vol. 4, pp. 161-173.
41 Isaacson W. *Einstein: his life and universe.* London, Pocket Books 2008. p. 105.
42 Ibid. p. 106.
43 POBF. Letter from Benjamin Franklin, Passy, to Jan Ingen Housz, Vienna, 2 September 1783.
44 POBF. Letter from Benjamin Franklin, Passy, to Sir Joseph Banks, London, 30 August 1783.
45 Ibid.
46 Gillispie C. *The Montgolfier brothers, and the invention of aviation 1783-1784.* Princeton, Princeton University Press 1983. p. 21 passim.
47 Schiff S. *Dr. Franklin goes to France.* London, Bloomsbury 2005. p. 345.
48 POBF. Letter from Jan Ingen Housz, Vienna, to Benjamin Franklin, Passy, 19 November 1783.
49 POBF. Letter from Jan Ingen Housz, Vienna, to Benjamin Franklin, Passy, 2 January 1784.
50 HVTS 2244. Letter from Jan Ingen Housz, Vienna, to Jacob van Breda, Delft, 15 December 1783.
51 POBF. Letter from Jan Ingen Housz, Vienna, to Benjamin Franklin, Passy, 2 January 1784.
52 IHLD. Abstract of a letter from Jan Ingen Housz, Vienna, to Achille-Guillaume Bègue de Presle, Paris, 2 January 1784. Abstract of a letter from Jan Ingen Housz, Vienna, to Jacques

Charles, Paris, 2 January 1784. Abstract of a letter from Jan Ingen Housz, Vienna, to Jean Baptiste Le Roy, Paris, 2 January 1784.
53   POBF. The Papers of Benjamin Franklin. Letter from Benjamin Franklin, Passy, to Jan Ingen Housz, Vienna, 16 January 1784.
54   http://www.wydera.de/balloon/history1.htm (Last accessed 4 April 2011).
55   POBF. Letter from Jan Ingen Housz, Vienna, to Benjamin Franklin, Passy, 10 February 1784.
56   POBF. Letter from Benjamin Franklin, Passy, to Jan Ingen Housz, Vienna, 16 January 1784.
57   HVTS 2244. Letter from Jan Ingen Housz, London, to Jacob van Breda, Delft, 18 February 1784.
58   POBF. Letter from Abbé Johann Nekrep, Vienna, to Benjamin Franklin, Passy, 12 June 1784.
59   http://www.wydera.de/balloon/history1.htm (Last accessed 11 April 2011).
60   HVTS 2244. Letter from Jan Ingen Housz, Vienna, to Jacob van Breda, Delft, 2 April 1785.
61   HVTS 2244. Letter from Jan Ingen Housz, Vienna, to Jacob van Breda, Delft, 9 April 1785.
62   HVTS 2244. Letter from Jan Ingen Housz, Vienna, to Jacob van Breda, Delft, 2 April 1785.
63   Beales D. *Joseph II. Against the world, 1780-1790*. Cambridge, Cambridge University Press 2009. p. 321.
64   Alexander Du Toit, 'Keith, Sir Robert Murray, of Murrayshall (1730–1795)', *Oxford Dictionary of National Biography*, Oxford University Press, 2004; online edn, Jan 2008. http://www.oxforddnb.com/view/article/15272 (Last accessed 14 October 2010).
65   Actually saved by an unexpected but very generous pension.
66   Joan Lane, 'Stokes, Jonathan (1755?–1831)', *Oxford Dictionary of National Biography*, Oxford University Press, 2004; online edn, May 2010. http://www.oxforddnb.com/view/article/54048 (Last accessed 14 October 2010).
67   Uglow J. *The Lunar Men*. London, Faber & Faber 2002.
68   HKB 24. Letter from Jonathan Stokes, Kidderminster, to Jan Ingen Housz, Vienna 31 October 1787.
69   Philadelphia, American Philosophical Society Library. Fabbroni papers, BF113. Letter from Jan Ingen Housz, Vienna, to Giovanni Fabbroni, Florence, 27 December 1784.
70   The King had promoted Lord Shelburne to the first Marquis of Lansdowne in December 1784 as reward for having served as 'Prime Minister' from July 1782 to April 1783.
71   Bowood. Shelburne Papers. Letter from Sir Robert Murray Keith, Vienna, to the first Marquis of Lansdowne, London 21 February 1785.
72   Bowood. Shelburne Papers. Letter from Sir Robert Murray Keith, Vienna, to the first Marquis of Lansdowne, London 31 December 1785.
73   Bowood. Shelburne Papers. Letter from Lord Wycombe, Vienna, to the first Marquis of Lansdowne, London 22 March 1786.
74   Mitchison R. *Agricultural Sir John. The life of Sir John Sinclair of Ulbster 1754-1835*. London, Bles 1962. p. 54 (opposite).
75   Ibid. Preface.
76   British Library. BM Add MS 8096. fol. 463. Letter from Jan Ingen Housz, Vienna, to Sir Joseph Banks, London, 27 October 1786.
77   Vienna. Österreichische Nationalbibliothek 50/5-1. Letter from Joseph Priestley, Birmingham, to Jan Ingen Housz, Vienna, 2 November 1787.
78   Ibid.
79   POBF. Letter from Jan Ingen Housz, Vienna, to Benjamin Franklin, Passy, 29 April 1783.
80   POBF. Letter from Benjamin Franklin, Passy, to Jan Ingen Housz, Vienna, 1 June 1783.
81   London, University College. Lord Odo Russell Collection. Letter from Dominic Beck, Salzburg, to Jan Ingen Housz, Vienna, 3 July 1783.
82   HVTS 2244. Letter from Jan Ingen

Housz, Vienna, to Jacob van Breda, Delft, 22 October 1784.
83 HVTS 2244. Letter from Jan Ingen Housz, Vienna, to Jacob van Breda, Delft, 28 August 1784.
84 HVTS 2245. Letter from Jan Ingen Housz, Vienna, to Jacob van Breda, Delft, 24 November 1786.
85 POBF. Letter from Jan Ingen Housz, Vienna, to Benjamin Franklin, Passy, 12 June 1782 enclosing one to Samuel Wharton, Philadelphia, of same date.
86 POBF. Letter from Jan Ingen Housz, Vienna, to Benjamin Franklin, Passy, 29 August 1781.
87 POBF. Letter from Jan Ingen Housz, Vienna, to Benjamin Franklin, Passy, 22 April 1782.
88 IHLD. Abstract of a letter from Jan Ingen Housz, Vienna, to Samuel Wharton, Philadelphia, 8 May 1783.
89 Ibid.
90 POBF. Letter from Jan Ingen Housz, Vienna, to Benjamin Franklin, Passy, 12 June 1782 enclosing one to Samuel Wharton, Philadelphia, of same date.
91 IHLD. Abstract of a letter from Jan Ingen Housz, Vienna, to Samuel Wharton, Philadelphia, 24 April 1782.
92 POBF. Letter from Jan Ingen Housz, Vienna, to Benjamin Franklin, Passy, 26 February 1783.
93 POBF. Letter from Jan Ingen Housz, Vienna, to Benjamin Franklin, Passy, 27 November 1782.
94 POBF. Letter from Jan Ingen Housz, Vienna, to Benjamin Franklin, Passy, 28 January 1783.
95 POBF. Letter from Jan Ingen Housz, Vienna, to Benjamin Franklin, Passy, 20 August 1782.
96 POBF. Letter from Jan Ingen Housz, Vienna, to Benjamin Franklin, Passy, 27 November 1782.
97 POBF. Letter from Jan Ingen Housz, Vienna, to Benjamin Franklin, Passy, 8 April 1783. John Williams senior.
98 IHLD. Abstract of a letter from Jan Ingen Housz, Vienna, to Francis Coffyn, Dunkerque, 29 April 1783.
99 IHLD. Abstract of a letter from Jan Ingen Housz, Vienna, to Samuel Wharton, Philadelphia, 8 May 1783.
100 POBF. Letter from Jan Ingen Housz, Vienna, to Benjamin Franklin, Passy, 19 November 1783.
101 Ibid.
102 IHLD. Abstract of a letter from Jan Ingen Housz, Vienna, to Francis Coffyn, Dunkerque, 2 January 1784.
103 IHLD. Abstract of a letter from Jan Ingen Housz, Vienna, to Samuel Wharton, Philadelphia, 9 October 1784.
104 IHLD. Abstract of a letter from Jan Ingen Housz, Vienna, to Jonathan Williams, Nantes, 9 October 1784.
105 POBF. Letter from Benjamin Franklin, Passy, to Jan Ingen Housz, Vienna, 29 April 1785.
106 POBF. Letter from Jan Ingen Housz, Vienna, to Benjamin Franklin, Passy, 11 June 1785.
107 IHLD. Copy of a letter from Jan Ingen Housz, Vienna, to Benjamin Franklin, Passy, 11 June 1785.
108 POBF. Letter from Benjamin Franklin, Philadelphia, to Jan Ingen Housz, Vienna, 27 June 1786.
109 POBF. Letter from Jan Ingen Housz, Vienna, to Benjamin Franklin, Philadelphia, 1 January 1787. POBF. Letter from Jan Ingen Housz, Vienna, to Benjamin Franklin, Philadelphia, 20 May 1787.
110 Ibid.
111 POBF. Letter from Benjamin Franklin, Philadelphia, to Jan Ingen Housz, Vienna, 27 June 1786.
112 POBF. Letter from Benjamin Franklin, Philadelphia, to Jan Ingen Housz, Vienna, 2 September 1786.

Notes to Chapter 17

1 Brands, H. *The First American*. New York, Doubleday 2000. p. 202.
2 Shelley, M. *Frankenstein, or the modern Prometheus*. London, Lackington, Hughes, Harding 1818.
3 Pierre Bertholon. *De l'électricité des végétaux*. Paris, Didot le jeune 1783.
4 Ingen-Housz, J. Letter to Mr. N.C.

Molitor. *Journal de Physique 1788*, Vol. 32, p. 322.
5   Ibid, passim.
6   Ibid. p. 322.
7   Ibid.
8   Ibid.
9   Ibid. p. 323.
10 Schwankhard. Letter to Mr. Ehrmann, *Journal de Physique 1785*, Vol. 27, December 1785, p. 467.
11 Ibid. pp. 462-468.
12 Ibid.
13 Ingen Housz, J. Letter to Mr. N.C. Molitor, *Journal de Physique 1788*, Vol. 32, May 1788, p. 324.
14 HVTS 2245. Letter from Jan Ingen Housz, Vienna, to Jacob van Breda, Delft, 2 May 1786.
15 British Library. BM.Add. Mss. 8097, ff. 83,84. Letter from Jan Ingen Housz, Vienna, to Sir Joseph Banks, London, 10 May 1786.
16 Schwankhard. Letter to Mr. Ehrmann, *Journal de Physique 1785*, Vol. 27, pp. 462-468.
17 Ingen Housz, J. Letter to Mr. N. C. Molitor, *Journal de Physique 1786,* Vol. 28, pp. 81-92.
18 Schwankhard. Letter to Mr. Ehrmann, *Journal de Physique 1785*, Vol. 27, pp. 462-468.
19 HVTS 224. Letter from Jan Ingen Housz, Vienna, to Jacob van Breda, Delft, 1 March 1786.
20 Duvarnier, D. Observations relative to the letter of M. Schwankhard inserted in the Journal de Physique, December 1785, p. 462. *Journal de Physique 1786*, Vol. 28, pp. 93,94.
21 Ingen Housz, J. Letter to Mr. N. C. Molitor. *Journal de Physique 1786*, Vol. 28, pp. 81-92.
22 Cited in: Ingen Housz, J. Letter to Mr. N.C. Molitor, *Journal de Physique 1788*, Vol. 32, p. 322.
23 Gardini, F. *The influence of atmospheric electricity on plants*. Turin, Briolus 1784.
24 Ingen Housz, J. Letter to Mr. N.C. Molitor, *Journal de Physique 1788*, Vol. 32, p. 325.
25 HTVS 2245. Letter from Jan Ingen Housz, Vienna, to Jacob van Breda, Delft, 3 June 1786.
26 Ingen Housz, J. Letter to Mr. N.C. Molitor, *Journal de Physique 1788*, Vol. 32, pp. 328-333 passim.
27 HVTS 2245. Letter from Jan Ingen Housz, Vienna, to Jacob van Breda, Delft, 3 August 1786.
28 Ibid, passim.
29 Ingen Housz, J. Letter to Mr. N.C. Molitor, *Journal de Physique 1788*, Vol. 32, p. 329.
30 Ibid. p. 333.
31 HVTS 2245. Letter from Jan Ingen Housz, Vienna, to Jacob van Breda, Delft, 6 July 1786.
32 Ingen Housz, J. Letter to Mr. N.C. Molitor, *Journal de Physique 1788*, Vol. 32, p. 335.
33 Ibid. p. 336.
34 Bertholon P. *De l'éléctrité de météores*. Paris, Croullebois 1787.
35 Bertholon P. *De l'éléctrité de météores. Volume 2*. Paris, Croullebois 1787. p. 371.
36 Ibid. p. 371 passim: quoted in: Ingen Housz, J. Letter to Mr. N.C. Molitor, *Journal de Physique 1788*, Vol. 32, p. 326.
37 Ingen Housz, J. Letter to Mr. N.C. Molitor, *Journal de Physique 1788*, Vol. 32, p. 333 passim.
38 HVTS 2245. Letter from Jan Ingen Housz, Vienna, to Jacob van Breda, Delft, 2 June 1786.
39 Ingen Housz, J. Letter to Mr. N. C. Molitor, *Journal de Physique 1788*, Vol.32, p. 333 passim.
40 Ibid. p. 334.
41 HVTS 2245. Letter from Jan Ingen Housz, Vienna, to Jacob van Breda, Delft, 6 February 1787.
42 Ingen Housz, J. Letter to Mr. N.C. Molitor, *Journal de Physique 1788*, Vol. 32, p. 334.
43 Ingen Housz J. *Nouvelles Expériences et observations sur divers objets de physique, tome second*. Paris, Barrois le jeune 1789.
44 Ingen Housz, J. Letter to Mr. N.C. Molitor, *Journal de Physique 1788*, Vol. 32, p. 334.
45 Schama S. *Patriots and liberators: revolution in the Netherlands 1780-1813*.

London, Fontana 1992. p. 64.
46 HVTS 2245. Letter from Jan Ingen Housz, Vienna, to Jacob van Breda, Delft, 16 January 1788.
47 Ibid.
48 Ibid.
49 Schama S. *Patriots and liberators: revolution in the Netherlands 1780-1813.* London, Fontana 1992. p. 63.
50 Ibid. p. 100.
51 Ibid. p. 83.
52 Ibid. p. 104.
53 Ibid. p. 129.
54 HVTS 2245. Letter from Jan Ingen Housz, Vienna, to Jacob van Breda, Delft, 16 January 1788.
55 Ibid.
56 HVTS 2245. Letter from Jan Ingen Housz, Vienna, to Jacob van Breda, Delft, 6 February 1787.
57 HVTS 2245. Letter from Jan Ingen Housz, Vienna, to Jacob van Breda, Delft, 16 January 1788.
58 HVTS 2245. Letter from Jan Ingen Housz, Vienna, to Jacob van Breda, Delft, 27 October 1787.
59 HVTS 2245. Letter from Jacob van Breda, Delft, to Jan Ingen Housz, Vienna, 30 January 1788.
60 Ibid.
61 Amsterdam. Amsterdam University Library Hs. Died. 11R. Letter from Hendrik Hoogeveen, Delft, to Jan Ingen Housz, Vienna, 11 February 1788.
62 Ibid.
63 Ibid.
64 HKB. 26. Letter from Salomon de Monchy, Rotterdam, to Jan Ingen Housz, Vienna, 14 February 1788.
65 Philadelphia. American Philosophical Society Library. Richard Price papers. B. P93. Film 54-61, frame 38. Letter from Jan Ingen Housz, Vienna, to Richard Price, London, 2 April 1787.
66 POBF. Letter from Benjamin Franklin, Philadelphia, to Jan Ingen Housz, Vienna, 11 February 1788. Jonathan Williams junior.
67 POBF. Letter from Benjamin Franklin, Philadelphia, to Jan Ingen Housz, Vienna, 2 September 1786.
68 BGA. A 13. Letter from Jan Ingen Housz, Vienna, to Louis Ingen Housz, Breda, 2 April 1788.
69 Ibid.
70 POBF. Letter from Jan Ingen Housz, Vienna, to Benjamin Franklin, Passy, June 11 1785.
71 HVTS 2245. Letter from Jan Ingen Housz, Vienna, to Jacob van Breda, Delft 4 May 1788.
72 Ingen Housz J. *Expériences sur les vegétaux, spécialement sur la propriété qu'ils possedent à un haut degré, soit d'amèliorer l'air quand ils sont au soleil, soit de le corrompre la nuit, ou lorsqui'ils sont à l'ombre: nouvelle edition, revue et augmentée.* Paris, Barrois le jeune 1787.
73 Ibid.
74 Ingen Housz J. *Versuche mit pflanzen, hauptsachlich über die eigenschaft, welche sie in einen hohen grade besitzen, die luft im Sonnenlichte zu reinigen, und in der nacht und im schatten zu verderben; nebst einer neuen methode, den grad der reinheit und heilsamkeit der atmosphärischen luft zu prüfen. Aus dem Franzosischen überfesetzt.* Translated from the French by Johann Andreas Scherer. Vienna, Wappler 1786.
75 Ingen Housz J. *Expériences sur les vegétaux, spécialement sur la propriété qu'ils possedent à un haut degré, soit d'amèliorer l'air quand ils sont au soleil, soit de le corrompre la nuit, ou lorsqui'ils sont à l'ombre: nouvelle edition, revue et augmentée.* Paris, Barrois le jeune 1787. p. 385.
76 Ingen Housz J. *Versuche mit pflanzen, hauptsachlich über die eigenschaft, welche sie in einen hohen grade besitzen, die luft im Sonnenlichte zu reinigen, und in der nacht und im schatten zu verderben; nebst einer neuen methode, den grad der reinheit und heilsamkeit der atmosphärischen luft zu prüfen. Aus dem Franzosischen überfesetzt. Zweyter Band.* Translated from the French by Johann Andreas Scherer. Vienna, Wappler 1788.
77 Ingen Housz J. *Versuche mit pflanzen, hauptsachlich über die eigenschaft, welche*

sie in einen hohen grade besitzen, die luft im Sonnenlichte zu reinigen, und in der nacht und im schatten zu verderben; nebst einer neuen methode, den grad der reinheit und heilsamkeit der atmosphärischen luft zu prüfen. Aus dem Franzosischen überfesetzt. Dritter Band. Translated from the French by Johann Andreas Scherer. Vienna, Wappler 1786.
78 Ingen Housz J. *Nouvelles Expériences et observations sur divers objets de physique. Tome second.* Paris, Barrois le jeune 1789.
79 HVTS 2245. Letter from Jan Ingen Housz, Vienna, to Jacob van Breda, Delft, 4 May 1788.
80 Ibid.
81 HVTS 2245. Letter from Jan Ingen Housz, Vienna, to Jacob van Breda, Delft, 19 June 1788.
82 Personal communication. FB.
83 Personal communication. Recently acquired at an antiquarian sale in Amsterdam.
84 British Library. BM Add MS 51735, ff 138,139. Letter from third Lord Holland, Plympton, to Lady Caroline Fox, Bowood, 14 September 1799.
85 BGA. A 13. Letter of credit to Jan Ingen Housz from Stametz & Co., bankers, Vienna, 2 July 1788.
86 Ibid.
87 Hussey A. *Paris, the secret history.* New York, Viking 2006. p. 166.
88 HKB 28. Letter from Jan Ingen Housz, London, to Josef Jacquin, Paris, 2 March 1790.
89 Ibid.
90 Hussey A. *Paris, the secret history.* New York, Viking 2006. pp. 188,189.
91 Hayes K. *The road to Monticello.* Oxford, Oxford University Press 2008. p. 293.
92 POBF. Letter from Jan Ingen Housz, Paris, to Benjamin Franklin, Philadelphia, 9 August 1788.
93 Ibid.
94 Ibid.
95 HVTS 2245. Letter from Jan Ingen Housz, Paris, to Jacob van Breda, Delft, 16 September 1788.
96 Ingen Housz J. *Expériences sur les végétaux, spécialement sur la propriété qu'ils possedent à un haut degré, soit d'amèliorer l'air quand ils sont au soleil, soit de le corrompre la nuit, ou lorsqui'ils sont à l'ombre. Tome second.* Paris, Barrois le jeune 1789.
97 It never did.
98 Ingen Housz J. *Nouvelles Expériences et observations sur divers objets de physique. Tome second.* Paris, Barrois le jeune 1789.
99 HVTS 2245. Letter from Jan Ingen Housz, Paris, to Jacob van Breda, Delft, 16 September 1788.
100 POBF. Letter from Benjamin Franklin, Passy, to Jan Ingen Housz, Vienna, 12 June 1782.
101 Philadelphia. American Philosophical Society Library: miscellaneous manuscripts collection. Letter from Jan Ingen Housz, Paris, to Jean Fabbroni, Florence, 2 April 1789.
102 Poirier J-P. *Lavoisier: chemist, biologist, economist.* Philadelphia, Penn 1998. pp. 140–143 passim.
103 Ibid. pp. 150,151.
104 De Morveau L, Lavoisier A, Bertholet C, de Fourcroy A. *Méthode de nomenclature chimique (un nouveau système).* Paris, Cuchet 1787.
105 Ibid.
106 HVTS 2245. Letter from Jan Ingen Housz, Vienna, to Jacob van Breda, Delft, 16 January 1788.
107 Ingen Housz J. *Nouvelles Expériences et observations sur divers objets de physique, tome second.* Paris, Barrois le jeune 1789. pp. 466 passim.
108 BGA. A 13. Letter from Jan Ingen Housz, Paris, to Jean-Henri Hassenfratz, Cambrai, 19 September 1788.
109 Schofield R. *The enlightened Joseph Priestley.* University Park PA, Pennsylvania State University Press 2004. p. 361.
110 HVTS 2245. Letter from Jan Ingen Housz, Paris, to Jacob van Breda, Delft, 19 March 1789.
111 Letter from Jan Ingen Housz, Bath, to William Falconer, Bath, 25 November 1791, which appeared in: Falconer W. *An account of the efficacy of the aqua*

NOTES TO PAGES 380-402

mephitica alkalina. London, Cadell 1792. p. 141.
112 Ibid. p. 140.
113 HVTS 2245. Letter from Jan Ingen Housz, Paris, to Jacob van Breda, Delft, 19 March 1789.
114 GBA. A 13. Letter from Jan Ingen Housz, London, to British Inland Revenue, 1789 or 1799.
115 HKB 28. Letter from Jan Ingen Housz, London, to Josef Jacquin, Paris, 2 March 1790.
116 Schama S. *Citizens*. London, Viking 1989. p. 305 passim.
117 Fraser A. *Marie Antoinette*. London, Weidenfeld & Nicolson 2001. p. 135.
118 Ibid. p. 44.
119 Ingen Housz J. *Nouvelles Expériences et observations sur divers objets de physique. Tome second*. Paris, Barrois le jeune 1789. Dedication.
120 Horace, Odes 2/17.
121 Schama S. *Citizens*. London, Viking 1989. p. 327 passim.
122 BGA. A 13. Letter of credit to Jan Ingen Housz from Stametz & Co., bankers, Vienna, 2 July 1788.
123 Schama S. *Citizens*. London, Viking 1989. p. 378 passim.
124 Ibid. p. 385.
125 BGA. A 8. Jan Ingen Housz, handwritten notes on a dream. passim.
126 BGA. A 13. Letter from Jan Ingen Housz, London, to British Inland Revenue, 1789 or 1799.
127 BGA. B 9. Letter from H. Crumpipen, Brussels, to Jan Ingen Housz, Paris, 16 July 1789.
128 GF. p. 25.
129 Ibid. p. 26.

Notes to Chapter 18

1 Rehm G. *De Bredase apothekers van de 15e tot het begin van de 19e eeuw. (The apothecaries of Breda from the 15th century to the beginning of the 19th century.)* Breda, Geschied- en Oudheidkundige Kring van Stad en Land van Breda 'De Oranjeboom' 1961. p. 104.
2 BGA. A 31. Doctorate (MD) certificate presented Henry Ferdinand Ingen Housz by University of Leiden, 26 September 1793.
3 BGA. B 21. Letter from nephew of Jan Ingen Housz, Breda, to Jan Ingen Housz, London, 21 November 1790. The Marquis of Lansdowne arranged for an agent from Amsterdam to call and value them for an English saleroom.
4 HNRH. Letter from Jan Ingen Housz, The Haig, to Martin van Marum, Haarlem, 5 September 1789.
5 Ibid.
6 Forbes R. (Ed.) *Martinus van Marum: life and work*. Haarlem, Tjeenk, Willink 1969. Vol. 1. p. 7 passim.
7 Ibid. p. 5.
8 Van Troostwyk A, Deiman J. Schets der nieuwe ontdekkingen omtrent het water. *Algemeen Magazin van Wetenschappen* 1790, Vol. 4, pp. 909-941.
9 Poirier J-P. *Lavoisier: chemist, biologist, economist*. Philadelphia, University of Pennsylvania Press 1996. p. 150.
10 Ibid.
11 HVTS 2246. Letter from Jan Ingen Housz, London, to Jacob van Breda, Delft, late 1789 or early 1790.
12 HKB 46. Letter from Jan Ingen Housz, London, to Josef Jacquin, Vienna, 16 January 1794.
13 Pearson G. Experiments and observations made with the view of ascertaining the nature of the gaz produced by passing electric discharges through water. *Journal of Natural Philosophy and the Arts 1791*. Vol. 1, p. 242.
14 HRNH. Letter from Jan Ingen Housz, Breda, to Martin van Marum, Haarlem, 15 September 1789.
15 Schama S. *Patriots and Liberators: revolution in the Netherlands 1780-1813*. London, Fontana 1992. p. 138.
16 Falconer W. *An account of the efficacy of the aqua mephitica alkalina in calculous disorders. Fourth Edition*. London, Cadell 1792. p. 142.
17 Falconer W. *An account of the efficacy of the aqua mephitica alkalina in calculous disorders. Third Edition*. London, Cadell

1789.
18. Falconer W. *An account of the efficacy of the aqua mephitica alkalina in calculous disorders. Fourth Edition*. London, Cadell 1792. p. 132.
19. Ibid. p. 141.
20. Ibid.
21. A drug that, literally, chases away sorrow; supposedly given to Helen of Troy.
22. Falconer W. *An account of the efficacy of the aqua mephitica alkalina in calculous disorders. Fourth Edition*. London, Cadell 1792. p. 142.
23. HVTS 2246. Letter from Jan Ingen Housz, London, to Jacob van Breda, Delft, Winter 1789/90.
24. Falconer W. *An account of the efficacy of the aqua mephitica alkalina in calculous disorders. Fourth Edition*. London, Cadell 1792. p. 143.
25. Ibid.
26. HRNH. Letter from Jan Ingen Housz, London, to Martin van Marum, Haarlem, 7 December 1789.
27. HVTS 2246. Letter from Jan Ingen Housz, London, to Jacob van Breda, Delft, 23 March 1790.
28. ÖNB. Fol. 19. Letter from Jan Ingen Housz, London, to Hendrick Hoogeveen, Delft, 27 March 1767.
29. HKB 30. Letter from Jan Ingen Housz, London, to Josef Jacquin, Paris, 7 June 1790.
30. HRNH. Letter from Jan Ingen Housz, London, to Martin van Marum, Haarlem, 31 May 1790. One and a half guineas = £88.00; one shilling = £2.80 (whence multiples).
31. Islington, London. The archived ledgers of Drummonds Bank at the Royal Bank of Scotland. DR/427/158, 1431.
32. Ibid.
33. Ibid.
34. HRNH. Letter from Jan Ingen Housz, London, to Martin van Marum, Haarlem, 7 December 1789.
35. Personal communication.
36. Lack W. *Franz Bauer: the painted record of nature*. Wien, Verlag des Naturhistorischen Museums 2008. p. 33 passim.
37. Royal Society Council Minutes 1782-1810, Vol. 7, p. 216. 5 November 1789.
38. Ibid.
39. Royal Society Club. Meeting books 1788-1800. Book 1.
40. Ibid.
41. Royal Society Journal Book, copy (33) 1787-1790. p. 461.
42. Ibid. Jacquin's book was '*Collectanea ad Botanicum, Chemiam, et Historiam Naturalem Spectantia, Vol II*. Wien, Wappler 1788.
43. Lack W. *Franz Bauer: the painted record of nature*. Wien, Verlag des Naturhistorischen Museums 2008. p. 42.
44. Desmond R. *The history of the Royal Botanic Gardens, Kew*. London, Harvill 1995. p. 410.
45. Greenwich. Caird Library, National Maritime Museum, MS80/031. Herschel, William and Caroline, Visitors Book at Slough. 21 November 1789.
46. HRNH. Letter from Jan Ingen Housz, London, to Martin van Marum, Haarlem, 7 December 1789.
47. Ibid.
48. Bowood Archives. Lansdowne House Dinner Guests, January 1788 - June 1792.
49. Ibid.
50. Ibid.
51. British Library. Add MS 51744; ff 135,136. Letter from Lady Caroline Fox, London, to third Lady Holland, Holland House, London, 30 December 1799.
52. British Library. Add MS 51744; f. 115. Letter from Lady Caroline Fox, Bowood, to third Lady Holland, Holland House, London, 8 September 1799.
53. BGA. A 13. Jan Ingen Housz; handwritten notes on the basis of a simple cypher, 25 December 1789. In London or, possibly, at Warwick Castle.
54. BGA. A 8. Jan Ingen Housz: commonplace book, loose-leaf.
55. HRNH. Letter from Jan Ingen Housz, London, to Martin van Marum, Haarlem, 23 March 1790.

56 BGA. A13. Letter from Josef Jacquin, Paris, to Jan Ingen Housz, London, 28 February 1790.
57 HKB 29. Letter from Jan Ingen Housz, London, to Josef Jacquin, Paris, 15 March 1790.
58 HKB 28. Letter from Jan Ingen Housz, London, to Josef Jacquin, Paris, 2 March 1790.
59 Smith C, Arnott R. *The genius of Erasmus Darwin*. Aldershot, Ashgate Publishing 2005. p. 199.
60 HVTS 2246. Letter from Jan Ingen Housz, London, to Jacob van Breda, Delft, 21 July 1791.
61 HVTS 2246. Letter from Jan Ingen Housz, Hertford, to Jacob van Breda, Delft, 25 July 1791.
62 Priestley J. *Experiments and observations relating to various branches of natural philosophy, third volume*. Pearson and Rollason, Birmingham, for Johnson, London 1786.
63 BGA. A 13. Letter from Jan Ingen Housz, London, to Joseph Priestley, Birmingham, 7 July 1790.
64 BGA. A 13. Letter from Joseph Priestley, Birmingham, to Jan Ingen Housz, London, 16 July 1790.
65 HKB 53. Letter from Jan Ingen Housz, Hertford, to Josef Jacquin, Vienna, 21 August 1790.
66 HKB 29. Letter from Jan Ingen Housz, London, to Josef Jacquin, Paris, 15 March 1790.
67 HKB 28. Letter from Jan Ingen Housz, London, to Josef Jacquin, Paris, 2 March 1790.
68 HRNH. Letter from Jan Ingen Housz, London, to Martin van Marum, Haarlem, 31 May 1790.
69 Brands H. *The first American*. New York, Doubleday 2000. p. 711.
70 HVTS 2246. Letter from Jan Ingen Housz, London, to Jacob van Breda, Delft, 8 July 1790.
71 HKB 34. Letter from Jan Ingen Housz, Hertford, to Josef Jacquin, Paris, 21 August 1790.
72 Horace, Satires II, 1,57.
73 Beales D. *Joseph II: against the world 1780-1790*. Cambridge. Cambridge University Press 2009. p. 635.
74 HKB 29. Letter from Jan Ingen Housz, London, to Josef Jacquin, Paris, 15 March 1790.
75 Wheatcroft A. *The Habsburgs: embodying empire*. Harmondsworth, Penguin 1996. p. 237.
76 HRNH. Letter from Jan Ingen Housz, London, to Martin van Marum, Haarlem, 31 May 1790.
77 Ibid.
78 Forbes R. (Ed.) *Martinus van Marum: life and work*. Haarlem, Tjeenk, Willink 1969. Vol. II. p. 266 passim.
79 HRNH. Letter from Jan Ingen Housz, Hertford, to Martin van Marum, London, 17 August 1790.
80 BGA. A 16a. Jan Ingen Housz: address book in London.
81 HKB 32. Letter from Jan Ingen Housz, London, to Josef Jacquin, Paris, 9 July 1790.
82 HRNH. Note from Jan Ingen Housz, London, to Martin van Marum, London, 24 July, 1790.
83 Personal communication.
84 Montefiore S. *Prince of princes: the life of Potemkin*. London, Weidenfeld & Nicolson 2000. p. 305.
85 Cross A. (Ed.) *An English lady at the Court of Catherine the Great: the journal of Baroness Elizabeth Dimsdale 1781*. Cambridge, Crest Publications 1989. p. 21.
86 Personal communcation.
87 Ibid.
88 HKB 34. Letter from Jan Ingen Housz, Hertford, to Josef Jacquin, Paris, 21 August 1790.
89 Ibid.
90 The 'main' house of Bowood was demolished in 1956. However, the dining room was carefully dismantled and re-assembled as the Board Room of the new Lloyds of London building. In a sense it has, therefore, survived.
91 Medd P. *Romilly: a life of Sir Samuel Romilly, lawyer and reformer*. London, Collins 1968. p. 89.
92 BGA. A 13. Letter from Lord Henry Petty, Bowood, to Jan Ingen Housz, London, undated.

93 Medd P. *Romilly: a life of Sir Samuel Romilly, lawyer and reformer.* London, Collins 1968. p. 90. We use the term 'college' advisedly. It is not clear whether the Marquis meant, by the term, his close relatives only or the expanded household at Bowood every autumn.
94 Bowood Archives. Sale catalogue of furniture and other valuable effects, Bowood House, 2 September 1805. Elaboratory, p. 44.
95 Britton J. *The Beauties of England and Wales: Wiltshire. Vol. XV.* London, Harris 1814. p. 549.
96 Garthshore M. Biographical account of Dr. Ingenhousz. *Annals of Philosophy 1817.* Vol. 10, pp. 161-165.   Bowood Archives. Sale catalogue of furniture and other valuable effects, Bowood House, 2 September 1805. Elaboratory, p. 44.
97 British Library. Add MS 51685; ff 60,61. Letter from Lord Wycombe, Margate, to the third Lord Holland, London, 15 September 1799.
98 Geneva, Biblioth. Publique & Universitaire. Dumont MSS, F1463. Letter from Étienne Dumont, Bowood, to Samuel Romilly, London, 19 September 1790.
99 BGA. A 13. Letter from Hendrick Hoogeveen, Delft, to Jan Ingen Housz, London, 7 December 1790.
100 BGA. A 13. Letter from Jan Ingen Housz, Bowood, to Anne Garbett, Birmingham, 5 November 1796.
101 Jervis T. *In Literary recollections of Rev. Richard Warner.* London. R Hunter 1831. p. 13.
102 BGA. A 8. Jan Ingen Housz: commonplace book (loose-leaf). p. 33.
103 BGA. A 16a. Jan Ingen Housz: address book in London.
104 Ibid.
105 British Library. Add MS 51967; f. 17. Letter from Jan Ingen Housz, Bath, to Lady Caroline Fox, Warwick, 17 November 1791.
106 Freeman J. *Jane Austen in Bath.* Alton, Mills 1969. p. 4 passim.
107 Hudson K. *The Bath and West. A bicentenary hisory.* Bradford-on-Avon, Moonraker Press 1976. p. 24.
108 Maehle A-H. Dissolving the stone: the search for lithontriptics. *Clio Medical/ The Wellcome Series in the history of medicine.* Vol. 55, 1999. 55-125.
109 Hulme N. *A safe and easy remedy for the relief of the stone and gravel, the scurvy, gout &c.* London, Robinson and Elmsley 1778. Hulme N. (Ingen Housz, J, translator) *Nova, tuta, facilisque methodus curandi calculum, scorbutum, podagram.* Leyden, Luzac and van Damme 1778.
110 HVTS 2246. Letter from Jan Ingen Housz, London, to Jacob van Breda, Delft, 21 June 1791.
111 BGA. A 8. Jan Ingen Housz: commonplace book, loose-leaf, p. 11.
112 Haarlem, Ryksarchief Noord-Holland, 529. Letter from Jan Ingen Housz, Bath, to Martin van Marum, Haarlem, 2 January 1791.
113 BGA. A 13. Letter from Count Orsini-Rosenberg, Vienna, to Jan Ingen Housz, London, 15 May 1790.

Notes to Chapter 19

1 Bowood House Archive. Lansdowne House Dinner Guests January 1788-June 1792.
2 Ibid.
3 Ibid.
4 Royal Society Club. Meeting books 1788-1800. Book 2.
5 BGA. A 13. Jan Ingen Housz; commonplace book, loose-leaf.
6 HVTS 2246. Letter from Jan Ingen Housz, London, to Jacob van Breda, Delft, 25 July 1791.
7 Royal Society Club. Meeting books 1788-1800. Book 2.
8 HRNH. Letter from William Herschel, Slough, to Martin van Marum, Haarlem, 11 February 1791.
9 HRNH. Letter from Jan Ingen Housz, London, to Martin van Marum, Haarlem, 11 March 1791.
10 Royal Society Council Minutes, copy, 1782 – 1810, Vol. 7; p. 237.
11 BGA. A 16a. Jan Ingen Housz; address

book in London.
12. Van Breda (ed). *Scheikundige Bibliotheek (Chemical Library)*, Delft 1792 (Vol. 1) & 1798 (Vol. 2).
13. HVTS 2246. Letter from Jan Ingen Housz, London, to Jacob van Breda, Delft, 23 March 1790.
14. Ingen Housz J. Aqua Mephitica Alcalina of Loogzoutig Luchtzuur Water. Een nieuw ontdekt en uitmundtend geneesmiddel in het graveel en den steen. (Alkaline soda water – a new discovery and an excellent cure for the gravel and the stone.) *Scheikundige Bibliotheek 1792*. Vol. 1, pp. 41-61.
15. Ingen Housz J. Tweede brief van den Heer J. Ingenhousz aan den Heere Deckers betreffende het Loogzoutig Luchtzuur Water. (Second letter from J. Ingenhousz to Mr. Deckers concerning the alkaline soda water.) *Scheikundige Bibliotheek 1792*. Vol. 1, pp. 95-114.
16. Ingen Housz J. Derde brief van den Heer J. Ingenhousz aan den Heere Deckers betreffende het Loogzoutig Luchtzuur Water. (Third letter from J. Ingenhousz to Mr. Deckers concerning the alkaline soda water.) *Scheikundige Bibliotheek 1792*. Vol. pp. 175-193.
17. Lindeboom G. *Dutch Medical Biography*. Amsterdam, Rodopi 1984.
18. Ibid.
19. Snelders H. *De geschiedenis van de scheikunde in Nederland (History of chemistry in the Netherlands), Deel 1*, Delft, Delft University Press 2008.
20. Ingen Housz J. Brief van Heer Johann Schmeisser aan Heer Jan Ingen Housz betreffende het Naphtha Vitrioli Martialis.(Letter from Mr. Johann Schmeisser to Mr. Jan Ingen Housz concerning Naphtha vitrioli martialis.) *Scheikundige* Bibliotheek 1798. Vol. 2, pp. 47-49.
21. Royal Society: Election Certificates, 1793, 20.
22. Ingen Housz J. Brief van den heere J. Ingenhousz and den Heere Jacob van Breda behelzende eenige proeven, met worm-doodende vogten genomen. (Letter of Mr. J. Ingenhousz to Mr. J. van Breda regarding some experiments made with worm-killing fluids.) *Scheikundige Bibliotheek 1798*. Vol. 2, pp. 153-163.
23. Ingen Housz J. Brief van den Heere J. Ingenhousz, aan den Heere Henric. Ferdin. Ingenhousz MD over eene nieuwe manier om kanker en verouderde verzweeringen te geneezen. (A letter from Mr. J. Ingenhousz to Mr. Hendrick Ferdinand Ingenhousz concerning a new way to heal cancer and inveterate ulcers.) *Scheikundige Bibliotheek 1798*. Vol. 2. pp. 201-208. Personal communication.
24. BGA. A 16a. Jan Ingen Housz; address book in London.
25. Vienna. Österreichische Nationalbibliothek, 6/97-6. Letter from Jan Ingen Housz, London, to the Imperial Librarian, Vienna, 18 April 1791.
26. Russell P. *A treatise of the plague, containing an historical journal and medical account of the plague at Aleppo*. London, Robinson 1791.
27. Vienna. Ost. Nationalbibliothek, 6/97-6. Letter from Jan Ingen Housz, London, to the Imperial Librarian, Vienna, 18 April 1791.
28. Ibid.
29. Breda, Gemeentearchief IV, (Van Hal), 39. Letter from Theophile Barrois, Paris, to Jan Ingen Housz, London, 22 April 1791.
30. British Library. Add MS 33541, f. 258. Letter from Jan Ingen Housz, Covent Garden, to Jeremy Bentham, Hendon, 27 May 1791.
31. Ibid.
32. He was elected on 3 August 1791: Haarlem. Election certificate of Jan Ingen Housz to Hollandsche Maatschappij der Wetenschappen, 3 August 1791. BGA. B 18b.
33. HNRH. Letter from Jan Ingen Housz, London, to Martin van Marum, Haarlem, 22 June 1791. Bierens de Haan J. *De Hollandsche Maatschappij der Wetenschappen 1752-1952*. Haarlem, Tjeenk Willink & Zoon 1970. pp. 335-336 & p. 363.

34 HRNH. Letter from Jan Ingen Housz, London, to Martin van Marum, Haarlem, 22 June 1791.
35 British Library. Add MS 51731, f. 45. Letter from Lady Caroline Fox, Bowood, to third Lord Holland, London, 20 June 1791.
36 HVTS 2246. Letter from Jan Ingen Housz, London, to Jacob van Breda, Delft, 29 May 1791.
37 BGA. A 8. Jan Ingen Housz; commonplace book, loose-leaf, p. 25.
38 Newman A. *The Stanhopes of Chevening*. London, Macmillan 1969. p. 111 passim.
39 Royal Society: Election Certificates 1772.
40 BGA. A 8. Jan Ingen Housz; common place book, loose-leaf, p. 25.
41 Ibid. p. 43.
42 HVTS 2246. Letter from Jan Ingen Housz, London, to Jacob van Breda, Delft, 28 July 1791.
43 BGA. A 8. Jan Ingen Housz; commonplace book, loose-leaf, p. 27.
44 HVTS 2246. Letter from Jan Ingen Housz, London, to Jacob van Breda, Delft, 14 September 1791.
45 BGA. A 8. Jan Ingen Housz; commonplace book, loose-leaf, p. 27.
46 www.delft.digitalstamboom.nl (last accessed 8 April 2011).
47 HVTS 2246. Letter from Jan Ingen Housz, London, to Jacob van Breda, Delft, 14 September 1791.
48 Horace, *Odes, Book 1*, 24.19 – to Virgil.
49 HVTS 2246. Letter from Jan Ingen Housz, London, to Jacob van Breda, Delft, 14 September 1791.
50 Personal communication.
51 HKB 53. Letter from Jan Ingen Housz, London, to Josef Jacquin, Vienna, 21 August 1790.
52 Schofield R. *The enlightened Joseph Priestley*. Pennsylvania, Pennsylvania State University Press. p. 284 passim.
53 HTVS 2246. Letter from Jan Ingen Housz, London, to Jacob van Breda, Delft, 14 September 1791.
54 Perusinus, Germanica Act 4, Scene 4.
55 Geneva. Biblioth. Publique & Universitaire. Dumont MSS, F1463, f. 62. Letter from Etienne Dumont, Bowood, to Samuel Romilly, London, 17 October 1791.
56 British Library. Add MS 51966, ff. 74,75. Letter from Lady Ann Upper Ossory, Oundle, to Lady Caroline Fox, Bowood, 28 October 1791.
57 Bowood Archive, Spool 961. Letter from the first Marquis of Lansdowne, London, to Miss Elizabeth Vernon, Bowood, undated but 1790s era.
58 British Library. Add MS 51967 f. 17. Letter from Jan Ingen Housz, Bath, to Lady Caroline Fox, Bowood, 17 November 1791.
59 BGA. A 13. Letter from the first Marquis of Lansdowne, Bowood, to Jan Ingen Housz, London, 24 June 1793.
60 British Library. Add MS 51966 ff. 74,75. Letter from Lady Ann Upper Ossory, Oundle, to Lady Caroline Fox, Bowood, 28 October 1791.
61 Bowring J.(ed). *The works of Jeremy Bentham*. Edinburgh, Tate, 1843. Vol. 3, p. 265. Letter from Jeremy Bentham, Hendon, to Lady E.G.(?), 27 November 1791.
62 Fitzmaurice E. *Life of William, Earl of Shelburne, afterwards First Marquis of Lansdowne, Vol 3., 1776-1805*. London, Macmillan 1876. p. 447.
63 British Library. Add MS 51967 f. 17. Letter from Jan Ingen Housz, Bath, to Lady Caroline Fox, Bowood, 17 November 1791.
64 Falconer W. *An account of the efficacy of the aqua mephitica alkalina in calculous disorders. Fourth edition*. London, Cadell 1792. Letter from John Ingen-Housz, pp. 132-148.
65 Roy Porter, 'Graham, James (1745–1794)', *Oxford Dictionary of National Biography*, Oxford University Press, 2004; http://www.oxforddnb.com/view/article/11199 (Last accessed 13 April 2011).
66 HVTS 2246. Letter from Jan Ingen Housz, Bath, to Jacob van Breda, Delft, 23 November 1795.
67 HRNH. Letter from Jan Ingen Housz, London, to Martin van Marum, Haarlem, 2 January 1791.
68 Now Glasshouse Street, opening

directly into Piccadilly Circus.
69 Walford E. *Old and new London.* London, Cassel 1891. Vol. 4. p. 235.
70 HKB 43. Letter from Jan Ingen Housz, Bowood, to Count Woronzoff, Vienna, 16 October 1793.
71 www.delft.digitalstamboom.nl (last accessed 8 April 2011).
72 BGA. A 13. Letter from Jan Hoogeveen, Delft, to Jan Ingen Housz, London, 8 October 1792.
73 HKB 38. Letter from Jan Ingen Housz, London, to Agatha Ingen Housz, Vienna, 10 February 1792.
74 Ibid.
75 Ibid.
76 HKB 40. Letter from Jan Ingen Housz, London, to Josef Jacquin, Vienna, 31 April 1792. (sic).
77 Ibid.
78 BGA. A 13. Letter from Josef Jacquin, Vienna, to Jan Ingen Housz, London, 3 April 1790.
79 BGA. A 13. Letter from Jan Ingen Housz, London, to Antoine Lavoisier, Paris, 15 March 1792.
80 Fitzmaurice E. *Life of William, Earl of Shelburne, afterwards First Marquis of Lansdowne, Vol. 3., 1776-1805.* London, Macmillan 1876. p. 505.
81 Schama S. *Citizens.* London, Viking 1989. p. 619.
82 Ibid.
83 HVTS 2246. Letter from Jan Ingen Housz, London, to Jacob van Breda, Delft, 10 June 1792.
84 BGA. A 8. Commonplace book, loose-leaf, pp. 42-46.
85 British Library. Add MS 51967 f. 17. Letter from Jan Ingen Housz, Bath, to Lady Caroline Fox, Bowood, 17 November 1791.
86 Ibid.
87 Jay M. *The atmosphere of heaven.* New Haven and London. Yale University Press 2009. p. 13.
88 BGA. A 8. Commonplace book, loose-leaf, p. 46.
89 Dickinson H. *Matthew Boulton.* Leamington Spa, Tee 1999, p. 142.
90 BGA. A 8. Commonplace book, loose-leaf, p. 46.
91 Uglow J. *The lunar men.* London, Faber 2002.
92 BGA. A 13. Letter from the first Marquis of Lansdowne, Bowood, to Jan Ingen Housz, London, 7 September 1792.
93 HVTS 2246. Letter from Jan Ingen Housz, Bowood, to Jacob van Breda, Delft, 29 September 1792.
94 HVTS 2246. Letter from Jan Ingen Housz, Bowood, to Jacob van Breda, Delft, 5 November 1792.
95 Ibid.
96 HVTS 2246. Letter from Jan Ingen Housz, Bath, to Jacob van Breda, Delft, 25 December 1792.
97 Personal communication.
98 HVTS 2246. Letter from Jan Ingen Housz, Bath, to Jacob van Breda, Delft, 15 November 1792.
99 Schama S. *Citizens.* London, Viking 1989. p. 668.
100 Ibid.
101 Haig W. *William Pitt the younger.* London, Harper Perennial 2004. p. 332.
102 Schama S. *Citizens.* London, Viking 1989. p. 688.
103 HKB 41. Letter from Jan Ingen Housz, London, to Josef Jacquin, Vienna, 15 March 1793.
104 Ibid.
105 Excess body fluid in the tissues – mostly in the abdomen in the case of liver disease.
106 HVTS 2246. Letter from Jan Ingen Housz, London, to Jacob van Breda, Delft, 27 June 1793.
107 HKB 42. Letter from Jan Ingen Housz, London, to Josef Jacquin, Vienna, 27 July 1793.
108 HKB 44. Letter from Jan Ingen Housz, London, to Josef Jacquin, Vienna, 22 October 1793.
109 HKB 42. Letter from Jan Ingen Housz, London, to Josef Jacquin, Vienna, 27 July 1793.
110 Ibid.
111 www.delft.digitalstamboom.nl (last accessed 8 April 2011).
112 BGA. A 13. Letter from Jan Hoogeveen, Delft, to Jan Ingen Housz, London, 8 October 1792.

113 Ibid.
114 Ibid.
115 Ibid.
116 Leonard W. Cowie, 'Parr, Samuel (1747–1825)', *Oxford Dictionary of National Biography*, Oxford University Press, Sept 2004. http://www.oxforddnb.com/view/article/21402 (Last accessed 13 April 2011).
117 Ibid.
118 Venn J (ed.) *Alumni Cantabrigienses Part II, 1752-1900.* Vol. II. p. 460.
119 HVTS 2246. Letter from Jan Ingen Housz, London, to Jacob van Breda, Delft, 23 April 1793.
120 BGA. A 13. Letter from Jan Ingen Housz, London, to Jan Hoogeveen, Delft, 5 May 1793.
121 Ibid.
122 BGA. A 13. Letter from the first Marquis of Lansdowne, Bowood, to Jan Ingen Housz, London, 24 June 1793.
123 Ibid.
124 BGA. A 13. Letter from Jan Ingen Housz, London, to the first Marquis of Lansdowne, Bowood 26 June 1793.
125 Ibid.
126 HVTS 2246. Letter from Jan Ingen Housz, London, to Jacob van Breda, Delft, 10 September 1793.
127 HKB 42. Letter from Jan Ingen Housz, 'in the country', to Josef Jacquin, Vienna, 27 July 1793.
128 Ibid.
129 HVTS 2246. Letter from Jan Ingen Housz, London, to Jacob van Breda, Delft, 10 September 1793.
130 Darwin E. *The botanic garden; a poem, in two parts. Part 1: The economy of vegetation. Part 2: The loves of plants.* London, Johnson 1791.
131 HKB 44. Letter from Jan Ingen Housz, 'in the country', to Josef Jacquin, Vienna, 22 October 1793
132 Ibid.
133 Porter R. *Doctor of society. Thomas Beddoes and the sick trade in late-enlightenment England.* London, Routledge 1992. p. 17.
134 Ibid. p.13
135 Jay M. *The atmosphere of heaven.* New Haven and London, Yale University Press 2009. p. 161.
136 Porter R. *Doctor of society. Thomas Beddoes and the sick trade in late-enlightenment England.* London, Routledge 1992. p. 15.
137 Jay M. *The atmosphere of heaven.* New Haven and London, Yale University Press 2009. p. 79.
138 Beddoes T. *Observations on the nature and cure of calculus, sea scurvy, consumption, catarrh and fever.* London, Murray 1793.
139 BGA. A 13. Letter from Thomas Beddoes, Hotwells, to Jan Ingen Housz, London, 8 June 1794.
140 BGA. A 13. Letter from Thomas Beddoes, Hotwells, to Jan Ingen Housz, London, 22 December (probably) 1794.

Notes to Chapter 20

1 Gest H. A misplaced chapter in the history of photosynthesis research: the second publication (1796) on plant processes by Dr. Jan Ingen-Housz MD, discoverer of photosynthesis. *Photosynthesis Research 1997*; 53, 65-72.
2 HKB 53. Letter from Jan Ingen Housz, London, to Josef Jacquin, Vienna, 16 January 1794.
3 Ibid.
4 Ibid.
5 Sargent M. (Ed.) *The outline of music.* Feltham, Middx., Newnes Books 1968. p. 161 passim.
6 Ibid.
7 BGA. A 31. Henry Ferdinand Ingen Housz, certificate of graduation, MD, from Leiden University, 26 September 1793.
8 HVTS 2246. Letter from Jan Ingen Housz, London, to Jacob van Breda, Delft, 21 February 1794.
9 HKB 53. Letter from Jan Ingen Housz, London, to Josef Jacquin, Vienna, 27 July 1793.
10 BGA. A 13. Letter from Johann Scherer, Vienna, to Jan Ingen Housz, London, 7 April 1794.
11 Ibid.
12 BGA. A 13. Letter from Josef Jacquin, Vienna, to Jan Ingen Housz, London, 8

April 1794.
13 BGA. A 13. Letter from Johann Scherer, Vienna, to Jan Ingen Housz, London, 7 April 1794.
14 Ibid.
15 Ingen Housz J. Brief van den heere J. Ingenhousz and den Heere Jacob van Breda behelzende eenige proeven, met worm-doodende vogten genomen. *Scheikundige Bibliotheek 1798*. Vol. 2, pp. 153-163.
16 HVTS 2246. Letter from Jan Ingen Housz, London, to Jacob van Breda, Delft, 16 May 1794.
17 BGA. A 8. Jan Ingen Housz, commonplace book, loose-leaf, p. 133.
18 Poirier J-P. *Lavoisier: chemist, biologist, economist*. Philadelphia, University of Pennsylvania Press 1996. p. 381.
19 BGA. A 8. Jan Ingen Housz, commonplace book, loose-leaf, p. 133.
20 Ingen Housz J. (Scherer J, ed) *Miscellanea Physico-medica (A miscellany of medical treatments)*. Vienna, Patzowsky 1795.
21 BGA. A 13. Letter from Jan Ingen Housz, London, to Sir John Sinclair, London, 2 December 1794.
22 BGA. A 13. Letter from Jean Henri Hassenfratz, Cambrai, to Jan Ingen Housz, Paris, 26 September 1788.
23 BGA. A 13. Letter from Jan Ingen Housz, London, to Sir John Sinclair, London, 2 December 1794.
24 Ibid.
25 HRNH. Letter from Jan Ingen Housz, London, to Martin van Marum, Haarlem, 16 January 1795.
26 Ibid.
27 Schama S. *Patriots and Liberators: revolution in the Netherlands 1780-1813*. London, Fontana 1992. p. 191.
28 BGA. A 13. Letter from Josef Jacquin, Vienna, to Jan Ingen Housz, London, 28 February 1795.
29 Ibid.
30 HKB 51. Letter from Jan Ingen Housz, London, to Josef Jacquin, Vienna, 8 March 1795.
31 Ingen Housz J. (Scherer J, ed) *Miscellanea Physico-medica (A miscellany of medical treatments)*. Vienna, Patzowsky

1795.
32 Ibid. p. 163.
33 HKB 52. Letter from Jan Ingen Housz, London, to Josef Jacquin, Vienna, 2, December 1795.
34 HKB 53. Letter from Jan Ingen Housz, London, to Josef Jacquin, Vienna, 7 March 1796.
35 HKB 52. Letter from Jan Ingen Housz, London, to Josef Jacquin, Vienna, 2, December 1795.
36 HKB 53. Letter from Jan Ingen Housz, London, to Josef Jacquin, Vienna, 7 March 1796.
37 BGA. B 8. Letter from Josef Jacquin, Vienna, to Jan Ingen Housz, London, 20 May 1796.
38 Mitchinson R. *Agricultural Sir John: the life of Sir John Sinclair of Ulbster*. London, Geoffrey Bles 1962. p. 120.
39 Ibid. p. 112.
40 Ibid. p. 146 passim.
41 Ibid. p. 139.
42 Ibid. p. 140.
43 HRNH. Letter from Jan Ingen Housz, London, to Martin van Marum, Haarlem, 16 January 1795.
44 BGA. A 13. Letter from Jan Ingen Housz, London, to Sir John Sinclair, London, 2 December 1794.
45 HRNH. Letter from Jan Ingen Housz, London, to Martin van Marum, Haarlem, 16 January 1795 .
46 Ibid.
47 Ibid.
48 Ibid.
49 HVTS 2246. Letter from Jan Ingen Housz, London, to Jacob van Breda, Delft, 16 February 1796.
50 Ingen Housz J. *An essay on the food of plants and the renovation of soils. Article no. III in Additional appendix to the outlines of the fifteenth chapter of the proposed general report from the Board of Agriculture: on the subject of manures*. London, Bulmer 1796. p. 17.
51 BGA. A 13. Letter from third Lord Stanhope, Chevening, to Jan Ingen Housz, London, 21 August 1796.
52 Ingen Housz J. *An essay on the food of plants and the renovation of soils. Article no. III in Additional appendix to the outlines of*

the fifteenth cahpter of the proposed general report from the Board of Agriculture: on the subject of manures. London, Bulmer 1796. p. 17.
53 Reed H. Jan Ingenhousz, plant physiologist. *Chronica Botanica 1949*, Vol. II, p. 290. Letter from Jan Ingen Housz, Bowood, to first Marquis of Lansdowne, London, 1 December 1795.
54 BGA. A 8. Jan Ingen Housz. Handwritten manuscript, 'On manures'.
55 Bath, Bath University Library. *Agricultural surveys Vol. 11, 1795*. Ingen Housz J. The chemistry of plants.
56 Actually, we now know the bulk to be provided by marine phytoplankton.
57 Bath, Bath University Library. *Agricultural surveys Vol. 11, 1795*. Ingen Housz J. The chemistry of plants. p. 7.
58 BGA. A 8. Jan Ingen Housz. Handwritten manuscript, 'On manures'. p. 10.
59 Ingen Housz J. *An essay on the food of plants and the renovation of soils. Article no. III in Additional appendix to the outlines of the fifteenth chapter of the proposed general report from the Board of Agriculture: on the subject of manures.* London, Bulmer 1796. p. 8.
60 BGA. A 8. Jan Ingen Housz. Handwritten manuscript, 'On manures'. p. 13.
61 Ingen Housz J. Some farther considerations on the influence of the vegetable kingdom on the animal creation. *Philosophical Transactions of the Royal Society 1782*, Vol. 72. pp. 438-9.
62 BGA. A 13. Letter from Jan Ingen Housz, Bath, to Sir John Sinclair, London, 10 January 1796.
63 Personal communication.
64 BGA. A 13. Letter from William Matthews, Secretary of Bath and West of England Society, Bath, to Jan Ingen Housz, Bath, 14 January 1796.
65 Bath, Bath University Library. *Agricultural surveys Vol. 11, 1795*. Ingen Housz J. The chemistry of plants.
66 Virgil. Georgics book 1.
67 BGA. A 13. Letter from Jan Ingen Housz, Bath, to Sir John Sinclair, London, 10 January 1796.
68 HTVS 2246. Letter from Jan Ingen Housz, London, to Jacob van Breda, Delft, 16 February 1796.
69 On the nature of things – Lucretius Book IV.
70 Ingen Housz J. *An essay on the food of plants and the renovation of soils. Article no. III in Additional appendix to the outlines of the fifteenth cahpter of the proposed general report from the Board of Agriculture: on the subject of manures.* London, Bulmer 1796.
71 BGA. A 13. Letter from Sir John Sinclair, London, to Jan Ingen Housz, London, 17 February 1796.
72 BGA A 13. Letter from Sir John Sinclair, London, to Jan Ingen Housz, London, 24 February 1796.
73 HKB 53. Letter from Jan Ingen Housz, London, to Josef Jacquin, Vienna, 7 March 1796
74 Ibid.
75 HVTS 2246. Letter from Jan Ingen Housz, London, to Jacob van Breda, Delft, 4 May 1796.
76 BGA. A 13. Letter from third Lord Stanhope, Chevening, to Jan Ingen Housz, London, 21 August 1796.
77 BGA. A 13. Letter from Jan Ingen Housz, London, to Ramsay Robinson, Kew, 4 July 1796.
78 Gest H. A misplaced chapter in the history of photosynthesis research: the second publication (1796) on plant processes by Dr. Jan Ingen-Housz MD, discoverer of photosynthesis. *Photosynthesis Research 1997*. Vol. 53, pp. 65-72.
79 HKB 53. Letter from Jan Ingen Housz, London, to Josef Jacquin, Vienna, 7 March 1796.
80 Ibid.
81 HRNH. Letter from Jan Ingen Housz, London, to Martin van Marum, Haarlem, 28 March 1796.
82 BGA. A 13. Letter from Johann Deckers, 's-Hertogenbosch, to Jan Ingen Housz, London, 10 June 1796.
83 BGA. B 8. Letter from Josef Jacquin, Vienna, to Jan Ingen Housz, London, 20 May 1796.
84 British Library, Add MS 51821, ff.

61,62. Letter from Jan Ingen Housz, London, to first Marquis of Lansdowne, Bowood, 29 July 1797.
85 BGA. A 13. Letter from Jan Ingen Housz to officers in his London parish regarding income tax, circa 1798.
86 BGA. A 13. Letter from Dr. Samuel Parr, Cambridge, to Jan Ingen Housz, London, 4 September 1794.
87 BGA. A 13. Letter from Dr. William Pearce, Cambridge, to Jan Ingen Housz, London, 31 December 1794.
88 BGA. A 13. Letter from Dr. Samuel Parr, Cambridge, to Jan Ingen Housz, London, 4 September 1794.
89 Ibid.
90 BGA. A 13. Letter from Dr. Samuel Parr, Cambridge, to Jan Ingen Housz, London, 29 December 1794.
91 Ibid.
92 BGA. A 13. Letter from Dr. William Pearce, Cambridge, to Jan Ingen Housz, London, 31 December 1794.
93 BGA. A 13. Letter from Dr. William Pearce, Cambridge, to Jan Ingen Housz, London, 16 December 1798.
94 BGA. A 13. Letter from Dr. Samuel Parr, Cambridge, to Jan Ingen Housz, London, 18 January 1795.
95 BGA. A 13. Letter from Jan Hoogeveen, Delft, to Jan Ingen Housz, London, 6 March 1798.
96 Cambridge University Library. Reports of the Syndics of the Cambridge University Press. Printing Office Records 1798: 29 June.
97 Cambridge University Library. Reports of the Syndics of the Cambridge University Press. Minutes 1801: 2 November.
98 BGA. A 15. Letter from Jan Ingen Housz, London, to Arnold Ingen Housz, Breda, 10 June 1796.
99 BGA. B 7a. Letter from Jan Ingen Housz, London, to Stametz Bros., Vienna, 24 July 1798.
100 Josey D. *Broome Park, Betchworth, Surrey*. Sutton, Surrey, Spring Gardens Trust 2005. p. 7.
101 BGA. A 13. Letter from James Petty, Betchworth, to Jan Ingen Housz, London, 26 February 1797.
102 HRNH. Letter from Jan Ingen Housz, London, to Martin van Marum, Haarlem, 7 July 1796.
103 BGA. A 13. Letter from Sir John Sinclair, Whitehall, to Jan Ingen Housz, London, 28 September 1796.
104 Ibid.
105 Sinclair J. (Ed.). *Hints on vegetation and questions regarding the nature and principles thereof.* London, McMillan 1796.
106 Medd P. *A life of Sir Samuel Romilly, lawyer and reformer.* London, Collins 1968. p. 104.
107 Ibid. p. 105.
108 BGA. A 13. Letter from Jan Ingen Housz, Bowood, to Anne Garbett, Birmingham, 5 November 1796.
109 Medd P. *A life of Sir Samuel Romilly, lawyer and reformer.* London, Collins 1968. p. 90.
110 Fitzmaurice E. *Life of William, Earl of Shelburne, afterwards First Marquis of Lansdowne, Vol 3., 1776-1805*. London, Macmillan 1876. p. 447.   Xanthippe was the wife of Greek philosopher, Socrates. She had a reputation (unjustifiable according to many) of being a domineering hag.
111 Medd P. *A life of Sir Samuel Romilly, lawyer and reformer.* London, Collins 1968. p. 109.
112 Ibid. p. 107.
113 BGA. A 13. Jan Ingen Housz: advise to a young nobleman cursorily written out, October 1796.
114 BGA. A 13. Letter from Jan Ingen Housz, Bowood, to Anne Garbett, Birmingham, 5 November 1796.
115 BGA. A 13. Letter from Jan Ingen Housz, Bowood, to Nathaniel Dimsdale, London, 9 November 1796.
116 British Library, Add. Mss. 51734, ff. 182,183. Letter from Lady Caroline Fox, Bowood, to third Lord Holland, London, 6 October 1797.
117 Geneva. Biblioth. Publique & Universitaire. Dumont Mss, F1463, ff. 44,45. Letter from Etienne Dumont, Bowood, to Samuel Romilly, London, 19 September 1791.
118 Bowood Archives, 962,114. Letter from Lord Henry Petty, Edinburgh, to

first Marquis of Lansdowne, Bath, 24 December 1796.
119 Jervis T. In *Literary recollections of Rev. Richard Warner*. London. R Hunter 1831. p. 13.
120 BGA. A 13. Jan Ingen Housz: commonplace book (loose-leaf). p. 33.
121 BGA. A 11. Handwritten note of Jan Ingen Housz entitled '*Casus phthiseos pulmonaris curata*. October 1796 at Bowood.' – passim.
122 Wiltshire History Centre 2083/20. St Mary the Virgin, Calne: burial record 1799.
123 Jay M. *The atmosphere of heaven*. New Haven and London, Yale Univeristy Press 2009. p. 98.
124 British Library, Add MS 51821, ff. 61,62. Letter from Jan Ingen Housz, Wandsworth, to first Marquis of Lansdowne, Bowood, 29 July 1797.
125 Jay M. *The atmosphere of heaven*. New Haven and London, Yale University Press 2009. p. 98.
126 British Library, Add MS 51821, ff. 61,62. Letter from Jan Ingen Housz, Wandsworth, to first Marquis of Lansdowne, Bowood, 29 July 1797.
127 HKB 54. Letter from Jan Ingen Housz, London, to Josef Jacquin, Vienna, 5 February 1797.
128 Alexander J. *Catherine the Great: life and legend*. Oxford, Oxford University Press 1989. p. 324.
129 HKB 54. Letter from Jan Ingen Housz, London, to Josef Jacquin, Vienna, 5 February 1797.
130 BGA. A 13. Letter from Josef Jacquin, Vienna, to Jan Ingen Housz, London, 20 May 1797.
131 Ibid.
132 HKB 56. Letter from Jan Ingen Housz, London, to Josef Jacquin, Vienna, 26 February 1797.
133 HKB 57. Letter from Jan Ingen Housz, London, to Josef Jacquin, Vienna, 23 November 1797.
134 HKB 55. Letter from Jan Ingen Housz, London, to Josef Jacquin, Vienna, 18 February 1797.
135 HKB 56. Letter from Jan Ingen Housz, London, to Josef Jacquin, Vienna, 26 February 1797.
136 BGA. A 13. Letter from Jan Ingen Housz, London, to James Petty, Betchworth, 28 February 1797.
137 Darwin E. *Zoonomia: the laws of organic life. Volume 2*. London, Johnson 1796.
138 HVTS 2246. Letter from Jan Ingen Housz, London, to Jacob van Breda, Delft, 24 February 1797.
139 HRNH. Letter from Jan Ingen Housz, London, to Martin van Marum, Haarlem, 28 February 1797.
140 BGA. A 13. Letter from Jan Ingen Housz, London, to James Petty, Betchworth, 28 February 1797.
141 HVTS 2246. Letter from Jan Ingen Housz, London, to Jacob van Breda, Delft, 24 February 1797.
142 Ibid.
143 HRNH. Letter from Jan Ingen Housz, London, to Martin van Marum, Haarlem, 28 February 1797.
144 Ibid.
145 HRNH. Letter from Jan Ingen Housz, London, to Martin van Marum, Haarlem, 25 June 1797.
146 London, Royal Society Certificates of Election, EC/1798/06.
147 Desmond R. *Kew: the history of the Royal Botanic Gardens*. London, Harvill Press 1995. p. 1 passim.
148 BGA. A 13. Letter from Jan Ingen Housz, London, to Alexander Ramsay Robinson, Kensington, 4 July 1796.
149 HVTS 2246. Letter from Jan Ingen Housz, London, to Jacob van Breda, Delft, 25 March 1797.
150 Ibid.
151 BGA. A 11. Jan Ingen Housz: various manure experiments.
152 Ibid.
153 Wilson D. *The British Museum: a history*. London, British Museum Press 2002. p. 29.
154 Peacock G. *Miscellaneous works of the late Thomas Young MD, FRS. Vol. 2*. London, John Murray 1855, p. 502.
155 Breda. Breda Gemeentearchief. Van Hal collection no. 40. Letter from first Marquis of Lansdowne, Bowood, to Jan Ingen Housz, London, 5 July 1797.
156 British Library, Add. MSS 51821 ff

61,62. Letter from Jan Ingen Housz, London, to first Marquis of Lansdowne, Bowood, 29 July 1797.
157 BGA. A 16a. Jan Ingen Housz: address book in London.
158 British Library, Add. MSS 51821 ff 61,62. Letter from Jan Ingen Housz, London, to first Marquis of Lansdowne, Bowood, 29 July 1797.
159 BGA. A 13a. Letter from William Herschel, Slough, to Jan Ingen Housz, London, 29 July 1797.
160 Keppel S. *The sovereign lady.* London, Hamish Hamilton 1974. p. 62 passim.
161 BGA. A 4. Book of prescriptions copied from Sir John Pringle.
162 Bowood Archives: 'Dr. Ingenhousz and his Austrian servant' (soft ground etching, pencil original).
163 Ibid.
164 British Library. Add. MS 51734, ff. 184,185. Letter from Lady Caroline Fox, Bowood, to third Lord Holland, Holland House, 29 October 1797.
165 HKB 57. Letter from Jan Ingen Housz, London, to Josef Jacquin, Vienna, 23 November 1797.
166 Ibid.
167 British Library. Add. Mss. 51734, fol. 194. Letter from Lady Caroline Fox, London, to third Lord Holland, Holland House, 12 November 1797.
168 British Library. Add. Mss. 51821, fol. 85. Letter from Jan Ingen Housz, London, to third Lord Holland, Holland House, 24 December 1797.
169 British Library. Add. Mss. 51845, fol. 190. Letter from Jan Ingen Housz, Wandsworth, to third Lord Holland, Holland House, 6 May 1798.

Notes to Chapter 21

1 Haig W. *William Pitt the younger.* London, Harper Perennial 2005. p. 433.
2 BGA. A 13. Letter from Jan Ingen Housz to officers in his London parish regarding income tax, circa 1798.
3 BGA. B 5. Letter from Agatha Ingen Housz, Vienna, to Jan Ingen Housz, London, 21 November 1798.
4 Islington, London. The archived ledgers of Drummonds Bank at the Royal Bank of Scotland. DR/427/158, 1430.
5 BGA. A 13. Letter from Jan Ingen Housz to officers in his London parish regarding income tax, circa 1798.
6 BGA. B 6. Letter from Jan Ingen Housz, Bowood, to Josef Jacquin, Vienna, 1 June 1799.
7 BGA. A 13. Letter from Nikolaus Jacquin, Vienna, to Jan Ingen Housz, London, 9 January 1798.
8 BGA. A 13. Letter from Agatha Ingen Housz, Vienna, to Jan Ingen Housz, London, 9 January 1798.
9 Ibid.
10 BGA. B 6. Letter from Jan Ingen Housz, Bowood, to Josef Jacquin, Vienna, 1 June 1799.
11 Ibid. BGA. B 4. Letter from Jan Ingen Housz, London, to Agatha Ingen Housz, Vienna, 24 July 1798.
12 BGA. A 13. Letter from Josef Jacquin, Vienna, to Jan Ingen Housz, London, 27 August 1798.
13 BGA. A 13. Letter from Count de Sauran, Vienna, to Jan Ingen Housz, London, 21 August 1798.
14 BGA. B 5. Letter from Agatha Ingen Housz, Vienna, to Jan Ingen Housz, London, 24 August 1798.
15 Jenner E. *An inquiry into the causes and effects of the variolae vaccinae.* London, Sampson Low, 1798.
16 Ibid.
17 Fisher R. *Edward Jenner (1749–1823).* London, Andre Deutsch 1991. p. 12 passim.
18 Jenner E. *Further observations on the variolae vaccinae, or cow pox.* London, Sampson Low 1799.
19 Royal Society of London, Certificates of Election V, 112, 1789.
20 Hunter J. *Letters from the past: from John Hunter to Edward Jenner.* London, Royal College of Surgeons 1976.
21 Fisher R. *Edward Jenner (1749–1823).* London, Andre Deutsch 1991. pp. 66–7. The folklore that cowpox protected against smallpox was long established. The theory had been tested as early as 1774 by Jesty, a farmer in Dorset who

'vaccinated' his wife and child during a smallpox epidemic but there can be no doubt that it was Jenner who first applied critical observation to the theory.
22. Baxby D. Edward Jenner's unpublished cowpox Inquiry and the Royal Society: Everard Home's report to Sir Joseph Banks. *Med. Hist.*, 1999, Vol. 43, pp. 108–10.
23. Ibid.
24. Jenner E. *An inquiry into the causes and effects of the variolae vaccinae.* London, Sampson Low 1798.
25. Fisher R. *Edward Jenner (1749–1823).* London, Andre Deutsch 1991. p. 75.
26. BGA. B 4. Letter from Jan Ingen Housz, London, to Agatha Ingen Housz, Vienna, 24 July 1798.
27. Greenwich. Caird Library, National Maritime Museum, MS80/031. Sir William Herschel, C Herschel: Visitors Book (1783–1846).
28. British Library. Add. Mss. 51966, f. 88. Letter from Lady Upper Ossory, Oundle, to Lady Caroline Fox, Bowood, 4 Oct. 1798.
29. BGA. A 13. Letter from Lady Caroline Fox, 77 Upper Guildford Street, London, to Jan Ingen Housz, 35 Marylebone Street, London, 31 Dec. 1798.
30. Breda Gemeentearchief IV, (Van Hal), 5–38, 55. Letter from Jan Ingen Housz, Bowood, to Edward Jenner, Berkeley, 12 Oct. 1798.
31. Henry Stiles had been born at Whitcombe Farm, Hilmarton, near Calne, Wiltshire in September 1745. From about 1783 he farmed at Whitley Farm in Bremhill, a nearby parish that neighboured Bowood. It must have been, therefore, at Whitcombe that Allsup had inoculated the 23-year old Stiles—in December 1768. His father, Samuel Stiles, was buried in Compton Bassett, the parish immediately to the east of Hilmarton, on 4 February 1769.
32. Breda Gemeentearchief IV, (Van Hal), 5–38, 55. Letter from Jan Ingen Housz, Bowood, to Edward Jenner, Berkeley, 23 October 1798.
33. Warren Hastings, retired Governor-General of India.
34. Horace (BC 65–8), *The art of poetry,* lines 352–3. Rushton Fairclough H. *Horace: satires, epistles and ars poetica.* London, Heinemann 1926, pp. 478–9.
35. Saunders P. *Edward Jenner, the Cheltenham years, 1753–1823.* Hanover NH, and London, University Press of New England 1982. p. 26.
36. Ibid.
37. Baron J, *The life of Edward Jenner, M.D., LL.D., F.R.S.* London, Henry Colburn 1838. Vol. 1, pp. 293–5.
38. Ibid.
39. Ibid.
40. Breda, Gemeentearchief IV, (Van Hal), 5–38. Letter from Edward Jenner, Cheltenham, to Jan Ingen Housz, London, (undated).
41. Van der Pas P. The Ingenhousz–Jenner correspondence. *Janus,* 1964, Vol. 51, pp. 217–18.
42. Baron J, *The life of Edward Jenner, M.D., LL.D., F.R.S.* London, Henry Colburn 1838. Vol. 1, pp. 298.
43. Ibid.
44. Gemeentearchief, Breda IV, (Van Hal), 5, 38. Letter from Jan Ingen Housz, London, to Edward Jenner, Berkeley, 20 Dec. 1798.
45. Jenner E. *Further observations on the variolae vaccinae, or cow pox.* London, Sampson Low 1799.
46. Beale N, Beale E. Evidence-Based Medicine in the Eighteenth Century: the Ingen Housz–Jenner Correspondence Revisited. *Medical History 2005,* Vol. 49, pp. 79–98.
47. Jenner E. *The origin of vaccine inoculation.* London, Shury 1801, p. 8.
48. Fenner F, Henderson D, Arita I, Ježek, Z, Ladnyi I. *Smallpox and its eradication.* Geneva, World Health Organisation 1988. p. 1062.
49. Bowood. Shelburne MSS 3,44. Letter from first Marquis of Lansdowne, London, to Lord Henry Petty, Cambridge, 11 March 1799.
50. Finally published as: Hoogeven H. *Dictionarium Analogicum Linguae Graecae.* Cambridge, Cambridge University

Press 1801.
51 BGA. A 13. Letter from John Cuthbertson, London, to Jan Ingen Housz, London, 15 February 1799. BGA. A 8. Jan Ingen Housz: commonplace book, loose-leaf 1790-1797. p. 172.
52 British Library, Add. Mss. 51967, ff. 64,65. Letter from Jan Ingen Housz, London, to Lady Caroline Fox, London, 3 January 1799.
53 Ibid.
54 HRNH. Letter from Jan Ingen Housz, Bowood, to Martin van Marum, Haarlem, 1 August 1799.
55 BGA. B 6. Letter from Jan Ingen Housz, Bowood, to Josef Jacquin, Vienna, 1 June 1799.
56 British Library, Add. Mss. 51821. f. 218. Letter from Étienne Dumont, Bowood, to third Lord Holland, before September 1799.
57 Bowood. Shelburne MSS 3,44. Letter from first Marquis of Lansdowne, London, to Lord Henry Petty, Cambridge, 11 March 1799.
58 Home Secretary in the Pitt administration.
59 BGA. B - 6. Letter from Jan Ingen Housz, Bowood, to Josef Jacquin, Vienna, 1 June 1799. A Heller was a small Viennese coin of little value.
60 Ibid.
61 BGA. A 13. Letter from first Marquis of Lansdowne, Bowood, to Arnold Ingen Housz, Breda, 9 September 1799.
62 Bath Record Office: Catalogue of Leases for Bath City, 793.
63 BGA. A 13. Letter from first Marquis of Lansdowne, Bowood, to Arnold Ingen Housz, Breda, 9 September 1799.
64 BGA. A 8. Jan Ingen Housz: commonplace book, loose-leaf 1790-1798. p. 164.
65 BGA. A 13. Letter from Jan Ingen Housz, Bowood, to William Falconer, Bath, 24 August 1799.
66 Fitzmaurice E. *Life of William, Earl of Shelburne: afterwards first Marquis of Lansdowne. Volume 3, 1776-1805.* London, Macmillan 1876. p. 449.
67 BGA. A 13. Letter from first Marquis of Lansdowne, Bowood, to Arnold Ingen Housz, Breda, 9 September 1799.
68 BGA. B 6. Letter from Jan Ingen Housz, Bowood, to Josef Jacquin, Vienna, 1 June 1799.
69 British Library. Add. Mss. 51950, f. 7. Dinner Books for Holland House, Vol. 1; 1799. The book also shows that Ingen Housz slept at Holland House those two nights.
70 Had taken over the tailoring business and premises from the Williams family sometime before January 1796. (Islington, London. The archived ledgers of Drummonds Bank at the Royal Bank of Scotland. DR/427/158, 1431.)
71 British Library. Add Mss. 51744, ff. 109-110. Letter from Lady Caroline Fox, Bowood, to third Lady Holland, Holland House, 23 July 1799.
72 Battle of the Trebbia, 17-19 June 1799.
73 BGA. A 4. Jan Ingen Housz, book of prescriptions copied from Sir John Pringle.
74 British Library. Add. Mss. 51735, f. 119. Letter from Lady Caroline Fox, Bowood, to third Lord Holland, London, 3 August 1799.
75 HVTS 2247. Letter from Jacob van Breda, Delft, to Jan Ingen Housz, Bowood, 17 July 1799.
76 BGA. B 5. Letter from Agatha Ingen Housz, Vienna, to Jan Ingen Housz, Bowood, 13 August 1799.
77 Ilchester, Earl of. (Ed). *Journal of Elizabeth, Lady Holland. Volume 2: 1799-1811.* London, Longmans 1908. p. 11.
78 Medd P. *Romilly: a life of Sir Samuel Romilly.* London, Collins 1968. p. 111.
79 Geneva. Biblioth. Publique & Universitaire. Dumont MSS, F1463. Letter from Étienne Dumont, Bowood, to Samuel Romilly, London, 7 September 1799.
80 British Library. Add. Mss. 51744, fol.115. Letter from Lady Caroline Fox, Bowood, to third Lady Holland, Saltram, 8 September 1799.
81 BGA. A 13. Letter from first Marquis of Lansdowne, Bowood, to Arnold Ingen

Housz, Breda, 9 September 1799.
82 BGA. B 18b. The will of Jan Ingen Housz in London, 10 June 1796.
83 Beale N, Beale E. Looking for Dr. Ingen Housz: the evidence for the site and nature of the burial of the famous Dutch physician and scientist. *Wiltshire Archeological and Natural History Magazine (2000)*; Vol 92, pp. 120-130.
84 Ibid.
85 Litten J. *The English way of death*. London, Robert Hale 1991. p. 101.
86 BGA. A 13. Letter from first Marquis of Lansdowne, Bowood, to Arnold Ingen Housz, Breda, 9 September 1799.
87 Ibid.
88 *Gentleman's Magazine October 1799*, Vol. 69. pp. 900-901.
89 Bath. Bath Library, The Podium. *The Bath Chronicle* of Thursday 19 September 1799.
90 Wiltshire History Centre 2083/20. St Mary the Virgin, Calne: burial record 1799.
91 Marsh A. *A history of the borough and town of Calne*. Calne, Heath 1903. p. 144.
92 BGA. A 13. Letter from Drummonds, bankers in London, to Arnold Ingen Housz, Breda, 12 September 1799.
93 Marsh A. *A history of the borough and town of Calne*. Calne, Heath 1903. p. 364.
94 Beale N. *Is that the doctor? A history of the Calne GPs*. Bradford on Avon. Ex Libris Press 1998. p. 21.
95 British Library. Add. Mss. 51965, ff. 5,6. Letter from Mrs. Anne Romilly, London, to Lady Caroline Fox, Bowood, 16 September 1799.
96 British Library. Add. Mss. 51744, ff. 119-120. Letter from third Lady Holland, Saltram, to Lady Caroline Fox, Bowood, 16 September 1799.
97 British Library. Add. Mss. 51735, ff. 140-141. Letter from Lady Caroline Fox, Bowood, to third Lord Holland, Saltram, 16 September 1799.

98 Ibid.
99 WJ. pp. 236-237.
100 BGA. A 13. Letter from the first Marquis of Lansdowne, Bowood, to Dominque Tede, Bowood, as a testimonial, 10 October 1799.
101 Islington, London. The archived ledgers of Drummonds Bank at the Royal Bank of Scotland. DR/427/158, 1430.
102 Ibid.
103 's-Hertogenbosch, Rijksarchief in Noord-Brabant. Cat. Family van Lanschot, 1262.
104 *Nederland's Patriciaat. 59e Jaargang 1973*. The Hague, Centraal bureau voor Genealogie, p. 112.
105 Rehm G. *De Bredase apothekers van de 15e tot het begin van de 19e eeuw. (The apothecaries of Breda from the 15th century to the beginning of the 19th century.)* Breda, Geschied- en Oudheidkundige Kring van Stad en Land van Breda 'De Oranjeboom' 1961. p. 104.
106 BGA. A 18. The Hague. Passport issued to Petrus Ingen Housz, 8 October 1799.
107 Ibid.
108 BGA. A 13. Letter from first Marquis of Lansdowne, Bowood, to Arnold Ingen Housz, London, 28 November 1799.
109 Ibid.
110 BGA. A 18. The Hague. Passport issued to Petrus Ingen Housz, 8 October 1799.
111 Rehm G. *De Bredase apothekers van de 15e tot het begin van de 19e eeuw. (The apothecaries of Breda from the 15th century to the beginning of the 19th century.)* Breda, Geschied- en Oudheidkundige Kring van Stad en Land van Breda 'De Oranjeboom' 1961. p. 104.
112 Vienna. Stadt/Landesarchiv, 12505. p. 64.
113 Ibid.
114 Ibid.
115 Vienna. Imperial & National Court. Ingen Housz will and probate documents at Vienna, 2318 (1799).
116 Ibid.

Ingen Housz Publications

(in chronological order and including translations of his works).

Ingen Housz J. *Lettre de Monsieur Ingen Housz, Docteur in Medicine, à Monsieur Chais, Pasteur de L'Église Wallonne de la Haye, au sujet d'une brochure contenant sa lettre à Monsieur Sutherland, et une réponse de Monsieur Sutherland à Monsieur Chais, sur la nouvelle méthode d'inoculer la petite vérole. (Letter from Mr. Ingen Housz, Doctor in Medicine, to Mr. Chais, Pastor of the Walloon Church of The Hague on the subject of a brochure containing his letter to Mr. Sutherland, and a response of Mr. Sutherland to Mr. Chais, on the new method of inoculating the small pox.)* Amsterdam, van Harrevelt 1768.

Ingen Housz J. Extract of a letter to Sir John Pringle, Bart. P.R.S. containing some experiments on the Torpedo, made at Leghorn, January 1, 1773 *Philosophical Transactions of the Royal Society 1775.* Vol. 65, pp. 1-4.

Ingen Housz J. Easy methods of measuring the diminution of bulk, taking place upon the mixture of common air and nitrous air; together with experiments on platina. *Philosophical Transactions of the Royal Society 1776.* Vol. 66, pp. 257-267.

Hulme N. (Ingen Housz J, translator). *Nova, tuta, facilisque methodus curandi calculum, scorbutum, podagram, ci addita est methodus extemporanea impregnandi aquam aliosque liquores aëre fixo. (A safe and easy remedy proposed for the relief of the stone and gravel, the scurvy and gout, and for the destruction of worms in the human body, together with an extemporaneous method of impregnating water with fixed air).* Leiden, Luzac and van Damme 1778.

Ingen Housz J. A ready way of lighting a candle, by a very moderate electrical spark. *Philosophical Transactions of the Royal Society 1778.* Vol. 68, pp. 1022-1026

Ingen Housz J. Electrical experiments, to explain how far the phenomena of the electrophorus may be accounted for by Dr. Franklin's theory of positive and negative electricity. (Royal Society Bakerian Lecture 1778). *Philosophical Transactions of the Royal Society 1778.* Vol. 68, pp. 1027-1048.

Hulme N. (Veirac J, translator of the Latin version of Ingen Housz J.) *Nieuwe, veilige en gemaklijke manier om den steen, den scheurbuik, het voeteuvel, enz. te geneezen, en de wormen uit het menschlijk ligchaam te verdrijven. (A safe and easy method for the relief of the stone and gravel, the scurvy and gout, and for the destruction of worms in the human body). Rotterdam, Arrenberg 1778.

Ingen Housz J. *A letter from Dr. Ingenhousz FRS to Dr. Priestley, on the effect of a new species of inflammable air or vapour, 1 March 1779*. In: Priestley J. *Experiments and observations relating to various branches of natural philosophy with a continuation on the observations on air*. London, Johnson 1779.

Ingen Housz J. Account of a new kind of inflammable air or gass, which can be made in a moment without apparatus, and is as fit for explosion as other inflammable gasses in use for that purpose; together with a new theory of gunpowder. *Philosophical Transactions of the Royal Society 1779*. Vol. 69, pp. 376-418.

Ingen Housz J. On some new methods of suspending magnetical needles. *Philosophical Transactions of the Royal Society 1779*. Vol. 69, pp. 537-546.

Ingen Housz J. Improvements in electricity. (Royal Society Bakerian Lecture 1779). *Philosophical Transactions of the Royal Society 1779*. Vol. 69, pp. 661 – 673.

Ingen Housz J. *Experiments upon vegetables; discovering their great power of purifying the air in the sunshine and of injuring it in the shade and at night to which is joined, a new method of examining the accurate degree of salubrity of the atmosphere*. London, Elmsly and Payne 1779.

Ingen Housz J. On the degree of salubrity of the common air at sea, compared with that of the sea-shore, and that of places far removed from the sea. In a letter from John Ingen Housz MD, FRS, to Sir John Pringle, Bart. FRS, dated Paris Jan. 22, 1780. *Philosophical Transactions of the Royal Society 1780*. Vol. 70, pp. 354-377.

Ingen Housz J. Exposition de plusieurs Lois qui paroissent s'observer constatement dans les divers mouvements du fluide électrique et auxquelles les physiciens n'avoient pas fait suffisante attention. (Exposition of the many laws which appear to be seen, consistently, in the many movements of electrical fluid and to which physicists had not paid sufficient attention). *Journal de Physique 1780*. Vol. 16, pp. 117-126.

Ingen Housz J. (Van Breda J, translator). *Proeven op plantgewassen: ontdekkende derzelver zeer aanmerkelyk vermogen om de lucht des Dampkrings te zuiveren, geduurende den Dag, en in de Zonne-schyn; en om gemeene lucht des nachts, en wanneer zy in de schaduw zyn, te bederven; beneevens eene niewe manier om den graad van gezontheid des dampkrings nauwkeurig te toetfen.* (Experiments upon vegetables, discovering their great power of purifying the air in the sunshine and of injuring it in the shade and at night to which is joined a new method of examining the accurate degree of salubrity of the atmosphere). Delft, Van der Zmout & De Groot 1780.

Ingen Housz J. *Expériences sur les végétaux, spécialement sur la propriété qu'ils possèdent a un haut degré, soit d'améliorer l'air quand ils son au soleil, soit de le corrempre la nuit, ou lorsqu'ils sont à l'ombre; auxquelles on a joint une méthode nouvelle de juger du degré de salubrité de l'atmosphère.* (Experiments upon vegetables discovering their great power of purifying the air in the sunshine and of injuring it in the shade and at night to which is joined a new method of examining the accurate degree of salubrity of the atmosphere). Paris, Didot le jeune 1780.

Ingen Housz J. (Probably translated by Molitor N.) *Versuche mit Pflanzen: wodurch entdeckt worden, dass sie die Kraft besitzen, die atmosphärische Luft beim Ssonnenschein zu reinigen, und im Schatten und des Nachts über zu verderben, und nebst einer Methode, die Reinigkeit der Atmosphäre genau abzumessen.* (Experiments upon vegetables discovering their great power of purifying the air in the sunshine and of injuring it in the shade and at night to which is joined a new method of examining the accurate degree of salubrity of the atmosphere). Leipzig, Weygand 1780.

Ingen Housz J (Lippert J. translator). *Neue sichere Methode der Heilung des Steinskorbuts, Podagra und Vernichtung der im menschlichen Körper enststehenden Würmer. Aus dem Englischen des Nath. Hulme ins Lateinische übersetzt von J. Ingen-housz.* (A safe and easy method for the relief of the stone and gravel, the scurvy and gout, and for the destruction of worms in the human body). Wien, Kurzbeck 1781.

Ingen Housz J. (Van Breda J, translator). Verhandelingen over de gedéphlogisteerde lucht, en de manier hoe men dezelve kan bekomen en tot de ademhaaling doen dienen. (Dissertation on the dephlogisticated air and the way to obtain it and make it useful for breathing.). *Verhandelingen van het Bataafsch Genootschap der proefondervindelyke wysbegeerte te Rotterdam 1781.* Vol. 6, pp. 107-159.

Ingen-Housz J. (Molitor N, translator). *Anfangsgründe der Electricität, hauptsächlich in Beziehung auf den Elektrophor; nebst einer leichten Art, vermittelst eines elektrischen Funkens das Licht anzuzünden, und einem Briefe in Betref einer neuen entzündbaren Knalluft, mit Anmerkungen.* (The rudiments of electricity, principally in relation to the electrophore; with a means of making a flame using an electric spark and a letter concerning a new inflammable explosive air, with notes.) Wien, Wappler 1781.

Ingen Housz J. Some farther considerations on the influence of the vegetable kingdom on the animal creation. *Philosophical Transactions of the Royal Society 1782.* Vol.72, pp. 436-439.

Ingen-Housz J. (Molitor N, translator.) *Vermischte Schriften, physisch-medizinisch inhalts.* (Various memoires on physical and medical topics). Wien, Krauss 1782.

Ingen Housz J. Observations sur la vertu de l'eau imprégnée d'air fixe, de différents acides, et de plusieurs autres substances, pour en obtenir, par le moyen des plantes et de la Lumière du soleil, de l'air déphlogistiqué. (Observations on the virtue of water impregnated with fixed air, and with different acids, in obtaining dephlogisticated air by means of plants and sunlight.) *Journal de Physique 1784*. Vol. 24, pp. 337-348.

Ingen Housz J. Réflexions sur l'économie des végétaux. (Reflections on the economy of plants). *Journal de Physique 1784*. Vol. 24, pp. 443-455.

Ingen Housz J. Remarques sur l'origine et la nature de la matière verte de M. Priestley, sur la production de l'air déphrogistiqué par le moyen de cette matière, et sur le changement de l'eau en air déphrogistiqué. (Remarks on the origin and nature of the green material of Mr. Priestley, on the production of dephlogisticated air by this material, and on the changes in water in dephlogisticated air.) *Journal de Physique 1784*. Vol. 25, pp. 3-12.

Ingen Housz J. Lettre de M. Ingen Housz à M. J. van Breda au sujet de la quantité d'air déphlogistiqué que les végétaux répandent dans 1'atmosphère pendant le jour: au sujet des raisons de l'inexactitude de la quantité de l'air déphlogistiqué qu'on obtient par les végétaux, exposés au soleil dans l'eau imbibée d'air fixe, ainsi que sur la véritable cause de l'influence méphetique nocturne des végétaux dans l'air. (Letter from Mr. Ingen Housz to Mr. van Breda on the subject of the quantity of dephlogisticated air that plants release into the atmosphere during the day: on the subject of the reasons for the imprecision of the quantity of dephlogisticated air that one obtains from plants that are exposed to the sun in water impregnated with fixed air, as well as the true cause of the mephitic influence of plants on the air at night.) *Journal de Physique 1784*. Vol. 25, pp. 437-450.

Ingen-Housz J. (Molitor N, translator.) *Vermischte Schriften, physisch-medizinisch inhalts. Erster Band. (Various memoires on physical and medical topics. First volume).* Wien, Wappler 1784.

Ingen-Housz J. (Molitor N, translator.) *Vermischte Schriften, physisch-medizinisch inhalts. Zweiter Band. (Various memoires on physical and medical topics. Second volume).* Wien, Wappler 1784.

Ingen Housz J. *Nouvelles experiences et observations sur divers objets de physique. (New experiments and observations on a variety of topics in physics.)* Paris, Barrois le jeune 1785.

Ingen Housz J. (Van Breda J, translator). *Verzameling van verhandelingen, over verschillende natuurkundige onderwerpen. (Collection of treatises on various topics in physics).* 's-Gravenhage (The Hague), van Cleef 1785.

Ingen Housz J. Observations sur la construction et l'usage de 1'eudiomètre de M. Fontana, et sur quelques propriétés particulières de l'air nitreux, adressée à M.

Dominique Beck.(Observations on the construction and use of the eudiometer of Mr. Fontana, and on some peculiar properties of nitrous air, addressed to Mr. Dominique Beck.) *Journal de Physique 1785.* Vol. 26, pp. 339-359.

Schwankhardt J. (Ingen Housz J). Lettre de M. J. D. Schwankhardt à M. Ehrmann, Professor de physique a Strasbourg, au sujet de l'influence de l'eIectricité sur la végétation. (Letter from Mr. J. D. Schwankhardt on the subject of the influence of electricity on plants.) *Journal de Physique 1785.* Vol. 27, pp. 462-468.

Ingen Housz J. Lettre de M. Ingen Housz à M. Molitor au sujet de l'effet particulier qui ont sur la germination des sémences, et sur l'accroissement des plantes formées, les différentes espèces d'air, les différents degrés de Iumière et de la chaleur et l'électricité. (Letter from Mr. Ingen Housz to Mr. Molitor on the subject of the particular effects, over the germination of seeds and over the growth of seedlings, exerted by different kinds of air, different degrees of light and by heat and electricity.) *Journal de Physique 1786.* Vol. 28, pp. 81-92.

Ingen Housz J. (Scherer J, translator). *Versuche mit Pflanzen, hauptsächlich über die Eigenschaft, welche sie in einem hohen Grade besitzen, die Luft im Sonnenlichte zu reinigen, und in der Nacht und im Schatten zu verderben; nebst einer neuen Methode, den Grad der Reinheit und Heilsamkeit der atmosphärischen Luft zu prüfen. Aus dem Französischen übergesetzt.* (Experiments with plants, especially on the ability, which they possess to a high degree, to improve the air in the sunshine, but which they spoil at night or in the shade; together with a new method for grading the salubrity of atmospheric air. *Translated from the French.).* Vienna, Wappler 1786.

Ingen Housz J. (Scherer J, translator). *Versuche mit Pflanzen, hauptsächlich über die Eigenschaft, welche sie in einem hohen Grade besitzen, die Luft im Sonnenlichte zu reinigen, und in der Nacht und im Schatten zu verderben; nebst einer neuen Methode, den Grad der Reinheit und Heilsamkeit der atmosphärischen Luft zu prüfen. Aus dem Französischen übergesetzt. Zweyter Band.* (Experiments with plants, especially on the ability, which they possess to a high degree, to improve the air in the sunshine, but which they spoil at night or in the shade; together with a new method for grading the salubrity of atmospheric air. *Translated from the French. Second volume.)* Vienna, Wappler 1786.

Ingen Housz J. *Expériences sur les vegétaux, spécialement sur la propriété qu'ils possedent à un haut degré, soit d'amèliorer l'air quand ils sont au soleil, soit de le corrompre la nuit, ou lorsqui'ils sont à l'ombre; auxquelles on a joint une méthode nouvelle de juger du degré de salubrité de l'atmosphère.* (Experiments on plants, especially on the ability which they possess, to a high degree, to improve the air when they are in sunshine, but which they corrupt by night or in the shade: to which is appended a new way of judging the degree of salubrity of the atmosphere.) Nouvelle edition, revue et augmentée. Paris, Barrois le jeune, 1787.

Ingen Housz J. Lettre de M. Ingen Housz à M. Molitor au sujet de l'influence de l'éIectricité atmosphérique sur les végétaux. (Letter from Mr. Ingen Housz to Mr. Molitor on the subject of the influence of atmospheric electricity on plants.) *Journal de Physique 1788*. Vol. 32, pp. 321-337.

Ingen Housz J. (Scherer J, translator). *Versuche mit Pflanzen, hauptsächlich über die Eigenschaft, welche sie in einem hohen Grade besitzen, die Luft im Sonnenlichte zu reinigen, und in der Nacht und im Schatten zu verderben; nebst einer neuen Methode, den Grad der Reinheit und Heilsamkeit der atmosphärischen Luft zu prüfen. Aus dem Französischen übergesetzt. Dritter Band. (Experiments with plants, especially on the ability, which they possess to a high degree, to improve the air in the sunshine, but which they spoil at night or in the shade; together with a new method for grading the salubrity of atmospheric air. Translated from the French. Third volume.)* Vienna, Wappler 1788.

Ingen Housz J. *Nouvelles Expériences et observations sur divers objets de physique. Tome second. (New experiments and observations on a variety of topics in physics. Second volume.)* Paris, Barrois le jeune 1789.

Ingen Housz J. Lettre de M. Ingen Housz à M. de la Metherie sur les métaux comme conducteurs de la chaleur. (Letter from Mr. Ingen Housz to Mr. de la Metherie on metals as conductors of heat.) *Journal de Physique 1789*. Vol. 34, pp. 68-69.

Ingen Housz J. *Expériences sur les végétaux, spécialement sur la propriété qu'ils possèdent à un haut degré, soit d'améliorer l'air quand ils sont au soleil, soit de le corrompre la nuit, ou lorsqui'ils sont à l'ombre: auxquelles on a joint une méthode nouvelle de juger du degré de salubrité de l'atmosphère. (Experiments on plants, especially on the ability which they possess, to a high degree, to improve the air when they are in sunshine, but which they corrupt by night or in the shade: to which is appended a new way of judging the degree of salubrity of the atmosphere.)* Tome Second. Paris, Barrois le jeune 1789.

Ingen Housz J. Expériences qui prouvent: 1e que les plantes évaporent une quantité plus grande d'air vital pendant le jour à l'air libre, que nous en voyons répandre étant couvertes d'eau pure; 2e que leur évaporation nocturne d'une air méphitique, qui est très petite lorsqu'elles sont couvertes d'eau, est très considérable dans l'état naturel; qu'il y a un mouvement et déplacement continuel du fluide aérien dans les végétaux. (Experiments that prove : firstly, that plants evaporate a larger quantity of vital air during the day to common air than they do whilst surrounded by pure air; secondly, that their night-time evaporation of mephitic air, which is very modest whilst they are under water, is very considerable in the natural state; that there is a continual movement and displacement of arial fluid in plants.) *Journal de Physique 1789*. Vol. 34, pp. 436-446.

Ingen Housz J. Effet de l'éIectricité sur les plantes. Reflexions ulterieures sur le contenu du memoire de M. Ingen Housz, publié dans le cahier de ce journal de mois de mai 1788. (The effect of electricity on plants. Final reflections on the contents of the memoire of Mr. Ingen Housz, published in this journal in May 1788.) *Journal de Physique 1789.* Vol. 35, pp. 81-83.

Ingen Housz J. A letter from John Ingen-Housz, Body Physician to their Imperial and Royal Majesties, to William Falconer MD, 25 November 1791. In: Falconer W. *An account of the efficacy of the aqua mephitica alkalina in calculous disorders. Fourth Edition.* London, Cadell 1792.

Ingen Housz J. Aqua Mephitica Alcalina of Loogzoutig Luchtzuur Water. Een nieuw ontdekt en uitmundtend geneesmiddel in het graveel en den steen. (Alkaline soda water – a new discovery and an excellent cure for the gravel and the stone.) *Scheikundige Bibliotheek 1792.* Vol. 1, pp. 41-61.

Ingen Housz J. Tweede brief van den Heer J. Ingenhousz aan den Heere Deckers betreffende het Loogzoutig Luchtzuur Water. (Second letter from J. Ingenhousz to Mr. Deckers concerning the alkaline soda water.) *Scheikundige Bibliotheek 1792.* Vol. 1, pp. 95-114.

Ingen Housz J. Derde brief van den Heer J. Ingenhousz aan den Heere Deckers betreffende het Loogzoutig Luchtzuur Water. (Third letter from J. Ingenhousz to Mr. Deckers concerning the alkaline soda water.) *Scheikundige Bibliotheek 1792.* Vol. pp. 175-193.

Ingen Housz J. (Scherer J, ed) *Miscellanea Physico-medica.* Vienna, Patzowsky 1795.

Ingen Housz J. *An essay on the food of plants and the renovation of soils. Article no. III in additional appendix to the outlines of the fifteenth chapter of the proposed general report from the Board of Agriculture: on the subject of manures.* London, Bulmer 1796. p.17.

Ingen Housz J. (van Breda J. translator.) *Proeve over het Voedzel der Planten, en de vrugtboormaking van Landereijen. (An essay on the food of plants and the renovation of soils.)* Delft 1797.

Ingen Housz J. (Fischer G, translator.) *Über Ernährung der Pflanzen und Fruchtbarkeit des Bodens. (On the food of plants and the refurbishment of soils.)* Leipzig, Schaferischen 1798.

Ingen Housz J. Brief van Heer Johann Schmeisser aan Heer Jan Ingen Housz betreffende het Naphtha Vitrioli Martialis. (Letter from Mr. Schmeisser to Mr. Ingenhousz concerning the Naphtha Vitrioli Martialis.) *Scheikundige Bibliotheek 1798.* Vol. 2, pp. 47-49.

Ingen Housz J. Brief van den heere J. Ingenhousz and den Heere Jacob van Breda behelzende eenige proeven, met worm-doodende vogten genomen. (Letter of

Mr. J. Ingenhousz to Mr. J. van Breda regarding some experiments made with worm-killing fluids.) *Scheikundige Bibliotheek 1798.* Vol. 2, pp. 153-163.

Ingen Housz J. Brief van den Heere J. Ingenhousz, aan den Heere Henric. Ferdin. Ingenhousz MD over eene nieuwe manier om kanker en verouderde verzweeringen te geneezen. (A letter from Mr. J. Ingenhousz to Mr. Hendrick Ferdinand Ingen Housz concerning a new way to heal cancer and inveterate ulcers.) *Scheikundige Bibliotheek 1798.* Vol. 2. pp. 201-208.

Bibliography

1
UNPUBLISHED MATERIAL

For these sources please see 'Bibliography Key' on page 527 (although Godefroi and Wiesner are published material).

2
PUBLISHED MATERIAL

a. Previous significant biographies and biographical articles on Jan Ingen Housz.

Beale N, Beale E. Looking for Dr. Ingen Housz: the evidence for the site and nature of the burial of the famous Dutch physician and scientist. *Wiltshire Archeological and Natural History Magazine (2000)*; Vol. 92, pp. 120-130.

Beale N, Beale E. *Who was Ingen Housz, anyway?* Calne, Calne Town Council, 1999.

Burzbach, C. *Biographisches Lexikon des Kaiserthums Oesterreich, Volume 10*, pp. 206-208. (Ingenhouss, Johann).

Dolezal H. Ingen-Housz, Jan: naturforscher und arzt. *Neue Deutsche Biographie 1974*, Vol. 10, pp. 171-172.

Flint, V. *Jan Ingen Housz: MSc dissertation 1950*. University of London Library, Senate House, London.

Garthshore M. Biographical account of Dr. Ingenhousz. *Annals of Philosophy 1817*. Vol. 10, pp. 161-165.

Gest H. Bicentenary homage to Dr. Jan Ingen Housz, MD (1730-1799), pioneer of photosynthesis research. *Photosynthesis Research 2000*, Vol. 63, pp 183-190.

Godefroi M. Dr Jan Ingen-Housz. Geheimraad en Lijfarts van Z.M.Kaiser Jozef II van Oostenrijk (Private counsellor and personal physician to His Majesty Emperor Joseph II). *Records of the Provincial Association of Arts and Sciences of North-Brabant.* 's-Hertogenbosch, July 1875.

Heinsius H. Jan Ingen-Housz. *Album der Natuur 1897*. pp 1-15.

Hoogenboom J. Dr. Jan Ingen-Housz. *Uit de geschiedenis van Breda en omgeving (From the history of Breda and its surroundings.)* Deeltje II. pp. 63-66.

Ingen Housz, J-M, Beale N, Beale E. The life of Dr Jan Ingen Housz (1730-1799) Private Counsellor and Personal Physician to HM Emperor Joseph II of Austria. *Medical Biography 2005*, Vol. 13: pp. 15 – 21.

Jenkins J. The English inoculator. *Journal of the Royal Society of Medicine 1999*, Vol. 92, pp. 534 – 537.

Magiels G. *From sunlight to insight. Jan Ingen Housz, the discovery of photosynthesis and science in the light of ecology.* Brussels, Brussels University Press 2010.

Reed, H.S. Jan Ingen Housz, physiologist. *Chronica Botanica 1949*, 11; pp. 285 – 396.

Smit, P. Jan Ingen-Housz (1730-1799): some new evidence about his life and work. *Janus 1980*, Vol. 68, pp. 125-139.

Treub, M. J. Ingen-Housz. *De Gids (The Guide) 1880*, Vol. 18, pp. 478-500.

Van der Aa, A. *Levensschets van J. Ingenhousz (Life sketches of J. Ingenhousz).* Herinneringen uit het gebied der Geschiedenis (Reminders from the domain of history), Amsterdam 1835. pp. 5-12.

Van der Pas, P. Ingen-Housz, Jan. *Dictionary of Scientific Biography.* Vol VII; pp 11-16. Gillespie C (Ed). Charles Scribner's Sons, New York.

Wap. Doctor Jan Ingen-Housz. *Noord-Brabandsche Volks-Almanak (North Brabant People's Almanac) 1843*, pp. 35-45.

Wiesner J. *Jan Ingen-Housz: sein leben und sein wirken als naturforscher und artz (Jan Ingen Housz, his life and works as a natural historian and doctor).* Vienna, Carl Konegen 1905.

b. Other secondary sources relating to the life of Ingen Housz (see, also, list of publications by Jan Ingen Housz, on pages 585-592).

Ackroyd P. *London: the biography.* London, Chatto and Windus 2000.

Acton H. *The Bourbons of Naples.* London, Methuen 1956.

Alexander J. *Catherine The Great: life and legend.* New York, Oxford University Press 1989.

Arneth A, Geffey M. *Correspondence secrete entre Marie-Thérèse et le Cte. de Mercy-Argenteau.* Paris, Mesnil 1874.

Arneth A. (Ed.) *Maria Theresia und Joseph II. Ihre correspondenz sammt Briefen Joseph's an seinen Bruder Leopold. Ester Band 1761 – 1772.* Wien, von Carl Gerold's Sohn 1867.

Baron J, *The life of Edward Jenner, M.D., LL.D., F.R.S.* London, Henry Colburn 1838.

Baxby D. Edward Jenner's unpublished cowpox Inquiry and the Royal Society: Everard Home's report to Sir Joseph Banks. *Med. Hist.*, 1999, Vol. 43, pp. 108–10.

Baynes J. *The Jacobite rising of 1715.* London, Cassell 1970.

Beale N, Beale E. Evidence-Based Medicine in the Eighteenth Century: the Ingen Housz–Jenner Correspondence Revisited. *Medical History 2005*, Vol. 49, pp. 79–98.

Beale N, Beale E. Looking for Dr. Ingen Housz: the evidence for the site and nature of the burial of the famous Dutch physician and scientist. *Wiltshire Archeological and Natural History Magazine 2000.* Vol. 92, pp. 120-130.

Beale N, Beale E. Sunlight at Southall Green: Dr. Ingen Housz discovers photosynthesis. *Perspectives in Biology and Medicine* 2001, Vol. 44, pp. 333-341.

Beale N. *Is that the doctor? A history of the Calne GPs.* Bradford on Avon. Ex Libris

Press 1998.
Beale N. *Joseph Priestley in Calne*. East Knoyle, Salisbury, Hobnob Press 2008.
Beales D. *Joseph II. Against the world, 1780-1790*. Cambridge, Cambridge University Press 2009.
Beales D. *Joseph II. In the shadow of Maria Theresa 1741-1780*. Cambridge, Cambridge University Press 1987.
Black J. *The grand tour in the eighteenth century.* Stroud, The History Press 2009.
Bowring J.(ed). *The works of Jeremy Bentham*. Edinburgh, Tate 1843. .
Brands H. *The first American: the life and times of Benjamin Franklin*. Doubleday, New York 2000.
Braunbehrens V. *Mozart in Vienna 1781-1791*. London, André Deutch 1990.
Brenni P. *Volta's electric lighter and its improvements*. Proceedings of a conference held in Pognano sul Lario, Italy June 2003. http://sci-ed.org/Conference-Pognana/Brenni.pdf (last accessed 4 June 2011).
Brighton T. *The discovery of the Peak District*. Chichester, Phillimore 2004.
Britton J. *The Beauties of England and Wales: Wiltshire. Vol. XV.* London, Harris 1814.
Brown R. A brief account of microscopical observations made on the particles contained in the pollen of plants. *Edinburgh Philosophical Magazine 1828*, Vol. 4, pp. 161-173.
Buranelli V. *The wizard from Vienna*. London, The Scientific Book Club 1977.
Burnby J. *A study of the English Apothecary from 1660 to 1760*. London, Wellcome 1983.
Caygill M. *The story of the British Museum*. London. British Museum Press 2002.
Churchill, W. S. *Marlborough: his Life and Times.* London, Harrap 1938.
Cohen E. Wie heeft de verbranding van een horlogeveer in suurstofgas het eerst uitgevoerd? (Who was the first to publish on the burning of a watch spring in oxygen?) *Chemisch Weekblad 1907,* Vol. 48, pp. 787-798.
Constantine D. *Fields of fire. A life of Sir William Hamilton*. London, Weidenfeld and Nicolson 2001.
Cooke A. *America*. London, British Broadcasting Corporation 1973.
Cooley A, Cooley M. *Pompeii: a source book*. London, Routledge 2004.
Crankshaw E. *Maria Theresa*. London, Constable 1983.
Cross A. *An English Lady at the court of Catherine the Great: the journal of Baroness Elizabeth Dimsdale, 1781*. Cambridge, Crest Publications 1989.
da Mosto F. *Venice*. London, BBC Books 2004.
Desmond R. *The history of the Royal Botanic Gardens, Kew.* London, Harvill 1995.
Dickinson H. *Matthew Boulton*. Leamington Spa, Tee 1999.
Dimsdale T. *The present method of inoculating for the small-pox*. London, Owen 1767.
Draper T. *A struggle for power: the American Revolution*. London, Abacus 1997.
Dunlop I. *Marie Antoinette: a portrait*. Sinclair-Stevenson, London 1993.
During M, Didi-Huberman G, Poggesi M. *Encyclopaedia anatomica*. Koln, Taschen 2004.
Falconer W. *An account of the efficacy of the aqua mephitica alkalina. London,* Cadell 1792.
Farrand M. *(A restoration of) The autobiography of Benjamin Franklin*. Berkeley CA,

University of California Press 1949.
Fisher R. *Edward Jenner (1749–1823)*. London, Andre Deutsch 1991.
Fitzmaurice E. *Life of William, Earl of Shelburne, afterwards First Marquis of Lansdowne*. London, Macmillan 1876.
Fontana F. *Descrizione e usi di alcuni stromenti per misurare la salubrita del aria. (Description and uses of some instruments for measuring the salubrity of the air.)* Firenze, Gaetano Cambiagi Stampatore granducale 1775.
Fontana F. Experiments and observations on the inflammable air breathed by various animals. *Philosophical Transactions of the Royal Society 1779*. Vol. 69, pp. 337-361.
Forbes R. (Ed.) *Martinus van Marum: life and work*. Haarlem, Tjeenk, Willink 1969.
Fraser A. *Marie Antoinette: the journey*. London, Weidenfeld & Nicolson 2001.
Freeman J. *Jane Austen in Bath*. Alton, Mills 1969.
French R. *Robert Whytt, the soul, and medicine*. London, Wellcome Institute of the History of Medicine 1969.
Geikie A. *Annals of the Royal Society club*. London, Macmillan 1917.
Gest H. Sun-beams, cucumbers and purple bacteria. Historical milestones in early studies of photosynthesis. *Photosynthesis Research 1988*, Vol. 19, pp 287-308.
Gillespie C (Ed. in chief). *Dictionary of Scientific Biography*. New York, Scribner's 1970-1980.
Gillispie C. *The Montgolfier brothers, and the invention of aviation 1783-1784*. Princeton, Princeton University Press 1983.
Girard G. *Correspondence entre Marie-Thérèse et Marie-Aintonette*. Paris 1933.
Glynn I, Glynn J. *The life and death of smallpox*. London, Profile Books 2004.
Golinski J. *Science as public culture: chemistry and enlightenment in Britain, 1760-1820*. Cambridge. Cambridge University Press 1992.
Grundy I. *Mary Wortley Montagu: comet of the enlightenment*. Oxford, Oxford University Press 1999.
Gubbins D. *Encyclopedia of Geomagnetism and Paleomagnetism*. Dordrecht, Springer Press 2007.
Hackmann W. *Electricity from glass. The history of the frictional electrical machine 1600-1850*. Alphen aan den Rijn, Sijthoff & Noordhoff 1978.
Haig W. *William Pitt the younger*. London, Harper Perennial 2008.
Halsband R (Ed). *The complete letters of Lady Mary Wortley Montagu*. Oxford, Clarendon Press 1965.
Hartmann P. *Elfenbeinkunst (Micropictures)*. Wien, Hartmann 1999.
Hayes K. *The road to Monticello: the life and mind of Thomas Jefferson*. Oxford, Oxford University Press 2008.
Heath C. *The book of Hertford*. Chesham, Barracuda 1975.
Hibbert C. *George III. A personal history*. London, Penguin Books 1998.
Hoff H. Jan Ingen-Housz and the cover slip. *Bulletin of the History of Medicine 1962*. Vol.36, pp. 365-8.
Hoogeven H. *Dictionarium Analogicum Linguae Graecae*. Cambridge, Cambridge University Press 1801.
Hopkins D. *The greatest killer. Smallpox in history*. Chicago, University of Chicago Press 2002.

Hudson K. *The Bath and West. A bicentenary history*. Bradford-on-Avon, Moonraker Press 1976.
Hughes R. *The fatal shore. A history of the transportation of convicts to Australia, 1787–1868*. London, Collins Harvill 1987.
Hulme N. *A safe and easy remedy proposed for the relief of stone and the gravel, scurvy and gout*. London, Robinson and Elmsly 1778.
Hussey A. *Paris, the secret history*. New York, Viking 2006.
Ilchester, Earl of. (Ed). *Journal of Elizabeth, Lady Holland*. London, Longmans 1908.
Isaacson W. *Einstein: his life and universe*. London, Pocket Books 2008.
Japikse C. *Fart proudly: writings of Benjamin Franklin you never read in school*. Berkeley, California, Frog Books 2003.
Jay M. *The atmosphere of heaven*. New Haven and London, Yale University Press 2009.
Jenner E. *An inquiry into the causes and effects of the variolae vaccinae*. London, Sampson Low, 1798.
Jenner E. *Further observations on the variolae vaccinae, or cow pox*. London, Sampson Low 1799.
Johnson, S. *The invention of air*. London, Penguin 2009.
Josey D. *Broome Park, Betchworth, Surrey*. Sutton, Surrey, Spring Gardens Trust 2005.
King-Hele D. *Erasmus Darwin: a life of unparalleled achievement*. London, Giles de la Mare 1999.
Kugler G. *Schönbrunn Palace. The State Apartments. Schloss Schönbrunn Kultur- und Betriebsges*. Vienna, Christian Brandstätter 1995.
Lack W. *Franz Bauer: the painted record of nature*. Wien, Verlag des Naturhistorischen Museums 2008.
Leppmann W. *Winckelmann*. London, Victor Gollancz 1971.
Litten J. *The English way of death*. London, Robert Hale 1991.
Luyendijk-Elshout A.M. (editor) *Wandelen met Boerhaave in en om Leiden (Walking with Boerhaave in Leiden)*. Caecilia Foundation, Leiden 1994.
Macaulay T. (Firth C. Ed.) *History of England*. London, Macmillan 1914.
Maehle A-H. Dissolving the stone: the search for lithontriptics. *Clio Medical/The Wellcome Series in the history of medicine*. 1999. Vol. 55, pp. 55-125.
Magiels G. *From sunlight to insight. Jan Ingen Housz, the discovery of photosynthesis and science in the light of ecology*. Brussels, Brussels University Press 2010.
Maloney W. *George & John Armstrong of Castleton*. Edinburgh and London, Livingstone 1954.
Marsh A. *A history of the borough and town of Calne*. Calne, Heath 1903.
Maurice D. *The Marlborough Doctors*. Stroud, Alan Sutton 1994.
Maurois A. *The life of Sir Alexander Fleming*. Harmondsworth, Middlesex, Penguin 1959.
McDonald D, Hunt L. *A history of platinum and its allied metals*. London, Johnson Matthey 1982.
McLure R. *Coram's Children*. New Haven CT, Yale University Press 1981.
Medd P. *Romilly: a life of Sir Samuel Romilly, lawyer and reformer*. London, Collins 1968.

Mesmer A. *Mémoire sur la découverte du magentism animal. (Memoire on the discovery of animal magnetism.)* Geneve et Paris, Didot le jeune 1779.

Mitchison R. *Agricultural Sir John. The life of Sir John Sinclair of Ulbster 1754-1835.* London, Bles 1962.

Montefiore S. *Prince of princes: the life of Potemkin.* London, Weidenfeld & Nicolson 2000.

Nederland's Patriciaat. 59e Jaargang. The Hague, Centraal bureau voor Genealogie 1973.

Newman A. *The Stanhopes of Chevening.* London, Macmillan 1969.

Nichols J. *Biographical and literary anecdotes.* London, Bowyer 1782.

Petz-Grabenbauer M. *Der Botaniker Nikolaus Joseph Freiherr von Jacquin.* Diplomarbeit (Dissertation). Wien, Archiv der Universitat. 885/94.

Pezzl, J. *Skizze von Wein – Sketch of Vienna, six instalments 1786 –1790.* As edited and reproduced in Robbins Landon, H. *Mozart and Vienna.* London, Thames and Hudson 1991.

Phillips D. *The Great Road to Bath.* Newbury, Countryside Books 1983.

Pichler C. *Denkwürdigkeiten aus meinem leben (Memorable events of my life).* Wien, Stabt 1844.

Plumb J. *Sir Robert Walpole. The King's minister.* London, The Cresset Press 1960.

Poirier J-P. *Lavoisier: chemist, biologist, economist.* Philadelphia, University of Pennsylvania Press, 1996.

Porter R. *Doctor of society. Thomas Beddoes and the sick trade in late-enlightenment England.* London, Routledge 1992.

Porter R. *Enlightenment: Britain and the creation of the modern world.* London, Penguin Books 2001.

Porter R. *The greatest benefit to mankind. A medical history of humanity from antiquity to the present.* London, Harper Collins 1997.

Pottle F (Ed). *Extracts from Boswell's London Journal 1762 - 1763.* Heinemann, London, 1950.

Prebble J. *Culloden.* London, Secker and Warburg 1961.

Priestley J. *Experiments and observations on different kinds of air.* Vol. II. London, Johnson 1776. (Second edition).

Priestley J. *Experiments and observations relating to various branches of natural philosophy.* London, Johnson 1779.

Priestley J. *Experiments and observations relating to various branches of natural philosophy; with a continuation of the observations on air. The second volume.* Birmingham, Pearson and Rollason, for J. Johnson, London 1781.

Pringle J. *Six Discourses delivered before the Royal Society on the annual assignments of the Copley Medal.* London, Strahan & Cadell 1783.

Rae P. *The History of the Late Rebellion, rais'd against King George by the Friends of the Popish Pretender.* London, Millar 1746.

Rehm G. *De Bredase apothekers van de 15e tot het begin van de 19e eeuw. (The apothecaries of Breda from the 15th century to the beginning of the 19th century.)* Breda, Geschieden Oudheidkundige Kring van Stad en Land van Breda 'De Oranjeboom.' 1961.

Robbins Landon H. *Mozart and Vienna.* London, Thames and Hudson 1991.

Russell J. *Nelson and the Hamiltons.* Harmondsworth, Middlesex, Penguin 1972.
Rutledge, J-J. *The Englishman's fortnight in Paris: or the art of ruining himself there in a few days.* London, Durham and Kearsly 1777.
Saunders P. *Edward Jenner, the Cheltenham years, 1753–1823.* Hanover NH, and London, University Press of New England 1982.
Schama S. *Citizens: a chronicle of the French Revolution.* London, Viking 1989.
Schama S. *Patriots and Liberators. Revolution in the Netherlands 1780 – 1813.* London, Collins 1977.
Schama S. *The embarrassment of riches. An interpretation of Dutch culture in the Golden Age.* London, Collins 1987.
Schiebinger L. *The mind has no sex? Women in the origins of modern science.* Cambridge, Mass., Harvard University Press 1989.
Schiff S. *Dr. Franklin goes to France.* London, Bloomsbury 2005.
Schmidt-Görg J, Schmidt H. *Ludwig van Beethoven.* London, Pall Mall Press 1970.
Schofield R. (Ed.) *A scientific autobiography of Joseph Priestley: selected scientific correspondence.* Cambridge, Mass., MIT Press 1966.
Schofield R. *The enlightened Joseph Priestley: a study of his life and work from 1773-1804.* University Park PA, Pennsylvania State University Press 2004.
Schofield R. *The enlightenment of Joseph Priestley. A study of his life and work from 1733-1773.* University Park PA, Pennsylvania State University Press 1997.
Seager J. *Hoogeveen's particles: abridged and translated into English.* London 1829.
Senebier J. *Expériences sur l'action de la lumière solaire dans la végétation.* Geneva and Paris. Briand 1788.
Senebier J. *Memoires physico-chymiques, sur l'influence de la lumière solaire pour modifier le êtres des trois règnes de la nature, et surtout ceux du règne végétal.* Geneva, Barthelemi Chirol 1782.
Singer D. Sir John Pringle and his circle: part 1. Life. *Annals of Science 1949,* Vol. 6: pp. 127 – 180.
Singer D. Sir John Pringle and his circle: part 2. Copley discourses. *Annals of Science 1950;* Vol. 6, pp. 248 – 261.
Sonntag O (Ed). *John Pringle's correspondence with Albrecht von Haller.* Basel. Schwabe-Verlag 1999.
Sonntag O. *The correspondence between Albrecht von Haller and Charles Bonnet.* Huber, Bern 1983.
Spaethling R. (Ed.) *Mozart's letters, Mozart's life.* London, Faber and Faber.
Spoehr H. The development of conceptions of photosynthesis since Ingen-Housz. *The Scientific Monthly 1919, 9: 32 – 46.* The Science Press. New York.
Sprigge T. (Ed.). *Collected works and correspondence of Jeremy Bentham.* London, Athlone Press 1968.
Stoll H. *Tod in Triest.* Berlin, Union Verlag 1968.
Thomas, D. *Collected Poems 1934 – 1952.* London, Dent 1952.
Thompson C. *The mystery and art of the apothecary.* Detroit, Singing Tree Press 1971.
Till N. *Mozart and the enlightenment: truth, virtue and beauty in Mozart's operas.* London, Faber 1992.
Tomalin C. *Samuel Pepys. The unequalled self.* London, Viking 2002.

Trease G. *Pharmacy in History*. London, Baillière, Tindall and Cox 1964.
Uglow J. *The Lunar Men*. London, Faber & Faber 2002.
Van Barneveld W. Proeve van onderzoek omtrent de hoeveelheid van bederf, t'welk in onzen dampkring ontstaat, nevens deszelfs verbetering door den groei der plantgewassen. (Investigation on the amount of pollution of the air and its restoration by the growth of plants.) *Verhandelingen Provinciaal Utrechts Genootschap van Kunsten en Wetenschappen 1781*, pp. 408-472.
Van Boven M. (ed.). *Ach lieve tijd. (Ah, good old times). Part 11, The people of Breda and their education*. Zwolle, Waanders 1985.
Van Breda (ed.). *Scheikundige Bibliotheek (Chemical Library)*, Delft 1792 (Vol. 1) & 1798 (Vol. 2).
Van der Korst J. *Een dokter van formaat. (A doctor of stature.)* Amsterdam, Bert Bakker 2003.
Van der Pas P. The discovery of the Brownian motion. *Scientarium Historia 1971*, Vol. 13, pp. 27-35.
Van der Pas P. The Ingenhousz–Jenner correspondence. *Janus*, 1964, Vol. 51, pp. 217-18.
Vickery A. *Behind closed doors. At home in Georgian England*. New Haven. Yale University Press 2009.
Wassink E. On the discovery of the light factor in photosynthesis. *Mededelingen van de Landbouwhogeschool te Wageningen/Nederland 1958*, Vol. 58, p. 5.
Webster N. *The great north road*. Bath, Adams and Dart 1974.
Wedd K. *The Foundling Museum*. London, The Foundling Museum 2004.
Wheatcroft A. *The Habsburgs. Embodying empire*. London, Viking 1995.
Whitworth R. *William Augustus, Duke of Cumberland*. London, Leo Cooper 1992.
Wiesner J. *Jan Ingen-Housz: Sein Leben und sein Wirken als Naturforscher und Artz*. Vienna, Carl Konegen 1905.
Wilson D. *The British Museum. A History*. London, British Museum Press 2002.
Zegers, R. Mozart and smallpox. *Clinical and Experimental Ophthalmology 2007*, Vol. 35, pp. 372–373.
Zwanenberg D. The Suttons and the business of inoculation. *Medical History 1978*; Vol. 22, pp. 71 – 82.

Index of Names

Apart from the members of various European royal families, all the subjects are English unless otherwise stated.
FRS = Fellow of the Royal Society.

Adam, Robert 1728 – 1792: neoclassical architect, 417, 420.
Adams, George (jnr) 1750 – 1795: scientific instrument maker, 409, 417.
Adams, John 1735 – 1826: American, lawyer, diplomat, second President of United States, 373.
Aeneae, Henricus 1743 – 1810: Dutch scientist and mathematician, 247, 264.
Aiton, William (junior) 1766 – 1849: head gardener at Kew from 1793, 486.
Aiton, William 1731 – 1793: head gardener at Kew, 276.
Albani, Alessandro 1692 – 1779: Italian Cardinal, nephew of Pope Clement XI, 120 – 122, 145, 146.
Albert of Saxony 1738 – 1822: Duke of Teschen, German aristocrat, 100.
Albinus, Bernhard Siegfried 1697 – 1770: German anatomist working at Leiden, 27.
Allsup, Christopher 1731 – 1816: Calne surgeon-apothecary, Bowood doctor, 280, 423, 481, 500, 508, 515.
Ann, wife of second earl of Upper Ossory – see Liddel.
Anne 1665 – 1714: Queen of Great Britain from 1702, 4, 72.
Anne 1709 – 1759: Princess Royal and Princess of Orange, 115.
Anselm, Karl 1733 – 1805, fourth Prince of Thurn and Taxis: Austrian noble, 234.
Arcangeli, Francesco, died 1768: Italian, assassin of Johann Winckelmann, 140, 141.
Arenberg, Duke of, see Engelbert.
Armstrong, George 1719 – 1789: London surgeon-apothecary, pioneer of paediatrics, 88.
Augusta 1719 – 1772, Princess of Saxe-Gotha: Mother of George III, 42.

Bache, Benjamin 1769 – 1798: American, grandson of Benjamin Franklin, 241.
Bacon, Francis 1561 – 1626: philosopher, statesman, lawyer, pioneer of scientific method, 41, 341.
Baker, George 1722 – 1809: FRS, royal physician, 44, 62, 63, 65, 88, 110, 170, 183, 209, 403.
Baker, Henry 1698 – 1774: FRS, Naturalist and microscopist, 235.
Bancroft, Edward 1745 – 1821: American physician, secretary to Benjamin Franklin in France, 290, 291, 357 – 360, 378.
Banks, Joseph Sir 1743 – 1820: Botanist, circumnavigated with James Cook. President of Royal Society, 260, 266, 317, 318, 330, 344, 346, 348, 354, 367, 406, 407, 418, 428, 429, 435, 466, 473, 483, 485, 498.
Barbier de Tinant, Jean-Jacques, died 1791: Amateur physicist of Strasbourg, 287, 292.
Barletti, Carlo 1735 – 1800: Italian. Professor of physics at Pavia, 320, 321.
Barneveld, Willem van 1747 – 1826: Amsterdam apothecary, natural scientist, 323, 324, 326, 327.
Barrois, Pierre Theophile le jeune 1752 – 1820: Publisher and bookseller in Paris, 382, 386, 389, 390, 433.
Barruel, Augustin 1741 – 1820: French priest, author on French revolution, 484, 485.

Bauer, Franz 1758 – 1840: Austrian, botanical artist employed at Kew, 405 – 407, 486.
Baye, Baron de, dates uncertain: friend of first Marquis of Lansdowne, 468.
Beccaria, Giovanni 1716 – 1781: Italian, professor of physics, Turin, 369.
Beck, Dominique, dates uncertain: German physicist, Salzburg, 356.
Beckers, (Gouban), Allegonda, dates uncertain: maternal grandmother of Jan Ingen Housz, 10, 12.
Beckers, Abraham, dates uncertain: maternal grandfather of Jan Ingen Housz, 10.
Beckers, Jan, dates uncertain: maternal uncle of Jan Ingen Housz, 12.
Beckers, Maria 1698 – 1731: wife of Arnold Ingen Housz, mother of Jan, 10 – 14.
Beddoes, Thomas 1760 – 1808: physician and chemist who pioneered inhalation therapies, 452, 454, 481 – 483.
Beethoven, Ludwig van 1770 – 1827: German pianist and composer, 294.
Bégue de Presle, Achille-Guillaume 1735 – 1807: French physician, scientist, writer; friend and doctor to Rousseau, 307, 308, 349, 379, 383.
Beman, Mr., dates uncertain: dairy farmer in Gloucestershire, 501, 505.
Bentham, Jeremy 1748 – 1832: jurist, philosopher, social reformer, 410, 433, 434, 439.
Bentinck, William Cavendish, 1738 – 1809: Whig politician, 507.
Bentley, Thomas 1731 – 1780: business partner of Josiah Wedgwood, 261.
Bernini, Gian Lorenzo 1598 – 1680: Roman sculptor and architect, 136.
Bertholet, Claude-Louise 1748 – 1822: French chemist, 384.
Bertholon, Saint-Lazare de, Pierre 1741 – 1800: French physicist, electricity expert, 364, 365, 367, 370.
Berthoud, Ferdinand 1727 – 1807: FRS. Swiss chronometer and clockmaker, Paris, 182.
Bicker, Lambert 1732 – 1801: Dutch physician, Rotterdam, 57, 247, 301, 402.
Biheron, Marie Marguerite 1719 – 1795: French anatomical model maker, 192.
Black, Joseph 1728 – 1799: Scottish physician, natural scientist, 425.
Blagden, Charles 1748 – 1820: FRS, scientist, secretary of Royal Society, 406, 418.
Blair, John, died 1782: FRS, clergyman, 170.
Boerhaave, Hermann 1668 – 1738: Dutch, revolutionary Leiden professor of medicine, pioneered bedside clinical teaching, 20, 25, 27, 28, 128.
Bonn, Andreas 1738 – 1817: Dutch, professor of anatomy, Amsterdam, 247.
Bonnet, Charles 1720 – 1793: FRS, Swiss naturalist, 271, 272, 275, 328, 329.
Bosch, Jan, dates uncertain: Dutch, printer of Haarlem, 399.
Bosch, Joanna, 1740-1821: wife of Martin van Marum, 399, 417, 418.
Bosch, Maria Anna von 1724 – 1790: Austrian, wife of Anton Mesmer, 223.
Boswell, James 1740 – 1795, ninth Lord Auchinleck: biographer of Samuel Johnson, 44.
Boulton, Matthew 1728 – 1809: manufacturer and partner of James Watt, 190, 413, 419, 444, 482, 488.
Bowman, Benjamin, dates uncertain: church warden, St. Mary's Church, Calne in the 1790s, xv, xvi.
Bowyer, William 1699 – 1777: London publisher, printer to Parliament and Royal Society, 49, 50, 60, 61, 63.
Brady, Terence, dates uncertain: physician to Austrian army and court at Brussels, 101.
Braschi, Giovanni Angleo 1717 – 1799: Italian, Pope Pius VI from 1775, 309, 339, 340.
Breda, Jacob van 1743 –1818: Dutch, physician and town official, Delft; linguist and editor of a chemical journal, 286 – 288, 301, 308, 311, 314, 323, 326 – 328, 333, 339, 342, 349, 371, 372, 374, 375, 379, 380, 383, 386, 398, 415, 430, 431, 432, 436 – 438, 440, 458, 467, 474, 485, 511.
Bridgewater, Duke of – see Egerton.
Brindley, James 1716 – 1772: canal engineer; cut Bridgewater canal, 187.
Brocklesby, Richard 1722 – 1797: FRS, London physician, 487.
Brown, Lancelot 'Capability' 1716 – 1783:

INDEX OF NAMES

landscape architect and gardener, 420, 468.
Brown, Robert 1773 – 1858: FRS, Scottish botanist and microscopist, 344 – 346.
Bruyn, Jacob de. Dates uncertain: Dutch, lawyer at Breda, 34.
Bulmer, William 1757 – 1830: London publisher and printer, 471, 473.
Burcot, Doctor, dates uncertain: Elizabethan physician, possibly of German extraction, 71.
Burke, Edmund 1729 – 1797: Irish politician, writer and orator, 444.
Bute, Earl of – see Stuart.
Butler, James 1665 – 1745, second Duke of Ormonde: Irish statesman, soldier; Jacobite, 8.

Caldani, Leopoldo Marco 1725 – 1813: Italian, professor of anatomy and medicine, Padua, 169.
Campen, Anna Elsenera, van, died 1791: wife of Jacob van Breda, 431, 436.
Canton, John 1718 – 1772: FRS, electrical expert, scientist, 186, 189.
Carbury, Count, dates uncertain: English, personal physician to Count D'Artois, 387, 402.
Caroline of Ansbach, Princess 1683 – 1737: wife of George II, 76, 485.
Catherine, the Great 1729 – 1796: Russian Empress from 1762, 142 – 145, 150, 151, 336, 337, 419, 483.
Cato, Marcus Porcius 234 BC - 149 BC: Roman statesman and author, 15.
Cavaceppi, Bartolomeo 1716? – 1799: Italian, sculptor in Rome, 118, 121 – 123.
Cavallo, Tiberius 1749 – 1809: FRS, from Naples; settled in England, electrical expert and scientist, 235, 263, 311, 314 – 316, 324, 326, 327, 330, 379.
Cavendish, Henry 1731 – 1810: FRS, chemist, discovered hydrogen, 183, 184, 252, 346, 384, 434.
Cavendish, William 1748 – 1811, fifth Duke of Devonshire: English aristocrat, 190.
Caxton, William 1422 - 1492: thought the first printer in the English language, 67.
Chais, Charles 1743 – 1790: Dutch, pastor at the Walloon church, The Hague, 104 – 106, 110, 114, 115.

Charlemagne 768 – 814: King of the Franks, 239.
Charles Alexander, Prince of Lorraine 1712 – 1780: Brother-in-law of Maria Theresa. Viceroy, Austrian Netherlands, 117, 185.
Charles II, 1630 – 1685: King of England from 1660, 4, 55.
Charles III, 1716 – 1788: King of Spain from 1759, 159.
Charles Philippe 1757 – 1836, Comte D'Artois: younger brother of Louis XVI, 387.
Charles, Alexander (VI) 1685 – 1740: Emperor of Austria, 19.
Charles, Jacques 1746 – 1823: French chemist and balloonist, 346 – 349, 384, 412.
Charlotte of Mecklenburg-Strelitz, 1744 – 1818: Queen of England from 1761, 42, 407, 485, 486.
Chaulnes, Duke de – see D'Albert.
Chaumont, Jacques-Donatien le Ray de 1725 – 1803: French entrepreneur, 240, 241, 285, 290, 382.
Chiswell, Sarah, died 1726: friend of Mary Wortley Montagu, 74.
Churchill, John, first Duke of Marlborough 1650 – 1722: soldier and statesman, 2, 3.
Churchill, Winston 1874 – 1965: statesman and politician, 483.
Cicero, Marcus Tullius 106 BC - 43 BC: Roman statesman and philosopher, 15, 27.
Clement XIII, Pope – see Rezzonico.
Clive, Robert ('of India') 1725 – 1774: Founder of Empire of British India, 204.
Cobenzl, Johann Karl Philipp von 1712 – 1770: chief minister, Austrian Netherlands at Brussels, 117.
Coffyn, Francis, dates uncertain: Belgian, American agent at Dunkerque, 290, 291, 357 – 359, 378.
Colborne, Benjamin, died 1793: general practitioner, Bath, expert on kidney and bladder stones, 423 – 426, 431, 439, 446.
Collard, Isaac, dates uncertain: secretary of the Dutch ambassador to London, 47, 61, 89, 110, 113, 183, 219.
Collin, Heinrich Joseph, 1731 – 1784:

Austrian, chief physician of the Vienna hospital, 131.

Colonna, Odo 1368 – 1431: Italian, Pope Martin V from 1417, 25.

Comus – see Ledru.

Cook, James 1728 – 1779: FRS, explorer, circumnavigator, 191.

Coram, Thomas 1668 – 1751: philanthropist who created London Foundling Hospital, 81 – 83.

Corsellis, Frederick, dates uncertain: Haarlem printer, 67.

Cullen, William 1710 – 1790: FRS, Scottish physician, teacher of medicine, 42.

Cumberland, Duke of – see William.

Cunaeus, Andreas 1712 – 1788: Dutch, scientist, joint inventor of Leyden Jar, 28.

Cunego, Domenico 1726 – 1803: Italian, engraver of Rome, 157.

Cushing, Thomas 1725 – 1788: American, politician and merchant, 216.

Cuthbertson, John 1743 – 1821: English scientific instrument maker, worked in Amsterdam, 247, 264.

D'Albert d'Ailly, Louis Joseph, 1741 – 1792: seventh Duke of Chaulnes, FRS, French aristocrat and scientist, 244.

D'Arcet Jean 1724 – 1801: French chemist, director of Sèvres porcelain works, 412.

D'Arlandes, François Laurent 1742 –1809: French ballooning pioneer, 347, 348.

D'Arpajon, Anne Claudine Louise 1710 – 1794: Comtesse de Noailles, French, Lady-in-Waiting to Marie Antoinette, 180.

D'Artois, Count – see Charles Philippe.

Dalrymple, Alexander 1737 – 1808: FRS, hydrographer and meterologist, 406.

Dalrymple, John 1673 – 1747, second Earl Stair: Scottish soldier and diplomat, 19, 21, 39, 41.

Dartmouth, Lord, see Legge.

Darwin, Erasmus 1731 – 1802: physician and naturalist, 413, 451, 484.

Daun, Leopold von 1705 – 1766: high-ranking Austrian soldier, 99.

Davis, Revd., dates uncertain: patient of Jan Ingen Housz at Calne, 481.

Davy, Humphry 1778 – 1829: FRS, chemist and inventor, 452.

Deblands, Josepha, dates uncertain: inoculation patient of Jan Ingen Housz at Florence, 196.

Deckers, Johann Hendrik, born 1729: Dutch physician at 's-Hertogenbosch, 27, 29, 46, 87, 95, 131, 155, 167, 170, 208, 209, 220, 430, 431, 460, 464, 474.

Degenfeld-Schomburg, Count Frederick von 1721 – 1781: Dutch ambassador at Vienna, 126.

Deiman, Johann Rudolf 1743 – 1808: German, physician, scientist in Amsterdam, 400, 401.

Der Pfalz, Karl Theodor von 1724 – 1799: Elector Palatine, Duke of Bavaria, 113, 238, 314.

Devonshire, Duke of – see Cavendish.

Didot, François-Ambroise 1730 – 1804: Paris printer, publisher, 287.

Dietrichstein, Prince – see Xaver.

Dietrichstein-Proskau-Leslie, Johann Karl, Prince 1728 – 1808: Czech aristocrat, 474.

Dimsdale, Ann, 1715 – 1779: second wife of Thomas Dimsdale, 94.

Dimsdale, Elizabeth, 1732 – 1812: third wife of Thomas Dimsdale, 337, 419, 420, 446, 470, 477, 481, 490, 510, 515.

Dimsdale, Nathaniel 1748 – 1811: FRS, physician and politician, 142, 145, 150, 151, 419, 467, 468, 480, 517.

Dimsdale, Thomas 1712 – 1800: FRS, Quaker, physician, smallpox inoculator, 90, 91; attends a fatal smallpox case, 92, 93; youth and medical training, 94, 95; performs a mass inoculation with Jan Ingen Housz, 95, 96, 97; 103; practice defended by Ingen Housz, 107; 108, 109, 114, 132, 141; inoculates Catherine the Great in Russia, 142, 143, 145 and 150, 151; 148, 171, 173, hosts a visit from Ingen Housz, 191; 200, 202, returns to Russia to inoculate royal children, 337; is subject of an Ingen Housz book dedication, 389; reunited with Ingen Housz 405 and 419, 420; hosts visit from Ingen Housz at Bath 423; at Hertford, 435; in Bath, 439 and 446; at Hertford, 450, 451; at Hertford, 467 and 477; 470, 481, 490, 510; hears of death of Ingen Housz from Marquis of Lansdowne, 515; 517.

INDEX OF NAMES

Dominique, Jan Ingen Housz's manservant – see Tede.

Douglas, James 1702 – 1768: fourteenth Earl Morton, President of the Royal Society, astronomer, 33, 65.

Drew Dr. dates uncertain: personal physician to third Lord Holland and Lady Holland, 508.

Drummond, Alexander, died 1782: physician to Sir William Hamilton at Naples, 203, 205.

Du Bourg, Jacques-Barbeu 1709 – 1799: French physician, inventor, writer, 241.

Dumont, Pierre Étienne 1759 – 1829: Swiss pastor and writer, secretary to Lord Lansdowne, 438, 512, 513, 515.

Durfort, Marquis de 1715 – 1789: French ambassador in Vienna, 177.

Egerton, Francis 1736 – 1803: Duke of Bridgewater, entrepreneur, 187.

Eginton, Francis 1737 – 1805: Birmingham glass painter, 444.

Ehrmann, Frederic Louis 1741 – 1800: German, professor of physics, Strasbourg, 295, 366.

Einstein, Albert 1879 – 1955: German, theoretical physicist and mathematician, 345, 346.

Elizabeth I, 1533 – 1603: Queen of England from 1558, 71.

Elmsly, Peter 1735/6 – 1802: London bookseller, 185, 190, 279, 280, 409, 433.

Engelbert, Louis 1750 – 1820: sixth Duke of Arenberg, ruler of a German Duchy, 263, 269, 286.

Erasmus, Desiderius 1466-1536: Dutch, Renaissance humanist, 15.

Esterhazy, Prince Nikolaus 1714 – 1790: Hungarian prince, 457.

Euripides 480 BC – 406 BC: Greek playwright, 15.

Fabbroni, Giovanni 1752 – 1822: Florentine naturalist, economist, chemist, 352.

Fagel, Count, dates uncertain: Dutch, clerk to the Netherlands ambassador in London, 461.

Fagel, Hendrik 1706 – 1790: Dutch, chief minister of States General, 113, 114.

Falconer, William 1744 – 1824: FRS, Bath physician, 402, 424 – 427, 431, 439, 508.

Farmer, Richard 1735 – 1797: Master of Emmanuel College, Cambridge, Shakespearean scholar, 449.

Ferdinand 1751 – 1825: King of the two Sicilies (Naples), 100, 155, 156, 159 – 162, 165, 166.

Ferdinand 1754 – 1806: Archduke of Austria, ruler of Breisgau, 137, 138, 146, 147, 168, 178, 356.

Ferdinand 1769 – 1824, Archduke, became Grand Duke of Tuscany, 159.

Firmian, Karl Joseph von 1712 – 1782: Austrian aristocrat, minister in Milan, 199.

Fitzpatrick, John, 1719 – 1758: first Earl of Upper Ossory, 411.

Fitzpatrick, Louisa, 1755 – 1789: Lady Shelburne, then Marchioness of Lansdowne, 410, 411.

Fitzpatrick, Mary 1751 – 1778: wife of second Lord Holland, 411.

Folkes, Martin 1690 – 1754: President of the Royal Society, antiquarian, 40.

Fontana, Felice Abbé 1730 – 1805: Florentine physiologist and naturalist, 168 – 170, 189, 199, 211, 212, 232, 242, 256, 261- 266, 269, 271, 272, 279, 286, 296, 302, 304, 305, 313, 315, 334, 341, 379, 522.

Foster, John, dates uncertain: Headmaster of Eton College (1765 – 1773), 62.

Fothergill, John 1712 – 1780: FRS, Quaker physician in London, 90.

Fourcroy, Antoine François 1755 – 1809: French chemist, 384.

Fox, Charles James 1749 – 1806: Whig politician, 411.

Fox, Henry, 1773 – 1840: third Lord Holland, 411, 488 – 491, 507, 508.

Fox, Lady Caroline 1765 – 1845: sister of third Lord Holland, 411, 438, 439, 450, 477, 480, 488 – 490, 500, 506, 511, 513, 516.

Fox, Stephen 1745 – 1774: second Lord Holland, 411.

Francis II, 1768 – 1835: Grand Duke of Tuscany and Emperor of Austria, 159, 199, 200 – 202, 493, 494.

Francis Stephen 1708 – 1765: Austrian Emperor, 98, 99, 101, 117, 331.

Francois, Elizabeth, dates uncertain: inoculation patient of Jan Ingen Housz,

Florence, 196.
Franklin, 'Frankie' 1732 – 1736: son of Benjamin Franklin, 95.
Franklin, Benjamin 1706 – 1790: FRS, American printer, scientist, diplomat; 32; re-arrives in London 54; youth and middle age summarised 55; gives letters to Ingen Housz for translation into Latin 56, 57; 59; holidays in Holland and Germany with Sir John Pringle 61; obtains printer dies from Haarlem 63, 64; performs electrical experiments with Ingen Housz 65, 66; 68; regrets not having his son inoculated 95; 110; supports ballot to have Ingen Housz elected FRS 170; 172; reunited with Ingen Housz in London 183; serves on government advisory committee re. lightning conductors 184; the 'Franklin stove' 185; tours cities in north of England with Ingen Housz et al 186, 189, 190; dines with Ingen Housz, Pringle and Mlle. Biheron (c/f) 191, 192; 194; reports to Royal Society on torpedo fish 204; 205, 209, 210; involved in tension between England and American colonies 216, 217, 218; returns to Philadelphia 219; in talks about declaring American independence 221; 222; returns to Europe – to Paris 231; 232, 234, 236, 237, 238; visited at Passy by Ingen Housz 240, 241; advises Ingen Housz on investing in America 242; 243; organises a safe passage to England for Ingen Housz 244; consults Sir John Pringle on his skin problems (by letter) 245; elected (with Ingen Housz) to the Batavian Science Society 246; 248, 257, 258; positive and negative electricity 259; 260, 266; reunited with Ingen Housz at Passy 283; 284; his popularity in France 285; applies to return home to America 285; receives copy of Ingen Housz book 285; corresponds with Ingen Housz on electricity 286; 288, 289, 290, 291, 293; conduction of heat in metals 297, 299, 319; 305, 307, 308, 312, 318, 319; plans (unfulfilled) trip to Italy and Germany 320; writes on lightning strike on church at Cremona 320, 321; discusses electric shocks with Ingen Housz 322; 323, 335, early ballooning at Paris 346 – 50; 356, 357, 358; writes to Ingen Housz on latter's investments in America, 359, 360; returns to Philadelphia 360; 361; lobbies to have Ingen Housz elected fellow of American Philosophical Society 362; 364, 367; informs Ingen Housz further on his American investment failures 377, 378; 382; observes the first melting of platinum 383; 390, dies 415; 416, 521.

Franklin, Deborah 1708 – 1774: wife of Benjamin Franklin, 216.

Franklin, Rosalind 1920 – 1958: x-ray crystallographer, 521.

Franklin, William 'Temple' 1760 - 1823: grandson of Benjamin Franklin, 241.

Franklin, William 1730 - 1813: son of Benjamin Franklin, Governor of New Jersey, 216.

Frederick II 1712 – 1786: King of Prussia, 'Frederick the Great', 336, 374.

Frederick William 1744 - 1797: King of Prussia, nephew of Frederick the Great, 374.

Frederick William, 1768 – 1816: Prince of Nassau-Weilburg, nephew of Prince of Orange, 375.

Frederick, Prince of Wales 1707 – 1751: father of George III, 80.

Fries, Johann von 1719 – 1785: Alsatian, industrialist and banker, Vienna, 155, 172, 195.

Fürstemberger, dates uncertain: amateur physicist, Basel; invented 'lighter' after meeting Volta, 292, 295, 296, 356.

Fürstenberg, Prince - see Nepomuk.

Gage, Thomas, General 1719 – 1787: led British Army in America, 219.

Galiani, Ferdinando 1728 – 1787: Italian economist and diplomat, 164.

Galileo, Galilei 1564 – 1642: Italian physicist, mathematician and astronomer, 25, 169.

Garampi, Guiseppe 1725 – 1792: Papal Nuncio in Vienna, 339.

Garbett, Anne. 1773 – 1818: wife of Sir Samuel Romilly, 478, 480, 516.

Garbett, Samuel 1717 – 1803: Birmingham industrialist and financier, 444, 478.

Gardini, Francesco Guiseppe 1740 – 1816:

INDEX OF NAMES

Italian, professor of physics at Alba, 367, 368, 369.
Garth, Samuel, Sir 1661 – 1719: FRS, royal physician, 75.
Gatti, Angelo 1730 – 1798: Italian, professor of medicine at Pisa, smallpox inoculator, 220.
Gaubius, Hieronymus David 1705 – 1780: German, professor of medicine at Leiden, 27, 183.
George I 1660 – 1727: King of England from 1714, 4, 5, 76.
George II 1683 – 1760: King of England from 1727, 39, 76, 82, 115, 485.
George III, 1738 – 1820: King of England from 1760, 42, 53, 65, 102, 103, 108, 146, 155, 162, 245, 246, 403, 408, 428, 473, 485, 486.
Gest, Howard: professor emeritus of microbiology, University of Indiana, USA, 318.
Gräffer, Rudolf, died 1817: Vienna publisher, bookseller, 172, 198.
Graham, James 1745 – 1794: mountebank, sexologist, 439, 440, 441.
Greenwood, Thomas, dates uncertain, Vicar of Calne (1782 – 1821), xv, xvi, 515, 518.
Grenville, George 1712 – 1770: First Lord of the Treasury (Prime Minister), 163.
Grenville, Lady Louisa 1758 – 1829: wife of third Earl Stanhope, 435.
Greville, George 1746 – 1816: second Earl of Warwick, 411, 443, 511.
Grey, James, Sir 1708 – 1773: diplomat, 162.
Gruber, Abbé Leonhard 1740 – 1811: Viennese priest, 405 – 407.
Guillotin, Joseph-Ignace 1738 – 1814: Parisian physician, advocate of 'machine' for executions, 443.
Guyton de Morveau, Louis-Bernard 1737 – 1816: French chemist, 384.

Haen, Anton de 1704 – 1776: Dutch, pupil of Boerhaave, professor of medicine, Vienna, 103, 111, 130, 150, 171, 173, 206, 222 – 224, 293, 303.
Hahn, Friedrich von 1742 – 1805: German astronomer, 237.
Hales, Stephen 1677 – 1761: FRS, priest, scientist and inventor, 189, 272, 273.
Haller, Albrecht von 1708 – 1777: Swiss anatomist and physiologist, 169, 182.
Halley, Edmond 1656 – 1742: FRS, Astronomer Royal, 268.
Hamilton, Catherine 1737/8 – 1782: first wife of Sir William Hamilton, 162, 163.
Hamilton, William, Sir 1730 – 1803: FRS, diplomat, archaeologist, 153, 161 – 166.
Hancock, John 1737 – 1793: American statesman, 221, 222.
Handel, George Frideric 1685 – 1759: German-born composer, 83.
Harrach, Count Ernst von, dates uncertain: Bohemian aristocrat, 126.
Harrach, Count Ferdinand von 1708 – 1778: Bohemian aristocrat, 126.
Harvey, William 1578 – 1657:physician who discovered blood circulation, 25, 38, 272.
Hassenfratz, Jean-Henri 1755 – 1827: French chemist, 384, 386, 461, 469, 470.
Hastings, Warren 1732 – 1818: Governor-General, East India Company, 501, 505.
Haugwitz, Friedrich Wilhelm von, 1702 – 1765: Austrian politician, 99.
Haydn, Joseph 1732 – 1809: Austrian musician and composer, 457.
Heberden, William 1710 – 1801: FRS, London physician, 44, 63, 65, 110, 170, 183.
Hell, Maximilian 1720 – 1792: Hungarian astronomer, 224, 226.
Heneage family: prominent family of Compton Bassett House, near Calne, 423.
Henry Petty 1780 – 1863: third Marquis of Lansdowne, Chancellor of the Exchequer, 420, 477, 479, 480, 507, 515.
Herschel, Caroline 1750 – 1848: sister of Herschel, astronomer in own right, 407, 408, 428, 429.
Herschel, William 1738 – 1822: FRS, astronomer, 407, 408, 419, 428 – 430, 488.
Hewson, Mary 1734 – 1795: daughter of Margaret Stevenson, Franklin's London landlady, 285.
Highett, family: family of a Calne carpenter, many of whom suffered from tuberculosis, 481, 482.
Hodges, George 1757/8 – 1768: smallpox victim of Berkhamsted, Hertfordshire, 92, 93, 96, 200.

Hoff, Hebbel, dates uncertain: Twentieth century Texan physiologist, 343.
Hogarth, William 1697 – 1764: London artist, 85.
Holland, third Lady, see Webster, Elizabeth.
Holland, third Lord – see Fox, Henry.
Home, Everard 1756 – 1832: FRS, London surgeon, 498.
Home, James 1719 – 1833: Scottish physician, 352.
Homer, dates uncertain: Greek poet, 15.
Hoogeveen, Hendrik 1712 – 1791: Dutch, schoolteacher at Breda and latterly head teacher of the Delft Latin School; classics scholar; birth and youth 16; marriage 17; 21, 27; asks Ingen Housz to organise publication of his *Particles* in London 30, 59 – 66; 35, 49, 50, 67, 68, 109; becomes head teacher at Delft Latin School 116; visited by Ingen Housz 116; *Particles* offered for sale, by Ingen Housz, in Austria and Germany 124, 140, 145, 172, 198; delayed publication of *Particles* in London 185; 207; hears that Ingen Housz has married in Vienna 229, 230; visited by Ingen Housz 248, 286; introduces Ingen Housz to Jacob van Breda 286; 287; is asked to correct Ingen Housz's Latin texts 288; 294, 372; suffers in the Patriot Revolt 376; revisited by Ingen Housz 398; 422; dies 441; his posthumous *Analogical Dictionary* published, with help of Ingen Housz, in London 448 – 450, 456, 467, 474, 475, 506.
Hoogeveen Jan, dates uncertain:.son of Hendrik Hoogeveen, 441, 448 - 450, 467, 475, 485.
Hoogeveen Theodorus, died 1787: son of Hendrik Hoogeveen, 376.
Hoogeveen, Theophilus, died 1787: son of Hendrik Hoogeveen, 376.
Horace 65 BC – 8 BC: Roman poet, 416, 422, 436.
Houlston, Thomas, dates uncertain: English physician, inoculator sometime at Vienna, 131.
Huck, Richard (later Huck-Saunders) 1720 – 1785: FRS, physician, 44, 65, 88, 111, 126, 128, 130, 170, 171, 183, 184, 219, 226, 230, 279, 314, 317.
Hulme, Nathaniel 1732 – 1807: FRS, physician to Charterhouse, 258, 259, 387, 425, 431, 432.
Hulst, Pieter Teyler van der 1702 – 1778: Haarlem banker, merchant, 399.
Hume, David 1711 – 1776: Scottish philosopher and historian, 64.
Hunter, John 1728 – 1791: Scottish surgeon working in London, 497, 498.
Hunter, William 1718 – 1783: Scottish anatomist, medical teacher in London, 65, 88, 192, 196 – 198.
Hussel, Major, dates uncertain: officer in Dutch Army, 1, 8.
Hutchinson, Thomas 1711 – 1780: Governor of Massachusetts, loyalist, 216, 218.

Ingen Housz Arnold. 1693 – 1764: father of Jan Ingen Housz, 1 - 3, 5 - 14, 17 - 21, 23, 27, 29, 31.
Ingen Housz, Agatha: wife of Jan Ingen Housz - see Jacquin, Agatha.
Ingen Housz, Arnold Joseph 1766 – 1859: son of Louis, 116, 183, 397, 475, 476, 517, 518.
Ingen Housz, Caspar born 1697: younger brother of Arnold Ingen Housz, 12, 19.
Ingen Housz, Cornelis 1690 – 1710: elder brother of Arnold Ingen Housz, 3.
Ingen Housz, Hendrica – see Van Beest.
Ingen Housz, Henry Ferdinand 1768 – 1838: son of Louis, 183, 397, 432, 457.
Ingen Housz, Jan (junior) 1769 – 1839: son of Louis, 183.

INGEN HOUSZ, JAN 1730 – 1799, birth, 12; death of mother, 13; at school, 14 – 17; meets Hendrik Hoogeveen, 16; meets John Pringle 20; university, Louvain 25; MD degree, 27; postgraduate studies, Leiden, 27, 28; doctor at Breda, 29 – 31; gives up practice, 31; father dies, 31; arrives in London, 33; reunited with Pringle, 35; to Edinburgh, 44; returns to London, 46; medical refresher with Pringle, 52; meets Benjamin Franklin, 54; translating works, 53 – 57; electrical machine, 58, 59; publishing Hoogeveen's *Particles*, 59 – 65; visits Oxford, Cambridge, 66; to London Foundling Hospital with William Watson, 80; inoculating at

INDEX OF NAMES

Foundling Hospital, 84 – 87; attending sick children with George Armstrong, 88; meets Daniel Sutton, 89; travels to Hertford, 90; meets Thomas Dimsdale, 93; attends smallpox case with Dimsdale, 92 – 94; inoculates villagers of Berkhamsted, 96 – 98; proposed as inoculator of royal family at Vienna, 103; called back from Hertford, 104; writes letter on inoculation in Netherlands, 107; returns to Hertford, 107; appointed at Vienna, 108, 109; leaves London for Vienna, 111; travels to Vienna, 113 – 118 & 123 – 125; meets William V of Orange, 115; meets Johann Winckleman, 123; meets van Swieten, 126; meets Kaunitz, 127; meets Maria Theresa and her younger children, 136; meets Joseph II, 138; starts inoculating (poor children) at Meidling, 139; inoculates royal children, 146; celebrations in Vienna, 147; inoculates at Castle St. Vitus, 148; travels to Italy, 155; in Rome, 157; at Naples, 159; inoculates King of Naples, 159 – 160; meets William Hamilton, 161; visits Pompeii and Herculaneum, 165; to Florence, 166; inoculates Grand Duke of Tuscany, 166; meets Felice Fontana, 169; elected Fellow of Royal Society, 170; demonstrates inoculation in Italy and southern Austria, 171 – 172; returns to Vienna, 172; befriends Nikolaus Jacquin, 173; travels to Bohemia and Moravia to demonstrate inoculation, 174; returns to Vienna, 176; leaves Vienna for London, 176; at Versailles, 179 – 182; meets von Haller, 182; at Breda, 183; returns to London, 183; admitted FRS, 184; travels around England with Franklin, 186 – 190; meets Joseph Priestley, 188; meets James Cook, 191; returns to Versailles 194; returns to Vienna, 195; travels to Italy again, 199; inoculations in Florence, 199 – 200; torpedo experiments, 205; returns to Vienna, 206; returns to Prague, 206; begins scientific experimenting, 209; invents eudiometers, 212; platinum experiments, 215; challenges Mesmer, 225; marries Agatha Jacquin, 230; travels to Prague and Berlin, 232; electrophore experiments, 232; candle-lighting, 235, travels to Bavaria, 236; travels in Germany and Netherlands, 237; meets Elector Palatine, 238; to Paris, reunites with Franklin, 240 – 245; reunites with Fontana at Paris, 242; to Breda, 246; travels in Holland, 246 – 249; returns to England, 257; reunites with Lord Shelburne and Joseph Priestley, 257; translates Hulme's book on kidney stones, 258; 1778 Bakerian Lecture of Royal Society, 259; air experiments with Fontana in London, refining the eudiometer, 261; refining the air pistol, 264; 1779 Bakerian Lecture, 266; paper on marine compasses, 267; to Southall Green, 273; experiments on plants, 273 – 280; publishes *Vegetables*, 280; travels to Bowood and Bath, 280; says goodbye to John Pringle, 280; sails for the continent, 281; air experiments at sea, 282; returns to Paris, 284; writes French version of *Vegetables*, 285; says goodbye to Benjamin Franklin, 286; returns to Vienna, 287; investments in America, 288 – 291; writes *Vermischte Shriften*, 294; writes *Nouvelles Experiences,* 294; improves the Fürstemberger lighter, 295; heat conduction in metals experiments, 297; publishes *Rudiments of Electricity*, 300; oxygen as therapy, 300 – 304; vacuum apparatus, 304; burning metals in oxygen, 305 – 307; conflict with Priestley, 311; more plant experiments, 315; *Further Considerations* to Royal Society, 317; concussed by electric shock, 321; conflict with van Barneveld, 323; conflict with Senebier, 326; advises Emperor on Freemasonry, 330; entertains Joseph II in laboratory, 335; entertains Grand Duke, and Grand Duchess of Russia in laboratory, 338; Pope Pius VI in Vienna, 339; the 'Green Matter', 340; invents cover slip, 342; new edition of *Vermischte*, 342; Brownian motion, 344; ballooning, 346 – 350; tries to freeze mercury, 351; visitors, 352 - 356; entertains Emperor and Grand Duke of Tuscany in laboratory, 356; meets Volta, 356; American investments fail, 356 – 362; elected fellow of American Philosophical Society, 362; experiments on electricity in plants, 364 – 372; revolution in Netherlands, 372 – 377; new editions of *Vegetables*, in French and in German, 379; brother, Louis, dies, 380; leaves Vienna for Paris, 381; meets Jefferson, 382; 'new chemistry' in Paris, 383 – 386; severely ill in Paris, 387; new edition of *Nouvelles Experiences*, 387; the French Revolution, 388 – 395; escapes from Paris 15 July 1789, 395; at Breda, 397; meets van Marum in Haarlem, 398; in Amsterdam, 400; returns to England, 402; treated for kidney stones with alkaline soda water, 403; reunites with many old London friends, 404 – 407; reunites with Josef Jacquin in London, 406; visits Herschel, 407, reunites with Lord Shelburne (now Marquis of Lansdowne) and his circle, 409 – 411; introduces van Marum in London, 417 – 419; revisits Hertford, Dimsdales, 419; Bowood and Bath, 420 – 427; meets William Falconer, 424; contributes to *Scheikundige Bibliotheek*, 430 – 432; elected to Royal Society Council, 430; elected member of Science Society of Holland, 434; visits Chevening, 434; Lady Caroline Fox, 439; meets Tallyrand, 443; defers returning to Vienna, 443; visits Birmingham, 444; arranges publication of Hoogeveen's *Analogical Dictionary*, 448 – 450; visits Thomas Beddoes in Bristol, 452, oxygen therapy in Bristol, 454; to Twickenham Common to write new book, 459, begins manure experiments, 460; new book – *Miscellanea physico-medica* published in Vienna, 462; writes paper for Board of Agriculture, 466; *On the food of plants* published, 471; elected member of Board of Agriculture, 472; 'busy' autumn at Bowood 1796, 477; treats patients in Calne, 481; further visit to Beddoes at Bristol, 482; arranges election of van Marum to Royal Society fellowship, 485; soil experiments at Kew and Hertford, 486; financial problems in London, withholding of Vienna pension, 492 – 496; dispute with Edward Jenner, 496 – 506; health deteriorates, 506; cared for at Bowood, 508; consults Dr. Falconer at Bath – terminal illness diagnosed, 508; last journey - to London and back to Bowood, 509; lapses into coma and dies 6/7 September 1799, 512; buried under St. Mary's Church, Calne, 515; Agatha Ingen Housz dies at Vienna, October 1800, 518: subsequent historical appraisal of Jan Ingen, 519.

Ingen Housz, Josina Agatha 1776 – 1832: 232.

Ingen Housz, Louis (junior) 1770 – 1774: 183.

Ingen Housz, Louis , died 1711: father of Arnold Ingen Housz, 3.

Ingen Housz, Louis 1729 – 1788: brother of Jan Ingen Housz, leather merchant and apothecary; birth 12; schooling 13; enters family business 19, 23; 29; marries 31; handles Jan's affairs while latter is abroad 34, 35; takes over family business on death of father 35; visited by Jan 116; revisited by Jan 183; 209; receives medallions from Jan 220; hears that Jan is married 230; 231, 239, 242; wife very ill, visited urgently by Jan 246; 286, 308; feels threatened by Patriot Revolt 372, 374; loses money invested in Amsterdam 377; 378; dies after an accident 380; 381, 397, 457.

Ingen Housz, Maria - see Stuyck.

Ingen Housz, Maria Catherina 1765 – 1812: daughter of Louis, 116, 183.

Ingen Housz, Martin 1685 - 1710: oldest brother of Arnold Ingen Housz, 3.

Ingen Housz, Peter William 1767 – 1848: son of Louis, 116, 183, 517.

Isabella of Parma 1741 – 1763: first wife of Joseph II, 98, 99.

Ismay, Joseph, dates uncertain: vicar of Mirfield, near Huddersfield, 45, 46.

Ismay, Thomas, dates uncertain: medical student, Edinburgh, 45.

Jacquet-Malzet, Louise-Sebastien de 1715 – 1800: French, priest, teacher settled at Vienna, 230.

Jacquin, Agatha 1735 – 1800: wife of Jan Ingen Housz; living with her brother

INDEX OF NAMES

and family in Vienna 219; meets Jan Ingen Housz 229; marries Jan Ingen Housz, 230; sets up home with Ingen Housz 230 – 232; 244, 269, 291; prominence in Vienna society 293, 295; moves home in Vienna 317; 352, 355; hostess to visiting scientists 356; 404; problems with servants 441; forced to move home again 447; 460; under threat of having to leave Vienna 494; 496; receives last letter from husband 511; 515, 516; dies 518.

Jacquin, Franziska 1769 – 1830: daughter of Nikolaus Jacquin, pupil of Mozart, 219, 293, 442.

Jacquin, Gottfried 1767 – 1792: son of Nikolaus Jacquin, friend of Mozart, 219, 293, 352, 442.

Jacquin, Josef 1766 – 1839: professor of chemistry, Vienna after his father; 170; childhood 219; 228, 232; tours Europe 405; meets Jan Ingen Housz in London by chance 405; 406; at Royal Society 407; in Paris 412, 414, 415; corresponds with Ingen Housz from Vienna 442, 484, 490; marries; helps Agatha Ingen Housz move apartment 447; publishes book on chemistry 451; 458, 462, 464; distributes *Miscellanea* in Vienna 473, 474, 483; tries to help with Ingen Housz financial problems in Vienna 493, 494, 496; 513; administers the Ingen Housz estate at Vienna 518.

Jacquin, Katharina – see Schreibers.

Jacquin, Nikolaus Josef von 1727 – 1817: Dutch, professor of botany and chemistry, Vienna; meets Jan Ingen Housz 173; publishes *Hortus Botanicus Vindobonensis* 176; 182, 190; reassures Ingen Housz that he has friends at Vienna 191; 195; helps Ingen Housz buy gemstones for William Hunter at Vienna 196; publishes *Flora Austriacae* 199; helps Ingen Housz buy mining shares 219, 228, 229, 232; witnesses marriage of Ingen Housz 230; 231, 232; 246, 279, 293, 310, 327, 352; helps train botanical artist, Franz Bauer 405; has a book presented to the Royal Society by Ingen Housz 407; widowed 436; 474, 484; writes to Ingen Housz about latter's financial crisis at Vienna 494; 515; is commemorated, 1905, in Vienna University 524.

James II 1633 – 1701: King of England from 1685 (James VII of Scotland), 4, 5, 7.

James III 1688 – 1766: (James VIII of Scotland), 'The old pretender', 4, 7.

Jamin, Jules 1818 – 1886: French physicist, 299.

Janety, Marc Étienne 1738 – 1820: French, jeweller in platinum, 383, 412.

Jantzon, Nicolaas, dates uncertain: Dutch, lawyer at Breda, 66 – 68, 90, 109.

Jay, John 1745 – 1829: American politician, 221, 222.

Jean (Jong), Ferdinand de 1731 – 1797: Dutch, physician, amateur musician, 294.

Jefferson, Thomas 1743 – 1826: third President of the United States, 222, 320, 382.

Jenner, Edward 1749 – 1823: FRS, physician who developed smallpox 'vaccination', 76, 496 – 506, 519, 522.

Joanne Gabriella 1750 – 1762: Archduchess of Austria, 98, 100, 159.

John Henry 1765 – 1809: Lord Wycombe, second Marquis of Lansdowne, 353, 354, 410.

Johnson, Samuel 1709 – 1784: writer and dictionary editor, 33, 449.

Joseph II 1741 – 1790: Emperor of Austria; with his father when latter dies in the street 98; widowed 99; remarries 99; widowed for second time 100; meets Jan Ingen Housz, asks him to inoculate his daughter 138; 142, 146, 148; leaves for Italy 155; at Naples 159; 160, 165, 166, 167, 171, 190; with his brother at Florence when latter is inoculated 194, 260; asks Ingen Housz to install lightning conductors at Vienna 209; 215, 234, 235, 236, 260, becomes full Emperor when mother dies 293; 295; 309, 317; and Freemasonry 331 – 333; visits Ingen Housz at his apartment/laboratory 334, 335; meets with Catherine the Great in Russia 336; welcomes Grand Duke Paul to Vienna 337; 338; welcomes the Pope to Vienna 339; 340, 349; commissions Allgemeine Krankenhaus 351;353, 355; revisits Ingen Housz in his apartment 356; ends free lodgings of court officials 358; 360, 371, 379, 389, 405, fights the Turks 416; health deteriorates and dies 416; 419, 433, 474, 521.

Kanter, Marianne de, dates uncertain: Dutch, nursemaid of Louis and Jan Ingen Housz, 13, 14, 220.

Kaunitz, Count Ernst Christoph, von 1737 – 1797: Austrian, Imperial ambassador at Naples, 161.

Kaunitz-Rittberg, Wenzel Anton, von 1711 – 1792: Austrian statesman, Chancellor of State and minister of foreign affairs, 102, 126 - 128, 148, 193, 223, 309, 329, 331, 339, 340, 353.

Keith, George 1692 – 1778: tenth Earl Marischal, Scottish army officer, served in Prussia, 8.

Keith, Robert, Murray- 1730 – 1795: British ambassador in Vienna, 236, 316, 317, 352, 353.

Kier, James 1735 – 1820: Scottish chemist, member of Lunar Society, 445.

King of Spain – see Charles.

Kippis, Andrew 1725 – 1795: FRS, Presbyterian minister in London, 44, 310.

Klaproth, Martin 1743 – 1817: German chemist, 432.

Klinkock, Josef Thaddeus 1734 – 1778: Bohemian, professor of anatomy, Prague, 232.

Knight, Gowin 1713 – 1772: FRS, expert on magnetism, 170, 224, 268.

Krauss, Johann Paul, dates uncertain: Vienna publisher and bookseller, 307, 308.

L' Epée, Charles de 1712 - 1789: French philanthropist, developed signing for the deaf, 243, 244.

Landriani, Marsilio 1751 – 1815: Italian, professor of experimental physics, Milan, 211, 329.

Lane, Anna Louisa, dates uncertain: portraitist, sister of William Lane, 157.

Lane, William 1746 – 1819: portraitist, 157.

Lansdowne, Marchioness – see Fitzpatrick, Louisa.

Lansdowne, Marquis of – see William Petty-Fitzmaurice.

Laugier Alexandre-Louis 1719 – 1774: French, royal physician to Habsburgs, 156, 157, 159, 166, 167, 171, 172, 199.

Laugier, Robert Francois 1722 – 1793: French, one time professor of botany, Vienna, 156, 173.

Launay, Marquis de 1740 – 1789: French, governor of Bastille prison, 392.

Lavoisier, Antoine 1743 – 1794: French, chemist and economist, 383 – 385, 394, 395, 400, 412, 414, 442, 456, 459.

Le Roy, Jean Baptiste 1720 – 1800: French academician and electrical expert, 242, 349.

Ledru, Nicholas-Philippe 1731 – 1807: Parisian showman-scientist, 243, 364, 365.

Legge, William 1731 – 1801: Lord Dartmouth, FRS, statesman, 217.

Leigh, Revd. dates uncertain: Vicar of Adlestrop in Gloucestershire in 1798, 503, 505.

Leopold Clement, Prince 1707 – 1723, 101.

Leopold, Peter 1747 – 1792: Grand Duke of Tuscany, Emperor of Austria, 98, 139, 149, 155, 166 – 172, 196, 199, 201, 232, 474.

Levett, Richard, dates uncertain: rector in Berkhamsted, Hertfordshire, 92.

Lewis, William 1708 – 1771: FRS, chemist, 214.

Lichtenberg, Georg Christoph 1744 – 1799: FRS, German, professor of physics, Göttingen, 306.

Liddel Ann 1745 – 1804: wife of second Earl of Upper Ossory, 439.

Linne, Carl von (Linneaus) 1707 – 1778: Swedish, developed hierarchical taxonomy, 341, 342.

Locherer, dates uncertain: Austrian, head of Vienna maternity hospital, 131.

Louis (August) XVI 1754 – 1793: grandson of Louis XV, King of France from 1775, 177 - 180, 243, 347, 387, 389, 394, 446.

Louis XIV 1638 – 1715: King of France from 1643 (1661), 179.

Louis XV 1710 – 1774: King of France from 1723, 177, 178, 180.

Lucretius 99 BC – 55 BC: Roman poet, 422.

Lyonet, Pierre 1708 – 1789: Dutch naturalist, artist and engraver, 247.

INDEX OF NAMES

Macauley, Thomas Babington 1800 – 1859: historian, 71.
MacGregor, Robert Roy, 'Rob Roy' 1671 – 1734: Scottish Highland outlaw, 1, 7, 8.
Macquer, Pierre Joseph 1718 – 1784: French chemist, 384.
Magellan, Jean Hyacinth 1722 – 1790: FRS, mathematician, 409.
Mahon, Charles, third Earl Stanhope 1753 – 1816: FRS, aristocrat and scientist, 434, 435.
Maitland, Charles, dates uncertain: physician, smallpox inoculation pioneer, 73, 75, 76, 77.
Maria Anna 1770 – 1809: Archduchess of Austria, 196, 201.
Maria Antonia (Marie Antoinette) 1755 – 1793: Archduchess of Austria, Queen of France, 99, 175 - 181, 194, 260, 347, 389, 394, 451, 452.
Maria Carolina 1752 – 1814: Archduchess of Austria, wife of King of Naples, 156, 159, 160, 161, 177.
Maria Elisabeth 1743 – 1808: Archduchess of Austria, 100.
Maria Fedorovna, formerly Princess Sophia Dorothea of Würtemberg 1759 – 1828: wife of Czar Paul of Russia, 336 – 338.
Maria Josepha 1751 – 1767: Archduchess of Austria, 100, 159, 161.
Maria Josepha of Bavaria 1739 – 1767: second wife of Joseph II, 99, 100.
Maria Luisa 1745 – 1792: Infanta of Spain, wife of Leopold, Grand Duke of Tuscany and Emperor, 98, 159, 196, 201.
Maria Theresia 1762 – 1770: Archduchess of Austria, daughter of Joseph II, 138, 146, 147.
Marie Christine 1742 – 1798: Archduchess of Austria, wife of Albert of Saxony, 99, 100.
Marie Theresa 1717 – 1780: Empress of Austria from 1740; 19, 20; widowed at Innsbruck 98; loses ten of her family/household, many to smallpox 98, 99; 100; origin of her marriage 101; determines to bring a smallpox inoculator to Vienna 101; 102, 103; writes to her ambassador in London 108; 114, 115, 121, 123; spends summers at Schönbrunn Palace 124; recruits Gerard van Swieten to Vienna 128, 129; 130, 132, 133; gives audience to Jan Ingen Housz 136; 137 –140; observes Ingen Housz inoculating children at Meidling 141; 142, 143, 145, 146; arranges celebrations after safe inoculation of her children 147; arranges an inoculation hospital for Ingen Housz 148; rewards Ingen Housz handsomely 148, 149; presents Ingen Housz with medallions 151; 154; sends Ingen Housz to Italy to inoculate 155, 159; writes personally to Ingen Housz to thank him 167; 175, 177, 178; sends Ingen Housz to Versailles to investigate apparent infertility of Marie Antoinette 179; 180, 181, 193, 194; mourns death of van Swieten 198; 202; presents Ingen Housz with micropicture jewel 207; offers Ingen Housz succession to van Swieten 207; 208, 219, 222; arranges marriage for Ingen Housz 229; 236, 289; has suffered serious fall and is ill 292; in heart failure and dies 293; 331, 335, 336, 339, 356, 358, 416, 495, 521, 524.
Marischal – see Keith, George.
Marlborough, Duke of, see Churchill, John.
Martin, Benjamin 1704 – 1782: London optician, scientific instrument maker, 282.
Marum, Martin van 1750 – 1837: Dutch, FRS, first director, Teyler's Museum, Haarlem, 398, 399, 400, 404, 409, 412, 417, 418, 429, 430, 434, 461, 474, 476, 485, 511.
Mary II 1662 – 1694: Queen of England from 1689 with her husband, William, 4, 72.
Mary, 1542 – 1587; Queen of Scots, 71.
Maskelyne, Neville 1732 – 1811: FRS, Astronomer Royal, 34, 183, 184.
Mather, Cotton 1663 – 1728: Puritan minister at Boston, promoter of smallpox inoculation, 95.
Mathis, Louis, dates uncertain: Ingen Housz's servant in Vienna, 125, 195.
Maty, Matthew 1718 – 1776: Dutch, lived London, FRS, writer, editor, secretary Royal Society, 33, 34, 47, 61, 62, 66, 104, 106, 107, 110, 170, 183, 407, 487.

Maurice, Adrien, 1678 – 1766: third Duke of Noailles, French aristocrat and soldier, 41.
Maximilian 1756 – 1801: Archduke of Austria, 137, 138, 146, 147, 168.
Maximilian Joseph 1756 – 1825: nephew of Elector Palatine, 238.
Mayer, Christian 1719 – 1783: Czech astronomer at Mannheim, 238.
McLellan, Robert, dates uncertain: apothecary, London Foundling Hospital, 85.
Mead, Richard 1673 – 1754: FRS, physician and antiquarian, 40, 75.
Meade family: of Blacklands House, Calne, 423.
Meerman, Gerard, dates uncertain: Dutch bibliophile, 67, 68, 109, 238.
Meissner, Carl 1800 – 1874: professor of Botany at Basle, 523.
Mercy-Argenteau, Compte de 1727 – 1794: Austrian ambassador at Versailles, 180, 181, 194.
Mesmer, Franz Anton 1734 – 1815: German physician and illusionist, 223 – 227, 241, 333, 440.
Metherie de la, Jean-Claude 1743 – 1817: French, naturalist and geologist, 367, 384.
Middleton, David, dates uncertain: surgeon to the King, 102.
Milman, Francis 1746 – 1821: FRS, President, Royal College of Physicians, 209.
Mochet, Franz, dates uncertain: Ingen Housz's servant in Vienna, 126, 195.
Molitor, Niklas 1754 – 1826: German, professor of chemistry, Mainz, 287, 300, 307, 308, 314, 379.
Monchy, Salomon de 1716 – 1794: Dutch, professor of medicine, Rotterdam, 183, 246, 248, 301, 372, 376, 402.
Money-Kyrle family: prominent landowning family near Calne, 423.
Monge, Gaspard 1746 – 1818: French mathematician and physicist, 384.
Monro, Alexander ('secundus') 1733 – 1817: Scottish physician, medical teacher, 45, 46.
Monro, Donald 1728 – 1802: FRS, Scottish physician working in London, 44.
Montgolfier, Jacques-Etienne 1745 – 1799: French balloonist, 347, 348.
Montgolfier, Joseph-Michel 1740 – 1810: French balloonist, 347, 348.
Morton, Charles 1716 – 1799: FRS, British Museum librarian, 487.
Morton, Lord – see Douglas.
Morveau Louis-Bernard, Guyton de 1737 – 1816: French chemist, 384.
Mozart, Leopold 1719 – 1787: Austrian, father of Wolfgang Amadeus Mozart, 100, 223.
Mozart, Maria Anna ('Nannerl') 1751 – 1829: sister of Mozart, 100.
Mozart, Wolfgang Amadeus 1756 – 1791: Austrian, musician and composer, 100, 174, 223, 293, 294, 330, 331, 333, 442.
Musin-Pushkin, Aleksey, Count, dates uncertain: Russian ambassador in London, 142, 145.
Musschenbroek van, Pieter 1692 – 1761: Dutch scientist, electrical expert, 27, 28.

Nairne, Edward 1726 – 1806: London optician, scientific instrument maker, 264, 265.
Naples, King of – see Ferdinand.
Nash, Richard 'Beau' 1674 – 1762: master of ceremonies at Bath, 423.
Necker, Jacques 1732 – 1804: French financier, politician from Geneva, 389, 391.
Needham, John Turberville 1713 – 1781: FRS, naturalist, 208.
Nekrep, Johann d.1785: Viennese priest and teacher, 349, 350.
Nelmes, Mary, dates uncertain: cowpox sufferer 1796, patient of Edward Jenner, 498.
Nepomuk, Joseph Wenzel Johann, 1728 – 1783: second Prince Fürstenberg, Czech, 232.
Newton, Isaac 1643 – 1727: President of Royal Society, mathematician and scientist, 26, 28, 34, 264, 271, 272.
Noailles, Comtesse de - see D'Arpajon.
Nollet, Jean-Antoine 1700 – 1770: French, professor of physics, Paris, expert on electricity, 367.
Nostitz-Rieneck, Franz Anton, Count 1725 – 1794: Bohemian, builder of Estates Theatre, Prague, 206.

INDEX OF NAMES

Oesterlin, Franziska, dates uncertain: Austrian, cousin of wife of Mesmer, Mesmer's patient, 223, 225, 226.

Oliver, Andrew 1706 – 1774: Massachusetts politician, 216, 218.

Oliver, Charlotte, died 1754: wife of John Pringle, 41.

Oliver, William 1695 – 1764: FRS, Bath physician, inventor of the 'Bath Oliver' biscuit, 41.

Orange, Prince of – see William.

Orsini-Rosenberg, Franz, Prince 1723 – 1796: Austrian, courtier, 268, 427, 461, 462, 474, 484.

Ovid 43 BC – 17/18 AD: Roman poet, 462.

Owen, Reverend Humphrey 1702 – 1768: Principal of Jesus College, Oxford, 66.

Pacassi, Nikolaus 1716 – 1790: Italian, final architect of Schönbrunn palace, 148.

Palatine, Elector – see Der Pfalz.

Paradise, John 1743 – 1795: FRS, mathematician and classicist, 263, 382.

Parr, Samuel 1747 – 1825: schoolmaster, minister of religion and writer, 449, 474, 475.

Parry, Caleb 1755 – 1822: FRS, Bath physician, 499.

Parsons, James 1705 – 1770: FRS, physician, foreign secretary of the Royal Society, 170.

Pas, Peter van der, dates uncertain: Dutch engineer, author on Jan Ingen Housz, 344.

Patzowsky, A, dates uncertain: Viennese publisher, 464.

Paul, Grand Duke 1754 – 1801: son of Catherine the Great. Russian Czar from 1796 – 1801, 142 – 145, 150, 151, 336 – 338, 419, 483.

Paytherus Thomas, dates uncertain: English physician, friend of Edward Jenner, 504.

Pearce, William, dates uncertain: Fellow, Jesus College, Cambridge, Dean of Ely Cathedral, 475.

Pearson, Thomas, dates uncertain: English physician, 354, 355.

Pelletier, Bertrand 1761 – 1797: French chemist, 383, 412, 442.

Pepys, Samuel 1633 – 1703: FRS, civil servant, MP and diarist, 43.

Percival, Thomas 1740 – 1804: Manchester physician, 425.

Petty, James died 1822: FRS, Surrey landowner, 476, 484, 485.

Phipps, James born around 1788: son of Edward Jenner's gardener, subject in first vaccination experiment, 498.

Pickel, Georg 1751 – 1838: German, professor of chemistry, Würzburg, 296, 297, 303, 306.

Pitt, William (the younger) 1759 – 1806: Prime Minister, 446, 465, 492.

Pitt, William, the elder 1708 – 1788: Lord Chatham, politician and 'Prime Minister', 435.

Pitters family: Ingen Housz's landlords in London, 65, 183, 261.

Plaggenborg, John, dates uncertain: Ingen Housz's last landlord in London, 510.

Pliny, Gaius (the younger) 61 AD - 112 AD: Roman lawyer and author, recorded the 79 AD eruption of Vesuvius, 163.

Pope Martin – see Colonna.

Pope Pius VI - see Braschi.

Pope, Alexander 1688-1744: satirical poet, 73.

Portland, Duke of - see Bentinck.

Priestley, Joseph 1733 – 1804: Unitarian minister, religious reformer, educationalist, scientist; meets Ingen Housz at Leeds 188; youth and development 188, 189; originator of eudiometry 210, 211; 252; employed by Lord Shelburne 257; experiments with Jan Ingen Housz in London 257; 258, 261, 263; publishes Ingen Housz ether gun details 264; 265, 269; shows that plants sometimes revitalise the air 270, 271; pneumatic chemistry 271; 272, 274, 275, 277; re-united with Ingen Housz at Bowood 280; 285, 302; conflict with Ingen Housz 311 – 314; 315, 316, 318, 319, 323, 326, 327, 330; dispute with Ingen Housz on the 'green matter' 340, 341; 352, 354; continues to assert his precedence over Ingen Housz's discoveries 355; 379; persists with phlogiston theory 386, 400, 413, 414; 422, 425; flees the riots at Birmingham 436, 437; 445, 520; features in *Biographie Universelle*. 523.

Pringle, Charlotte, see Oliver.

Pringle, Francis 1699 – 1746: professor of Greek, University of St. Andrews, 20.

Pringle, John, Sir 1708 – 1782: President of Royal Society, royal physician and army doctor; meets Ingen Housz family in Breda 20; youth and education 21; befriends and patronises Jan Ingen Housz 21; 23, 24, 25, 32, 35; invites Ingen Housz to London 36; 38; miltary career 39, 40; elected FRS 40; famous books published 40, 41; conceives sanctity of battle wounded 41; marries 41; widowed 41, 42; royal physician 42; weekly intellectual soirées at his house 43, 44; arranges further medical training for Ingen Housz 44; 46, 47, 50; gives Ingen Housz access to all his prescriptions 51, 52; 53 – 57; knighted 61; tours Holland and Germany with Benjamin Franklin 61; 62, 64; electrical experiments with Franklin and Ingen Housz 65, 66; invites the King to subscribe to *Particles* 65; 68, 80, 87; advises Ingen Housz to widen his medical experience 88; supports Dr. George Armstrong; 88, 90, 94; writes to Dr. Brady at Brussels re. smallpox inoculation 101; recommends Ingen Housz for Vienna 103; tells Ingen Housz of his selection as royal inoculator at Vienna 104; 107, 109; gives Ingen Housz letter of introduction to Gerhard van Swieten 110; 111; receives fractious letter from Ingen Housz at Vienna 126; advises and cautions Ingen Housz 128, 130; 131; receives letter of thanks from van Swieten 141, 148; updated on inoculation at Vienna by Ingen Housz 154; 165; supports ballot for Ingen Housz to be elected FRS 170, 182; re-united with Ingen Housz in London 183; hosts Ingen Housz at Royal Society club 184; 190, 191; dinner with Mlle. Biheron (c/f) 192; 194, 198, 199, 201; writes to Ingen Housz at Florence 202; elected President of Royal Society 203; encourages Ingen Housz in Torpedo experiments 203, 205, 206; invites Ingen Housz to be secretary to Royal Society 209; 210, 211, 214, 225, 226, 231; invites Ingen Housz to give 1778 Bakerian lecture of Royal Society 235; 245, 256, 257; unwell after an accident 258; advises Benjamin Franklin on his skin problems (by letter); 259; resigns as PRS 260; 271; hears of Ingen Housz's 'sunlight' discovery 275; 276 – 280; encourages Ingen Housz to do eudiometry at sea 282; 283, 300; retires fully from medical practice and removes to Scotland 310; returns to London 310; suffers a stroke and dies 310; 367, 379, 404, 406, 449, 521.

Pushkin. see Musin.

Quarin, Joseph von 1733 – 1814: Viennese physician, 351.

Querini, Angelo 1721– 1796: Venetian senator, 370.

Radcliffe, John 1652 – 1714: royal physician to William and Mary, 72.

Ramsden, Jesse 1735 – 1800: London scientific instrument maker, 59, 266.

Ranby, John 1703 – 1773: FRS, royal surgeon, 103.

Razumovsky, Andrey Kirillovich 1752 – 1836: Russian ambassador at Vienna, 483.

Reischach, Thaddäus 1698 – 1782: Austrian ambassador at The Hague, 109, 114, 115.

Reveillon, Jean-Baptiste 1725 – 1811: Parisian paper manufacturer, 347, 390.

Reynolds, Joshua 1723 – 1792: FRS, English society portrait painter, 64.

Rezzonico, Carlo Torre di, 1693 – 1769: Italian, Pope Clement XIII from 1758, 146, 157.

Rob Roy – see MacGregor.

Robert, Aine-Jean 1758 – 1820: French pioneering balloonist, 346, 347.

Robertson, James, dates uncertain: Scottish physician, 356.

Robertson, John 1712 – 1776: FRS, librarian of the Royal Society, 184.

Robinson, Alexander Ramsay d.1824: manager of the King's farm at Kew, 473, 486, 487.

Rockingham, Lord – see Watson-Wentworth.

Romilly, Anne –see Garbett, Anne.

Romilly, Samuel 1757 – 1818: Solicitor General, reforming lawyer, 410, 478,

INDEX OF NAMES

479, 512.
Rosenberg, Count – see Orsini-Rosenberg.
Rousseau, Jean-Jacques 1712 – 1778: French philosopher, 64.
Rozier, Jean-François, Pilâtre de 1754 – 1785: French chemist and balloonist, 347.
Rucker, John Anthony 1719 – 1804: City of London merchant, 487, 488, 504, 506.
Russell, Patrick 1727 – 1805: FRS, London physician, 432, 433.

Sachs, Julius, von 1832 – 1897: German, botanist, plant physiologist, 523.
Salomon, Johann Peter 1745 – 1815: German musician and impresario, 457.
Saunders, William 1743 – 1817: FRS, royal physician in London, 425.
Sauran, Francis, Count, dates uncertain: Austrian, minister of finance at Vienna, 496.
Saussure, Horace-Bénédict de 1744 – 1799: Swiss physicist and alpine mountaineer, 182.
Saussure, Nicolas-Theodore de 1767 – 1845: Swiss chemist, pioneer of plant physiology, 470.
Savory, William, dates uncertain: Calne guild steward (mayor) in 1799, 515.
Scheele, Carl Wilhelm 1742 – 1786: Swedish chemist, 244, 269 – 272, 302.
Scherer, Johann Anton b. 1730: Viennese physician, 379, 380, 451, 458, 460 - 463, 474.
Schikaneder, Emanuel 1751 – 1812: German impresario, 330.
Schitar, Antoni, dates uncertain: servant of Jan Ingen Housz, Vienna, London, 177, 183, 194, 199, 441.
Schmeisser, Johann 1767 – 1837: FRS, German chemist, 432.
Schreiber family: in-laws of the Jacquin family, 447.
Schreibers, Katharina 1735 – 1791: wife of Nikolaus Jacquin, 352, 436.
Schwankhard, dates uncertain: student of natural history, Vienna 366, 367.
Schwediaur, Franz Xavier 1748 – 1824: German physician, venereologist, 235 – 237, 239, 240, 246, 247, 257, 260, 281, 316.
Schwenke, Thomas 1694 – 1767: Dutch, professor of anatomy, The Hague, 106.
Secker, Thomas 1693 – 1768: Archbishop of Canterbury, 64.
Seilern, Christian August von 1717 – 1801: Austrian ambassador in London, 102, 108, 109.
Senebier, Jean 1742 – 1809: Genevan priest and natural philosopher, 292, 309, 326 – 330, 379, 470.
Shelburne, Lord – see William Petty-Fitzmaurice.
Sheremeteva, Anna, died 1768: Russian, personal adviser to Catherine the Great, 144.
Shuldham, Molyneux. Baron 1717 – 1798: FRS, naval officer and politician, 430.
Sibthorp, John 1758 – 1796: botanist, professor at Oxford, 405, 418.
Siedentopf, Henry 1872 – 1940: German physicist, 346.
Sinclair, John, Sir 1754 – 1835: FRS, Scottish politician, agriculturalist, 354, 464 – 467, 471, 472, 477.
Sloane, Sir Hans 1660 – 1753: President of the Royal Society, physician and collector; collection was foundation of British Museum, 76, 171.
Smeaton, John 1724 – 1792: FRS, civil engineer, scientific instrument maker, 183, 418.
Socrates 469 BC – 399 BC: Greek philosopher, 27.
Sombreuil – see Virot.
Sonnenfels, Josef von 1733 – 1817: Austrian economist, adviser to the Emperor, 495.
Sophie Charlotte of Mecklenburg-Strelitz – see Charlotte.
Stadion, Johann Philipp, 1763 – 1824: Count von Warthausen, Austrian ambassador in London, 432.
Stair, Earl of – see Dalrymple.
Stanhope – see Mahon.
Stanhope, Lady – see Grenville, Louisa.
Starhemberg, Georg Adam, Prince 1724 – 1807: Austrian statesman and diplomat, 126.
Starhemberg, Ludwig von 1762 – 1833: Austrian ambassador in London, 462, 483, 494.
Stevenson, Margaret 1706? – 1783: Benjamin Franklin's landlady in London, 285.

Stiles, Henry, dates uncertain: farmer at Calne, patient of Christopher Allsup, 500, 501.
Stokes, Jonathan 1755? – 1831: physician, botanist; member of Lunar Society, 352.
Stoll, Maximilian 1742 – 1787: Viennese royal physician, 303, 304, 433.
Störck, Anton 1731 – 1803: Viennese royal physician, 130, 131, 224, 293.
Stuart, John 1713 – 1792: third Earl of Bute, Scottish, Prime Minister, 42.
Stubenrauch, Johann Georg 1732 – 1802: Viennese balloonist, 350.
Stuyck, Maria 1737 – 1812: wife of Louis Ingen Housz, 31, 116, 183, 231, 246, 395, 397, 402, 457, 517.
Sutherland, Alexander, dates uncertain: physician who took the Suttonian method of inoculation to Netherlands and Germany, 105 – 107, 110, 114.
Sutton, Daniel 1735 – 1819: smallpox inoculator, son of Robert Sutton: 89, 90.
Sutton, Robert 1707 – 1788: smallpox inoculator in East Anglia: 88, 89, 95, 103, 106, 110.
Swieten, Gerard van 1700 – 1772: Dutch physician trained under Boerhaave; personal physician to Maria Theresa, 20, 103, 110, 111, 126 - 142, 146 – 148, 156, 173, 182, 191.
Swieten, Gottfried van 1733 – 1803: son of Gerard, diplomat, musical impresario, 294.
Swinden, Jan-Hendrik van 1746 – 1823: Dutch, natural philosopher, 328, 329.

Tacitus, Publius 56 AD - 117 AD: Roman historian, 243.
Talleyrand-Périgord, Charles Maurice de 1754 – 1838: French diplomat, 442, 443.
Tanucci, Bernardo 1698 – 1783: Italian, diplomat in Naples, 161.
Tassie, James 1735 – 1799: Scottish, engraver in London, 261.
Tede, Dominique, dates uncertain: Italian, long-serving personal servant of Jan Ingen Housz; engaged by Jan Ingen Housz 196; travels to Italy 199; fishes for Torpedo specimens 203; travels to England 236; 239; helps with plant experiments 250 – 256, 273; leaves England 281; 283, 284; returns to Vienna 291; leaves Vienna 381; nurses Ingen Housz in Paris 387, 388, 396; is in Breda with Ingen Housz 397; arrives back in England 402; 418; to Bowood 420; helps Ingen Housz with demonstrations 422; 423; in Marylebone Street 441; saves Ingen Housz from an assault 451; with Ingen Housz at Twickenham Common 461; 483; at Bowood 489; present when Ingen Housz dies 492; 496, 509; attends Ingen Housz's funeral 515; grieves for his master 516; receives testimonial from Lord Lansdowne 516.
Teyler – see Hulst.
Timoni, Emanuele, dates uncertain: Italian, physician in Constantinople, 75.
Toaldo, Giuseppe 1719 – 1797: FRS, Italian, mathematician, astronomer, Padua, 370.
Tour, de la, Louise-François, dates uncertain: distinguished Paris printer, 243.
Troostwyk, Adriaan Paets van 1752 – 1837: Amsterdam merchant and scientist, 400, 401.
Turner, John, dates uncertain: surgeon-apothecary in Debenham, Suffolk, 89.
Tyson William 1763 – 1831: fellow of Emmanuel College, Cambridge, 475.

Upper Ossory, Lord – see Fitzpatrick.

Van Beest van Renoy, Hendrica 1661 – 1742: wife of Louis Ingen Housz at Bommel, 2, 3, 19.
Vaughan, Benjamin 1751 – 1835: diplomat, agriculturalist; American sympathiser, 410.
Vaughan, Samuel d. 1802: American merchant, close friend of Benjamin Franklin, 356.
Vermond, Mathieu Jacques de, Abbé 1735 – 1806: French, tutor to Marie Antoinette, 180.
Vernon, Caroline born 1761: married Robert (Bobus) Smith of 'Holland House set', 411.
Vernon, Elizabeth 1762? – 1830: life companion to Lady Caroline Fox, 411, 444, 488, 511, 513.
Vernon, Henrietta 1760 – 1838: married

INDEX OF NAMES

second Earl of Warwick, 411, 444, 511.
Vernon, Richard 1726 – 1800: politician and racehorse breeder, 411.
Vesalius, (van Wesel) Andreas 1514 – 1564: Belgian, physician, anatomist, 27, 38.
Virgil 70 BC – 19 BC: Roman poet, 27, 342, 422.
Virot de, Charles-François 1725 – 1794: Marquis de Sombreuil, governor, Les Invalides, Paris, 392.
Viveash, Samuel, dates uncertain: Calne guild steward (mayor) in 1799, 515.
Volta, Alessandro 1745 – 1827: Italian, professor of physics, Pavia, 232, 233, 242, 259, 292, 295, 320, 355, 356.
Voltaire – Francois-Marie Arouet 1694 – 1778: French writer and philosopher, 145, 177.

Walpole, Horace 1717 – 1797: art historian, man of letters, 164.
Walpole, Horatio 1678 – 1757: politician and diplomat, 5.
Walpole, Robert 1676 – 1745: Statesman and, effectively, first Prime Minister, 5, 72.
Walsh, John 1726 – 1795: FRS, diplomat and natural scientist, 203 – 205.
Wappler, Christian Friederich, dates uncertain: Viennese publisher, 300.
Warthausen – see Stadion.
Warwick, Earl of, - see Greville.
Warwick, Lady - see Henrietta Vernon.
Washington, George 1732 – 1799: first President, United States of America, 227.
Watson, William 1715 – 1787: FRS, London physician, inoculator, natural philosopher, 44, 65, 78, 83 – 85, 87, 95, 97, 103, 107, 110, 114, 170, 183, 189, 214, 215, 449.
Watson, William Sir (junior) 1744 – 1825: FRS, physician and astronomer, 170.
Watson-Wentworth, Charles 1730-1782: Lord Rockingham, Whig politician, Prime Minister, 190.
Watt, James 1736 – 1819: Scottish inventor and engineer, 413, 482.
Watzel, dates uncertain: a merchant travelling between Vienna and London, 456.
Wayte, Allin 1752 – 1816: surgeon-apothecary in Calne, 515.
Webster, Elizabeth 1771 – 1845: wife of third Lord Holland, 488 – 491, 511, 513, 516.
Webster, Godfrey, Sir 1747 – 1800: first husband of Elizabeth Webster, 489.
Wedderburn, Alexander 1733 –1805: Scottish, lawyer at English Bar; Solicitor-General, 218.
Wedgwood, Josiah 1730 – 1795: FRS, potter, member Lunar Society, 261.
Welderen Count van, dates uncertain: Dutch ambassador in London, 47, 113.
West, James 1703 – 1772: President of the Royal Society, barrister, politician and antiquary, 184.
Wharton, Samuel 1732 – 1800: Boston businessman, land developer: friend of Franklin, 289 – 291, 357 – 362, 377, 382.
Wharton, Samuel Lewis, dates uncertain: son of Samuel Wharton, 357 – 361, 377.
Whytt, Robert 1714 – 1766: FRS, Edinburgh physician, professor of medicine, 53, 54, 56, 68.
Widman Lazar 1697 – 1769: Austrian sculptor, 149.
Widmanstätten, Alois von Beckh 1753 – 1849: Austrian printer, 349.
Wiesner, Julius 1838 – 1916: professor of Botany at Vienna, 174, 522.
Wightman, Joseph, dates uncertain: royalist commander at Inverness in 1719, 8, 9.
Wilhelmina 1751 – 1820: Princess of Orange from 1767, 374.
William 1721 – 1765: Duke of Cumberland, younger son of George II, soldier, 39, 40, 80, 94.
William III 1650 – 1702: Prince of Orange and joint Monarch of England with wife, Mary. 4, 72.
William III, Prince of Orange – see Orange-Nassau.
William Petty-Fitzmaurice 1737 – 1805: second Lord Shelburne, later first Marquis of Lansdowne, Whig politician, Prime Minister; has Jan Ingen Housz as dinner guest 257; employs Joseph Priestley as adviser 257; visited by Ingen Housz at Bowood 280; 312, 313; arranges for his son to visit Vienna; 353; Ingen Housz a dinner guest 409, 410, 429, 442, 443,

507; marries Louisa Fiztpatrick 411;
employs Capability Brown at Bowood
420; introduces Ingen Housz to rest
of his 'family' 410; hosts large autumn
gatherings at Bowood 421 – 423, 438,
439, 444, 445, 450, 487, 488, 490; 468,
477, 478, 480, 481; pays for surgery for
a Calne patient 482; 492, 501; urges
Ingen Housz to live at Bowood during
illness 508; 512; writes to Ingen Housz
relatives after Ingen Housz death 513,
514, 515; invites Dominique Tede to
stay at Bowood 516; writes to Ingen
Housz bankers 517; offers to help Ingen
Housz nephews in London 517.

William V, 1748 – 1806: Prince of Orange,
114 – 116, 373 – 375, 462.

Williams family: Jan Ingen Housz's
landlords in Marylebone Street,
London, 441.

Williams, John, dates uncertain: American,
brother, Jonathan (senior, born in 1719),
243, 288 – 290, 357 - 360.

Williams, Jonathan 1750 – 1815: Franklin's
grand-nephew, American agent at
Nantes, 186, 188, 190, 288, 291, 357,
377.

Wilson, Benjamin 1721 – 1788: FRS,
electrical expert, 184, 245.

Winckelmann, Johann Joachim 1717 -
1768. German art expert, pioneer of
archaeology, 118 – 126, 140, 141, 145,
146, 157, 164, 165.

Withering, William 1741 – 1799:
Birmingham physician, discoverer of
the power of digitalis, 445.

Woodward, John 1665 – 1728: FRS (though
expelled in 1710), London physician, 75.

Wortley Montagu, Edward 1678 – 1761:
Whig politician, diplomat, 72, 73, 75.

Wortley Montagu, Mary 1689 – 1762: wife
of Edward Wortley Montagu, writer,
73 – 76, 95, 152.

Wurtzburg, Prince, dates uncertain: Ingen
Housz inoculation patient in Vienna,
151.

Wycombe, Lord – see John Henry.

Xanthippe, fifth century BC: wife of
Socrates, 478.

Xaver, Karl Maximilian Philipp 1702 –
1784: Prince Dietrichstein, Austrian
politician, 190, 215.

Young, Arthur 1741 – 1820: economist,
agriculturalist, 466, 469.

Zienmayer, Baron, dates uncertain: Austrian
aristocrat of Vienna, 321.

General Index

Aa, river, 10.
aanteken boekje (memo book), 272.
Achen, 237, 239.
acids, 384, 445, 460, 468, 486, 487, 490.
Adelphi, 417, 440.
Adlestrop, 501, 503.
Adrianople, 73, 75.
air analysis (see, also, eudiometry), 210, 211, 252, 261, 263, 271, 276, 277, 282, 286, 296, 300, 324, 412.
air pistol, 247, 264, 265, 292, 308, 334, 356.
Albion Mill, London, 418.
alchemy, alchemists, 299, 333.
Aleppo, 433.
algae, green – see 'green matter'.
alkaline soda water, 402, 403, 424, 425, 427, 430, 431, 434, 435, 439, 446, 451, 457, 460, 464, 481.
Allgemeine Krankenhaus, Vienna, 351.
Almanacks, 38.
America, American colonies, 54, 55, 57, 216 – 219, 227, 346, 357 – 361, 373, 415.
American business ventures of Jan Ingen Housz, 243, 288 – 291, 357 – 362, 377, 382.
American Philosophical Society, 285, 362.
American War of Independence, 219, 236, 241, 285, 290, 291, 334, 373, 377, 381, 399, 400, 410, 432.
Amsterdam, 65, 183, 197, 236, 247, 264, 282, 286, 291, 334, 373, 377, 381, 399, 400, 410, 432.
Analogical Dictionary, 448, 449, 450, 474, 475.
ancien régime, 522.
Anfangsgründe der Electricität – see Rudiments of Electricity.
animacules, 334.
animal magnetism, 224 – 227, 241.
Annagasse, Vienna, 447, 518.
Annals of Agriculture, 466.
Annonay, France, 347.
Ansbach, Germany, 76, 112.
anti-septic, 40.

Antwerp, 116, 282.
apothecary, apothecary shop, 17 – 20, 23, 29, 52, 79, 85, 165, 192, 277, 397.
aqua mephitica alkalina - see alkaline soda water.
archeology, origin of, 120.
Aschattenburg, Germany, 41.
Assembly Rooms, Bath, 423.
asthma, 52, 162, 452, 488.
astronomy, 33, 34, 407, 428.
Augustinerkirche, Vienna, 178, 198.
Austria, Austrians, 98 – 103, 111, 117, 122, 128 – 132, 150, 177, 178, 202, 333, 350, 371, 406, 416, 433, 447, 451, 474, 493, 494, 511.
Austrian National Library, Vienna, 56, 199.
Austrian Netherlands – see Belgium.
Austri-chienne, 389.

Bäckerstrasse, Vienna, 228.
bacteria, 78, 342.
Baden, Austria, 292.
Bakerian Lecture, 235, 258, 259, 265, 266, 267, 269, 287.
balloons, hot air and hydgrogen, 346 – 350, 390.
Barbary pirates, 161.
Bastille prison, Paris, 391, 392, 437.
Bataafsch Genootschap – Batavian science society, Rotterdam, 246, 247, 300, 301, 376, 431.
Bates Hotel, London, 417.
Bath (and West) Agricultural Society, 470.
Bath Oliver, biscuit, 41.
Bath Philosophical Society, 280, 423.
Bath, Avon, 275, 280, 402, 407, 420, 423, 424, 431, 439, 440, 446, 452, 470, 481, 508, 515 – 517.
Bayford, Hertfordshire, 96, 107.
Belgium (Austrian Netherlands), 3, 19, 24, 117, 128, 185, 281, 394, 406, 462.
Bengeo, Hertford, 94.
Berkeley Square, London, 441, 507.
Berkeley, Gloucestershire, 497, 502, 504.

Berkhamsted, Hertfordshire, 92, 94, 96, 107, 200.
Berlin, 105, 118, 121, 123, 125, 142, 145, 232, 337, 374.
Betchworth, Surrey, 476, 484.
Binnenhof, the, The Hague, 114, 373.
Biographie Universelle, 513.
Birmingham, 186, 190, 354, 419, 437, 444, 445, 478.
bladder stones, 258, 396, 402, 403, 424, 425, 430, 431, 451, 453.
blanket weed – see 'green matter'.
Bloemkool, de – The Cauliflower (Inn), 380
Board of Agriculture, 464 – 467, 471 – 473, 477, 487.
Board of Ordnance, London, 184.
Bohemia, 98, 173, 174, 176, 206, 232, 424.
Bologna, 118, 169.
Bommel (Zaltbommel), 3, 9.
Boston Tea Party, 218.
Boston, Massachusetts, 55, 81, 95, 216, 217, 227, 289, 357, 358, 377.
Bowood House, xv, 257, 280, 410, 411, 420 – 423, 434, 438, 444 – 446, 450 – 452, 464, 468, 469, 476 – 481, 487 – 490, 499 – 501, 508 – 517.
Brabant, 19, 20, 25, 49, 376.
brass pistol – see air pistol.
breathing oxygen, 300 – 304, 307, 308, 453, 482.
Breda: in early eighteenth century, 10; schools, 14 – 17; becomes main Ingen Housz residence, 19; English army camped near, 20; Jan Ingen Housz in medical practice in, 29 – 31; Ingen Housz passes through, 116; Ingen Housz passes through again, 183; Ingen Housz visits family, sister-in-law ill, 246; Patriot Revolt in Breda, 374; Ingen Housz flees to from Paris, 395, 397, 401, 402; Ingen Housz clothes, carriage stranded in, 443, 446, 447; French occupy town, 467; Ingen Housz and Breda Museum, 519, 522, 524.
Breda, Latin School, 14, 16, 66, 154.
Brenner Pass, 122, 206.
Brenta, River, 370.
Brescia, Italy, 172, 184, 209, 265.
Bridgewater canal, 187, 188.
Bristol, 304, 452, 481, 482, 483.
British army – see English army.
British Museum, 33, 47, 61, 66, 80, 104, 164, 165, 184, 418, 487.
British Wool Society, 464.
Broome Park, 476, 484.
Brownian motion, 344 – 346.
Brugstraat, Breda, 11, 12.
Brussels, 101, 109, 114 – 117, 129, 182, 183, 185, 194, 208, 241, 245, 246, 269, 282, 286, 394, 419.
bubonic plague, 73, 156, 246, 433.
Burg Trugenhofen (Schloss Taxis), Baden-Württemberg, 237.
Burgerspitalinshaus, Vienna, 518.

Cabalism, 333.
calcium sulphate, 468, 471, 473.
Calne, Wiltshire, xv, 257, 280, 312, 420, 423, 478, 481, 482, 500, 514, 515, 518, 524, 525.
calomel, 146.
Calvinists, 15, 398.
Cambridge University Press, 449, 467, 475.
Cambridge, Cambridge University, 63, 66, 449, 466, 475, 477, 507.
candles, 45, 94, 153, 183, 186, 235, 236, 270, 277, 300, 336, 421.
caoutchouc – see rubber.
carbon dioxide, 189, 278, 302, 329, 330, 425, 469, 470.
carbon, 304, 305, 344, 383, 469.
carbonated water, 258, 387, 425.
Carlsbad, Germany, 424.
Carlton House, London, 257.
Caserta Royal Palace, Naples, 153, 161, 179.
Castle House, Calne, 514.
Castleton, Derbyshire, 186.
Catholic, Roman, 3, 4, 7 – 12, 15, 25, 27, 83, 94, 103, 120, 128, 157, 239, 339, 398, 512.
Champ de Mars, Paris, 346.
Chancellor of the Exchequer, 72, 478.
charcoal – see carbon.
Charing Cross, London, 32, 56, 65, 183, 394, 404, 441, 475.
Chatsworth, Derbyshire, 377.
Chelsea, Chelsea Physic Garden, 80, 417, 487.
Cheltenham, 502, 503.
chemistry, 297, 384, 386; *see also* new chemistry.
Chemnitz, Germany, 219.
Cherhill, Wiltshire, 423, 478.
Chevening House, Kent, 434, 435.

GENERAL INDEX

chimneys, smoky, 390.
China, 95.
Chippenham, Wiltshire, 420, 423, 478.
chlorophytes, 342.
Christ Church College, Oxford, 353.
Church and King riots, Birmingham, 437.
Clarinet trio by Mozart (K498), 293.
Clerbessem, de, 10, 11, 13.
Cockpit, The, Whitehall, 218.
Cold Bath Fields, London, 69, 85, 86.
Colloquia familiaria, 15.
compass needles – see marine compasses.
Compiègne, France, 178.
Compton Bassett, Wiltshire, 423.
Concord, Massachusetts, 219, 227.
conferva rivularis (crow silk), 315, 316, 342.
Congress of the United States, 285, 286, 319, 320, 359, 377.
conjunctivitis, 339, 340.
Constantinople, 73, 75, 220.
consumption – see tuberculosis.
Copley medal of the Royal Society, 40, 183, 205, 215, 260.
copper, 28, 159, 251, 297, 299, 305- 307, 320, 321, 369, 370, 444.
Cornhill, London, 184, 405.
corona civica, 157.
Courier de Europe, 245.
Covent Garden, London, 429, 435, 441.
cover slip, 343, 344, 345, 346, 519,
Cowbridge House, Hertford, 419, 467.
cowpox, 496 – 506.
Crane Court – London, 33, 43, 170, 184, 406.
Craven Street, London, 56.
Cremona, 320, 321.
cress, 365, 369.
Critical Review, 314.
Crown and Anchor Tavern, London, 183, 406.
Culloden, battle of, 39.
customs houses, Paris, 395.

Danube, River, 117, 122 – 124, 193, 350, 363.
De civilitate morum puerilium libellus, 15.
Declaration of Independence of America, 55, 222.
Delft Latin School, 116, 248, 376, 441, 448.
Delft, the Netherlands, 59 – 62, 65, 67, 115, 116, 248, 286, 301, 308, 333, 371, 374 – 376, 398, 430, 431, 448, 474, 475.

Denmark, 517.
dephlogisticated air – see oxygen.
Derby, 61. 190.
Dettingen, Battle of, 39.
Devil's Arse, 186.
diabetes mellitus, 508.
diamond ring, 148, 338.
diffraction of light, 343.
digitalis – see foxglove.
Dischingen, Baden-Württemberg, 234, 237.
Discourses, Six, 271, 310.
Dissertation on dephlogisticated air, 301.
DNA, 520, 521.
Dr. Jan Ingen Housz Plein, Breda, 524.
Doctrina Particularum Graecae - see *Particles*.
Douai, France, 3, 25.
Drummonds Bank, London, 183, 404, 474, 476, 493, 496, 510, 516, 517.
Duchess Street, London, 432.
Duke Street, London, 404
Dunkerque, France, 290, 357.
Dunn and Taylor, linen merchants, Fleet Street, London, 47, 404.
Dutch army, 1, 2, 7, 432.
Dutch Embassy, London, 88, 89.
Dutch language, 15, 20, 21, 24, 54, 106, 259, 295, 301, 308, 430, 432, 492.

earth-bathing, 440, 441.
East India Company, 217.
Edgbaston Hall, Birmingham, 445.
Edinburgh Royal Infirmary, 45, 46.
Edinburgh, Edinburgh University, 20, 27, 35, 42, 44 – 46, 48, 53, 88, 151, 189, 310, 352, 439, 440, 452, 464, 477, 480.
Eilean Donan, castle of, Scotland, 8.
Eindstraat, Breda, 10, 11, 13, 23, 29, 165, 397, 402.
Elba, Island of, 170, 183.
electrical machine, 31, 58, 59, 79, 205, 209, 232, 235, 244, 295, 364, 390, 398, 400, 401, 447.
electricity (static), 28, 31, 54 – 59, 205, 209, 224, 232, 235, 242, 266, 286, 322, 355, 364 – 372, 390, 400, 401, 439.
electro-convulsive therapy, 323.
electrophore, 232, 233, 242, 259, 287, 295, 300.
electrostatic induction - see electrophore.
Emmanuel College, Cambridge, 449, 475.
England: fear of Catholicism, 4; Jan Ingen Housz arrives, 34; stronghold

of smallpox inoculation, 35, 46;
tension with American colonies, 57;
introduction of inoculation, 76, 77;
Marie Theresa requests inoculator from,
101, 102, 108; Ingen Housz leaves, 111;
Ingen Housz returns, 183; Ingen Housz
leaves again, 194; outbreak of hostilities
with America, 216 – 219; Ingen Housz
leaves, 281, 282; peace with America,
346; Ingen Housz longs to return,
354; Ingen Housz returns, 402; Ingen
Housz begins to settle, 414, 429; fear
of revolution, 445, 446, 487; war with
France, 446; Stadholder flees to, 462;
economic problems in, 492; Ingen
Housz nephews travel to, 517.
English army, 19, 21, 39, 41, 162, 165, 377.
English Channel, 5, 245, 257, 282, 348, 436.
English language, 42, 44, 89, 241, 278, 279, 287, 295, 310, 379, 457.
Enlightenment, The, 78, 163, 266, 277, 295, 299, 519.
Essex, 77, 94.
ether, 264, 308, 334, 356.
Eton College, 62.
eudiometer, eudiometry, 211 - 215, 242, 251 – 256, 261 – 263, 269, 271, 272, 274 – 279, 282, 286, 300, 302, 309, 310, 313, 315, 327, 328, 329, 334, 335, 341, 379, 390, 417, 520, 522.
evaporation, 343.
expeditionary force – see English Army.
Expériences sur les vegétaux . . . , 379.
Experiments and Observations . . . of Priestley, 264, 311, 413.

Falmouth, 415.
Farmers-General, 395, 459.
feminism, 73.
fertilisers – see manures.
First Lord of the Treasury – see Prime Minister.
Firth of Forth, 6.
fixed air – see carbon dioxide.
flagellates, 342.
Flanders, 19, 28, 39, 40, 474.
Fleet Street, London, 33, 36, 43, 47, 49, 50, 54, 80, 279, 282, 404, 406, 409, 441, 483, 488, 499.
Flora Austriacae, 199.
Florence, 155, 166, 169 – 171, 173, 196, 199, 202, 203, 205, 207, 211, 220, 266, 304, 352, 356, 489.
flute quartet of Mozart (K285b), 294.
Foundling Hospital (see London).
France, 3, 4, 24, 176, 178, 194, 234, 241, 245, 346, 364, 374, 378, 388, 404, 429, 442, 446, 459, 465, 467.
Frankenstein, Dr., 364.
Franziskanerplatz, Vienna, 317, 356, 363, 447.
Freemasonry, 330 – 333, 484, 512, 521.
French army, 19, 238, 378, 462, 511.
French language, 16, 24, 44, 103, 107, 117, 171, 177, 199, 287, 294, 295, 301, 310, 379.
French revolution, 333, 390 – 395, 414, 436, 443, 444, 446, 452, 484, 523.
Friesland, 17, 328.
Fürstemberger lighter, 292, 295, 296, 356.
Further observations . . . of Jenner, 505.

gallstones, 387, 388, 441, 446.
Garda, Lake, 169, 172.
Gardes Francaises, 390, 391.
gases, 169, 189, 210, 213, 244, 263, 265, 269, 271, 272, 288, 302, 314, 345, 352.
Geneva, Lake Geneva, 182, 271, 292, 309, 326 – 329, 389, 467.
Gentleman's Magazine, 102, 515.
German language, 42, 118, 259, 287, 294, 295, 379, 451, 456, 458.
Germany, Germans, 26, 42, 61, 112, 118, 121, 234 – 239, 320, 336, 391, 405, 432, 524.
Glauber's salt – see sodium sulphate.
Glen Shiel, battle of, 1, 7, 8.
Glorious Revolution, 4.
gold, 306.
'golden age' of the Netherlands, 17, 373.
Golden Square, London, 393, 441.
gout, 285, 318, 346, 349, 453.
Grand Canal, Venice, 156.
Grand Hotel de Toulouse, Paris, 382.
Grand Union Canal, 273.
Gravesend, 281, 282.
Greek language, 15, 16, 24, 49, 60, 62, 103, 119, 123, 124, 248, 385, 410, 448, 449.
green matter, the, 311, 312, 340 – 342, 344, 355, 379, 389.
green vitriol – see iron sulphate.
Greenwich, Greenwich Observatory, 418.
guillotine, 442, 443, 446, 452, 455, 459.
gunpowder, 264, 265, 301, 308, 334, 390.

gypsum – see calcium sulphate.

Haarlem, the Netherlands, 64, 67, 398 – 400, 417, 430, 434, 474.
Habsburgs, 25, 100, 101, 117, 135, 136, 159, 160, 196, 207, 211, 336, 337, 484.
Hague, The, 5, 33, 67, 104 – 106, 110, 113 – 116, 247, 308, 373, 376, 398.
Hamburg, Germany, 474, 483, 485, 487.
Hampton Court Palace, 71.
Handsworth, Birmingham, 444.
Hanover Square, London, 441.
Hanover, 407.
Harwich, 42, 111, 183, 402, 417.
heat conduction in metals, 297 – 300, 307, 308, 319.
Heathrow airport, 459.
Hemert, bankers in Amsterdam, 183.
henbane, 250, 253 – 256.
Herculaneum, Italy, 153, 164 – 166.
Hertford, 90 – 97, 101, 104, 107 – 109, 141, 142, 191, 419, 435, 436, 443, 450, 451, 467, 468, 477, 486, 487, 490.
Hints on Vegetables, 477.
History and Present State of Electricity, 189.
Hofburg Palace, Vienna, 100, 101, 156, 158, 159, 175, 177, 179, 193, 338, 494, 496.
Hofrat, 148, 155, 294.
Holland House, Kensington, 489 – 491, 506, 507, 509.
Holland science society, Haarlem, 398, 399, 431, 434.
Holland, Province of, 17, 20, 61 – 63, 105, 106, 373 – 376, 398, 430, 474 – 476, 517.
Hollandsche Maatschappij der Wetenschappen – see Holland Science Society.
'Honest Whigs', 50, 184.
horse chestnut, 250, 252, 253, 254.
Hortus Botanicus Vindobonensis, 176, 182, 190, 199.
Hortus Kewensis, 407.
hot air balloons – see balloons.
Hotel de Valentinois, Paris, 240.
Hotel Dieu, Paris, 351.
Hotel Dubouloir, Paris, 283.
Hotwells, Bristol, 452, 453, 481, 482.
Hounslow, 459, 490.
House of Commons, 80.
House of Lords, 63, 219, 507.
huisarts, 29, 44.
humoral theory of disease, 26.

Hungary, 98, 138, 219.
hydrogen, 263 – 265, 296, 297, 302, 308, 316, 334, 346, 347, 384, 385, 401.
hydrogen balloons – see balloons.
hydroponics, 244.

Île de Ré, 204.
immunology, 496, 506.
Imperial library, Vienna, 129, 146.
income tax, 493.
industrial revolution, 429.
inflammable air – see hydrogen.
influxus nocturnum, 252, 309, 328, 329.
'Ingen Housz Box', 297 – 299.
Ingenhouszia, 523.
Ingenhouszstrasse, Vienna, 524.
Innsbruck, 98, 99.
Inquiry of Jenner, 496, 499, 505, 506.
International Red Cross, 41.
Inverness, 1, 6, 7.
Investigation on pollution, 323.
iron sulphate, 435.
iron, 214, 306, 346.
Italy, 16, 35, 121, 124, 140, 145, 149, 155, 157, 193, 196, 211, 320.

Jacobins, Jacobinism, 442, 444, 480, 484, 507, 512.
Jacobite rebellion of 1715, 4.
Jacobite rebellion of 1719, 7.
Jacobite rebellion of 1745, 24, 39, 47, 94.
Jacobites, 4, 6, 7.
jasmine, 370.
jaundice, 387, 388.
Jermyn Street, London, 404, 415.
Jesuits, 129, 329, 385, 414, 484.
Jesus College, Oxford, 66.
Josefplatz, Vienna, 56.
Journal Britannique, 33.
Journal de Physique, 330, 364, 367, 372.
Journal de Voyage, 1777, 237.
Julianischer House, Vienna, 124.

Kasteel Plein, Breda, 14.
Kekkelstadt Trio – see clarinet trio.
Kew, 245, 276, 407, 418, 473, 485, 486, 487, 490.
kidney stones, 387, 424, 431, 453.
kinetic theory of atoms, 345.
kraam kloppertje, 11.

La Rochelle, 203, 204.

lagere school, 14.
lancets, for inoculation, 84, 86, 96, 235, 236, 400.
Landstrasse, Vienna, 223, 224, 225.
Lansdowne House, London, 257, 409 – 411, 429, 442, 443, 478, 507.
Latin language, 15, 16, 21, 24, 25, 27, 49, 53, 56, 60, 103, 123, 258, 288, 385, 431, 448, 458, 460, 461.
laudanum, 51, 167, 197.
leather, leather merchant, 10, 11, 17, 18, 19, 37, 63, 84, 121, 159, 165, 239, 452.
Leeds, Leeds Cloth Hall, 186, 189, 210, 257, 271, 425.
Leghorn (Livorgno), 203, 205, 206.
Leiden, 16, 26, 27, 28, 29, 45, 66, 128, 229, 230, 248, 258, 321, 333, 398, 432, 457.
Leipzig, 287, 351.
lemon, 254, 256, 274, 387.
Les Invalides, Paris, 391, 392.
letter-copying press, 413.
Lettre de M. Ingen Housz a M. Chais, 104, 105.
Levant, The, 95, 125, 433.
Lexington, Massachusetts, 219.
Leyden jar, 28, 204, 205, 215, 232, 233, 235, 264, 295, 321, 322, 334, 401.
libelles, 389.
Liège, 117, 239.
light - see sunlight.
lightning, lighting conductors, 172, 184, 205, 209, 235, 241, 245, 246, 260, 307, 318, 320, 321, 322, 351, 363, 364, 368, 370, 371.
lime, limewater, 303, 370, 371.
Lincoln's Inn Fields, 79. 80.
Linz, Austria, 131, 234.
lithontriptics, 424.
litterae patentes, 381.
Livorgno – see Leghorn.
Lombardy, 122, 170 – 172, 199, 320, 356.
London coffee houses, 46, 47, 50, 80, 184.
London Foundling Hospital, 80, 82 – 87, 97, 107, 138.
London: trials of the 1715 Jacobite rebels, 6; absence of a university, primacy of the Royal Society, 34; great fire of, and rebuilding, 36; population in mid-eighteenth century, 37; social situation in mid-eighteenth century, 37, 38; physicians in, 38; Jan Ingen Housz in society of, 43, 62; smallpox mortality rates, 72; child abandonment in, 80, 82; rents and living costs in 1770s, 183; lightning conductors in, 209; unpopularity of Franklin in, 217, 218; ballrooms and assembly rooms, 261; air quality in, 282; Freemasonry in, 330, 331; state of science in 1780s in, 383; room rents at 1789, 404; Ingen Housz living style in 1790s in, 404, 412, 506; population in 1790, 405; Ingen Housz scientific reputation in, as at 1790, 413; street dangers in, 434, 445; Ingen Housz social status and network in, 449.
Louvain, Louvain University, 24 – 27, 29.
Louvre, The, Paris, 384.
Lunar Society, 352, 445.
Lutheran religion, 120.
Luzac and van Damme, Amsterdam booksellers, 258.
lye soap, 424.
lymphoma, 509.
Lyon, Lyon Academy, 347, 367.

Maastricht, 239, 240.
Magic Flute, The, 330, 333.
magnets, magnetism, 215, 224 – 227, 241, 267, 268, 295, 308, 440.
Mainz, Germany, 67, 238, 287, 300, 314.
malum ischiadicum (backache), 195.
Manchester, 186, 187, 188.
manganese, 386, 390, 413, 488.
Mannheim, Germany, 117, 238, 294.
manures, 466 – 469, 471, 473, 476, 477, 485, 486, 490.
Maria Theresa Brooch, 207, 289.
marine compass, 267 – 269, 356, 390.
Martinique wax, 413.
Marylebone Street, London, 441, 451, 461, 484, 494, 504.
Massachusetts, 216 – 218, 227, 289, 358.
Mayfair, 36, 44, 47.
measles, 17.
medals, medallions, (see, also, Copley) 140, 149, 151, 158, 220, 242, 267, 525.
Mediterranean, 203, 220.
Meerle, The Netherlands, 380.
Meidling, Austria, 134, 139 – 142, 146 – 148.
Memoires . . . , of Barruel, 484.
mephitic water – see carbonated.
mercury, 304, 351.

Messiah, The, 83.
Meuse, River, 239, 240.
miasma, 210.
micro-picture – see Maria Theresa brooch.
microscope, microscope slide, 215, 261, 295, 338, 341 – 345, 356, 519.
Milan, 171, 199, 206, 211, 329, 356.
Mimosa pudica, 364, 365.
Ministry of Agriculture, 465.
mint – see peppermint.
Miscellanea Physico-medica, 462, 463, 473, 483.
Mitre coffee house – see London coffee houses.
Montague House – see British Museum.
moon, 34, 112, 118, 408, 422, 428, 445.
Moravia, 174, 318.
Moscow, 151, 336.
musculi falcati, 204.
Museo Pio-Clementino, Vatican, 122.
mustard, 365, 366, 369.

Nantes, France, 240, 243, 289.
naphtha vitriolis martialis, 432.
Naples, 100, 155, 156, 159 – 166, 177, 203, 206.
Narrenturm, the, Vienna, 351.
Nehalennia, the goddess, 248.
Netherlands: education in, 14; 'Golden Age' and subsequent slump, 17; middle classes in, 38; smallpox inoculation in, 80, 87 – 89, 106, 114; 'Royal' family of, 115; revolution in, 372 – 377; French invasion of, 446; morale-boosting postage stamps, 524.
new chemistry, 385, 386, 390, 400, 453, 485.
New York, 56, 221, 415.
Newgate Gaol, 76, 394.
nightshade, 254, 255.
Nijmegen, The Netherlands, 6, 197, 374.
nitre – see saltpetre.
nitric acid, 210, 212, 251, 253.
nitrous air (nitric oxide), 210 – 214, 251, 252, 255, 256, 262, 263, 269, 274, 277, 313, 323.
Nobel prize, 520.
North Sea, 4, 107, 109, 113, 248, 402, 417.
Northumberland Court, London, 65, 183, 257, 261.
Nouvelles Expériences, 294, 307, 308, 318, 342, 380, 383, 386, 388, 389, 407.

Observations . . . by Beddoes, 453.
Observations . . . by Pringle, 40.
oil of vitriol – see sulphuric acid.
onnozele schaapje, 11.
Orange-Nassau, House of, 4, 114, 115, 373, 374.
Ostende, Belgium, 5, 190, 237, 282.
Ottoman Empire, 73, 125.
Oxford, Oxford University, 42, 63, 209, 273, 405 – 407, 418, 452, 454.
oxygen: first isolation of, 265; in the power of gunpowder, 265; in the ether pistol, 265; discovery of release from leaves in sunlight, 275 – 277; as therapy for lung diseases, 300 – 304; burning metals in, 305 – 307; from the 'green matter', 340; in melting platinum, 383; in formation of water, 384; a phial of always carried by Jan Ingen Housz, 422; preventing access of to promote healing of ulcers, 432; in 'pneumatic medicine', 452 – 454; in photosynthesis, 470; and tuberculosis, 482; from 'air machines', 488.
oxygenated muriate of potash – see potassium chlorate.

Padua, 169, 370.
paediatrics, beginnings of, 88.
Palais de Justice, Paris, 391.
Palais Royale, Paris, 391.
Palazzo Pitti, Florence, 166, 169, 199.
Palazzo Sessa, Naples, 163.
Palazzo Torrigiani, Florence, 169.
Pall Mall, 32, 157, 257, 260, 261, 273, 310, 404, 440.
Panopticon, 433.
Pantheon, 38, 261.
pantograph, 413.
Papal antiquary, 120.
Papal Bull, 331.
Papal States, 120, 155.
Paragon Buildings, Bath, 423, 439.
Paris: extracting air from skin experiments, 284; salons, 297; ballooning at, 346 – 349; visit of Joseph II, 351; social situation in 1788, 382; 'new' chemistry in, 384 – 386; difficult winter in, 1788/9, 388; experiments on platinum in, 383; revolution in, 391 – 395; Jan Ingen Housz escapes from, 394, 395; guillotines in, 442, 443; the 'Terror' 443.

Particles, 49, 50, 59 – 66, 103, 116, 124, 140, 145, 146, 172, 185, 198, 248, 288, 448, 475.
passport, (see, also, *litterae patentes*), 199, 517.
Passy, 240 - 245, 259, 260, 283 – 285, 289 – 291, 297, 308, 322, 346, 382.
Patriot Revolt, the Netherlands, 372 – 377, 395, 401, 402, 446, 462.
Pavia, Italy, 171, 320, 356.
peach, 246, 254, 256, 274, 279.
Peak District, 186.
pearl ash, 486.
Peking, 243.
penicillin, 305.
Pennsylvania fireplace (Franklin Stove), 185.
Pennsylvania, 216, 221, 243, 359, 360, 378, 382.
pension of Jan Ingen Housz, 151, 208, 317, 331, 356, 358, 359, 404, 416, 474, 493 – 495, 499, 507.
Pensionary, 113, 114.
peppermint, 253, 255, 270, 313.
pharmacist – see apothecary.
Philadelphia, 54, 55, 216, 219, 221, 290, 359 – 362, 377, 382, 415, 417, 439, 446.
Philosophical Transactions of the Royal Society, 214, 236, 256, 268, 300, 316, 317, 330, 406, 498.
phlogiston, 271 (explanation of term), 302, 323, 384, 386, 400, 401, 521.
phosphate glass, 383.
phosphorus, 244, 336.
photosynthesis, 278, 312, 318, 342, 453, 469, 470, 472, 473, 519, 520, 523.
Physico-chemical memoirs of Senebier, 326, 328.
physics, 15, 27, 211, 295, 297, 306, 320, 329, 386.
Piccadilly, London, 310, 393, 404.
Pisa, 169, 220.
pistol – see air pistol.
Place de Concorde, Paris – see Place de la Revolution.
Place de la Révolution, Paris, 446, 455, 459.
Place de Lorraine, Brussels, 240.
plague – see bubonic.
plant physiology, 340, 372, 456, 487, 523.
planta animalis, 341.
platinum, 214, 215, 383, 390, 401, 412, 444, 518.
pneumatic chemistry – see air analysis.
Pneumatic institute, Bristol, 452, 454, 482.

pneumatic trough, 273, 274, 286, 489.
pollen, 344, 345.
Pompeii, 164 – 166.
Ponte Vecchio, Florence, 166.
porphyria, 72.
Portland Place, London, 430.
postage stamp, of Ingen Housz, 524.
post-mortem dissection, 26, 31, 104, 237, 247.
potassium carbonate, 434.
potassium chlorate, 442.
power of attorney, 359, 360, 361, 378, 382, 476.
Prague, 125, 174, 206, 232.
Prater, The, Vienna, 223, 350, 363.
Pressburg (Bratislava), 100, 495.
Prime Minister, 72, 73, 410, 445.
Princenhage, the Netherlands, 380.
Princes Street, Edinburgh, 45.
Priory, The, Hertford, 419.
Privy Council, 148, 155, 208, 218, 220.
'Project Ingen Housz', 523.
Protestant, Protestants, 4, 7, 25, 239, 326, 389.
Prussia, Prussian army, 118, 119, 173, 232, 336, 374, 375, 378, 391, 447, 462.
Pulteney Bridge, Bath, 423, 470.
Pump Room, Bath, 423.
Purfleet, 184, 218.

Quaker, Quakers, 90, 94, 419.
Quay des Augustines, Paris, 382.
Queen Square, Bath, 423.
quicksilver (see mercury).

'Ramsden' electrical machine, 59, 266.
Ranelagh. 38.
Ratisbon, 117, 118, 122, 123, 236.
red blood cells, 169.
Red Cross – see International.
Reformation, The, 25.
Reformed Churches, 15, 239, 512.
Regensburg – see Ratisbon.
Rennweg, Vienna, 228.
Rhine, River, 178, 182, 238, 239.
Rialto Bridge, Venice, 156.
rinderpest, 17.
Roman Catholic – see Catholic.
Roman Catholic Church, Breda – see Brugstraat.
Rome, 16, 35, 83, 118, 120 – 122, 124, 140, 155, 157 – 159, 173, 309, 339.

Rosicrucians, 332.
rotation of crops, 470, 473, 490.
Rotterdam, 6, 57, 67, 183, 246 – 248, 282, 286, 301, 303, 349, 402, 424, 476.
Royal Academy of Arts, London, 157.
Royal Academy of Sciences, Paris, 242, 284, 330, 383, 384, 388, 442.
Royal Botanical Gardens, Vienna, 176, 228, 310, 315, 327, 352, 368, 370 – 372, 494.
Royal College of Physicians of Edinburgh, 42, 53.
Royal College of Physicians, London, 42, 79, 88, 209.
Royal Crescent, Bath, 423.
Royal Institution, 422.
Royal Navy, 8, 357.
Royal Society Council, 42, 80, 258, 266, 406, 430.
Royal Society dining club, 80, 183, 191, 258, 406, 407, 429, 487, 521.
Royal Society: Ingen Housz first attends, as guest, 33, 34; meets to consider the rebuilding of London after the fire, 36; elects Benjamin Franklin as a fellow, 54; elects Jan Ingen Housz a fellow, 170; Ingen Housz admitted in person, 184; form a committee to advise on lightning conductors on arsenals, 184, 209, 245; elects John Pringle as President, 203; Ingen Housz's first paper in the Philosophical Transactions, 214; and platinum, 215; invites Ingen Housz to give 1778 Bakerian Lecture, 235, 236; Copley Medal of, 260; Joseph Banks elected President, 260; medallion presented by, 267; meetings schedule, 316; response to papers read, 330; politics in, 330; moves to Somerset House, 406; annual official visit to Greenwich Observatory, 418; Ingen Housz elected to Council of, 430; van Marum elected a fellow, 485; rejects paper on cowpox by Edward Jenner, 498; reputation hotly defended by Ingen Housz, 507.
rubber, 212.
Rudiments of Electricity, 300, 307.
Rue de Fauborg Saint-Antoine, Paris, 390.
Rue de Jardinet, Paris, 382.
Russia, Russians, 142 – 145, 150, 151, 173, 191, 336 – 338, 416, 483, 511.

St. Evangile Church, Geneva, 326.
St. Giles-in-the-Fields, 37.
St. James' (Parish, London), 36, 37, 471, 473, 493.
St. James' Church, Piccadilly, 310.
St. James' Palace, 102, 486.
St. James' Square, London, 310.
St. Mark's Cathedral, Venice, 156.
St. Martin in the Fields Church, London, 37.
St. Mary's Church, Calne, 514, 524, 525.
St. Paul's Cathedral, 50, 67, 184, 211, 345, 418, 449.
St. Paul's coffee house – see London coffee houses.
St. Petersburg, 142 – 145, 150, 191, 238, 336, 337, 389, 419, 483.
St. Pietersberg Mountain, the Netherlands, 239.
St. Stephen's Cathedral, Vienna, 195, 212, 223, 228.
St. Thomas' Hospital, 94, 142, 482.
salt of tartar, 425.
saltpetre, 301, 302.
Salzburg, 100, 356.
Sankt-Veit, Schloss, 148, 154, 173.
sans-culottes, 480.
Santa Maria Della Salute, Venice, 156.
Saturn, 408, 428, 429.
Scheikundige Bibliotheek, 430, 431, 432, 451, 458.
Schiedam, the Netherlands, 247.
Schonnbrunn Palace, 100, 124, 134 - 141, 146 – 148, 154, 173, 179, 195, 207, 246.
schutterij, 373.
Scotland, 1, 2, 5, 8, 20, 46, 47, 53, 66, 202, 310, 432, 464.
Scuir-Curan (Sgurr Uaran), 2, 8.
seeds, 365, 366, 467, 468, 487.
Seine, River, 240, 284, 346, 455.
self-lighting candles - see candles.
septic, 40.
Severn Bore, 163.
sexuality, 119, 121, 124, 140.
Sheffield, 189, 190.
Shelburne House – see Lansdowne House.
Sheriff Muir, 6.
's-Hertogenbosch, 3, 27, 29, 87, 220, 430, 474.
Shiel, River, 7.
shuldernsteuern, 195.
Sicily, 155.

signing language for the deaf, 244.
silver, 299, 306.
Six Discourses . . . of Pringle, 271, 310.
Skye, Isle of, 7.
slime moulds, 342.
Slough, 407, 408, 419, 488, 499.
smallpox inoculation: description of, 74; introduced into England, 76; particulars of process, 77; and Jan Ingen Housz's exposure, 80; at London Foundling Hospital, 84, 85; in the Netherlands, 88 – 90; book by Thomas Dimsdale, 93; in Hertford, 94; mass inoculation, 94; in America, 95; in Berkhamsted, Hertfordshire, 96 – 98; introduction into Austria, 102; Ingen Housz chosen to take to Vienna, 103; in the Netherlands, 104 – 107; opposition at Vienna, 130, 131; early attempts at Vienna, 131; inadequate techniques at Vienna, 132; 'rehearsal' inoculations at Meidling, Vienna, 137 – 142; Dimsdale in Russia, 142 – 145, 150, 151; of Habsburg children at Vienna, 146; inoculation house at Vienna, 148; at Naples, 159, 161; in Florence, 166, 167; teaching inoculation around Austria, 171; in Bohemia, 173; Ingen Housz has near-disaster at Florence, 199; and doctor-patient relationship, 203; finally accepted practice at Vienna, 293; versus 'vaccination' of Edward Jenner, 496 – 506.
smallpox: in London, 37; description of and effects of, 71, 72; in Royal households, 71, 72; mortality rate in London, 72; clinical features, 78; in servants and soldiers, 86; confluent case, 92; at Vienna, 98 – 101; Mozart and, 100, 294; scars of, on Maria Theresa, 136; in Russian Royal family, 143 – 145; eradication, 151, 506; and Marie Antoinette, 177; and cowpox, 496 – 499.
Smout and de Groot, publishers at Delft, 287.
snuffbox, gold, 149, 167, 176.
sodium sulphate, 239.
Soho, Birmingham, 190, 444.
Soho, London, 185, 441, 485.
Some further considerations, Priestley, 316.
Somerset House, London, 406, 407, 418, 485, 499.
Southall, Southall Green, 252, 273 – 277, 279, 311, 316, 326, 327, 341, 366, 461.
Southampton, 360.
Spain, Spanish army, 9, 159, 214, 461.
Specola, La, 169.
spectroscopy, 306.
spirit of nitre – see nitric acid.
Stadholder, 115, 373, 374, 375.
Stametz Bank, Vienna, 176, 183, 236, 291, 381, 474, 493, 496.
States-General, 3, 113.
steam engine, Boulton and Watt, 418.
steel, 96, 224, 268, 297, 299, 305, 306, 363, 378.
Stitchill House, Scotland, 20.
Strand, London, 59, 67, 183, 185, 311, 406, 407, 417.
Strasbourg, 178, 182, 287, 291, 292, 295, 366, 381, 388.
studium generale, 25.
sulphuric acid, 346, 425, 434, 468, 471, 477, 486.
sunlight, 256, 270, 275 – 278, 311 – 313, 319, 323 – 327, 329, 340, 342, 355, 366, 379, 390, 413, 414, 470, 520, 522, 524.
surgeon-apothecary, 29, 77, 89, 481, 497, 504.
Sydney, Street, Gardens, Bath, 423.
syphilis, 71.

taffeta, 347.
taking the waters, 239, 275, 280, 424, 453, 481.
tartar, salt of – see potassium carbonate.
Teddington, 271.
telescope, 295, 407, 408, 409, 419, 422, 428, 429, 447, 488.
Temple Bar, 47.
Temple Law Courts, 47.
Temple of Health, 440.
Temple of Hymen, 440.
Terheijden, the Netherlands, 20.
'Terror', the, 443.
Teyler's Museum, foundation, 398 – 400, 409, 417.
Thames, River and estuary, 5, 36, 162, 276, 281, 282, 417, 418, 485.
The Circus, Bath, 423, 508.
The food of plants, 469 – 473, 523.
Theatre Royal, Bath, 423.
thermometer, 351.

'Third Estate', 389.
Thirty-Years' War, 119.
Tilburg, 10, 19.
Tipton, Birmingham, 445.
torpedo fish, 203- 206.
touch screen, 233.
Tourton and Ravel (Bauer) bank, Paris, 433.
Tranquil Dale – see Broome Park.
Transylvania, 125, 176, 485.
Treasury, British, 72, 493.
Treaty of Utrecht, 117.
Treaty of Versailles, 1756, 177.
Trecktang, de, 10, 11, 116.
Tremella, 341.
Trento, Italy, 122.
Trieste, 122, 125, 140, 141, 172.
Trinity College, Cambridge, 478, 507.
Tsarskoe Selo, Russia, 145.
tuberculosis, 177, 416, 453, 481, 482.
Tuileries Palace, Paris, 348.
Turin, 171, 336, 367, 368, 369.
Turkey, 74, 75, 76, 77, 433.
Turks, 135, 416, 433.
Tuscany, 155, 159, 170, 196, 203, 211, 416.
Twickenham, 162, 459, 468.
typhus, 40, 98, 99.
Tyrol, Austria, 98, 199.

'uniform of Ingen Housz', 140, 141.
Unitarian sect, 189, 257.
United Provinces – see Netherlands.
University of Vienna, 129, 207, 294, 524.
Uranus, 408.
Utrecht Provincial Society of Science and Arts, 323, 324, 326, 431.
Utrecht, 16, 33, 324, 373, 374.

vaccination, 496, 498, 499, 503, 519, 522.
vacuum, 304, 305, 422.
variolae vacciniae, 498.
variolation (see, also, smallpox inoculation), 498, 499, 501, 502.
vases, Etruscan, 164.
Vatican, 120, 121, 122, 172, 331, 339.
Vauxhall, Vauxhall Gardens, London, 36, 38.
Vegetables, 280, 285, 287, 289, 300, 311, 312, 314, 323, 326 – 328, 341, 379, 409, 414, 521.
Vegetaux, Vegetaux II, 379, 386, 387, 390, 458.
Venice, 118, 125, 140, 156, 157, 206, 235, 241, 387.
Venus, 1761 & 1769 transits of, 34, 238.
Vermischte Shriften, 294, 307, 308, 314, 316, 342, 453.
Verona, Italy, 118, 172.
Versailles, 135, 136, 161, 177, 179, 180 – 182, 183, 192, 194, 240, 260, 347, 378, 391.
Versuche mit pflanzen, 379, 380.
Vesuvius, 153, 159, 163, 165.
Vienna General Hospital – see Allgemeine Krankenhaus.
Vienna Maternity Hospital, 131.
Vienna physicians' pension club, 230.
Vienna: smallpox epidemics, 98 – 101; Jan Ingen Housz sent as an inoculator, 109; Jan Ingen Housz arrives, 124; in the mid eighteenth century, 124 – 125; tensions in the medical fraternity, 130 – 132; celebrations of successful inoculation of Royal children, 147; prevalent eye problems in, 191; Ingen Housz socially isolated, 207; Ingen Housz erecting lightning conductors in, 209; Freemasonry in, 330 – 333; possible peace conference in, 335; Russian Grand Duke visits, 337, 338; Pope visits, 339, 340; ballooning at, 350; hard winters at, 350, 351; new hospital at, 351; introduction of alkaline soda water, 458; financial strictures at, 492 – 496.
Viennese Medical Gazette, 458.
Villa Poggio Imperiale, Florence, 166.
viper venom, 169.
Vizier, 338.

Waal, river, 3.
Walloon Church, The Hague, 105.
Waltham Cross, 90.
Wandsworth, 488, 504, 506.
War of Austrian Succession, 20, 24, 40.
War of Spanish Succession, 3.
Warsaw, 336, 354.
Warwick, 449, 513.
watch spring, 306.
water figwort, 250, 253, 254.
water lilies, 276.
water pepper, 251, 252.
watercress, 244.
Watt's air machine, 482, 488.
wax anatomical models, 192.
Weihburgasse, Vienna, 291, 317, 335, 338, 358.

Wentworth Woodhouse, 190.
West Hill, Wandsworth, 487, 488.
West Indies, 173.
Westminster Abbey, 310.
Westminster, 37, 45, 109, 217, 221, 273, 441.
Whigs, 72, 184, 410.
white horse at Cherhill, 423, 478.
Whitehall, 47, 217, 440.
whooping cough, 238, 246.
Wiener Neustadt Military Academy, Austria, 154.
willow, 254, 255.

Windmill Street, London, 196.
Windsor, 408, 428, 486.
Wipplingerstrasse, Vienna, 518.
wires, 297 – 299, 305 – 307, 319, 356, 368 – 371, 401.
Worcestershire, 352.
worms, intestinal, 51, 432, 481.

X-ray diffraction, 521.

Zur wahren Eintracht, 331.
Zur Wohltatigkeit, 331.

Rear cover image: Ingen Housz etched by the sun. A geranium plant is kept in the dark for two days; forced, thereby, to use up its energy stores – starch in the leaves. One leaf is then detached and sandwiched between glass plates. A slide negative of an engraving of Ingen Housz (1769) is then projected onto the leaf for an hour. During exposure, light falling on those parts of the leaf not obscured by the negative triggers photosynthesis and starch is produced. The leaf is then 'developed' – soaked in an iodine solution after blanching in alcohol. The iodine stains the starch-holding cells black and the 'father of photosynthesis' makes his appearance by the process he discovered: history, art, and science combine in a powerful demonstration.

Norman Beale MA, MD is a retired general practitioner and epidemiologist whose research on the effects of unemployment on family health was groundbreaking. His MD dissertation was awarded the 'best of the year' prize at Cambridge in 1989. He then developed a household deprivation marker for use in UK general practice. Retirement has allowed him the opportunity to publish a book, Joseph Priestley in Calne, and, latterly, to concentrate, with his wife, on writing 'the' biography of Jan Ingen Housz.

Elaine Beale is a retired schoolteacher who trained at Homerton College, Cambridge and whose main subjects were history and drama.

www.ingramcontent.com/pod-product-compliance
Lightning Source LLC
Chambersburg PA
CBHW021128230426
43667CB00005B/65